AF805887

Landslides in Sensitive Clays

Advances in Natural and Technological Hazards Research

Volume 36

For further volumes:
http://www.springer.com/series/6362

Jean-Sébastien L'Heureux • Ariane Locat
Serge Leroueil • Denis Demers • Jacques Locat
Editors

Landslides in Sensitive Clays

From Geosciences to Risk Management

Springer

Editors
Jean-Sébastien L'Heureux
Norwegian Geotechnical Institute
Trondheim, Norway

Serge Leroueil
Department of Civil Engineering
and Water Engineering
Laval University
Québec, QC, Canada

Jacques Locat
Laboratoire d'études sur les risques naturels
Department of Geology and Geological
Engineering
Laval University
Québec, QC, Canada

Ariane Locat
Department of Civil Engineering
and Water Engineering
Laval University
Québec, QC, Canada

Denis Demers
Service de la Géotechnique et de la géologie
Ministère des Transports du Québec (MTQ)
Québec, QC, Canada

ISSN 1878-9897 ISSN 2213-6959 (electronic)
ISBN 978-94-007-7078-2 ISBN 978-94-007-7079-9 (eBook)
DOI 10.1007/978-94-007-7079-9
Springer Dordrecht Heidelberg New York London

Library of Congress Control Number: 2013947095

© Springer Science+Business Media Dordrecht 2014
This work is subject to copyright. All rights are reserved by the Publisher, whether the whole or part of the material is concerned, specifically the rights of translation, reprinting, reuse of illustrations, recitation, broadcasting, reproduction on microfilms or in any other physical way, and transmission or information storage and retrieval, electronic adaptation, computer software, or by similar or dissimilar methodology now known or hereafter developed. Exempted from this legal reservation are brief excerpts in connection with reviews or scholarly analysis or material supplied specifically for the purpose of being entered and executed on a computer system, for exclusive use by the purchaser of the work. Duplication of this publication or parts thereof is permitted only under the provisions of the Copyright Law of the Publisher's location, in its current version, and permission for use must always be obtained from Springer. Permissions for use may be obtained through RightsLink at the Copyright Clearance Center. Violations are liable to prosecution under the respective Copyright Law.
The use of general descriptive names, registered names, trademarks, service marks, etc. in this publication does not imply, even in the absence of a specific statement, that such names are exempt from the relevant protective laws and regulations and therefore free for general use.
While the advice and information in this book are believed to be true and accurate at the date of publication, neither the authors nor the editors nor the publisher can accept any legal responsibility for any errors or omissions that may be made. The publisher makes no warranty, express or implied, with respect to the material contained herein.

Printed on acid-free paper

Springer is part of Springer Science+Business Media (www.springer.com)

Preface

Landslides in sensitive clays or quick clays represent a major hazard in the northern countries of the world because of their retrogressive potential and high mobility. Examples of catastrophic landslides in sensitive clays that impacted populations are numerous: e.g., Saint-Jean-Vianney in 1971, Rissa in 1978, Finneidfjord in 1996 and St-Jude in 2010. In order to respond to the societal demands, the scientific community has to expand its knowledge of landslide mechanisms in sensitive clays to assist authorities with state-of-the-art investigation techniques, hazard assessment methods, risk management schemes, mitigation measures and planning.

Scientific and engineering research on "Landslide in Sensitive Clays" has advanced significantly during the last decades owing to elevated public awareness and collaborations between the academia and the industry. This has led to major advances including the development of highly sophisticated geotechnical and geophysical investigation tools, hazard mapping procedures, risk management schemes and mitigation measures. Great progresses in numerical modelling have also lead to a better understanding of slope processes and stability assessment in sensitive clays.

This book gathers the latest scientific research made by international geoscientists in academia, industry and governmental agencies, and focuses on the full spectrum of challenges presented by landslides in sensitive clay terrain. The book is organized under the following topics: (1) Sensitive clays: source, nature and development; (2) Landslide characterization; (3) Geotechnical and geophysical site investigations; (4) Modelling, slope stability and progressive failure; and (5) Hazard assessment, risk management, regulations and policies.

Keywords: Sensitive clay, Quick clay, Landslide, Site investigation, Modelling, Hazard assessment, Risk assessment.

Trondheim, Norway — Jean-Sébastien L'Heureux
Québec, QC, Canada — Ariane Locat, Serge Leroueil, Denis Demers, Jacques Locat

Preface

Contents

1 **Landslides in Sensitive Clays – From Geosciences to Risk Management** 1
Jean-Sébastien L'Heureux, Ariane Locat, Serge Leroueil, Denis Demers, and Jacques Locat

Part I Sensitive Clay: Source, Nature and Development

2 **Chemistry, Sensitivity and Quick-Clay Landslide Amelioration** 15
J. Kenneth Torrance

3 **Nature of Sensitive Clays from Québec** 25
Jacques Locat and Daniel St-Gelais

4 **Three-Dimensional Quick-Clay Modeling of the Gothenburg Region, Sweden** 39
M.A. Persson

5 **Ion Exchange as a Cause of Natural Restabilisation of Quick Clay – A Model Study** 51
Pascal Suer, Hjördis Löfroth, and Yvonne Andersson-Sköld

6 **Potassium Chloride as Ground Improvement in Quick Clay Areas – A Preliminary Study** 63
Tonje E. Helle, Ingelin Gjengedal, Arnfinn Emdal, Per Aagaard, and Øyvind Høydal

Part II Landslide Characterization

7 **Inventory of Large Landslides in Sensitive Clay in the Province of Québec, Canada: Preliminary Analysis** 77
Denis Demers, Denis Robitaille, Pascal Locat, and Janelle Potvin

8 Characterization of Post-failure Movements of Landslides in Soft Sensitive Clays 91
V. Thakur, S.A. Degago, F. Oset, R. Aabøe, B.K. Dolva, K. Aunaas, T. Nyheim, E. Lyche, O.A. Jensen, M.B. Sæter, A. Robsrud, M. Viklund, Daniel Nigussie, and J.-S. L'Heureux

9 Controls on the Dimensions of Landslides in Sensitive Clays 105
Marten Geertsema and Jean-Sébastien L'Heureux

10 Prehistoric Sensitive Clay Landslides and Paleoseismicity in the Ottawa Valley, Canada 119
Gregory R. Brooks

11 Characterization and Post-failure Analysis of the 1980 Landslide in Sensitive Clays at Havre-St-Pierre, Québec, Canada 133
Pascal Locat, Serge Leroueil, Jacques Locat, and Denis Demers

12 The Evolution of Material Properties Within an *In Situ* Shear Zone in Sensitive Clay 145
Jean-Sébastien L'Heureux, Ragnar Moholdt, Vidar Gjelsvik, and Einar Lyche

Part III Integrated Geotechnical and Geophysical Site Investigations

13 The Use of Geophysics for Sensitive Clay Investigations 159
Shane Donohue, Michael Long, Jean-Sébastien L'Heureux, Inger-Lise Solberg, Guillaume Sauvin, Magnus Rømoen, Thomas Kalscheuer, Mehrdad Bastani, Lena Persson, Isabelle Lecomte, and Peter O'Connor

14 Applications of 2D Resistivity Measurements for Quick-Clay Mapping in Mid Norway 179
Inger-Lise Solberg, Louise Hansen, Jan Steinar Rønning, and Einar Dalsegg

15 An Integrated Approach to Quick-Clay Mapping Based on Resistivity Measurements and Geotechnical Investigations 193
Andreas Aspmo Pfaffhuber, Sara Bazin, and Tonje E. Helle

16 Geophysical and Geotechnical Investigations for a Major Highway in a Quick-Clay Area 205
Rolf Sandven and Inger-Lise Solberg

17 **Mapping of Quick Clay by ERT and CPT-R in the Göta Älv River Valley** 217
Torleif Dahlin, David Schälin, and Johannes Tornborg

18 **Geophysical Data Integration for Quick-Clay Mapping: The Hvittingfoss Case Study, Norway** 229
Guillaume Sauvin, Isabelle Lecomte, Sara Bazin, Jean-Sébastien L'Heureux, and Maarten Vanneste

19 **Joint Acquisition and Processing of Seismic Reflections and Surface Waves in a Sensitive Clay Deposit in the Outaouais Region (Québec), Canada** 241
Gabriel Fabien-Ouellet, Richard Fortier, and Bernard Giroux

20 **Empirical Geophysical/Geotechnical Relationships in the Champlain Sea Sediments of Eastern Ontario** 253
H.L. Crow, J.A. Hunter, A.J.M. Pugin, S.E. Pullan, S. Alpay, and M. Hinton

Part IV Slope Stability, Modelling and Progressive Failure

21 **The Effect of Deformation Rate in Progressive Slope Failure** 267
Anders Samstad Gylland

22 **Failure Mechanism of Spreads in Sensitive Clays** 279
Ariane Locat, Serge Leroueil, and Hans Petter Jostad

23 **How Well Do We Understand the Undrained Strain Softening Response in Soft Sensitive Clays?** 291
V. Thakur, H.P. Jostad, H.A. Kornbrekke, and S.A. Degago

24 **Effect of Strain-Softening in Design of Fills on Gently Inclined Areas with Soft Sensitive Clays** 305
Hans Petter Jostad, P. Fornes, and V. Thakur

25 **Effective Stress Based Stability Analysis of Normally Consolidated Clays** 317
Tim Länsivaara, Juho Mansikkamäki, and Ville Lehtonen

Part V Hazard Assessment, Risk Management, Regulations and Policies

26 **An Overview of the Mapping of Landslide-Prone Areas and Risk Management Strategies in the Province of Québec, Canada** 331
Janelle Potvin, Catherine Thibault, Denis Demers, and Chantal Bilodeau

27 **Regulatory Framework for Road and Railway Construction on the Sensitive Clays of Norway** 343
F. Oset, V. Thakur, B.K. Dolva, K. Aunaas, M.B. Sæter, A. Robsrud, M. Viklund, T. Nyheim, E. Lyche, and O.A. Jensen

28 **Risk Assessment for Quick Clay Slides – The Norwegian Practice** 355
Bjørn Kalsnes, Vidar Gjelsvik, Hans Petter Jostad, Suzanne Lacasse, and Farrokh Nadim

29 **Quaternary Geology as a Basis for Landslide Susceptibility Assessment in Fine-Grained, Marine Deposits, Onshore Norway** 369
Louise Hansen, Fredrik Høgaas, Harald Sveian, Lars Olsen, and Bjørn Ivar Rindstad

30 **Management of Quick Clay Areas in Slope Stability Investigations – The Göta River Valley** 383
Helen Åhnberg, Hjördis Löfroth, and Karin Lundström

31 **Safety Concepts for Slope Stability** 395
Tim Länsivaara and T. Poutanen

Author Index 409

Subject Index 411

Contributors

Roald Aabøe Norwegian Public Roads Administration (NPRA), Oslo, Trondheim, Norway

Per Aagaard Department of Geosciences, University of Oslo (UiO), Oslo, Norway

Helen Åhnberg Swedish Geotechnical Institute (SGI), Linköping, Sweden

Sam Alpay Natural Resources Canada, Geological Survey of Canada, Ottawa, ON, Canada

Yvonne Andersson-Sköld Swedish Geotechnical Institute (SGI), Linköping, Sweden

Department of Earth Science & COWI, University of Gothenburg, Göteborg, Sweden

Kristian Aunaas Geotechnical and Landslide Division, Norwegian Public Roads Administration (NPRA), Oslo, Norway

Mehrdad Bastani Geological Survey of Sweden (SGU), Uppsala, Sweden

Sara Bazin Norwegian Geotechnical Institute (NGI), Oslo, Norway

Chantal Bilodeau Service de la Géotechnique et de la géologie (Geotechnique and Geology Branch), Ministère de la Sécurité publique du Québec, Québec City, QC, Canada

Gregory R. Brooks Natural Resources Canada, Geological Survey of Canada, Ottawa, ON, Canada

Heather L. Crow Natural Resources Canada, Geological Survey of Canada, Ottawa, ON, Canada

Torleif Dahlin Engineering Geology, Lund University, Lund, Sweden

Einar Dalsegg Geological Survey of Norway (NGU), Trondheim, Norway

Samson A. Degago Norwegian Public Roads Administration (NPRA), Oslo, Trondheim, Norway

Denis Demers Service de la Géotechnique et de la géologie (Geotechnique and Geology Branch), Ministère des Transports du Québec (MTQ), Québec City, QC, Canada

Bjørn K. Dolva Geotechnical and Landslide Division, Norwegian Public Roads Administration, (NPRA), Oslo, Norway

Shane Donohue School of Planning, Architecture and Civil Engineering, Queen's University Belfast, Belfast, Northern Ireland, UK

Arnfinn Emdal Department of Civil and Transport Engineering, Norwegian University of Science and Technology (NTNU), Trondheim, Norway

Gabriel Fabien-Ouellet Département de géologie et de génie géologique, Université Laval, Québec City, QC, Canada

Petter Fornes Norwegian Geotechnical Institute (NGI), Oslo, Norway

Richard Fortier Département de géologie et de génie géologique, Université Laval, Québec City, QC, Canada

Marten Geertsema Ministry of Forests, Lands, and Natural Resource Operations, Prince George, BC, Canada

Bernard Giroux INRS-ETE, Québec City, QC, Canada

Vidar Gjelsvik Norwegian Geotechnical Institute (NGI), Trondheim, Norway

Ingelin Gjengedal Norwegian University of Science and Technology (NTNU), Trondheim, Norway

Norconsult, Molde, Norway

Anders Samstad Gylland Department of Civil and Transport Engineering, Norwegian University of Science and Technology, Trondheim, Norway

SINTEF Building and Infrastructure, Trondheim, Norway

Louise Hansen Geological Survey of Norway (NGU), Trondheim, Norway

Tonje E. Helle Department of Civil and Transport Engineering, Norwegian University of Science and Technology (NTNU), Trondheim, Norway

The Norwegian Public Roads Administration, Trondheim, Norway

M. Hinton Natural Resources Canada, Geological Survey of Canada, Ottawa, ON, Canada

Fredrik Høgaas Geological Survey of Norway (NGU), Trondheim, Norway

Øyvind Høydal Norwegian Geotechnical Institute (NGI), Oslo, Norway

J.A. Hunter Natural Resources Canada, Geological Survey of Canada, Ottawa, ON, Canada

Odd Are Jensen Landslides, Flood and River Management Division, Norwegian Water Resources and Energy Directorate (NVE), Trondheim, Norway

Hans Petter Jostad Norwegian Geotechnical Institute (NGI), Oslo, Norway

Thomas Kalscheuer ETH Zürich, Zürich, Switzerland

Bjørn Kalsnes Norwegian Geotechnical Institute (NGI), Oslo, Norway

H.A. Kornbrekke Norwegian University of Science and Technology, Trondheim, Norway

Suzanne Lacasse Norwegian Geotechnical Institute (NGI), Oslo, Norway

Tim Länsivaara Tampere University of Technology, Tampere, Finland

Isabelle Lecomte International Centre for Geohazards (ICG), Oslo, Norway

Department of Geosciences, University of Oslo (UiO), Oslo, Norway

NORSAR, Kjeller, Norway

Ville Lehtonen Tampere University of Technology, Tampere, Finland

Serge Leroueil Department of Civil Engineering and Water Engineering, Laval University, Québec City, QC, Canada

Jean-Sébastien L'Heureux Norwegian Geotechnical Institute (NGI), Trondheim, Norway

Ariane Locat Department of Civil Engineering and Water Engineering, Laval University, Québec City, QC, Canada

Jacques Locat Laboratoire d'études sur les risques naturels, Department of Geology and Geological Engineering, Laval University, Québec City, QC, Canada

Pascal Locat Service de la Géotechnique et de la géologie (Geotechnique and Geology Branch), Ministère des Transports du Québec (MTQ), Québec City, QC, Canada

Hjördis Löfroth Swedish Geotechnical Institute (SGI), Linköping, Sweden

Michael Long School of Civil, Structural and Environmental Engineering, University College Dublin, Dublin, Ireland

Karin Lundström Swedish Geotechnical Institute (SGI), Linköping, Sweden

Einar Lyche Landslides, Flood and River Management Division, Norwegian Water Resources and Energy Directorate (NVE), Trondheim, Norway

Juho Mansikkamäki Tampere University of Technology, Tampere, Finland

Ragnar Moholdt Norwegian Geotechnical Institute (NGI), Trondheim, Norway

Farrokh Nadim Norwegian Geotechnical Institute (NGI), Oslo, Norway

Daniel Nigussie Department of Civil and Transport Engineering, Norwegian University of Science and Technology (NTNU), Trondheim, Norway

Trude Nyheim Landslides, Flood and River Management Division, Norwegian Water Resources and Energy Directorate (NVE), Trondheim, Norway

Peter O'Connor Apex Geoservices, Gorey, Ireland

Lars Olsen Geological Survey of Norway (NGU), Trondheim, Norway

Forde Oset Geotechnical and Landslide Division, Norwegian Public Roads Administration (NPRA), Oslo, Norway

Lena Persson Geological Survey of Sweden (SGU), Uppsala, Sweden

Martin A. Persson Department of Earth Sciences, University of Gothenburg, Gothenburg, Sweden

Andreas A. Pfaffhuber Norwegian Geotechnical Institute (NGI), Oslo, Norway

Janelle Potvin Service de la Géotechnique et de la géologie (Geotechnique and Geology Branch), Ministère des Transports du Québec (MTQ), Québec City, QC, Canada

T. Poutanen Tampere University of Technology, Tampere, Finland

A.J.M. Pugin Natural Resources Canada, Geological Survey of Canada, Ottawa, ON, Canada

S.E. Pullan Natural Resources Canada, Geological Survey of Canada, Ottawa, ON, Canada

Bjørn Ivar Rindstad Geological Survey of Norway (NGU), Trondheim, Norway

Denis Robitaille Service de la Géotechnique et de la géologie (Geotechnique and Geology Branch), Ministère des Transports du Québec (MTQ), Québec City, QC, Canada

A. Robsrud Technology Division, Norwegian National Railways Administration (NNRA), Oslo, Norway

Magnus Rømoen Norwegian Geotechnical Institute (NGI), Oslo, Norway

Jan Steinar Rønning Geological Survey of Norway (NGU), Trondheim, Norway

M.B. Sæter Technology Division, Norwegian National Railways Administration (NNRA), Oslo, Norway

Rolf Sandven Multiconsult, Trondheim, Norway

Guillaume Sauvin International Centre for Geohazards (ICG), Oslo, Norway

Department of Geosciences, University of Oslo (UiO), Oslo, Norway

NORSAR, Kjeller, Norway

David Schälin Swedish Geotechnical Institute (SGI), Linköping, Sweden

Inger-Lise Solberg Geological Survey of Norway (NGU), Trondheim, Norway

Daniel St-Gelais DSG Management Group Inc, Vancouver, BC, Canada

Pascal Suer Swedish Geotechnical Institute (SGI), Linköping, Sweden

Harald Sveian Geological Survey of Norway (NGU), Trondheim, Norway

Vikas Thakur Geotechnical and Landslide Division, Norwegian Public Roads Administration (NPRA), Oslo, Norway

Catherine Thibault Service de la Géotechnique et de la géologie (Geotechnique and Geology Branch), Ministère des Transports du Québec (MTQ), Québec City, QC, Canada

Johannes Tornborg Skanska, Stockholm, Sweden

J. Kenneth Torrance Geography and Environmental Studies, Carleton University, Ottawa, ON, Canada

Maarten Vanneste International Centre for Geohazards (ICG), Oslo, Norway

Norwegian Geotechnical Institute (NGI), Oslo, Norway

M. Viklund Technology Division, Norwegian National Railways Administration (NNRA), Oslo, Norway

Chapter 1
Landslides in Sensitive Clays – From Geosciences to Risk Management

Jean-Sébastien L'Heureux, Ariane Locat, Serge Leroueil, Denis Demers, and Jacques Locat

1.1 Introduction

Landslides in sensitive clays represent a major hazard in the northern countries of the world such as Canada, Finland, Norway, Russia, Sweden and in the US state of Alaska. Examples of catastrophic landslides in sensitive clays that impacted populations are numerous: e.g., Saint-Jean-Vianney in 1971 (Tavenas et al. 1971; Potvin et al. 2001), Rissa in 1978 (Gregersen 1981; L'Heureux et al. 2012), Finneidfjord in 1996 (Longva et al. 2003), Kattmarka in 2009 (Nordal et al. 2009) and St-Jude in 2010 (Locat et al. 2012). In order to respond to the societal demands, the scientific community has to expand its knowledge of landslide mechanisms in sensitive clay to assist authorities with state-of-the-art investigation techniques, hazard assessment methods, risk management schemes, mitigation measures and planning.

This book gathers the latest scientific research made by international experts dealing with geological, geotechnical and geophysical aspects of slope failure in sensitive clays and focuses on understanding the full spectrum of challenges presented by landslides in such brittle materials. The book consists of a collection

J.-S. L'Heureux (✉)
Norwegian Geotechnical Institute (NGI), Trondheim, Norway
e-mail: jsl@ngi.no

A. Locat • S. Leroueil
Department of Civil Engineering and Water Engineering, Laval University, Québec City, QC, Canada
e-mail: ariane.locat@gci.ulaval.ca

D. Demers
Service de la Géotechnique et de la géologie (Geotechnique and Geology Branch), Ministère des Transports du Québec (MTQ), Québec City, QC, Canada

J. Locat
Laboratoire d'études sur les risques naturels, Department of Geology and Geological Engineering, Laval University, Québec City, QC, Canada
e-mail: Jacques.locat@ggl.ulaval.ca

J.-S. L'Heureux et al. (eds.), *Landslides in Sensitive Clays: From Geosciences to Risk Management*, Advances in Natural and Technological Hazards Research 36,
DOI 10.1007/978-94-007-7079-9_1, © Springer Science+Business Media Dordrecht 2014

of papers prepared for the 1st International Workshop on Landslides in Sensitive Clays (IWLSC), hosted by Laval University, Québec (Canada) in 2013. This is the first international effort focusing on landslides in sensitive clays since the Symposium on Slopes in Soft Clays held in Linköping, Sweden, in 1982 (Bergren and Lindgren 1983). Since then, over three decades have passed during which time the international community has done much work and advancement regarding the understanding of this type of natural hazard. An increased awareness has led to major advances in the development of highly sophisticated geotechnical and geophysical investigation tools as well as mapping procedures including state-of-the-art hazard assessment tools, risk management schemes and mitigation measures. Great progresses in numerical modelling techniques have also lead to a better understanding of slope processes and stability assessment in sensitive clays.

The main objective of the book is to provide an international perspective of landslides in sensitive clays and their consequences by bringing together state-of-the-art contributions from international researchers in academia, industry, governments, as well as planners working on land-use, hazard mapping and risk management. The book provides a collection of 30 papers presented at the 1st IWLSC which are broadly organized under the following topics: (1) Sensitive clays: source, nature and development; (2) Landslide characterization; (3) Geotechnical and geophysical site investigations; (4) Modelling, slope stability and progressive failure; and (5) Hazard assessment, risk management, regulations and policies.

1.2 Part I: Sensitive Clays: Source, Nature and Development

The first section of this book focuses on a fundamental aspect of sensitive clays which consists in understanding the source, nature and development of such materials. The term sensitivity (S_t) is defined as the ratio of the undrained shear strength of the intact clay (s_u) over the undrained shear strength of the remoulded clay (s_{ur}) at the same water content. It is thus a measure of the distance from the intact soil properties to those of the remoulded soil. Factors promoting high undisturbed shear strength and/or low remoulded shear strengths are thus necessary for the development of high sensitivities in fine-grained sediments. The essential factors for developing high sensitivities in clay are synthesized in Torrance (1983, 2012; Chap. 2, this volume). These factors relate to e.g. mineralogy (Chap. 3 by Locat and St-Gelais, this volume), pore water chemistry (Chap. 6 by Helle et al., this volume; Chap. 5 by Suer et al., this volume), stratigraphy (Chap. 4 by Persson, this volume), grain size and consolidation history. In turn, all of these factors link to depositional and/or post-depositional processes (Chap. 2 by Torrance, this volume).

The highly sensitive clays or quick clays of the World have all developed in a similar pattern during and following the most recent glacial periods (Chap. 2 by Torrance, this volume; Chap. 29 by Hansen et al., this volume; Chap. 3 by Locat and

St-Gelais, this volume). They first accumulated into the various postglacial seas during and following the last glaciation. As they settled to the seafloor, the silt and clay particles quickly flocculated into aggregates. Flocculation of silt and clay particles in marine waters produced an interlinked, random and "open" structure with a high water content (Chap. 12 by L'Heureux et al., this volume). Following the ice retreat in the northern countries the sea level rose slower than the local land surface, resulting in an apparent marine regression and the emergence of the old sea bottom (e.g. Gadd 1988; Andersen and Borns 1994). The clays became exposed to a flux of fresh groundwater which gradually leached the high salinity pore water from the sediment. Leaching is considered as the most important factor in the development of high sensitivity in the marine sediments as it leads to an increase in inter-particle repulsive forces (Rosenqvist 1953; Torrance 1975, 1983, Chap. 2, this volume). Because of this the leached material behaves more or less as a liquid following structural collapse; hence the term "quick" clay.

The role of mineralogy on the behaviour of sensitive clay sediments has been discussed by e.g. Quigley (1980), Torrance (1983), Torrance and Ohtsubo (1995), Locat (1995), and Locat et al. (2003). Locat and St-Gelais (Chap. 3, this volume) show that the clays from Québec are of low activity and contain little or no swelling clay minerals. As discussed by these authors, there are similar mineralogical trends between the sensitive clays from Québec and those found in areas which have experienced similar geological history (i.e. Ontario, British Columbia, Labrador, Scandinavia and parts of Russia and Alaska).

Chemical factors such as salinity and composition of ions in the pore water are important attributes in development of sensitive clays (Chap. 2 by Torrance, this volume; Chap. 6 by Helle et al., this volume; Chap. 5 by Suer et al., this volume). Changes in ions concentration and salinity may also contribute to strengthening the bonds between the minerals in clay. Such changes may be induced artificially through wells for soil stabilization (Chap. 6 by Helle et al., this volume), or naturally through groundwater flow and ion exchange (Chap. 5 by Suer et al., this volume).

1.3 Part II: Landslide Characterization

Landslide characterization includes all stages of slope movement referred to as pre-failure, onset of failure, post-failure and reactivation stages (Leroueil et al. 1996; Leroueil 2001; Vaunat and Leroueil 2002). Triggering of landslide in sensitive clay is usually attributed to natural factors (e.g. river erosion, rainfall, etc.), human activity (e.g. placement of fill), or a combination of both. Recent experience also shows that vibrations from blasting (e.g. Johansson et al. 2013) and from earthquake (e.g. Perret et al. 2011) can trigger landslide in sensitive clays. In this respect, Brooks (Chap. 10, this volume) shows that a large number of pre-historical landslides found in the Ottawa valley region could be linked to the paleoseismicity of the area.

Following the terminology of Cruden and Varnes (1996), the main types of landslides encountered in sensitive clays are usually referred to as flows and spreads (see also Tavenas 1984 and Karlsrud et al. 1984). Flows, also denoted as flowslides, are by far the most important landslide type in Scandinavia (L'Heureux 2012; Chap. 8 by Thakur et al., this volume). In comparison, they represent 57 % of the large landslides occurring in the Province of Québec, the rest being spread (38 %) and unidentified large landslides (5 %) (Chap. 7 by Demers et al., this volume).

Whereas at the pre-failure and failure stages, soil mechanics principles are needed to explain landslide initiation, the post-failure mechanisms may rather rely on a combination of soil mechanics and fluid dynamic principles (Locat and Demers 1988; Leroueil et al. 1996; Locat 1997; Thakur and Degago 2012). To avoid the difficulties in assessing the post-failure movements in terms of retrogression and run-out distance, hazard assessment methodologies in sensitive clay terrain are very much based on statistical data and historical records (Chap. 28 by Kalsnes et al., this volume; Chap. 26 by Potvin et al., this volume; Chap. 30 by Åhnberg et al., this volume). Detailed landslide inventories, such as those presented by Demers et al. (Chap. 7, this volume) and Thakur et al. (Chap. 8, this volume), are thus of great relevance for the hazard mapping methodologies used in eastern Canada and in Scandinavia, respectively. To this aim, the development and increase use of airborne light detection and ranging (LiDAR) in recent years has given us a better possibility to investigate the morphology and the failure mechanisms for historical and pre-historical landslides (Jaboyedoff et al. 2009; Chap. 10 by Brook, this volume, Chap. 7 by Demers et al., this volume).

As discussed by e.g. Mitchell and Markell (1974), Tavenas et al. (1983), Geertsema and L'Heureux (Chap. 9, this volume) and Thakur et al. (Chap. 8, this volume), important factors governing the retrogression of landslide in sensitive clays involve aspects of geometry, material properties and other external controls (e.g. network of trees). Many authors also agree to the fact that large retrogressive landslides in sensitive clays only occur when the remoulded shear strength of the soil is less than 1.0 kPa and when the liquidity index is larger than 1.2 (Lebuis et al. 1983; Tavenas et al. 1983; L'Heureux 2012; Chap. 8 by Thakur et al., this volume; Chap. 7 by Demers et al., this volume).

The mobility of the debris in a sensitive clay slide is a function of the potential energy available, the energy needed to remould the clay mass after failure, the flow capacity of the soil (i.e. its liquidity index) and the environment in which the landslide occurs (Chap. 11 by Locat et al. 2008, this volume). In support to empirical correlation, numerical analyses such as those presented for the 1980 Havre-St-Pierre landslide (Chap. 11 by Locat et al., this volume) can be used to obtain relatively good estimates of run-out distances for landslides in sensitive clays.

Landslide in sensitive clays may, in some instances, be limited to small displacements due to e.g. morphological constrains, soil properties and stratigraphy. For such landslides the mobilised shear strength along the newly formed slip surface will correspond to the remoulded strength of the soil. Mitigation measures are therefore difficult and expensive. L'Heureux et al. (Chap. 12, this volume) studied the evolution of material properties at the depth of a failure surface for a landslide from northern Norway. They show that, with time, dissipation of excess pore pressure and consolidation of the remoulded clay at the depth of failure slowly leads to an increase in shear strength; from remoulded strength and towards values slightly higher than typically found for normally consolidated clays (i.e. strengthening).

1.4 Part III: Integrated Geotechnical and Geophysical Site Investigations

The combination of geotechnical and geophysical methods for characterization of sensitive clay formations has advanced considerably over the last decade thanks to improve technology, field equipment and software. Collaboration between the academia and the industry has enhanced the use of several geophysical methods for investigations in sensitive clay terrain, and for estimation of soil parameters (Long and Donohue 2007, 2010; Long et al. 2012; Chap. 20 by Crow et al., this volume; Chap. 13 by Donohue et al., this volume). Electric Resistivity Tomography (ERT) and Resistivity Cone Penetrometer Testing (RCPT) appear to be the most powerful and popular geophysical techniques for sensitive clay investigation owing to the relatively strong relationship between resistivity and a number of geotechnical properties (Chap. 17 by Dahlin et al., this volume; Chap. 13 by Donohue et al. this volume; Chap. 15 by Pfaffhuber et al., this volume; Chap. 16 by Sandven and Solberg, this volume; Chap. 18 by Sauvin et al., this volume; Chap. 14 by Solberg et al., this volume). Nevertheless, the various authors in this section strongly stress the fact that resistivity measurements alone cannot be used to characterize the sensitivity and the geotechnical properties of a clay deposit. Resistivity measurements should be combined with traditional geotechnical methods, such as sounding and sampling. Moreover, the reader should be aware of the differences in the thresholds recommended in order to differentiate between leached and unleached clay materials (Chap. 17 by Dahlin et al., this volume; Chap. 15 by Pfaffhuber et al., this volume, Chap. 16 by Sandven and Solberg, this volume). Site-specific resistivity ranges are recommended to tell apart non-sensitive clays from quick clays.

Some of the limitations of the ERT methods lie in the geophysical data inversion and depth resolution, especially close to abrupt interfaces such as clay/bedrock (Chap. 15 by Pfaffhuber et al., this volume). An approach where data from different geophysical methods are integrated with geotechnical sounding results and where ultimately joint inversion is performed may be beneficial for proper imaging of sensitive clay terrain (Chap. 13 by Donohue et al., this volume; Chap. 19 by Fabien-Ouellet et al., this volume; Chap. 18 by Sauvin et al., this volume). Altogether, this section emphasizes the need for multidisciplinary studies and collaboration between geologists, geophysicists and geotechnical engineers for a better characterization and mapping in sensitive clay terrains.

1.5 Part IV: Slope Stability, Modelling and Progressive Failure

Sensitive clays may exhibit brittle strain-softening behaviour under shear with peak shear strength and lower shear strength at large deformation. The undrained strain-softening behaviour of clays is dependant, among other factors, of the shear induced pore water pressures, as explained by Thakur et al. (Chap. 8 by this volume). The understanding and characterization of the mechanical behaviour of sensitive clays during shear, through laboratory experiments (Chap. 21 by Gylland, this volume) or on the field (Chap. 12 by L'Heureux et al., this volume), is essential for analysing slope stability in these materials. This section presents the continuing work needed to accommodate this particular behaviour of sensitive clays in stability analysis regarding first time landslides (Lo 1972; Lo and Lee 1973; Lefebvre 1981; Chap. 24 by Jostad et al., this volume; Chap. 25 by Länsivaara et al., this volume) and in the understanding of failure mechanisms of large landslides (Bernander 2000, 2011; Quinn et al. 2011; Chap. 11 by Locat et al. 2011 and this volume; Locat 2012; Gylland 2012; Chap. 21 by Gylland this volume).

In engineering practice, the characterisation of strain-softening behaviour of clays and its implementation in stability analysis is not straightforward. Jostad et al. (Chap. 24, this volume) established, with a large number of finite element analyses of designed of fills on sensitive clays, a scaling factor giving the ratio between the calculated capacity of a soil to undrained loading without and with the effect of softening. This enables to indirectly take into account the strain-softening behaviour of sensitive clays in a regular limit equilibrium method (LEM) using the peak shear strength. Moreover, Länsivaara et al. (Chap. 25, this volume) presents methods to take shear induced pore water pressures into account in effective stress stability analyses, using LEM and finite element method (FEM). It is seen that both methods give acceptable results. However, LEM require the knowledge of the shear induced pore water pressures and the FEM require detailed knowledge of the yield surface of the clay.

Strain-softening behaviour may results in progressive failure, which can explain large landslides in sensitive clays (Bernander 2000, 2011; Locat et al. 2011; Gylland 2012). Recent developments in progressive failure modelling using FEM explain how a slope can fail under a small triggering load located in the slope, in case of downward progressive failure, and unloading located near the toe of the slope. Gylland (Chap. 21, this volume) showed the dependency to shear rate of shear band mechanical behaviour in clay using a modified triaxial cell. Locat et al. (Chap. 11, this volume) extended this mechanism to the initiation and extent of spreads, large landslides in sensitive clays, and the dislocation of the soil mass in blocks of more or less remoulded material having horsts and graben morphologies.

These studies generally conclude that brittleness of sensitive clays is the main governing aspect of progressive failure. Much work is still needed on the understanding of post-peak shear behaviour of sensitive clays and shear band formation (shear strain localization). Additional back calculation of well documented cases of large landslides is also necessary in order to increase our knowledge and understanding of these large landslides.

1.6 Part V: Hazard Assessment, Risk Management, Regulations and Policies

Landslides in sensitive clays pose a major socio-economic problem in the northern countries of the world because of their retrogressive potential and high mobility. Already in the 1970s and 1980s programs were established in the different countries to map large areas prone to landslide in sensitive clays (c.f. Viberg 1984). During the two last decades, the concepts of hazard, consequence and risk were efficiently incorporated into the different programs. The increase awareness has led to comprehensive hazard and risk assessment methodologies, management strategies, regulations and policies. In this section, authors present the different strategies adopted in Québec (Chap. 26 by Potvin et al., this volume), in Norway (Chap. 28 by Kalsnes et al., this volume; Chap. 27 by Oset et al., this volume) and in some parts of Sweden (Chap. 30 by Ånhberg et al., this volume) when dealing with the hazard and risk associated to landslides in sensitive clays. At this point in time, most of risk assessment methodologies are seen as qualitative/semi-quantitative (Chap. 28 by Kalsnes et al., this volume; Chap. 26 by Potvin et al., this volume).

Several differences exist between the various hazard mapping programs found in the province of Québec, Norway and Sweden (e.g. Viberg 1984; Lundström and Andersson 2008). For instance, cementation and overconsolidation generally lead to higher shear strength in the Canadian clays, compared with the Swedish and Norwegian ones, which makes it possible for steeper slopes to be stable. Therefore, the lowest slope inclination criterion used to identify areas to be mapped in the province of Québec is 14° while the slope inclination criterion used in Norway is 1:15 (~4°) and in Sweden 1:10 (~6°). In Norway, hazard and risk are evaluated only in areas where quick clay is found (i.e. clay with a remoulded shear strength less than 0.5 kPa). In Québec, delimitation of areas into different zones of susceptibility is influenced by the presence of sensitive soils but areas may get the highest degree of susceptibility even if sensitive soils are not present (e.g. Robitaille et al. 2002). In Sweden there is no requirement for investigations of quick clay occurrence or sensitivity (Lundström and Andersson 2008; Chap. 30 by Ånhberg et al., this volume). Furthermore, the Québec method divides the hazard depending on the type of mass movement that can occur; areas with prerequisites for initial rotational or superficial landslides, areas in which a large retrogression could occur and areas potentially affected by landslide debris downstream. In the Norwegian method, the area is divided in larger zones, each one corresponding to a potential landslide zone regardless of the type of movement. In the Swedish method, a hazard zone is not equivalent to a potential landslide zone, the results only indicate that more investigations are necessary (Lundström and Andersson 2008). In Québec, risk management covers several aspects: mandatory taking into account in municipalities regulations of mapping results; monitoring plans in landslide prone areas; preventive stabilization works in high risk zones; raising awareness by training sessions and public meetings (Chap. 26 by Potvin et al., this volume).

For each country, regional and local planning in sensitive clay terrain involves a detailed ground investigation program, laboratory tests and slope stability assessments. In Europe, the concepts of partial safety factors and reliability based design are used for slope stability evaluation in soft and sensitive clays (Chap. 31 by Lansivaara and Poutanen, this volume; Chap. 27 by Oset et al., this volume). This approach offers a framework to account for all the uncertainties as well as the consequences of failure. Methods for quick clay landslide stability calculations taking into account the brittle behaviour of the material are also under development in Norway (Chap. 24 by Jostad et al., this volume; Chap. 28 by Kalsnes et al., this volume).

Hazard and risks maps associated to landslide in sensitive clays are now available to the public through the web and other media in several of the Nordic countries. These maps are usually crude and will never replace detailed investigations and quantitative slope stability evaluation. In order to be effective, such maps require training and awareness-raising activities towards the public. The impacts and the associated regulations on the real estate sector need to be explored further (e.g. Chap. 26 by Potvin et al., this volume). In the future, such maps should also include the potential impact of climate changes.

1.7 Conclusions and Outlook

Scientific and engineering research on "Landslide in Sensitive Clays" has advanced significantly during the last decades owing to elevated public awareness and collaborations between the academia and the industry. Also, the use of geophysical methods for site characterization, state-of-the-art laboratory tests and the great strides made by modellers has given us new possibilities to better understand the development of sensitive clay, to perform more realistic slope stability analyses, and to better map the hazard posed by such landslides.

Despite these great advances, there are yet a number of areas in which further progresses are required. Firstly, there are some inconsistencies throughout the literature regarding the use of sensitivity scales and the definition of the term "quick clay". In Sweden, quick clay is defined as a clay with a sensitivity of 50 or more, and a fully remoulded shear strength of less than 0.4 kPa (Karlsson and Hansbo 1989; Chap. 30 by Åhnberg et al., this volume). In Norway, quick clays are defined as clays with a remoulded shear strength of less than 0.5 kPa and a with a sensitivity of 30 or more (NGF 1974; Thakur and Degago 2012). Other sensitivity scales have also been proposed by e.g. Rosenquist (1953), Broms and Stål (1980), Le Bihan and Leroueil (1981). In Canada, the description of the sensitivity is based on the classification found in the 4th Edition of the Canadian Foundation Engineering Manual (CGS 2006). Here, the sensitivity scale was recently modified to incorporate the descriptor "quick" when a clay presents a sensitivity of 16 or more (CGS 2013). All in all, different sensitivity scales are used worldwide and this often creates confusions in the literature, in particular in the domain of slopes in clay. In the future, the scientific community should preferably agree on a common scale.

At a first glance, other important questions and themes which need to be tackled in the upcoming years are: (1) How to account for strain-softening behaviour of clays in safety factors calculation; (2) How to account for the effect of shear strain localization in the laboratory and in the field; (3) Need for better understanding the physical processes involved in the transition from failure to post-failure for a better prediction of landslide mobility and retrogression; (4) Proper tools and/or methods to assess the retrogression potential in sensitive clay terrain; (5) More work on the relationships between geophysical and geotechnical parameters; (6) How to perform more quantitative hazard and risk assessment when dealing with landslides in sensitive clays; (7) Tools for early warning systems and stabilization methods. These points were raised as part of preparing for the 1st IWLSC. It is hope that as discussion takes place many more ideas will be generated toward a better understanding of the mechanisms involved in landslides in quick clays and how to strengthen our capabilities for quantitative risk assessment.

Finally, the overall goal of the 1st IWLSC is to bring together specialists from different geo-scientific fields, both from academia and industry, to share knowledge and common problems regarding landslide in sensitive clays. A main aim in the future is to continue this international arrangement so that geologists, geophysicists, geotechnical engineers, physicists, modellers, and others all join forces in understanding as much as possible the many aspects within the wide theme of landslide in sensitive clays.

References

Andersen BG, Borns HW (1994) The ice age world. Scandinavian Univ. Press, Oslo, 208 pp

Bergren B, Lindgren J (eds) (1983) Symposium on slopes on soft clays. Swedish Geotechnical Institute Report No. 17, Linkoping, pp 455–461

Bernander S (2000) Progressive landslides in long natural slopes. MSc thesis, Luleå University of Technology

Bernander S (2011) Progressive landslides in long natural slopes, formation, potential extension and configuration of finished slides in strain-softening soils. PhD thesis, Department of Civil and Mining Engineering, Luleå University of Technology

Broms BB, Stål T (1980) Landslides in sensitive clays. In: Proceedings: international symposium on landslides, vol 2. New Delhi, pp 39–66, 7–11 April 1980

Canadian Geotechnical Society (CGS) (2006) Canadian foundation engineering manual, 4th edn. c/o BiTech Publisher Ltd, Richmond, 506 p

Canadian Geotechnical Society (CGS) (2013) Errata – Canadian foundation engineering manual, 4th edn. http://www.cgs.ca/engineering-manual.php

Cruden DM, Varnes DJ (1996) Landslides types and processes. In: Turner AK, Schuster RL (eds) Landslides investigation and mitigation, special report 247, Transportation, Research Board, National Research Council, National Academy press, Washington, DC, pp 37–75

Gadd NR (1988) The basin, the ice, the Champlain Sea. The late quaternary development of the Champlain Sea basin. GAC special paper 35. Edited by Gadd NR. Geological Association of Canada. Ottawa, Ont. pp 15–24

Gregersen O (1981) The quick clay landslide in Rissa, Norway. NGI Publication, vol 135, pp 1–6

Gylland AS (2012) Material and slope failure in sensitive clays. PhD thesis, The Norwegian University of Sciences and Technology

Jabeyedoff M, Demers D, Locat J, Locat A, Locat P, Oppikofer T, Robitaille D, Turmel D (2009) Use of terrestrial laser scanning for the characterization of retrogressive landslides in sensitive clay and rotational landslides in river banks. Can Geotech J 46:1379–1390

Johansson J, Løvolt F, Andersen KH, Madshus C, Aabøe R (2013) Impact of blast vibrations on the release of quick clay slides. In: Proceedings of the 18th ICSMGE, Paris

Karlsrud K, Aas G, Gregersen O (1984) Can we predict landslide hazards in soft sensitive clays? Summary of Norwegian practice and experiences. In: Proceedings of the 4th international symposium on landslides, Toronto, vol 1, 16–21 September 1984. University of Toronto Press, Toronto, pp 107–130

Karlsson R, Hansbo S (1989) Soil classification and identification. Byggforskningsrådet. Dokument D8:1989. Stockholm

L'Heureux JS (2012) A study of the retrogressive behaviour and mobility of Norwegian quick clay landslides. In: Eberhardt E, Froese C, Turner AK, Leroueil S (eds) Landslides and engineered slopes – protecting society through improved understanding, vol 1. CRC Press, London, UK, pp 981–988

L'Heureux JS, Eilertsen RS, Glimstad S, Issler D, Solberg I, Harbitz CB (2012) The 1978 quick clay landslide at Rissa, mid-Norway: subaqueous morphology and tsunami simulations. In: Yamada Y et al (eds) Sub-marine mass movements and their consequences, vol 31, Advances in natural and technological hazards research. Springer, Dordrecht, pp 507–516

Le Bihan JP, Leroueil S (1981) The fall cone and the behaviour of remoulded clay. Terratech Ltd. Research Report, Montreal

Lebuis J, Robert JM, Rissmann P (1983) Regional mapping of landslide hazard in Quebec. In: Proceedings of the symposium on slopes on soft clays. Swedish Geotechnical Institute report no. 17, Linköping, Sweden, pp 205–262

Lefebvre G (1981) Fourth Canadian geotechnical colloquium: strength and slope stability in Canadian soft clay deposits. Can Geotech J 18(3):420–442

Leroueil S (2001) Natural slopes and cuts: movement and failure mechanisms. Géotechnique 51(3):197–243

Leroueil S, Vaunat J, Picarelli L, Locat J, Faure R, Lee H (1996) A geotechnical characterisation of slope movements. In: Senneset K (ed) Proceedings of the 7th international symposium on landslides, Trondheim, 1. Balkema, Rotterdam, pp 53–74

Lo KY (1972) An approach to the problem of progressive failure. Can Geotech J 9(4):407–429

Lo KY, Lee CF (1973) Analysis of progressive failure in clay slopes. In: Proceedings of the 8th international conference on soil mechanic and foundation engineering, vol 1, Moscow, August 1973. ICSMFE Publications, USSR, pp 251–258

Locat J (1995) On the development of microstructure in a collapsible soils. NATO workshop. In: Derbyshire E et al (eds) Genesis and properties of collapsible soils. Kluwer Academic Publishers, Dordrecht, pp 93–128

Locat J (1997) Normalized rheological behaviour of fine muds and their flow properties in a pseudoplastic regime. In: Chen CH (ed) Debris-flow hazards mitigation: mechanics, prediction, and assessment. ASCE, New York, pp 260–269

Locat A (2012) Rupture progressive et étalements dans les argiles sensibles. PhD thesis, Université Laval

Locat J, Demers D (1988) Viscosity, yield stress, remolded strength, and liquidity index relationships for sensitive clays. Can Geotech J 25:799–806

Locat J, Tanaka H, Tan TS, Dasari GR, Lee H (2003) Natural soils: geotechnical behavior and geological knowledge. In: Characterisation and engineering properties of natural soils, vol 1 (Proc. Singapore Workshop), Balkema, Swets and Zeitlinger, Lisse, pp 3–28

Locat P, Leroueil S, Locat J (2008) Remaniement et mobilité des débris de glissements de terrain dans les argiles sensible de l'est du Canada. In: Proceedings of the 4th Canadian conference on geohazards: from causes to management. Presse de l'Université Laval, Québec, pp 97–106

Locat A, Leroueil S, Bernander S, Demers D, Jostad HP, Ouehb L (2011) Progressive failures in Eastern Canadian and Scandinavian sensitive clays. Can Geotech J 48(11):1696–1712

Locat P, Demers D, Robitaille D, Fournier T, Noël F, Leroueil S, Locat A, Lefebvre G (2012) The Saint-Jude landslide of May 10, 2012, Québec, Canada. In: Eberhardt E, Froese C, Turner AK, Leroueil S (eds) Landslides and engineered slopes – protecting society through improved understanding, vol 1. CRC Press, London, UK, pp 635–640

Long M, Donohue S (2007) In situ shear wave velocity from multichannel analysis of surface waves (MASW) tests at eight Norwegian research sites. Can Geotech J 44:533–544

Long M, Donohue S (2010) Characterization of Norwegian marine clay with combined shear wave velocity and piezocone cone penetration test (CPTU) data. Can Geotech J 47:709–718

Long M, Donohue S, L'Heureux JS, Solberg IL, Rønning JS, Limacher R, O'Connor P, Sauvin G, Rømoen M, Lecomte I (2012) Relationship between electrical resistivity and basic geotechnical parameters for marine clays. Can Geotech J 49(10):1158–1168

Longva O, Janbu N, Blikra LH, Boe R (2003) The 1996 Finneidfjord slide: seafloor failure and slide dynamics. In: Locat J, Mienert J (eds) Submarine mass movements and their consequences. Kluwer, Dordrecht, pp 531–538

Lundström K, Andersson M (2008) Hazard mapping of landslides, a comparison of three overview mapping methods in fine-grained soils. In: J Locat, D Perret, D Turmel, D Demers, S Leroueil (eds) Proceedings of the 4th Canadian conference on geohazards: from causes to management. Presse de l'Université Laval, Québec

Mitchell RJ, Markell AR (1974) Flowsliding in sensitive soils. Can Geotech J 11:11–31

Nordal S, Alén C, Emdal A, Jendeby L, Lyche E, Madshus C (2009) Skredet i Kattmarkvegen i Namsos 13. mars 2009 – Rapport fra undersøkelsesgruppe satt ned av Samferdselsdepartementet. Tapir Uttrykk, Trondheim (In Norwegian)

Norsk Geoteknisk Forening (NGF) (1974) Retningslinjer for presentasjon av geotekniske undersøkelser. Oslo, 16 p (In Norwegian)

Perret D, Mompin R, Bossé F, Demers D (2011) Stop 2-5B: the Binette road earth flow induced by the June 23rd, 2010 Val-des-Bois earthquake. In: Russell HAJ, Brooks GR, Cummins DI (eds) Deglacial history of the Champlain Sea basin and implications for urbanization. Joint annual meeting GAC-MAC-SEG-SGA, Ottawa, Ontario, May 25–27, 2011, Field Guide Book, pp 72–74. Geological Survey of Canada, Open File 6947

Potvin J, Pellerin F, Demers D, Robitaille D, La Rochelle P, Chagnon J-Y (2001) Revue et investigation supplémentaire du site du glissement de Saint-Jean-Vianney. Paper presented at the 54th Canadian geotechnical conference, Calgary, vol 2, pp 792–800

Quigley RM (1980) Geology, mineralogy and geochemistry of Canadian soft soils: a geotechnical perspective. Can Geotech J 17:261–285

Quinn PE, Diederichs MS, Rowe RK, Hutchinson DJ (2011) A new model for large landslides in sensitive clay using a fracture mechanism approach. Can Geotech J 48(8):1151–1162

Robitaille D, Demers D, Potvin J, Pellerin F (2002) Mapping of landslide-prone areas in the Saguenay region, Québec, Canada. In: Proceedings of the international conference on instability – planning and management, Ventnor, Isle of Wight, UK, pp 161–168

Rosenqvist IT (1953) Consideration on the sensitivity of Norwegian quick-clays. Geotechnique 3:195–200

Tavenas F (1984) Landslides in Canadian sensitive clays – a state of the art. In: Proceedings of the 4th international symposium on landslides. Canadian Geotechnical Society, vol 1, pp 141–153

Tavenas F, Chagnon JY, La Rochelle P (1971) The Saint-Jean-Vianney landslide: observations and eyewitnesses accounts. Can Geotech J 8:463–478

Tavenas F, Flon P, Leroueil S, Lebuis J (1983) Remoulding energy and risk of slide retrogression in sensitive clays. In: Proceedings of the symposium on slopes on soft clays, Linköping, Sweden. Swedish Geotechnical Institute, SGI report no. 17, pp 423–454

Thakur V, Degago S (2012) Quickness of sensitive clays. Geotech Lett 2:81–88

Torrance JK (1975) On the role of chemistry in the development and behaviour of the sensitive marine clays of Canada and Scandinavia. Can Geotech J 12:326–335

Torrance JK (1983) Towards a general model of quick clay development. Sedimentology 30:547–555

Torrance JK (2012) Landslides in quick clay. In: Clague JJ, Stead D (eds) Landslides: types, mechanisms and modeling. Cambridge University Press, Cambridge

Torrance JK, Ohtsubo M (1995) Ariake bay quick clays: a comparison with the general model. Soils Found 35:11–19

Vaunat J, Leroueil S (2002) Analysis of post-failure slope movements within the framework of hazard and risk analysis. Nat Hazards 26:83–102

Viberg L (1984) Landslide risk mapping in soft clays in Scandinavia and Canada. In: Proceedings of the 4th international symposium on landslides, vol 1, Toronto

Part I
Sensitive Clay: Source, Nature and Development

Chapter 2
Chemistry, Sensitivity and Quick-Clay Landslide Amelioration

J. Kenneth Torrance

Abstract Quick clay's combination of sufficient undisturbed shear strength to be stable *in situ* and liquefaction upon structural failure derives from chemical factors that act during and after sediment accumulation. Flocculation of silt and clay particles in marine water produces an interlinked, random structure with a water content that approximates its high-salinity liquid limit. Co-sedimentation of cementing agents may augment its strength. Upon uplift above sea level, fresh water displacement of the high-salinity pore water decreases the liquid limit while the structure and water content remain almost constant, yielding a liquidity index in the range of about 1.5–4. The central premise of this paper is that development of solutions begins with consideration of how problems have arisen. Specifically, chemistry's dominating role in quick clay development provides insights into approaches that can be applied to amelioration of the quick clay landslide problem.

Keywords Quick clay • Origin • Landslides • Post-failure behavior • Chemical controls

2.1 Introduction

Sensitivity (St), the ratio of the undisturbed soil strength to its post-failure strength is a critical concept (Skempton and Northey 1952). For designation of a soil as 'sensitive', St must be >1. For designation as 'quick clay' (a particularly problematic sediment), the most important criterion is that the post-failure strength must be so low (<0.5 or 0.4 kPa (by fall cone test)) that the thoroughly disturbed soil behaves as a liquid. A minimum sensitivity criterion is also commonly applied, with St>30

J.K. Torrance (✉)
Geography and Environmental Studies, Carleton University,
Ottawa, ON, Canada K1S 5B6
e-mail: Ken.Torrance@carleton.ca

J.-S. L'Heureux et al. (eds.), *Landslides in Sensitive Clays: From Geosciences to Risk Management*, Advances in Natural and Technological Hazards Research 36,
DOI 10.1007/978-94-007-7079-9_2, © Springer Science+Business Media Dordrecht 2014

being commonly, but not universally, accepted. Sensitive clays are present in both sub-aerial and sub-aqueous (marine or lake) settings, while quick clays are almost entirely restricted to sub-aerial settings, even though their accumulation and early consolidation stages occurred in sub-aqueous (generally sub-marine) settings.

The quick clays of the world have developed in sediments that accumulated during the most recent late-glacial and early post-glacial periods when the weight of the continental ice sheets had depressed the underlying land surface to a greater extent than ice accumulation had lowered global sea level. The glaciers delivered debris, ranging in size from boulders to clay-sized rock flour, to the glacier/sea contact zone. Coarse material accumulated close to the glacial snout and the silt and clay particles were carried into the sea where they quickly flocculated into aggregates and settled to the sea floor. When ice sheets retreated from direct contact with the sea, their melt waters continued to deliver the fine particles to the sea; sand formed deltas at river mouths, and silt and clay particles were still carried seaward where flocculation and sediment accumulation continued. As the ice sheets retreated farther and farther, the local land surface rebounded faster than sea level rose, eventually lifting more and more of the old sea bottom above sea level. With exposure to subaerial processes, a weathering crust formed in the surface zone. Movement of fresh water through the sediment, whether by infiltration and downward movement from the surface or by upward movement driven by artesian pressures originating in nearby uplands, gradually leached the high-salinity pore water from the sediment. The flocculated microstructure remained intact and kept the water content almost constant during leaching, but the material properties changed. The original high-salinity sediment, if thoroughly disturbed, would behave as a plastic solid/extremely viscous liquid, with its water content being approximately at its high-salinity liquid limit. After leaching, the low-salinity sediment (at the same water content) behaves as a liquid. The high-salinity sediment was 'sensitive clay'; the leached sediment has become 'quick clay'.

2.2 Sensitivity Development

The essential factors for development of high sensitivities in fine-grained sediments have been synthesized into a general model for quick clay development (Fig. 2.1) (Torrance 1983, 2012). Chemical factors play critical roles in all four quadrants of the model, and they differ somewhat for fresh water and marine sediments.

2.2.1 Development of High Undisturbed Strength

2.2.1.1 Depositional Factors

Development of high undisturbed strength in high water content sediment requires the formation of an interconnected structure that is able to limit the extent of consolidation as further sediment accumulates. Inter-particle flocculation whereby

Fig. 2.1 The general model for quick clay development. [1]Essential in marine clays. [2]Essential in fresh water clays (Modified from Torrance 1983, 2012)

Factors producing high undisturbed strengths	
Depositional	Post-depositional
Flocculation [1,2] • Salinity [1] • Polyvalent cations dominant [2] • High suspension concentration	Cementation • Rapidly developed • Slowly developed Slow load increase Time-dependent diagenetic changes
Factors producing low remolded strengths	
Depositional	**Post-depositional**
Material properties • Domination by low activity minerals [1,2]	Decrease in liquid limit < decrease in water content • Leaching of salt [1] • Dispersants [2] Reducing conditions • Inhibition of high-swelling-smectite formation Minimal consolidation

[1]Essential in marine clays. [2]Essential in fresh water clays.

the net-negatively-charged clay and silt particles cluster into multi-particle floccules with random particle orientation produces high-water content sediment that is capable of developing increasing resistance to further consolidation as sediment accumulation slowly increases the load. In salty water, suppression of the repulsive forces between the negatively charged silt and clay particles allows flocculation to occur. In fresh-water, when incoming sediment concentrations are sufficiently high, calcium (Ca+2), as the dominant adsorbed cation, suppresses the repulsive forces and allows flocculation. In both case the result is that the sediment retains more water than if the particles had not flocculated before settling. Higher suspension concentrations speed up flocculation in both saline and fresh-water.

2.2.1.2 Post-depositional Factors

The undisturbed shear strength may be further enhanced by slow load increase, diagenetic change and post-depositional cementation. Slow sediment accumulation allows time for minor adjustments to the flocculated structure to accommodate the increasing load. Time-dependent diagenetic changes affecting particle mineralogy appear to be of little or no importance in most quick clays, but may play a role in the unusual quick clays in Kyushu, Japan (Egashira and Ohtsubo 1982; Torrance and Ohtsubo 1995). Cementation, which is present in many (not all) eastern Canadian quick clays, but is insignificant in Scandinavian and most other quick clays, is the most important post-depositional factor. Carbonates, iron oxides and hydrous oxides and 'amorphous' minerals have been proposed by a number of authors as being potential cementing agents (see Torrance 1983). Further research has determined that most of what was designated as 'amorphous' minerals is detrital, nano-sized, crystalline hematite and magnetite (Torrance et al. 1986) or was extracted from the illite and chlorite by excessively aggressive reagents (Torrance 1995). Carbonates, which are present in many quick clays, are not effective cementing agents because small crystals tend to dissolve, with reprecipitation onto the lower energy surfaces of larger carbonate crystals. This produces isolated concretions, but not general cementation. Torrance (2012) argued that the iron oxides are the only potential cementing agents that are present in sufficient quantity to be effective. There is reason to believe that the oxides preferentially accumulate at inter-particle contact points during flocculation.

2.2.2 Development of Low Remolded Strength

2.2.2.1 Depositional Factors

The essential depositional requirement for achieving low remolded strength is that minerals of low activity must dominate the sediment. Activity, the soil plasticity (liquid limit minus plastic limit) divided by the % clay (<2 μ) sized particles. The activities of Canadian and Scandinavian quick clays are in the 0.2–0.5 range; the Japanese quick clay's activity may slightly exceed 1. Only traces of high-swelling, high-activity smectite may be present in the Canadian and Scandinavian quick clays. The Japanese quick clay has a non-swelling, high-iron smectite of low activity (~1) as their dominant clay mineral (Egashira and Ohtsubo 1982). A large amount of ferrous iron in the octahedral sheet of this smectite leads to a high negative charge in each layer, strong attraction forces to the interlayer cations and the inability of water to penetrate into the interlayer. The abundant amorphous volcanic-ash particles act in a similar manner to the primary minerals in the Canadian and Scandinavian quick clays.

2.2.2.2 Post-depositional Factors

Displacement of the original salty pore water by fresh water is essential to the development of high sensitivity in the marine and brackish water sediments. Displacement by infiltration of fresh water through the surface commences when the sediment surface has risen above sea level. Such downward salt displacement dominates in the extensive, old-sea-bottom plains of eastern Canada. In Scandinavia, and in a few situations in Canada, artesian pressures at the base and edges of the sediment, originating from water entry into nearby uplands, displaces the salty pore water from most of the sediment laterally and upward to the surface drainage system; only in the near-surface zone is the salt displaced downward and laterally. In neither leaching situation do high sensitivities develop in the surface zone; rather oxidative weathering, wetting and drying and other pedological processes produce a soil profile and a stiff, strong weathering crust. Below the weathering crust, salinity reduction leads to an increase in inter-particle repulsive forces. In cemented sediment, the water content should remain constant during the leaching while uncemented sediment may experience a minor decrease. In both scenarios, the flocculated structure remains intact. What changes is that the liquid limit decreases as salt removal progresses; to the extent that, if the flocculated structure is destroyed after leaching, the increased inter-particle repulsion prevents re-flocculation. The liquidity index increases from approximately 1 at high salinity to 1.5 to >4 at salinities less than about 2 g/l. Upon structural collapse, the leached material behaves as a liquid; it has become 'quick clay'. In brackish water sediments, the liquid limit decrease accompanying salt displacement may not be sufficient to produce quick clay. Introduction of dispersing agents, such as organic compounds from organic soils, may decrease the liquid limit sufficiently for quick clay to be present. Inorganic dispersants, in modest amounts, would have the same effect.

The requirement for reducing conditions is introduced to the model for the first time in this paper. It is particularly applicable to Japanese quick clays in which a ferrous-iron-rich, low-swelling smectite is the dominant clay mineral. Under oxidizing conditions, this smectite is converted to ferric-iron-rich, high-swelling smectite which dramatically reduces the sensitivity. Likewise, reducing conditions inhibit swelling smectite formation in the Canadian and Scandinavian quick clays, a change that would increase their post-leaching liquid limit and decrease their sensitivity.

2.3 Synthesis

In all four quadrants of the general model for quick clay development, a chemical factor (salt flocculation, oxide cementation, low activity mineralogy, salt removal or dispersants) is important, even essential, to the development of high sensitivity. This relationship suggests that chemistry, most probably through the introduction of appropriate chemicals, by appropriate means, could be important to the management/amelioration of quick clay landslides.

2.4 Quick Clay Landslides

2.4.1 Classification

In the overall landslide classification scheme (Cruden and Varnes 1996), quick clay landslides classify as 'earth flows'. Quick clay 'earth flows' are described as being 'retrogressive' because the backscarp moves 'backward' (away from the river bank) as a sequence of failures occurs. For people unfamiliar with these landslides, 'retrogressive' can leave the impression that landslide enlargement occurs by an almost passive retreat of the backscarp, when in fact the landslide is rapidly and very aggressively advancing into the landscape beyond the riverbank (Torrance 2012). Carson and Lajoie (1981) presented a classification of landslides in sensitive sediments based on how they related to topographic constraints. It identified: (1) two-dimensional spreading failures (fill the adjacent river valley); (2) aborted failures (did not fill the adjacent valley); (3) excess retrogression (sediment exceeded the adjacent valley capacity and flowed downstream and upstream); (4) multidirectional retrogression (in which the opening to river is narrower than the scar width); and (5) flakeslides (in which the whole slide area commences sliding as a unit, and breaks up as it flows). This geomorphological classification is very useful in identifying how a landslide relates to its setting, but does not directly address the modes of failure. Torrance (2012) has grouped these landslides into two super-categories, STEPWISE (Type 4), and UNINTERRUPTED (the rest), according to how the failure process progresses. For both types of landslides, the normal first step is a failure of the fracture-laden weathered material of the valley slope. Exceptions to this 'normal' appear to be quick clay landslides triggered by earthquakes (Seed and Wilson 1967; Schaffer and Smith 1987) and landslides near Chelsea, Québec (May 8, 1973) and Småröd, Sweden (December 20, 2006) (Statens Haverikommission 2009) triggered by consolidation failures during road construction. To my knowledge, no one has observed, in action, a naturally caused, riverbank failure that was subsequently followed by a quick clay landslide. Natural factors are the triggers of most quick clay landslides (the Rissa landslide (Gregersen 1981) referred to later, was a spectacularly notable exception).

2.4.1.1 Stepwise Landslides

Stepwise landsliding requires that a sufficient proportion of the debris from each step must be liquid and capable of transporting itself, the surface crust and other semi-intact material from the landslide scar, leaving behind an unstable, unsupported slope which, in turn, fails in a similar manner. Only the multidirectional retrogressive landslides meet the stepwise criteria. Individual steps are of the simple rotational landslide character. The early stages of the 1978 Rissa landslide were of this character; small rotational slides collapsed from the side and back scarps of the developing scar, remolded to the liquid state as they fell, splashed and flowed away (NGI 2006). In this stepwise manner, a narrow scar advanced about 450 m diagonally into the slope over a 45 min period (Gregersen 1981). Stepwise failures normally cease when the soil properties change or when the failure plane gradually rises to a shallower depth and the backscarp is stable. At Rissa, the stepwise failure mode was followed by uninterrupted failures that expanded the slide area by 20 times in about 5 min. The common feature of stepwise failures is that very little landslide debris remains within the landslide scar. The flowing debris may cause further destruction for substantial distances downstream. In an extreme case at

Saint-Jean-Vianney, Québec, the debris flowed several kilometers downstream at high velocities (Tavenas et al. 1971). Stepwise landslides are the dominant landslide form in Scandinavia, and a substantial number have occurred in Canada. A stepwise landslide requirement is that the thickness of the quick clay zone that liquefies must be large compared to the thickness of the surface crust and other overlying soil that does not liquefy. Leaching by upward displacement of pore water by artesian pressures decreases the depth to which weathering agents penetrate from the surface thereby allowing for a thinner weathered crust and a lesser amount of non-liquid debris to transport out of the scar.

2.4.1.2 Uninterrupted Landslides

Uninterrupted landslides occur when the amount of liquefied debris is insufficient to transport the non-liquefied debris from the scar unless the momentum of that debris is maintained from start to finish (Torrance 2012). These landslides occur where the depth to the quick clay zone is comparable or greater than the thickness of the quick clay zone that liquefies. 'Grounding' of large, semi-intact chunks of crust and un-remolded debris within the scar obstructs further debris flow and rapidly closes down the landsliding process. A large amount of debris remains within the landslide scar, often forming a series of 'ribs' that are oriented perpendicular to the flow direction (Carson 1977). The uninterrupted landslide process is characterized by continuous advancement of the failure plane into the landscape from the riverbank, the breaking loose of non-quick-clay chunks as the underlying liquid debris moves forward and thins, and the creation of a new backscarp position which has already been passed by the rapidly advancing failure plane. Intact quick clay in the slumped unit will be remolded by gravitational and kinetic forces, but overlying clay that is not so sensitive may remain largely intact. As long as the failure zone advances rapidly and supplies sufficient liquefied quick clay to keep the chunky debris moving, the landslide continues to enlarge. As chunky debris becomes grounded in the river valley, debris removal slows down and the landsliding process rapidly ceases when such debris can no longer exit the landslide scar.

2.5 Strategies for Amelioration of Landsliding in Sensitive Clays

Two fundamentally different strategies are available for amelioration of the quick clay landslide problem: actions directed toward prevention of the triggering landslide in the fracture-laden, weathered material that constitutes the riverbank; and actions directed toward inhibiting/preventing advancement of the failure into the quick clay that lies beyond the weathered zone.

2.5.1 Prevention of the Initial Failure

Potential approaches to preventing riverbank failure include: building of earthworks at the base of the slope to prevent erosion and to provide support to the slope; reducing the slope height and the effective slope angle by soil removal from upper parts of the slope; improving drainage within the slope; installation of piles that would cross the potential failure planes. While one can envision such measures being feasible in selected locations, "the amount of earth material and disturbance involved, the slope height and steepness, and the valley narrowness eliminate these (physical) methods as practical solutions" for most situations that involve quick clay (Torrance 2012).

2.5.2 Prevention of Failure Zone Advancement

The alternative to prevention of the initial failure is to focus of how to prevent the riverbank failure (a relatively minor event) from being followed by a quick clay landslide. The best alternative appears to be to eliminate the critical portion of the quick clay by increasing its liquid limit sufficiently that its post-failure behaviour will be that of a plastic solid. That change will greatly enhance the probability that more of the debris from the initial failure will remain in the landslide scar and provide protection against subsequent backscarp failures. Since the time when the late Professor Ivan Rosenqvist demonstrated, on site, that the liquid debris of a quick clay landslide could be stabilized by the addition of common salt (personal communication, J. Moum), the potential of chemical additions to cure problems presented by quick clay has gradually gained recognition. Restoration of their pore-water salinity by NaCl addition transforms the quick clay to a material that is approximately at its plastic solid/liquid transition point. Moum et al. (1968) and Torrance (1975), using sensitive clays from Oslo and Ottawa, respectively, both found that KCl additions, at equivalent ionic concentrations, are about twice as effective as NaCl in increasing post-failure strengths and liquid limits. Because treatment is being carried out close to a potentially unstable slope, chemical introduction must not risk triggering a landslide. Hence, injection of NaCl or KCl brines under pressure, which would create excess pore pressures, must be rejected. The safest method appears to be that proposed by Moum et al. (1968) in which modest diameter bore holes would be filled with crystalline salts and water, and the salt would dissolve and slowly diffuse (a few years, dependent on bore hole spacing) throughout the surrounding soil. K has advantages over Na in that its diffusion coefficient is greater, its effect is greater, and that, if the salinity is reduced by future leaching, K-saturated clay at low salinity has a higher liquid limit than does Na-saturated clay. If either treatment is determined to have a 'side effect' of increasing the undisturbed strength of the treated quick clay, that will be a bonus.

2.6 Summary

Among landslide-prone earth materials, the quick clays are unique in their origin and in the range of ways in which chemical factors have been essential to the development of their landslide-prone character. The ease with which a landslide can be triggered, by nature and by human actions, their liquefaction during failure and the mobility of the landslide debris make physical intervention to decrease landslide risk a risky approach. In light of the extent to which their behaviour can be improved by chemical modification, and the feasibility of introducing common, generally non-problematic, chemicals (NaCl or KCl) in a relatively non-intrusive, almost passive manner, chemical modification would appear to be the safest, most effective approach to amelioration of the quick clay landslide problem. The time period for chemical modification may be several years, but once achieved, it should be long lasting – potentially for centuries or millennia.

Acknowledgments I would like to recognize the late Laurits Bjerrum, the late Johan Moum, and Tor Løken of the Norwegian Geotechnical Institute for introducing me to the quick-clay landslide problem, and the many students who, through research projects since 1970, have assisted me in gaining an understanding of quick clays and of how their properties reflect the influences of a wide range of chemical, physical and mineralogical factors. I also thank Dr. Yvonne Andersson-Sköld for her insightful comments that have improved this manuscript.

References

Carson MA (1977) On the retrogression of landslides in sensitive muddy sediments. Can Geotech J 14:582–602

Carson MA, Lajoie G (1981) Some constraints on the severity of landslides in sensitive deposits. Géog Phys Quat XXXV:301–306

Cruden DM, Varnes DJ (1996) Landslide types and processes. In: Turner AK, Schuster RL (eds) Landslides: investigation and mitigation, Transportation research board, special report 247. National Academy Press, Washington, DC, pp 36–75

Egashira K, Ohtsubo M (1982) Smectite in marine quick-clays of Japan. Clays Clay Miner 30:275–280

Gregersen O (1981) The quick clay landslide at Rissa, Norway, Publication Norwegian Geotechnical Institute 135. Norwegian Geotechnical Institute, Oslo, pp 1–6

Moum J, Sopp OI, Løken T (1968) Stabilization of Undisturbed quick clay by salt wells, Publication Norwegian Geotechnical Institute 81. Norwegian Geotechnical Institute, Oslo, pp 1–8

Norwegian Geotechnical Institute (2006) The Rissa landslide. DVD, Nor Geotek Inst, Oslo

Schafer CT, Smith JN (1987) Hypothesis for a submarine landslide and cohesionless sediment flows resulting from a 17th century earthquake-triggered landslide in Quebec, Canada. Geo-Mar Lett 7:1–37

Seed HB, Wilson SD (1967) The turnagain heights landslide, Anchorage, Alaska. Proc Am Soc Civ Eng 93:325–353

Skempton AW, Northey RD (1952) The sensitivity of clays. Géotechnique 3:30–53

Statens haverikommission (2009) Swedish Accident Investigation Board Jordskred vid väbygge E6 I Småröd, O län, den 20 December 2006. http://www.havkom.se/virtupload/reports/ro2009_01.pdf

Tavenas F, Chagnon J-Y, LaRochelle P (1971) The Saint-Jean-Vianney landslide: observations and eyewitness accounts. Can Geotech J 8(3):463–478

Torrance JK (1975) On the role of chemistry in the development and behaviour of the sensitive marine clays of Canada and Scandinavia. Can Geotech J 12:326–335

Torrance JK (1983) Towards a general model of quick clay development. Sedimentology 30:547–555

Torrance JK (1995) On the paucity of amorphous materials in the sensitive marine clays. Can Geotech J 32:535–538

Torrance JK (2012) Landslides in quick clay. In: Clague JJ, Stead D (eds) Landslides: types, mechanisms and modeling. Cambridge University Press, Cambridge

Torrance JK, Ohtsubo M (1995) Ariake Bay quick clays: a comparison on to the general model. Soils Found 35:11–19

Torrance JK, Hedges SW, Bowen LH (1986) Mössbauer spectroscopic study of the iron mineralogy of post-glacial marine clays. Clays Clay Miner 34:314–322

Chapter 3
Nature of Sensitive Clays from Québec

Jacques Locat and Daniel St-Gelais

Abstract Quantitative mineralogical analyses of sediment samples from 18 sites in Québec indicate that plagioclase is the most abundant mineral in all soils and that chlorite is generally the dominant clay mineral followed by illite and expandable clays consisting of mixed-layer clays minerals involving vermiculite. The study also illustrates how specific surface area and constitutive water content can be used to evaluate departures from average in the mineralogical composition of sensitive clays. The relationship between activity and specific surface area of sensitive clays shows that when compared to soils from different sedimentary basins that have different mineralogy, they are characterized by a much lower activity and specific surface area.

Keywords Sensitive clays • Quantitative mineralogy • Specific surface area • Activity • Québec • Oxides • Postglacial marine sediments

3.1 Introduction

Sensitive clays in eastern Canada are mostly found in Québec but also occur in Ontario and Labrador. In Eastern Canada, sources of sediment are mostly controlled by glaciations and de-glaciation patterns and by the trends of the major topographic features (Locat 1995). North of the St. Lawrence River, ice-flow directions were southerly while north of the water divide boundary in the Appalachian Mountains,

J. Locat (✉)
Laboratoire d'études sur les risques naturels, Department of Geology and Geological Engineering, Laval University, Québec City, QC, Canada
e-mail: Jacques.locat@ggl.ulaval.ca

D. St-Gelais
DSG Management Group Inc, Vancouver, BC, Canada

J.-S. L'Heureux et al. (eds.), *Landslides in Sensitive Clays: From Geosciences to Risk Management*, Advances in Natural and Technological Hazards Research 36,
DOI 10.1007/978-94-007-7079-9_3, © Springer Science+Business Media Dordrecht 2014

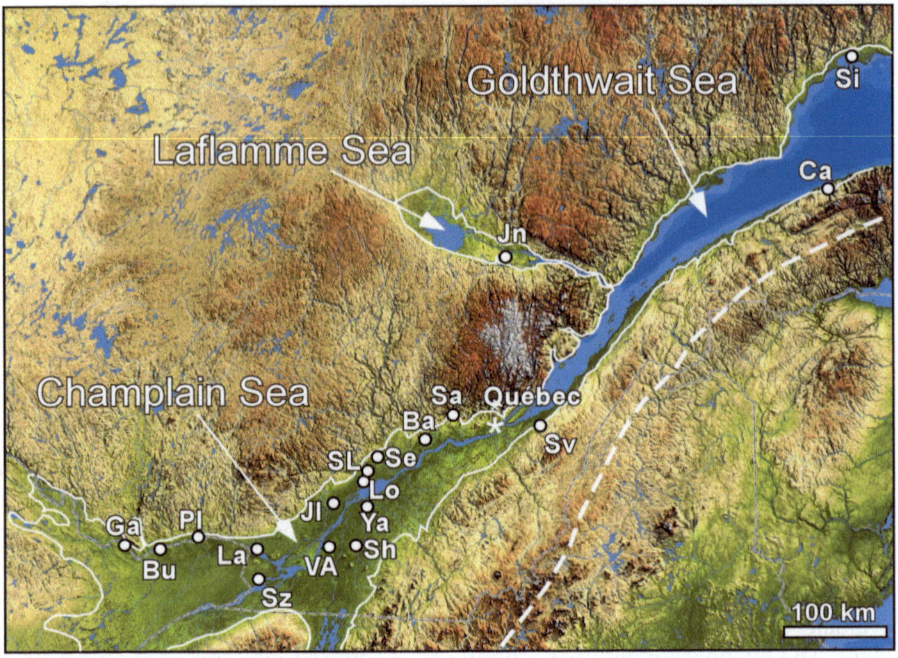

Fig. 3.1 The postglacial seas of the St. Lawrence valley and the location of the sensitive clay sample sites: (*Ba* Batiscan, *Bu* Buckingham, *Ca* Cap-Chat, *Ga* Gatineau, *Jl* Joliette, *Jn* Jonquière, *La* Lachute, *Lo* Louiseville, *Pl* Plaisance, *Si* Sept-Îles, *Sa* St. Alban, *Se* St. Étienne, *Sh* St. Hyacinthe, *SL* St. Léon, *Sv* St. Vallier, *Sz* St. Zotique, *Va* Varennes, *Ya* Yamaska). *Dashed line*: water divide where in the final de-glaciation phase ice-flow direction were to the north and south of that boundary

the ice-flow was towards the north near the end of de-glaciation (see dashed line in Fig. 3.1). In the St. Lawrence Basin in Québec, the source rocks can be roughly divided into three geological provinces: (1) the Precambrian shield (mostly metamorphic rocks), (2) the Appalachian (volcanic and sedimentary rocks) and (3) the Lower St. Lawrence Platform (mostly limestone and some shale). Most of the sediments coming into the various postglacial seas resulted from glacial grinding of the bedrock with little post-depositional weathering. The nature of the sensitive clays from Québec reflects this. Postglacial marine sedimentary basins, along the St. Lawrence River system (Fig. 3.1) consist of the Goldthwait Sea (actual Gulf of St. Lawrence, since 13,500 yBP), the Champlain Sea (St. Lawrence Valley, upstream of Québec City, 13,100–10,600 yBP), and the Laflamme Sea (at the northwestern tip of the Saguenay Fjord, 10,300–8,000 yBP).

Investigation of the mineralogy and chemistry of sensitive clays from Québec, are quite limited. Mineralogical analyses have been presented by Brydon and Patry (1961), Ballivy et al. (1971), Gillot (1971), Gravel (1974), Bentley (1976, 1980), Martel et al. (1978), Foscal-Mella (1976), Babineau (1977), Quigley (1980),

Lebuis et al. (1983), Locat et al. (1984), Torrance (1988, 1995), Locat 1995, and Berry and Torrance (1998). Most of these results can be described as semi-quantitative. We are taking the opportunity of this workshop to present the quantitative mineralogical analyses of sensitive clays from various sites in Québec (Fig. 3.1) that were conducted by St. Gelais (1990). We will commence by describing the methodology. It will not be possible to review the role of mineralogy (sediment source) on the behavior of sensitivity clay sediments. To that effect, the reader is referred to work Quigley (1980), Torrance (1983), Torrance and Ohtsubo (1995), Locat (1995), and Locat et al. (2003).

3.2 Methodology

Samples of sensitive clays were collected, using Shelby tubes, from 1.5 to 2.5 m depths to assure that they were from below the surface oxidizing zone. Great care was given to maximize analyses on the same sample (St. Gelais 1990). Physico-chemical properties, reported in Table 3.1, were acquired according to BNQ (Bureau de la normalisation du Québec) standards including grain size analysis on wet samples. The CEC was measured according to the method of Chapman (1965) using an ammonium acetate solution at a pH of 7, the organic content by calcination (Ball 1964), and the salinity by electric conductivity of the interstitial water (Torrance 1975). The specific surface area was determined using three methods: EGME (Diamond and Kinter 1958), and methylene blue (Tran 1977) for both internal and external surfaces, the BET (nitrogen method) for the external surfaces.

The main mineral phases for all samples were identified using X-ray diffraction. Samples were dried at 25 °C and lightly crushed for the powder diagrams and a wet subsample was put in a centrifuge to extract the <2 μm fraction used for natural, glycolated, and heated X-ray diffraction of oriented samples and to obtain the >2 μm for constitutive water analysis (H_2O_c). Water loss measurements, on both size fractions were determined using a thermogravimetric method (TGV) with temperature intervals as proposed by Jackson (1969), i.e. 105–110 °C for H_2O_h (hygroscopic water), 110–300 °C for interlayer water (as for montmorillonite and vermiculite), and from 300 to 900 °C for constitutive water (H_2O_c). As suggested by Jackson (1969), the water loss for constitutive water was computed with respect to the loss at 300 °C. By doing so, the relative amounts of H_2O_c for vermiculite, illite and chlorite are 5, 4.5, and 13 % respectively. For a Na-montmorillonite and K-montmorillonite the value are 2.7 and 6.2 % respectively (Bérend et al. 1995). For the TGV measurements, samples were pre-treated with hydrogen peroxide to remove the organic matter and the treatment lasted for as long as effervescence was visible (McKeague 1978). H_2O_c results are compiled in Table 3.1 for an average of three tests (St. Gelais 1990).

For the quantitative evaluation of the main mineral phases, the approach proposed by Engler and Tyengar (1987) was used and the results are presented in Table 3.2. It incorporates various methods in parallel. So, for quartz, K-feldspar, and

Table 3.1 Physico-chemical properties of selected sensitive clay sediments from Québec

Site	w	mo	wL	wp	Il	Ip	CF	CEC	Act	Sal	BI	BET	EG	H+F	noA	H+>F	n0B	A+B	H+to
Champlain Sea																			
Batiscan	63	0.8	64	21	0.97	42.6	78	12.6	0.53	0.8	77	29	112	4.3	3.5	3.4	0.7	4.1	4.0
Buckingham	39	0.6	42	17	0.86	25.1	50	12.3	0.55	0.2	61	19	101	5.2	2.6	2.6	1.3	3.9	4.0
Gatineau	60	1.0	68	25	0.81	42.2	75	16.0	0.56	0.1	90	32	131	5.2	3.9	2.6	0.6	4.5	4.6
Joliette	40	0.5	42	17	0.94	24.4	48	8.0	0.49	2.3	39	15	53	5.0	2.5	1.5	0.7	3.3	3.1
Lachute	47	0.8	57	24	0.69	33.4	78	14.6	0.42	0.6	93	31	106	5.6	4.2	3.8	1.0	5.2	5.1
Louiseville	73	0.8	76	24	0.94	52.5	80	13.0	0.66	1.9	91	29	97	5.1	4.1	2.2	0.4	4.5	4.5
Plaisance	70	0.9	77	25	0.86	52.3	80	16.6	0.65	0.4	102	36	133	5.9	4.7	3.0	0.6	5.3	5.3
St-Alban	74	0.9	55	21	1.57	33.8	78	12.6	0.42	0.5	98	32	119	5.6	4.5	2.6	0.5	5.0	4.9
St-Étienne	34	0.7	34	19	0.98	14.7	30	10.6	0.49	0.4	41	9	67	4.4	1.3	0.7	0.5	1.8	1.8
St-Hyacinthe	74	0.7	70	22	1.08	48.1	80	12.0	0.60	1.3	88	32	125	5.2	4.2	4.6	0.9	5.1	5.2
St-Léon	66	0.6	71	23	0.90	48.1	73	13.3	0.69	0.5	76	29	102	4.4	3.1	2.6	0.8	3.9	4.0
St-Zotique	86	1.2	71	23	1.32	47.6	83	15.6	0.56	0.9	98	32	112	5.9	5.0	4.2	0.6	5.6	5.7
Varennes	57	0.9	67	27	0.74	40.1	82	15.3	0.50	1.3	102	34	125	5.6	4.5	3.0	0.6	5.1	5.1
Yamaska	41	0.7	40	16	1.04	23.5	40	7.6	0.59	1.8	49	17	63	3.6	1.5	3.5	2.1	3.5	3.4
Goldthwait Sea																			
Cap Chat	32	0.4	40	22	0.58	17.2	30	6.3	0.57	1.5	26	10	35	6.4	1.9	3.8	2.7	4.6	4.5
Sept-Îles	48	0.6	63	27	0.60	36.1	75	12.0	0.48	11.8	38	24	73	7.0	5.2	2.9	0.7	6.0	5.9
St-Vallier	34	0.5	36	15	0.92	20.6	52	8.6	0.41	3.8	46	16	65	4.5	2.2	2.3	1.1	3.4	3.4
Laflamme Sea																			
Jonquière	32	1.0	41	20	0.58	20.5	48	11.3	0.41	0.3	48	18	88	5.5	2.5	1.1	0.6	3.0	3.1

Note: *w* water content (%), *mo* organic matter (%), *wL* liquid limit (%), *wp* plastic limit (%), *IL* liquidity index, *CF* clay fraction (%), *Act* activity, *Sal* salinity (g/L), *Bl* surface area (m^2/g), Bleu method (m^2/g), *BET* surface area, nitrogen method (m^2/g), *EG* surface area, ethylene glycol (m^2/g), *H+F* H_2O_c of clay fraction, *noA* H_2O_c normalized for the clay fraction, *H+>F* H_2O_c for non-clay fraction, *noB* H_2O_c normalized for non-clay fraction, *H+to* total constitutive water

Table 3.2 Mineralogical composition of selected sensitive clay sediments from Québec (all values are in %)

Site	Cal	Dol	Ca1	Ca2	Am	Qtz	KFd	Pla	I	V	M	C	aAl	aFe	aSi	aTo	CTo	PTo	MTo
Champlain Sea																			
Batiscan	2.0	1.0	3.0	9.3	12	10	11	34	8	5	0	9	0.2	0.5	0.2	0.9	23	70	93
Buckingham	1.0	0.0	1.0	0.5	9	20	7	36	8	4	0	11	0.2	1.2	0.4	1.8	22	73	97
Gatineau	1.0	0.0	1.0	1.0	8	11	12	36	8	9	0	9	0.2	1.0	0.4	1.6	27	68	96
Joliette	0.0	0.0	0.0	1.7	10	17	13	40	10	3	0	4	0.1	0.6	0.3	1.0	16	80	98
Lachute	2.0	1.0	3.0	8.4	12	11	11	33	9	5	0	14	0.2	0.7	0.4	1.3	28	70	100
Louiseville	1.0	0.0	1.0	3.6	9	9	15	39	5	5	0	16	0.2	0.9	0.4	1.5	26	72	100
Plaisance	1.0	1.0	2.0	4.8	10	13	11	32	10	6	0	13	0.1	0.9	0.4	1.5	29	67	98
St-Alban	2.0	0.0	2.0	7.3	8	10	14	34	8	7	0	12	0.2	0.7	0.4	1.2	27	69	98
St-Étienne	1.0	0.0	1.0	4.0	8	18	17	39	2	2	0	5	0.3	0.9	0.5	1.7	9	81	92
St-Hyacinthe	3.0	1.0	4.0	10.3	11	12	8	32	8	5	0	15	0.2	0.5	0.3	0.9	28	67	97
St-Léon	1.0	0.0	1.0	3.3	8	10	15	40	4	5	0	11	0.2	0.9	0.3	1.4	21	74	97
St-Zotique	3.0	0.0	3.0	9.8	10	10	9	31	10	7	0	15	0.2	0.8	0.4	1.4	32	62	95
Varennes	2.0	1.0	3.0	8.3	8	10	13	33	9	7	0	13	0.2	1.0	0.3	1.6	29	66	97
Yamaska	4.0	1.0	5.0	11.1	10	19	10	33	5	6	0	6	0.1	0.5	0.3	0.8	17	77	94
Goldthwait Sea																			
Cap Chat	5.0	2.0	7.0	8.3	16	3	28	20	2	2	0	20	0.1	0.6	0.2	0.9	24	74	99
Sept-îles	2.0	0.0	2.0	1.1	6	8	6	40	13	8	0	13	0.2	1.0	0.3	1.5	34	62	98
St-Vallier	1.0	1.0	2.0	5.3	8	21	10	37	10	3	0	5	0.1	0.5	0.3	0.9	18	77	96
Laflamme Sea																			
Jonquière	2.0	0.0	2.0	3.3	6	17	22	35	2	2	0	11	0.3	1.3	0.3	1.9	15	82	98

Note: *Cal* calcite, *Dol* dolomite, *Ca1* from Chittick, *Ca2* from chemical analysis (CO_2/0.44), *Am* amphibole as actinolite, *Qtz* quartz, *KFd* K-feldspar, *Pla* plagioclase, *I* illite, *V* vermiculite, *M* montmorillonite, *C* chlorite, *from DCB* aAl_2O_3, *aFe* Fe_2O_3, *aSi* SiO_2, *aTo* total, *CTo* total clay minerals, *PTo* total primary minerals, *MTo* all minerals (including minerals)

plagioclase, the pyrosulfate method was used (Kiely and Jackson 1965). Plagioclase was determined by assigning the Na_2O of the chemical analysis to anorthite (%An) indexed using the method proposed by Van der Plas (1966). Illite was determined by attributing the K_2O obtained from selective dissolution in HCl at 825 °C (Reynolds and Lessing 1962). Vermiculite and montmorillonite were determined using CEC after removing the iron oxides (Alexiades and Jackson 1965). Total carbonates (calcite and dolomite) were determined using the Chittick apparatus (Dreimanis 1962; Locat and Bérubé 1986). For chlorite and amphibole there are no quantitative dissolution methods. Therefore, amphiboles (referred to here as actinolite) were computed by attributing the residual CaO from chemical analysis after re-distribution to carbonates and plagioclases (calcite: 56 %, dolomite: 30.4 %). For plagioclase the distribution depends on the %An (percent anorthite). In the same way, the chlorite was determined by allowing the residual value of MgO after attribution to amphibole: 13.8 %, illite: 2.8 %, vermiculite: 22 % and dolomite: 21.9 %. This re-distribution of MgO is based on an average chemical composition of these minerals present in various Québec soils (Gravel 1974). Amorphous silica (Si), and Fe and Al oxides were determined using the dithionite-citrate-bicarbonate method (DCB, Mehra and Jackson 1960). For each sample, the mineralogical analysis was verified by chemical mass balance. For SiO_2 and Al_2O_3, differences in concentration computed from mineralogical composition and chemical analyses are on average -0.24 ± 0.8 and 1.1 ± 0.6 % respectively. For more details on the various procedure and test results (including chemical analyses), the reader is referred to St. Gelais (1990).

3.3 Results and Analysis

3.3.1 Physico-chemical Properties

Physico-chemical properties for the soils from 18 sites (Fig. 3.1) are presented in Table 3.1. Water contents values are mostly below or near the liquid limit. This reflects using samples been collected near the surface, within the crust (mechanical) but below the weathered crust. In that of the crust they were likely exposed to soluble weathering products coming from above. Three samples have clay fraction below 50 % (Cap Chat, St-Étienne, and Yamaska) and are consider as clayey silts. The organic matter content is less than 1.2 %. The activity of most soils varies between 0.4 and 0.7 which is within the range (0.25 and 0.75) noted by (Leroueil et al. 1983) for soils from Québec. The Sept-Îles, Joliette, and Saint-Vallier have a pore water salinity above 2 g/L which is also reflected by their higher Na^+ exchangeable cations but not clearly by their activity (St. Gelais 1990). Locat (1982) has shown that increasing the salinity to 35 g/L for 12 soils from Québec resulted in an average activity coefficient of 0.5 ± 0.09 and the increase was higher for soils with an initial activity below 0.3. Specific surface areas, for both internal and external surfaces, were measured by the Blue and EGME methods. They both show

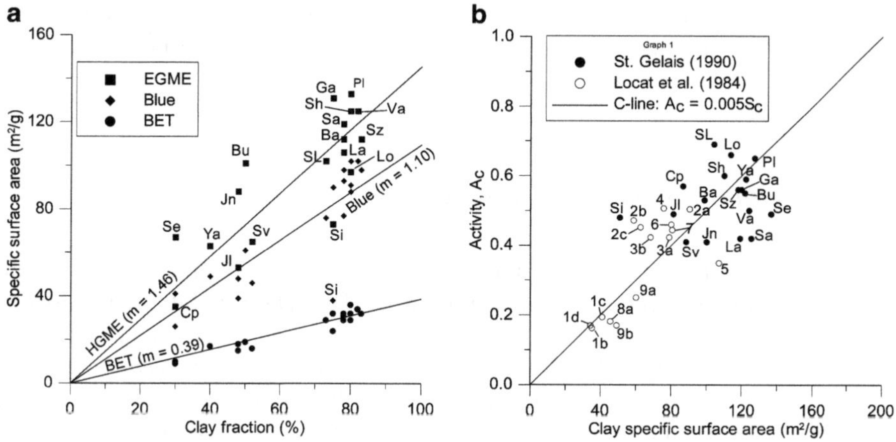

Fig. 3.2 (**a**) Specific surface areas and clay fraction for total (EGME and Blue methods) and external (BET method) surfaces. (**b**) Activity as a function of clay specific surface area (C-line from Locat et al. 2003, *C* stands for clay), sites with numerals are from Locat et al. (1984), *1* Grande-Baleine, *2* Olga, *3* St. Marcel, *4* St. Léon, *5* St. Alban, *6* St. Barnabé, *7* Shawinigan, *8* Chicoutimi, and *9* Outardes

a relationship between surface area and clay fraction, with the EGME values consistently higher than with the Blue method (Table 3.1, Fig. 3.2a). For a linear relationship going through the origin, the r^2 values are 0.97 for both. The BET method (external surfaces only) is less influenced by the mineralogy (St. Gelais 1990) and shows an even better relationship ($r^2=0.99$, Fig. 3.2). For the Blue method, results for the Sept-Îles sample (Si in Fig. 3.2) are much lower than the others when using the Blue method. It is believed that this may be due to its high pore water salinity (11.8 g/L) which may reduce the adsorption of the Blue on the particles (Pimol et al. 2008).

From these analyses, Locat et al. (2003) introduced a concept relating the Skempton Activity (*Ac*) to the clay specific surface area *Sc* (Blue or total surface area/<2 μm) to illustrate the effect of clay mineralogy on index and surface properties. The *Sc* value is equivalent to consider that all the surface area is due to the particles in the clay fraction. Results for sensitive clays from Québec are shown in Fig. 3.2b, also including data from Locat et al. (1984). The relationship has been used to introduce the C-line (Locat et al. 2003). As will be discussed below, the spread in Fig. 3.2b represents the variability of the mineralogical composition of sensitive clays from Québec characterized by a low activity and a relatively low value of clay specific surface area (*Sc*).

Locat et al. (1984) determined that the constitutive water (H_2O_c) could be a good indicator of the presence of clay minerals in sensitive clays from Québec. As part of the work by St-Gelais (1990), this approach was further developed and then tested on the clay fractions, the non-clay fractions and the bulk samples of the 18 soils, after the removal of organic matter. Results (Table 3.1) show that the method is quite robust with the total of constitutive water content of both clay and non-clay fraction being nearly the same as the one computed for the whole sample (standard deviation of ±0.08 %).

3.3.2 Primary Minerals (Non-clay)

Results of the mineralogical quantification are presented in Table 3.2. The suite of primary minerals, by decreasing concentration, is: plagioclase (19–40 %), K-feldspar (6–28 %), quartz (3–20 %), amphiboles (actinolite, 6–16 %), calcite (1–5 %) and dolomite (0–2 %). Taken globally, the amounts of primary minerals and their relative abundances are similar to what has been estimated by Brydon and Patry (1961), Locat et al. (1984) and by Berry and Torrance (1998). The primary mineral content ranges from 62 and 80 % while inversely the amount of clay minerals ranges from 9 to 34 %. The total amount of dithionite-citrate-bicarbonate (DCB) extractable material ranges from 0.82 to 1.87 % (Table 3.2), which is comparable to the range of 1.09–2.06 % reported by Berry and Torrance (1998).

3.3.3 Clay Minerals

Clay minerals are compiled in Table 3.2. The suite of clay minerals, by decreasing order of concentration is: chlorite (4–20 %), illite (2–13 %), and vermiculite (2–9 %) with an (illite + vermiculite)/chlorite ratio varying between 0.5 and 1.5. Only traces of montmorillonite was found (<0.1 %). The total amount of clay minerals varies from 9 % (St. Étienne) to 34 % (Sept-Îles). The highest amount of chlorite is found in Cap Chat (20 %), because of a large portion of chloritic shale in the non-clay fraction, and the lowest is in the Joliette sample (4 %). For illite, the maximum is found in the sample from Sept-Îles (13 %) and the lowest at St. Étienne, Cap Chat and Jonquière (2 %) with the first two having the lowest clay fraction (30 %). For vermiculite, inter-layered with illite and/or chlorite, the maximum is for the Gatineau sample (8 %) and the minimum again for Cap Chat and St. Étienne at 2 %. The abundance of clay minerals is clearly correlated with the clay-size fraction, except for Cap Chat (Cp, Fig. 3.3a). For the Cap Chat sample, the presence of clay minerals in the non-clay fraction is also in agreement with the higher values of H_2O_c in the non-clay fraction relative to the clay fraction (Table 3.1).

The amount of constitutive water (H_2O_c) in bulk samples, shows an excellent relationship to the abundance of the clay minerals, (Fig. 3.3b). However, when computed from the clay mineral composition, the estimated amount of constitutive water is significantly lower but still shows a positive relationship. The good relationship for the bulk samples, may be due to (St. Gelais 1990): (1) the (illite + vermiculite)/chlorite ratio only varies between 0.5 and 1.5; (2) the small amounts of hydrous oxides present; and (3) only small amounts of calcite are present (0–5 %).

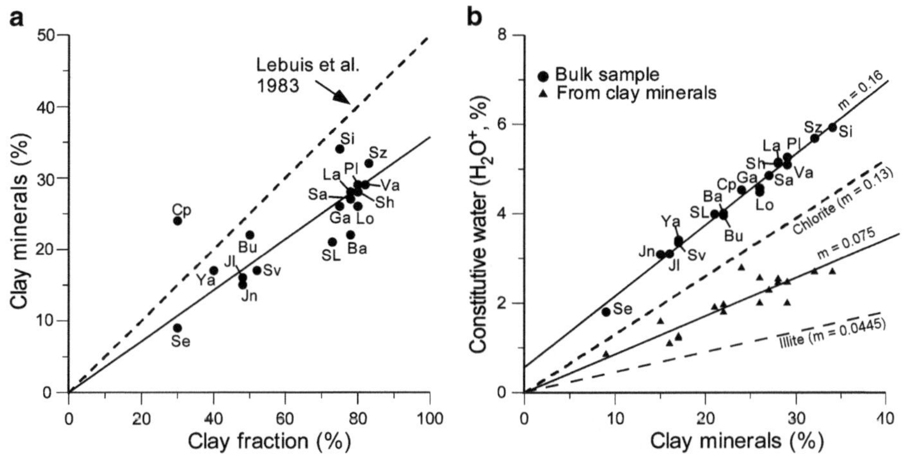

Fig. 3.3 (**a**) Clay minerals as a function of the clay fraction. Note here that results from Cap Chat (Cp) which contains clay minerals as part of shale particle in the silt fraction. (**b**) Amount of constitutive water as a function of the amount of clay minerals obtained from bulk samples measurements and computation from clay mineral composition

The relationship between constitutive water and mineral abundance can be a useful tool to evaluate if a clay sample in Québec has a mineralogical composition which departs from typical values.

Berry and Torrance (1998) note that both montmorillonite and vermiculite (in mixed-layer form) are expandable minerals. The impact of vermiculite, as an expandable clay, cannot have the same effect as that of montmorillonite at the same concentration (Locat et al. 2003). Still we consider, like Berry and Torrance (1998), that using the word 'expandable' is inclusive and should include mixed-layered minerals (illite/vermicyulite and/or chlorite/vermiculite).

When compared to other clays, it is clear that Québec clays have a very different signature than samples formed in different environments and containing a significant amount of swelling clays. As reported by Locat (1995) and Locat et al. (2003) the sedimentary environments found in Québec are also found in other regions like in Ontario, British Columbia (Geertsema and Torrance 2005), Labrador, Scandinavia and parts of Russia and Alaska. Results from various studies carried out at Laval University and reported by Locat et al. (2003) are presented in Fig. 3.4 which encompasses data for soils with very different mineralogical composition from kaolinite soils to smectite dominated soils. The C-line relationship, introduced above, presents both the plasticity index and the specific surface area normalized by the clay fraction (<2 μm). The C-line in Fig. 3.4 illustrates the overall link between these two parameters with a slope of 0.005. The C-line indicates that a sediment with an activity of 0.5, a clay specific surface area (*Sc*) of about 100 m^2/g should be expected and it would fall within the illite/chlorite dominated zone. Figure 3.4 puts the Québec sensitive clays in a broader perspective with respect to soils in Europe

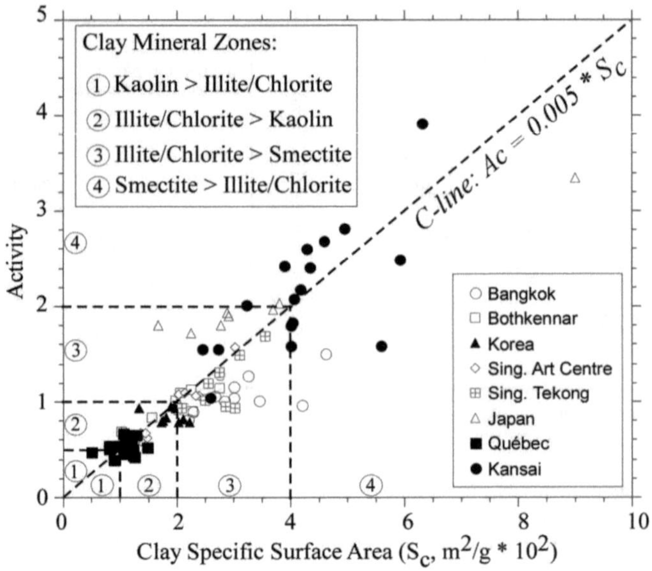

Fig. 3.4 The relationship between the Activity (*Ac*) of soils and their clay specific surface area (*Sc*). The *dashed line* from the origin represents the 'clay line' (C-line, modified after Locat et al. 2003)

and Asia where the geological history and source of sediment may be quite different, but similar to places in Europe and Asia where the geological history is similar. Ariake clay (Torrance and Ohtsubo 1995), which has a low activity for a Japanese clay (varying from 0.75 to 2.0), would fall in zone 3 but close also to boundary between zone 3 and 2. Figure 3.4 also illustrates the fact that sensitive clays from Québec are of low activity and contain little or no swelling clays. In this diagram, Québec clays fall within the illite/chlorite dominated zone, which for Québec could read as chlorite/illite.

3.4 Concluding Remarks

As indicated by Locat et al. (1984), it is difficult to develop a clear mineralogical signature of the soils as a function of their sedimentary basin but the sediments formed in these postglacial seas can be seen as a product of the concentration of fines from the till (Locat 1995). The end result is a clay soil with a low activity and a relatively low clay specific surface area (*Sc*). This may be due to various reasons including variability in the grain size distribution (e.g. in various profiles as shown by Berry and Torrance 1998) and also because some mixing may have taken place between sediment plumes coming from the various rivers entering the basins.

Although each measurement method has its own uncertainty, this paper has presented what we consider as the quantitative analysis of 18 samples from Québec, most of them within the Champlain Sea basin. The set of data can certainly be explored in more details and should provide a reference to help evaluate the performance of simpler methods.

As part of this work, we have underlined the interesting potential of the specific surface area measurements (all methods) and the amount of constitutive water (H_2O_c) to reflect on the clay mineralogy of the soils, even more when compared to soils from other environments. This work has also illustrated the use of the C-line to reflect on the nature of sensitive clays relative to other clays from around the World.

Acknowledgements The authors would like to thank Robert Ledoux and Marc-André Bérubé who greatly helped ensuring that the methods used for the mineralogical analysis were appropriate and used carefully. We also thank the support of the Ministère des transports du Québec for providing the samples. This research was supported by grants from the NSERC and Fonds FQRNT. Finally, we would like to thank Kenneth Torrance for his constructive comments that greatly helped the manuscript.

References

Alexiades CA, Jackson ML (1965) Quantitative determination of vermiculite in soils. Soil Sci Soc Am 29:522–527

Babineau P (1977) Minéralogie des argiles de St-Alban. MSc thesis, Université Laval, Ste.-Foy, Québec, 73 p

Ball DF (1964) Loss on ignition as an estimate of organic matter and organic carbon in non-calcareous soils. J Soil Sci 18:84–92

Ballivy G, Pouliot G, Loiselle A (1971) Quelques caractéristiques géologiques et minéralogiques des dépôts d'argile du nord-ouest du Québec. Can J Earth Sci 12:1525–1541

Bentley SP (1976) The contribution of primary mineral particles to the properties of sensitive clays from Eastern Canada. PhD thesis, University of Leeds, Leeds, 336 p

Bentley SP (1980) Significance of amorphous material relative to sensitivity in some Champlain clays: Discussion. Can Geotech J 17:632–634

Bérend I, Cases JM, François M, Uriot JP, Michot L, Masion A, Thomas F (1995) Mechanism of adsorption and desorption of water vapor by homoionic montmorillonites: 2. The Li+, Na+, K+, and Cs+ – exchanged forms. Clay Clay Miner 43:324–336

Berry RW, Torrance JK (1998) Mineralogy, grain-size distribution and geotechnical behavior of Champlain clay core samples, Quebec. Can Mineral 36:1625–1636

Brydon JE, Patry LM (1961) Mineralogy of Champlain Sea sediments and a Rideau clay soil profile. Can J Soil Sci 41:169–181

Chapman HD (1965) Cation exchange capacity. In: Black CA (ed) Methods of soil analysis. Agronomy, vol 9, part 2. American Society of Agronomy, Madison, pp 891–901

Diamond S, Kinter EB (1958) Surface area of clay minerals as derived from measurements of glycol retention. Clay Clay Miner 5:334–347

Dreimanis A (1962) Quantitative bazometric determination of calcite and dolomite by using Chittick apparatus. J Sediment Petrol 32:520–529

Engler P, Tyengar SS (1987) Analysis of mineral samples using combined instruments (XRD, TGA, ICOP) procedures for phase quantification. Am Mineral 72:832–838

Foscal-Mella OM (1976) Analyse mineralogique des argiles glaciaires. Masters thesis, Ecole Polytechnique, Montreal, Quebec, 148 p

Geertsema M, Torrance JK (2005) Quick clay from the Mink Creek landslide near Terrace, British Columbia: geotechnical properties, mineralogy, and geochemistry. Can Geotech J 42:907–918
Gillot JE (1971) Mineralogy of Leda clays. Can Mineral 10:797–811
Gravel JY (1974) Minéralogie de l'argile Champlain de St-Jean-Vianney. Master thesis, Université Laval, Ste.-Foy, Québec, 42 p
Jackson ML (1969) Soil chemical analysis, advanced course, 2nd edn. Parallel Press, Madison, 895 p
Kiely PY, Jackson ML (1965) Quartz, Feldspar, and Mica determination for soils using sodium pyrosulfate fusion. Soil Sci Soc Am 29:159–163
Lebuis J, Robert J-M, Rissmann P (1983) Regional mapping of landslide hazards. In: Proceedings of the international symposium on slopes in soft clays, Swedish Geotechnical Institute, Report No. 17, pp 205–256
Leroueil S, Tavenas F, Le Bihan JP (1983) Propriétés caractéristiques des argiles de l'est du Canada. Can Geotech J 20:681–705
Locat J (1982) Contribution à l'étude de l'origine de la structuration des argiles sensibles de l'Est du Canada. PhD thesis, University of Sherbrooke, Department of Civil Engineering, Québec, 512 p
Locat J (1995) On the development of microstructure in a collapsible soils. NATO workshop. In: Derbyshire E et al (ed) Genesis and properties of collapsible soils. Kluwer Academic Publishers, Dordrecht, pp 93–128
Locat J, Bérubé M-A (1986) L'influence de la granulométrie sur la mesure des carbonates par la méthode du chittick. Géogr Phys Quat 40:331–336
Locat J, Levebvre G, Ballivy G (1984) Mineralogy, chemistry, and physical properties interrelationships of some sensitive clays from Eastern Canada. Can Geotech J 21:530–540
Locat J, Tanaka H, Tan TS, Dasari GR, Lee H (2003) Natural soils: geotechnical behavior and geological knowledge. In: Characterisation and engineering properties of natural soils, vol 1 (Proc. Singapore Workshop), Balkema. Swets & Zeitlinger, Lisse, pp 3–28
Martel YA, De Kimpe CR, Laverdière MR (1978) Cation-exchange capacity of clay-rich soils in relation to organic matter, mineral composition and specific surface area. Soil Sci Soc Am 42:764–767
McKeague JA (1978) Manual on soil sampling and methods of analysis. Can Soc Soil Sci
Mehra OP, Jackson ML (1960) Iron oxide removal from soils and clays by a dithionite-citrate system buffered with sodium bicarbonate. In: Proceedings of the 7th national conference on clays and clay minerals. Pergamon Press, New York, pp 317–327
Pimol P, Khanidtha M, Prasert P (2008) Influence of particle size and salinity on adsorption of basic dyes by agricultural waste: dried Seagrape (*Caulerpa lentillifera*). J Environ Sci 20:760–768
Quigley RM (1980) Geology, mineralogy and geochemistry of Canadian soft soils: a geotechnical perspective. Can Geotech J 17:261–285
Reynolds RC, Lessing P (1962) The determination of dioctahedral mica and potassium feldspar in submicroscopic grain sizes. Am Mineral 47:979–982
St-Gelais D, (1990) La surface spécifique et l'eau de constitution comme indicateur de la composition minéralogique des sols argileux du Québec. MSc thesis, Department of Geology and Geological Engineering, Laval University, Québec, 124p
Torrance JK (1975) Pore water extraction and the effect of sample storage on the pore water chemistry of Leda clay. Soil Specimen Preparation Laboratory Testing. ASTM, STP 599, pp 147–157
Torrance JK (1983) Towards a general model of quick clay development. Sedimentology 30:547–555
Torrance JK (1988) Mineralogy, pore-water chemistry, and geotechnical behaviour of Champlain Sea and related sediments. In: Gadd NR (ed) The late quaternary development of the Champlain Sea Basin, Geological Association of Canada, special paper 35. Geological Association of Canada, St. John's, pp 259–275

Torrance JK (1995) On the paucity of amorphous minerals in the sensitive post-glacial marine clays. Can Geotech J 32:535–538
Torrance JK, Ohtsubo M (1995) Ariake bay quick clays: a comparison with the general model. Soil Found 35:11–19
Tran NL (1977) Un nouvel essai d'identification des sols: l'essai au bleu de méthylène. Bulletin de Liaison des Laboratoires des Ponts et Chaussées 88:136–137
Van der Plas L (1966) The identification of detrital feldspars. Elsevier Publishing Co, New York, pp 169–225

Chapter 4
Three-Dimensional Quick-Clay Modeling of the Gothenburg Region, Sweden

M.A. Persson

Abstract GIS Modeling of weighted quick-clay preconditions has previously resulted in clay-sensitivity predictions in SW Sweden (mainly the Västra Götaland county), but variability with depth has not been studied in connection with modeling in the area. Certain characteristics important for quick-clay distribution and related risk assessment change with stratigraphy and depth. These need to be modeled in 3D to better account for varying clay conditions. The possibilities for such work have improved with methodological advances and increased accessibility of geotechnical records. A three-step process is used to evaluate the distribution of sensitive clay within the region. First, geotechnical records are compiled and used to analyze clay sensitivity relative to their vertical distance from stratigraphically distinguishable units. Second, the results are combined with parts of an earlier quick-clay prediction model. Third, the combined results are tested against existing information and presented as 3D sensitivity. Suggestions for further refinements and applicability of findings are given based on the results.

Keywords Quick-clay development • Leaching • Stratigraphic modeling • Southwestern Sweden

4.1 Introduction

While commonly used quick-clay mapping methods (i.e. 1D geotechnical surveying and 2–2.5D resistivity profiling; cf. Lundström et al. 2009; Persson and Stevens 2012) are only practical in limited areas, modeling is, if the larger uncertainties can

M.A. Persson (✉)
Department of Earth Sciences, University of Gothenburg, Gothenburg, Sweden
e-mail: martin.persson.3@gvc.gu.se

J.-S. L'Heureux et al. (eds.), *Landslides in Sensitive Clays: From Geosciences to Risk Management*, Advances in Natural and Technological Hazards Research 36,
DOI 10.1007/978-94-007-7079-9_4, © Springer Science+Business Media Dordrecht 2014

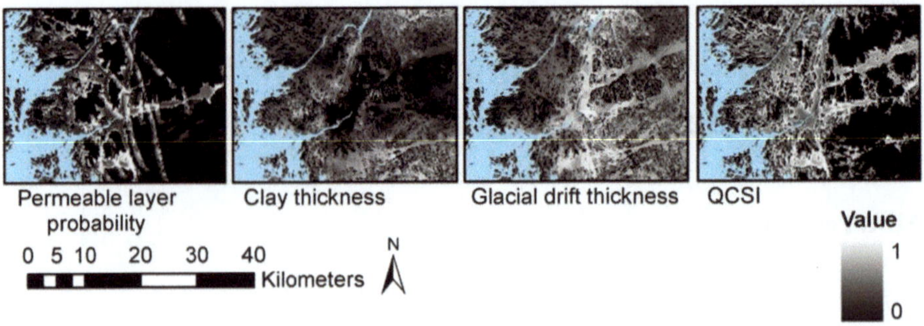

Fig. 4.1 Spatial distribution of stratigraphic criteria utility and QCSI within the Gothenburg region, SW Sweden as proposed by Persson et al. (2013). No potential for quick-clay formation is indicated by the value 0 while the maximum effect is expected at 1

be tolerated, nearly unlimited geographically and is feasible in three dimensions. Predictive modeling of favorable conditions for quick clay in SW Sweden has been previously applied to produce maps of maximum sensitivities, regardless of their stratigraphic positions (Persson et al. 2013). The quick-clay susceptibility index (QCSI; Fig. 4.1) is a weighted measure combining criteria (water flow accumulation, relative relief, clay thickness, aquifer thickness, permeable layer probability, distance to glacial drift, distance to bedrock, groundwater capacity and time available for post-depositional processes) that affect groundwater leaching, considered the chief cause for Swedish quick-clay formation. In addition to the criteria weights, utilities were assigned to account for local differences in criteria fulfillment and effects on leaching. Modeling quick-clay distribution allows a structured approach to the use of both empirical and conceptual information. This is especially important in the early stages of projects concerning the ground conditions where hazardous areas exists, but may also help identify knowledge and data gaps that are crucially important for subsequent understanding and safety. However, when data sources are scarce model result can be expected to frequently be ambiguous.

Knowledge of quick-clay distribution is a critical step when assessing if infrastructure, buildings and land may be affected by future landslide retrogression. The main objective of the present study is to combine sensitivity trends from existing geotechnical vertical documentation with earlier, largely conceptual results of the geographic distribution to present a 3D model of sensitivity to allow quick-clay prediction using varying degrees of input from new investigations. The primary intended use is in the early project stages where existing data are sparse but less precise analyzes may be motivated by the good geographical coverage in 3D.

A clay deposit is, by definition, quick if the undrained undisturbed to remolded shear strength ratio (i.e. sensitivity) is over 50 while the undrained remolded shear strength is <0.4 kPa (Karlsson and Hansbo 1989). High sensitive and quick clays are mostly dependent on low remolded shear strength, which is also, by magnitude, more variable than the undisturbed shear strength (Bjerrum 1954).

4.2 Quick-Clay Development Settings in the Gothenburg Area

The crystalline Precambrian bedrock is fractured in a pattern reflected by the orientation of today's valleys. Sediment architecture in the area (Figs. 4.1 and 4.2) generally follows the principles put forward by Stevens et al. (1991). Sandy till and glaciofluvium (hereafter glacial drift) are thickest in association with ice-front positions (i.e. where the ice-front remained stagnant for tens to hundreds of years during ice retreat, approximately 15,000 years ago; Hillefors 1969; Lundqvist and Wohlfarth 2001). Topographic features (e.g. fracture valley systems) have in places also promoted coarse-sediment deposition. Valleys oriented along glacial flow directions were favorable as glacial meltwater channels, which has led to relatively greater deposition of glaciofluvial sand and gravel (Hillefors 1969; Stevens and

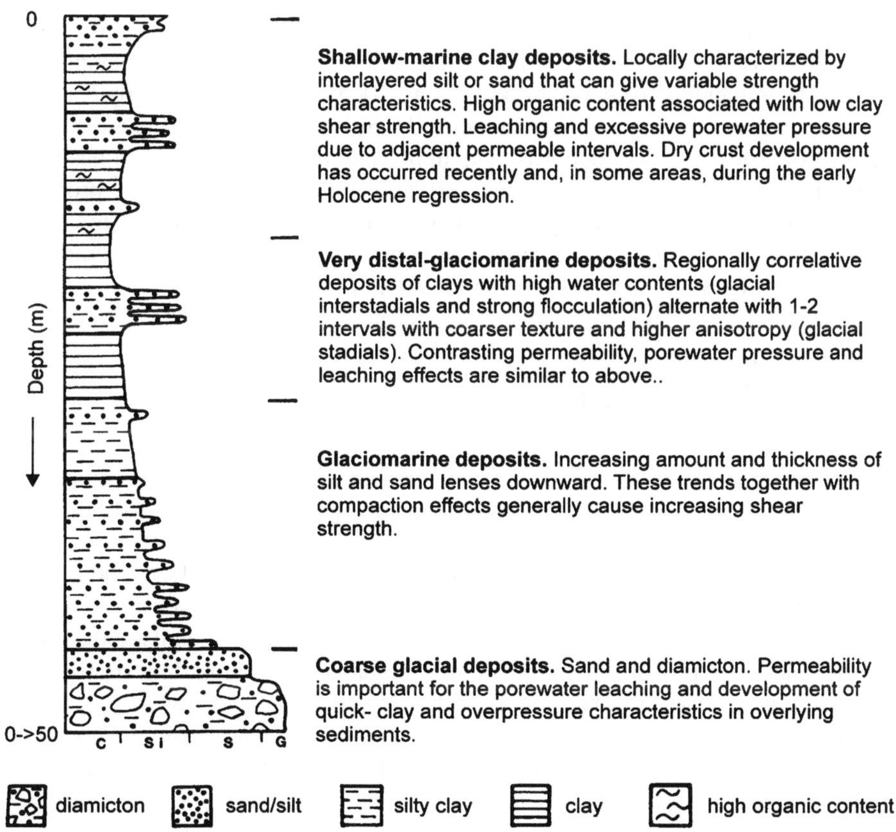

Fig. 4.2 Simplified stratigraphic sequence SW Swedish valleys (Modified from Stevens et al. 1991)

Hellgren 1990). Glacial drift thicker than 50 m have been interpreted locally along a reflection seismic survey in the southern Göta älv River (Klingberg et al. 2006).

Clay deposition during and after ice retreat has, in places, resulted in thicknesses in excess of 100 m (e.g. Klingberg et al. 2006). Shoreline regression and erosion have, in exposed areas, introduced permeable layers within the otherwise clayey sequences. These can be well connected to the source material deposits (i.e. glacial drift) along valley slopes where infiltration is likely and therefore allowing for effective groundwater leaching. The prevalent relief relationships, together with a slowing in the rate of land uplift, provided the greatest exposure to reworking by wave turbulence 7,000 years ago. At this time the sea level was ca. 25 m above the present sea level in the Gothenburg area. Since the low sedimentation rates following this time have been relatively uniform, the permeable layer level is often found ca. 15 m below the present ground level (Cato et al. 1982).

Quick-clay developments in Sweden have primarily occurred in the SW parts of the country where glaciomarine and marine clays compose a significant part of the valley stratigraphy. Mineralogical and textural quick-clay prerequisites are generally fulfilled. A large number of factors are known to affect clay sensitivity (reviewed e.g. by Rankka et al. 2004) although the most effect has often been attributed to leaching (Rosenqvist 1946). The pore-fluid cation concentration and composition are highly influential (Andersson-Sköld et al. 2005) and are changed by leaching. The highest sensitivities are often associated with permeable deposits either inter-layered or below the clay deposits. The upper few meters of clay have been subjected to a varying degree of dry-crust formation and other vadose zone processes causing a net strengthening by the release of divalent cations and compaction. The natural variability of sensitivities varies by at least one order of magnitude in different stratigraphic units at the same location (Talme et al. 1966).

4.3 Data and Methods

The Gothenburg study area (Fig. 4.3) was chosen both for well-known quick-clay areas and to benefit of several extensive geotechnical surveys (SGI 2012a, b) that were not available during the initial modeling (Persson et al. 2013).

In the parsing phase (Fig. 4.4), existing geotechnical data (SRA 2002, 2008; STA 2005–2009, 2006–2011) allowed definition of stratigraphic sensitivity trends. The data (fall cone and cone penetration test) were also used to identify and quantify the thickness and probability trends in the Gothenburg area regarding stratigraphic units used in the earlier QCSI model (Sect. 4.1; Persson et al. 2013).

The stratigraphy simplified to only three units (*Clay*, *Inter-layered silt or sand* and *Glacial drift*) together with geologic history gave four scenarios: (1) *Clay directly resting on bedrock*, (2) *Clay with glacial drift below*, (3) *Clay with inter-layered silt or sand at 15 m* and (4) *Clay with inter-layered silt or sand at 15 m and underlying glacial drift*. These *scenarios* were further specified using sub-clay glacial drift and silt or sand layer utility thresholds, in regard to leaching, of 0.73 (corresponding to 5 m glacial drift; cf. Persson et al. 2013) and 0.50 (permeable

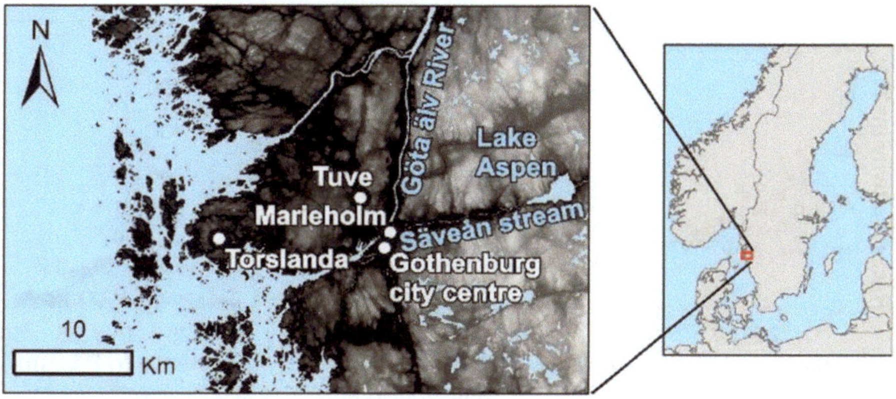

Fig. 4.3 Map over the study area with sites mentioned in text indicated

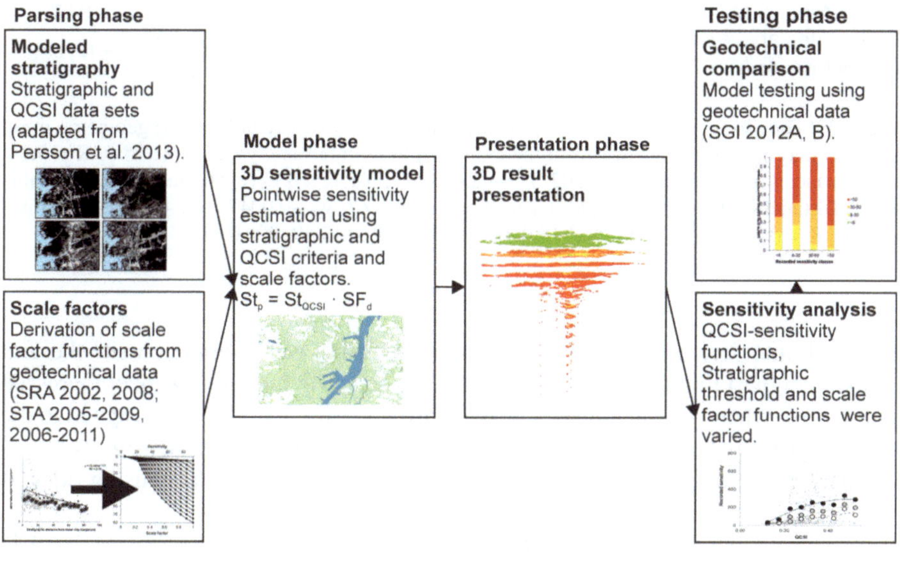

Fig. 4.4 Sequential phases and components used for model construction and analysis

layer probability), respectively. The thresholds were assumed to identify hydraulically effective units. The sensitivity trends in each scenario (Fig. 4.5) were interpreted from the geotechnical test data (specified in Fig. 4.4, lower left corner). Scale factor functions (SFFs; Fig. 4.6) for specific stratigraphic scenarios and clay thicknesses were constructed to get 0–1 scale factors that express each point's share of the maximum QCSI-derived sensitivity at a site (cf. Persson et al. 2013). This was achieved by rearranging terms in the line equations of Fig. 4.5 to resolve sensitivity at depth and compare to the QCSI-derived sensitivity. If several sensitivity-raising units

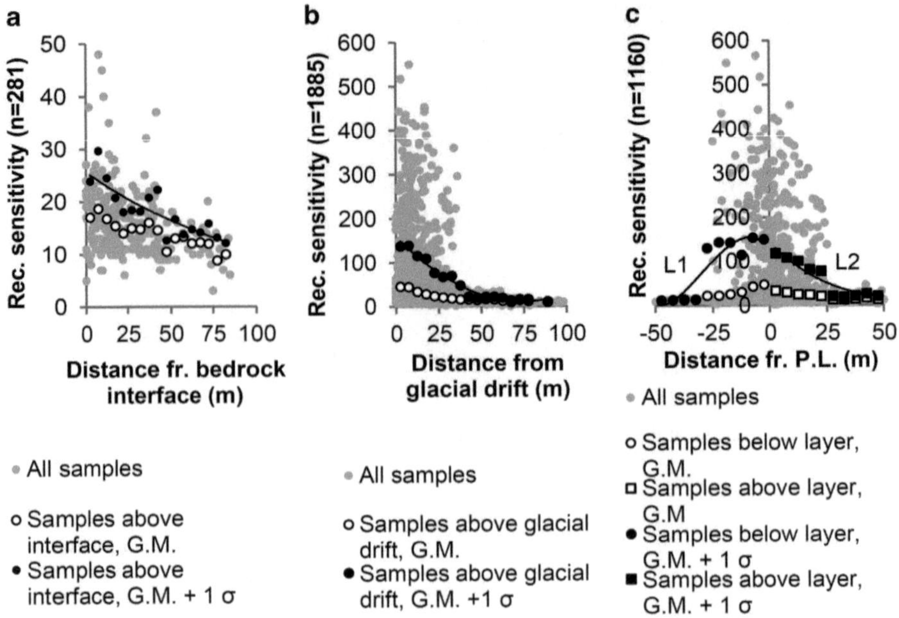

Fig. 4.5 Scenario sensitivity trends from geotechnical records where G.M. is the geometric mean per 5 m clay unit. (**a**) Distance from the clay – bedrock interface. $y=25.644e^{-0.009x}$, $R^2=0.74$. (**b**) Distance from glacial drift. $y=0.0271x^2-4.0063x+159.4$, $R^2=0.97$ (**c**) Distance from permeable layer. Distances >0 indicate samples taken above layers. *L1*: $y=-0.0047x^3-0.4144x^2-6.4286x+126.91$. $R^2=0.86$. *L2*: $y=-0.0004x^3+0.0806x^2-5.4931x+143.49$, $R^2=0.91$

(bedrock interface, permeable layer or sub-clay drift) were found at a site the unit with the highest impact at specific depths was used. The functions were compressed depth-wise when clay thicknesses were small. A sensitivity of 5 was assigned to the ground surface to account for near-surface processes lowering the sensitivity.

In the modeling phase (Fig. 4.4), the 3D sensitivity was calculated using Eq. 4.1.

$$St_p = St_{QCSI} \cdot SF_d \tag{4.1}$$

Where St_p is the predicted sensitivity at a specific depth and calculated for each point in point data sets (50 m horizontal and 5 m vertical point separation) covering the whole study area down to 80 m. St_{QCSI} is the maximum site-specific sensitivity interpreted by the QCSI-model. The regression line equation yielded when sensitivity was averaged for 0.05 QCSI classes and plotted against QCSI (Fig. 4.7) was used as primary QCSI-to-sensitivity function. SF_d is the scale factor unique for each depth given a stratigraphic scenario and a total clay thickness. All calculations combining QCSI, QCSI-sensitivity relations and scale factor functions (Eq. 4.1) were done using ESRI® ArcMap™ 10 (ArcInfo licence). The model results were presented using the depth-wise point sets and ESRI ArcScene 10®.

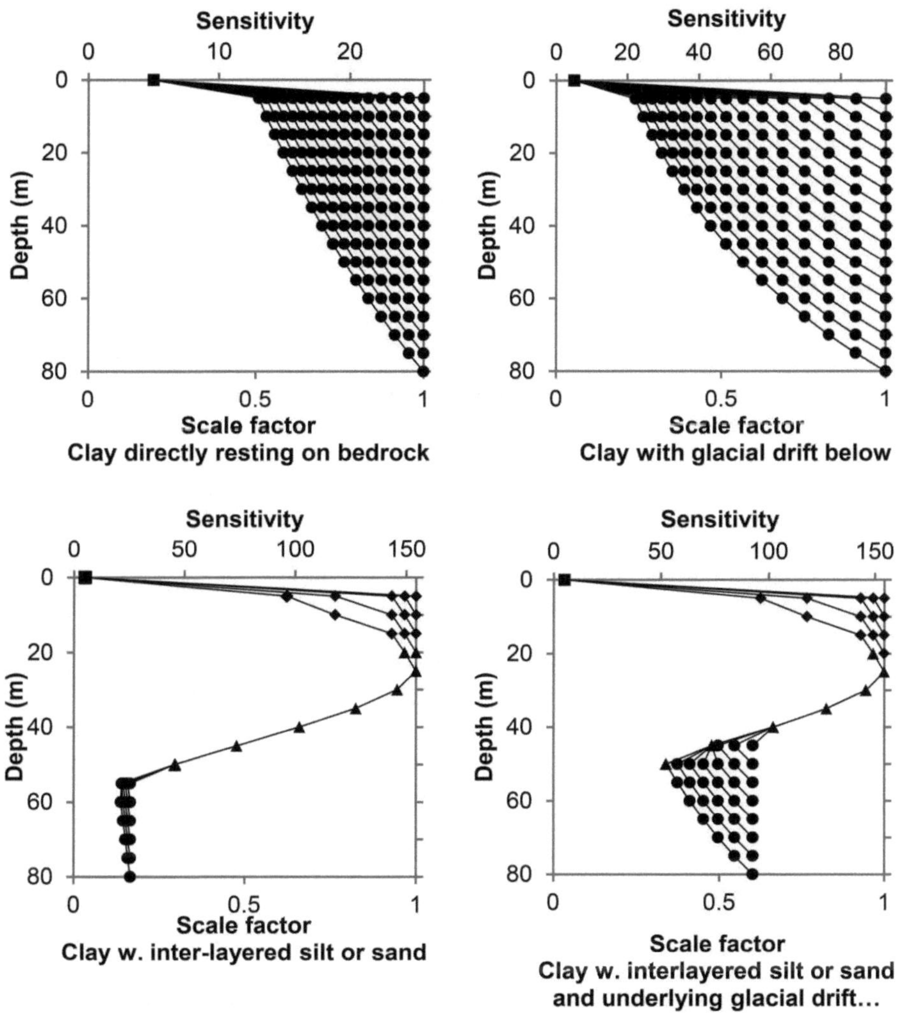

Fig. 4.6 Empirical SFFs, based on trends in Fig. 4.5. Separate functions were used for specific depths and clay thicknesses. Sensitivity scales are different between diagrams

The model performance was tested in two ways (Fig. 4.4). First, the model's sensitivity to changes in SFFs (Fig. 4.6), alternative QCSI-to-sensitivity functions (Fig. 4.7), and criteria thresholds was tested using three sets of SFF (original ± 25 %), three QCSI-to-sensitivity functions, and three criteria-threshold sets (original ± 25 %). Second, clay sensitivity maps were produced from the result (Fig. 4.8) and compared using ordinary kriging and ArcGIS standard settings (Fig. 4.9a, b) to independent geotechnical data (SGI 2012a, b). The use of alternative model components and functions did not give any significant improvements over the initially used, so these were not changed.

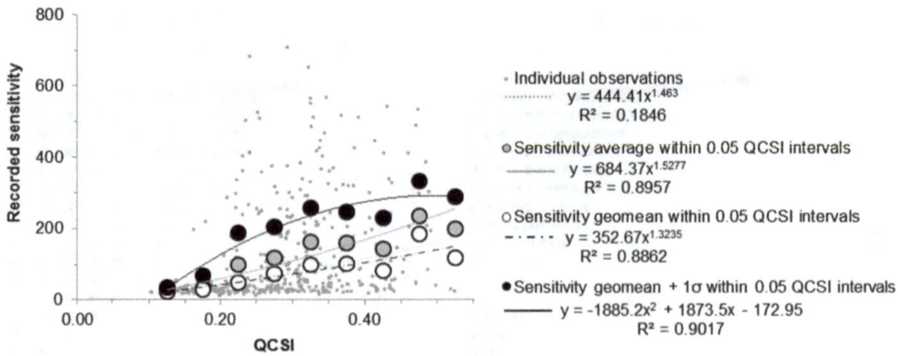

Fig. 4.7 Alternative functions for use in QCSI to sensitivity transformations

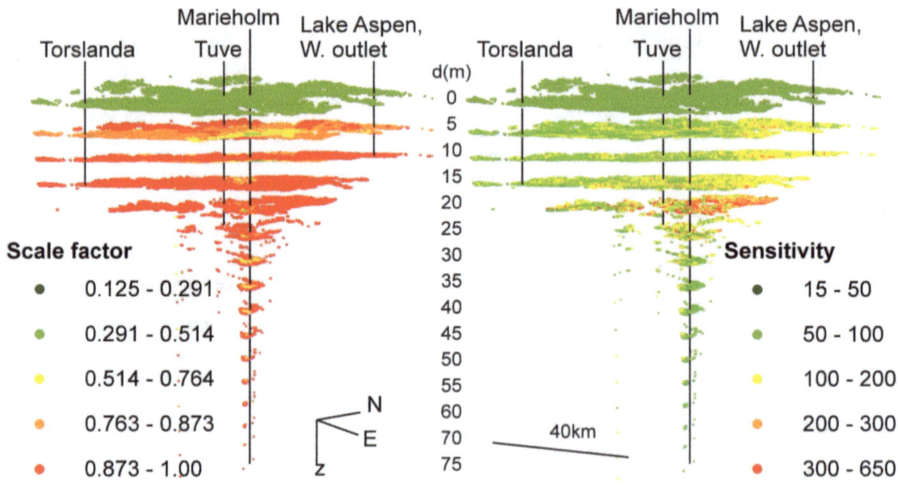

Fig. 4.8 Stacked maps of scale factors (from SFFs) and predicted sensitivity. The topography is exaggerated 10×. The spectator view is toward the northeast

4.4 Three-Dimensional Quick-Clay Predictions

Close to 8.5 km^3 of clay sediment have been investigated in the study area (Fig. 4.8). It is suggested, using the original QCSI-to-sensitivity transformation (Fig. 4.7), that approximately 70 % of the geographic points have an estimated sensitivity of >50 somewhere at depth. If the sample distribution of Fig. 4.9b, which has been used as a model test, is considered the proportion corresponds to ca. 50 % which still is, for reasons given below, an overestimate. If instead the function based on geometric mean of samples a more moderate 28 % of points are predicted to be quick. In either case, the upper 15 m is suggested by the model to contain 85 % of all quick data points.

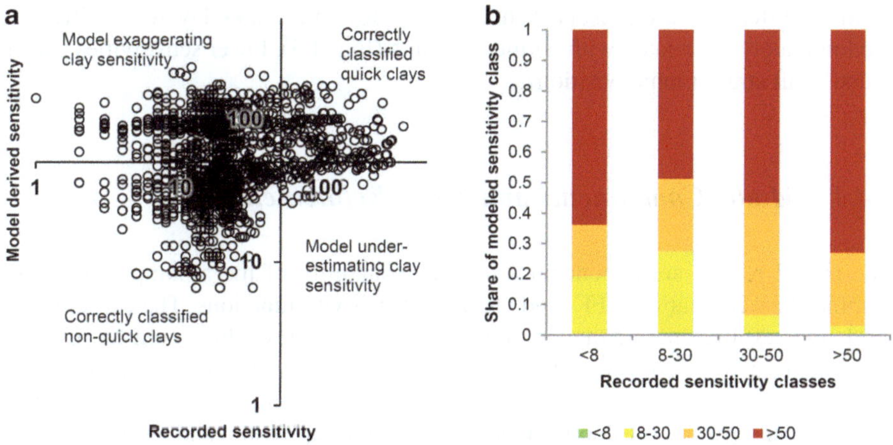

Fig. 4.9 (**a**) Comparison between documented and model-derived sensitivity where axis cross at a sensitivity of 50. (**b**) Distribution of modeled sensitivity within sensitivity classes

Practically, deeper lying quick clay poses little threat under most circumstances as they are below most slip surfaces.

The geotechnical data used for testing the results indicate that given favorable stratigraphic conditions sensitivity values are often more than ten times the original background sensitivity of unleached clay in the Gothenburg area (St ~ 7). This sensitivity is interpretable from the geotechnical comparison data and is comparable to what Bjerrum (1954) and Söderblom (1969) observed in samples from southern Norway and the Göta älv River valley respectively.

Areas with *Clay directly resting on bedrock* are characterized by only small deviations from the background sensitivity and are rarely quick. However, a sensitivity increase is evident near the clay – bedrock interface where sensitivities commonly reach 30. The *Clay with glacial drift below* scenario often has the thickest affected clay sequence (i.e. largest volume), where sensitivity trends can be traced from water bearing unit. Maximum, geometric means and range of recorded sensitivities before the sensitivity background is reached are here comparable to those in the *Clay with inter-layered sand or silt scenario*. The slightly higher geometric means related to permeable layering may be due to hydraulically more effective contact with infiltration sediments at the surface. The sensitivity maxima tend to be below the layer, where the pressure gradients are toward the layer and the clay may also be exposed to overpressure from the drift material beneath. The average values and standard deviations of recorded sensitivities (Fig. 4.5) decrease with distance from the permeable units, suggesting that a background value is reached.

If the other environmental criteria favoring quick clay (cf. Persson et al. 2013) are sufficiently fulfilled, quick clay may have formed even if glacial drift and permeable layering in the clay sequence are both absent. If many of these same criteria are unfulfilled low sensitive clay (St < 8) may occur even where significant permeable deposits are present. While high (>30) and quick-clay sensitivities are predominately associated with permeable units (Fig. 4.5), lower sensitivities can be found at all stratigraphic positions.

4.4.1 Model Components and Their Refinements

The model results are dependent on the reasonability of the assumptions, stratigraphic characterization, SFFs and QCSI-to-sensitivity functions. The earlier QCSI modeling resulted in an exaggeration of low sensitivities. This could be resolved using alternative transformation functions (Fig. 4.7) derived from geotechnical data comparisons. Modifications of these functions will result in changes of the estimated sensitivities at all depths. Changes in stratigraphic thresholds or SFFs will not affect the maximum sensitivity but will shift the stratigraphic sensitivity pattern. For example, a glacial drift utility threshold decrease by 25 % (i.e. from 5 to 3.4 m thickness) will increase the affected area from ca. 122 to 217 km^2 and thus give a different sensitivity distribution.

Several aspects of the model could benefit from further work. Specifically, the comparisons to geotechnical data indicate that the largest effort should involve improvement of the modeled stratigraphy. This may be achieved by further paleogeographic work better defining areas with permeable layers, geophysical work aimed at characterizing drift thicknesses in various settings and geomorphometric or geophysical work to better describe buried bedrock morphology.

Comparison to recent geotechnical data from the lower Göta älv River valley (SGI 2012a, b) show that the scale factors (from SFFs) are occasionally presented for deeper levels than the maximum recorded clay thicknesses. This is partly due to the 50 m geographic model resolution and the fact that clay thicknesses change rapidly. The earlier quantification of glacial drift thickness is significantly limited by that many geotechnical surveys focus on depth to bedrock or drift only. The empirical basis for the permeable layer predictions is better, but dropstones, isolated sand lenses and methodological limitations (cf. Larsson 2007) may falsely be interpreted as continuous layers. Further discussion on the parameterization is given by Persson et al. (2013). Other identifiable stratigraphical positions, with variable quick-clay preconditions (e.g. early glaciomarine clay with relatively high carbonate content or sand layers derived from glacial stadials; cf. Stevens et al. 1991), could be identified and recreated within future model revisions. The thresholds used for concluding the glacial drift or permeable layer presence have been arbitrary chosen knowing that some sites are incorrectly characterized. Leaching beyond quickness (cf. Solberg et al. 2008) or clay-sample disturbance cannot be excluded, contributing to samples in the upper left quadrant of Fig. 4.9a. The geotechnical data used in this study was originally developed mainly for infrastructure purposes and are therefore not normally distributed but skewed toward areas with problematic ground conditions. This is somewhat balanced by investigations at wide valley locations (e.g. at Marieholm). A more representative distribution of cored sites would arguably give overall lower 3D sensitivity estimates. Additional sensitivity records are available from municipality offices, consultant companies and authorities and could be digitalized, preferably into a national database that could serve multiple purposes.

4.4.2 Possible Model Applications

The 3D model results can be used to improve the relatively simplistic pilot study stage of Swedish landslide hazard zonation (MSB 2009), which does not consider quick-clay occurrences. Modeling could also be used in infrastructure feasibility studies for cost-effective planning. Since the model accuracy is highly dependent on input data density, implying that the areas with little or no documentation, i.e. those most highly benefited by modeling, are also where the results have lower reliability. Therefore, modeling needs to be repeated as more data becomes available. Since the QCSI modeling is focused on leaching, the results could also be used in specialized studies of quick-clay development, especially in areas where deviations between model results and observations occur.

The model is also useful where substantial geotechnical documentation exists (e.g. along road or railroad networks). Interpolation between sounding locations can be done more consistently and volumes can be calculated aided by the suggested modeling approach.

4.5 Conclusions

The current model is a first step towards incorporating 3D sensitivity estimations in quick-clay mapping applications. The results are based on a simplified stratigraphic model and cannot replace traditional mapping methods in survey applications. There is, nevertheless, a complementary relationship to empirical documentation that underscores a need for modeling and the possibilities for combined assessments with increasingly holistic (3D) character.

Acknowledgements The author is grateful to authorities who made data available and to Rodney Stevens at the Dept. of Earth Sciences, Univ. of Gothenburg who improved the manuscript language and contents. Hjördis Löfroth at the Swedish Geotechnical Institute is appreciated for reviewing the manuscript and suggesting improvements.

References

Andersson-Sköld Y, Torrance JK, Lind B, Odén K, Stevens RL, Rankka K (2005) Quick clay – a case study of chemical perspective in southwest Sweden. Eng Geol 82:107–118

Bjerrum L (1954) Geotechnical properties of Norwegian marine clays. Géotechnique 4(2):49–69

Cato I, Fredén C, Olausson E (1982) Summary of the investigation. In: Olausson E (ed) The Pleistocene/Holocene boundary in south-western Sweden, The Geological Survey of Sweden series C: 794. Sveriges geologiska undersökning, Uppsala, pp 253–268

Hillefors Å (1969) Västsveriges glaciala historia och morfologi. Meddelanden från Lunds universitets geografiska institution. Dissertation 60, Lund

Karlsson R, Hansbo S (1989) Soil classification and identification. Document D8:1989. Byggforskningsrådet, Stockholm

Klingberg F, Påsse T, Levander J (2006) Bottenförhållanden och geologisk utveckling i Göta älv. Geological Survey of Sweden report K43. Uppsala

Larsson R (2007) CPT-sondering, utrustning – utförande – utvärdering. En in-situ metod för bestämning av jordlagerföljd och egenskaper i jord. Swedish Geotechnical Institute Info 15. Linköping

Lundqvist J, Wohlfarth B (2001) Timing and east–west correlation of south Swedish ice marginal lines during the Late Weichselian. Quat Sci Rev 20:1127–1148

Lundström K, Larsson R, Dahlin T (2009) Mapping of quick clay formations using geotechnical and geophysical methods. Landslides 6:1–15

MSB (2009) Metod för kartering i finkorniga jordar. Swedish Civil Contingencies agency website retrieved January 26, 2013 from: https://www.msb.se/sv/Forebyggande/Naturolyckor/Oversiktlig-stabilitetskartering/Kartering-i-finkorniga-jordar/Metod-for-kartering/

Persson M, Stevens R (2012) Quick-clay formation and groundwater leaching trends in southwestern Sweden. In: Eberhardt E, Froese C, Turner AK, Leroueil S (eds) Landslides and engineered slopes – protecting society through improved understanding. CRC Press/Taylor & Francis Group, London, pp 615–620

Persson MA, Stevens RL, Lemoine Å (2013) Spatial quick-clay predictions using multi-criteria evaluation in SW Sweden. Landslides: Published online first Feb. 17

Rankka K, Andersson-Sköld Y, Hultén C, Larsson R, Leroux V, Dahlin T (2004) Quick clay in Sweden. Swedish Geotechnical Institute report 65, Linköping

SGI (2012a) "Borrhåltidigarepaketet", SGI/120801. Geotechnical survey results made available by the Swedish Geotechnical Institute, Linköping, Sweden. Retrieved October 16, 2012 from http://www.geodata.se/

SGI (2012b) "Borrhålpaketet", SGI/120801. Geotechnical survey results made available by the Swedish Geotechnical Institute, Linköping, Sweden. Retrieved October 16, 2012 from http://www.geodata.se/

Söderblom R (1969) Salt in Swedish clays and its importance for quick clay formation: results from some field and laboratory studies, no. 22. Swedish Geotechnical Institute, Stockholm, p 63

Solberg IL, Rønning JS, Dalsegg E, Hansen L, Rokoengen K, Sandven R (2008) Resistivity measurements as a tool for outlining quick clay extents and valley fill stratigraphy: feasibility study from Buvika, Central Norway. Can Geotech J 45:210–225

SRA (2002) Västra stambanan Göteborg–Torp, Partille–Jonsered section. Geotechnical data. The Swedish Rail Administration

SRA (2008) The Kil–Göteborg (Agnesberg–Marieholm) section. Geotechnical data. The Swedish Rail Administration

STA (2005–2009) Geotechnical data concerning E6 stretches between Hogstorp and Svinesund extracted from the Swedish Transport administration database CHAOS

STA (2006–2011) Geotechnical data concerning E45 stretches between Olskroken, Göteborg and Trollhättan extracted from the Swedish Transport administration database CHAOS

Stevens RL, Hellgren L-G (1990) A generalized lithofacies model for glaciomarine and marine sequences in the Göteborg area, Sweden. Geol Foren Stock Forh 112:89–105

Stevens RL, Rosenbaum MS, Hellgren L-G (1991) Origins and engineering hazards of Swedish glaciomarine and marine clays. In: Forster A, Culshaw MG, Cripps JC, Little JA, Moon CF (eds) Quaternary engineering geology. Engineering geology special publication 7, Geological Society, London, pp 257–264

Talme OA, Pajuste M, Wenner CG (1966) Secondary changes in the strength of clay layers and the origin of sensitive clay. Byggforskningsrådet Rapport 46, Stockholm

Rosenqvist I Th (1946) Om leirers kvikkagtighet. Statens Vegvesen. Veglaboratoriet meddelelse Nr 4. Oslo

Chapter 5
Ion Exchange as a Cause of Natural Restabilisation of Quick Clay – A Model Study

Pascal Suer, Hjördis Löfroth, and Yvonne Andersson-Sköld

Abstract Quick clay may be stabilised by increased magnesium concentrations in the pore water. Weathering has so far been cited as the source of magnesium, but is an unlikely process at 15 m depth in clays, where an increase was found at a test site in the south of Sweden. This study investigates how far ion exchange can explain the source of magnesium. A rough model in the program PHREEQC incorporating ion exchange, but not weathering, is used to model geochemical and transport processes since the latest ice age in a soil profile in the Göta River valley. Rain water, or rain water with added calcium or magnesium carbonate, is used as infiltrating solution. Calcium or magnesium could come from weathering in the dry crust. Advection, diffusion and ion exchange are sufficient to approximate concentrations in the pore water, if the infiltrating water contains calcium. Weathering at depths below 5 m is not included in the model, but observations of magnesium are reproduced nevertheless. The magnesium comes from the sea water at the end of the latest ice age and has been stored on the ion exchange sites in the clay. The magnesium is displaced into the pore water when calcium enters the soil, is transported downwards, and partly re-enters the exchange sites. The rough model supports the importance of ion exchange for the restabilisation of quick clay by natural magnesium at this site.

Keywords Quick clay • Desalinization • Ion exchange • Magnesium • Modelling • Leaching

P. Suer (✉) • H. Löfroth
Swedish Geotechnical Institute (SGI), Linköping, Sweden
e-mail: pascal.suer@swedgeo.se

Y. Andersson-Sköld
Swedish Geotechnical Institute (SGI), Linköping, Sweden

Department of Earth Sciences & COWI, University of Gothenburg, Göteborg, Sweden

J.-S. L'Heureux et al. (eds.), *Landslides in Sensitive Clays: From Geosciences to Risk Management*, Advances in Natural and Technological Hazards Research 36,
DOI 10.1007/978-94-007-7079-9_5, © Springer Science+Business Media Dordrecht 2014

5.1 Introduction

The sensitivity of clays in many areas where quick clay occurs is correlated to the concentrations of calcium and magnesium in the pore water. The source of the natural calcium and magnesium affects how quick clays will react to changes in amount and composition of precipitation, and affects the time frame for the development and restabilisation of quick clay.

When Løken (1968) reviewed the chemical processes that lead to quick clay development in Norway, weathering was cited as the dominant chemical process for restabilisation. Weathering in this context included mineralisation, ion exchange, and leaching processes. However, this definition is not useful to separate the importance of mineralisation from that of ion exchange, and identify the source of magnesium.

The term weathering in soil chemistry is usually restricted to the breaking down of soil particles. This is a slow process, unless speeded up by oxygen, organic matter, or acidity. Thus weathering is most important in the upper metre(s) of the soil profile. Weathering releases substances, e.g. ions, in the soil. These ions are then found in the pore water, on the exchange sites, or in new minerals (Appelo and Postma 2005).

Ion exchange is a simplified description of interactions between clay and pore water. Clay particles have a net negative charge that attracts cations. Exchange sites are used as a conceptual model to quantify the capacity that is available to cations. The cations on the exchange sites are no longer in the pore water, but will easily re-enter the pore water if the pore water composition changes. The majority of the cations in a clay soil is associated with the clay, and not with the pore water (Appelo and Postma 2005).

Talme (1968) discussed the influence of the salinity during deposition of the clays, and the calcium carbonate content of the clay. Solid calcium carbonate, e.g. from shells, dissolves until an equilibrium with the pore water is reached. This equilibrium is strongly influenced by pH and carbon dioxide pressure. Thalme did not pursue the issue further to separate the influence of the depositional environment from the weathering of minerals.

Moum et al. (1968) added potassium, iron, aluminium and calcium to the soil to increase stability by increasing the cations in the soil. The assumption was that adding cations decreased the sensitivity, but the results could be due *either* to the direct action of the cations that were added, *or* indirectly to the ion exchange processes caused by the addition, or to a combination of both.

Sensitive clays overlying quick clays were observed in the Göta River valley. For the clay to be classified as quick it should, according to the Swedish definition, have a sensitivity $S_t \geq 50$ and a remoulded undrained shear strength $\tau_R \leq 0.4$. Both sensitive and quick clays had low salt content, but the overlying sensitive clays (at 6–13 m depth) had higher concentrations of magnesium in the pore water (Löfroth et al. 2012a). This was accompanied by patterns of potassium that were reminiscent of the chromatographic effects that can be caused by ion exchange, while weathering is likely to be very slow at this depth in a clay (Appelo and Postma 2005).

Chromatographic effects are caused by the fact that different ions have different affinities for the exchange sites on the clay. For example, sodium has a weaker affinity for exchange sites than calcium or magnesium (Løken 1970; Appelo and Postma 2005). Such different affinities are approximated in exchange constants for the ions. The system gets complicated quickly by changes in pore water composition. This makes geochemical modelling an appropriate method to explore the interaction of the processes (Appelo and Postma 2005).

This paper studies the source of magnesium in the present day pore water in the Fråstad area in the Göta River valley. A geochemical model is used to test the hypothesis that a magnesium content is not caused by weathering in the deeper layers. Instead the depositional environment of the clay, combined with leaching and ion exchange, are used to model processes since the latest ice age.

5.2 The Study Area

The study area is near Fråstad, north of Lilla Edet, around 60 km north of Gothenburg and 10 km south of Trollhättan in Sweden. Geotechnical, geophysical and geochemical data of the site is presented in Löfroth et al. (2012a, b) and SGI (2011).

Bedrock in the area is more than 50 m below ground level, but it surfaces at the sides of the valley. From bedrock to ground level the main sediment is clay, with some coarser layers. The pore water pressures in the area are generally lower than hydrostatic indicating a downward groundwater flow (Löfroth et al. 2012a).

Borehole U07201Q is used in the model. It has a dry crust of 2–3 m overlying a 20 m thick clay layer. The clay is quick from around 14 to 22 m depth. At 22 m depth there is a 3 m thick sand layer. Below this friction layer, from 25 m down, is more clay. The clay below the sand layer is salt to brackish. All clay above the sand layer has a low salt content and resistivity over 10 Ωm. Bedrock is at more than 50 m depth (Fig. 5.1).

Pore water samples were taken at 8, 13 and 20 m depth. The clay at 8 and 13 m is highly sensitive but not quick, and the clay at 20 m is quick. The pore water ions are of magnesium-carbonate type at 8 and 13 m depth, and of sodium-carbonate type at 20 m. The sodium concentration in the pore water of the quick clay is 9 mM (20 m depth). Higher up in the profile, sodium is lower. The magnesium concentration is low at 20 m depth, and higher at 8 and 13 m (Table 5.1).

5.3 Modelling

The program used for modelling in this study is the PHREEQC code developed by USGS (Parkhurst and Appelo 2011). PHREEQC is primarily a geochemical program, with an extensive database of chemical reactions for speciation, ion exchange, mineral equilibria etcetera. The chemical reactions are defined for each cell in the

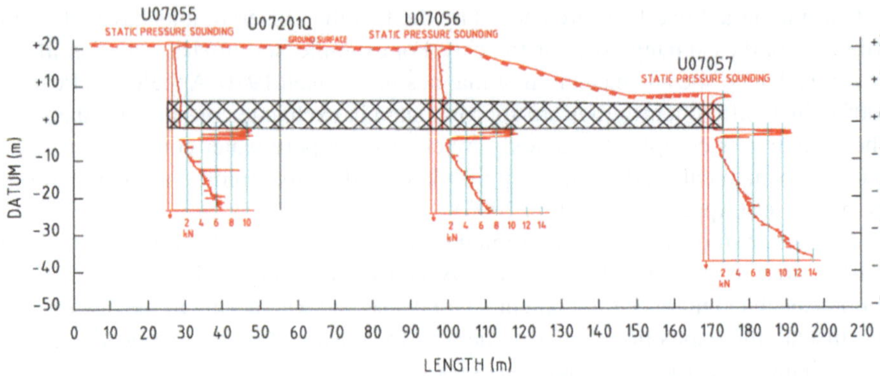

Fig. 5.1 Section parallel to the Göta River with profiles from static pressure sounding. A small stream lies to the right/east near borehole U07057. Depth after local reference, scale from +25 to –40 m. Borehole U07201Q is directly besides the centre sounding, U07056. The quick clay layer is marked with a *dashed interval*. Note the friction layer directly below (After SGI 2011)

Table 5.1 Sensitivity and concentrations in the pore water of borehole U07201Q, in mM[a]

	Sensitivity	Na	K	Mg	Ca	Cl	TIC[b]
8 m	50	0.80	0.75	0.94	0.66	0.23	4.33
13 m	60	1.47	1.04	1.73	0.35	0.18	7.44
20 m	>500	9.10	0.43	0.21	0.08	0.16	11.32

Löfroth et al. (2012a), SGI (2011)
[a]mM, also millimoles/litre, denotes the number of molecules of an ion in solution. 9 mM sodium is equivalent to 208 mg/l. 0.2 mM magnesium is 4.9 mg/l
[b]*TIC* Total Inorganic Carbon, a measurement of carbonate

model and related to the database phreeqc.dat dated 2010-09-30. The present model includes only ion exchange, and no mineral equilibria or kinetic reactions. The pore water volume in each cell is equilibrated with the exchange sites for each time step. The pore water then moves to the next cell and mixes with parts of the pore water from cells above and below to represent diffusion during the time step. Then pore water and exchange sites are equilibrated again. PHREEQC gives a reasonable representation of the water transport, and a very good calculation of the geochemistry. Relevant parts of the input file are shown in Fig. 5.2.

The 1D model starts at the time of the withdrawal of the sea up to present day, around 7,000 years. The model has a surface area of 1 m^2 and concerns the clay from 4 to 22 m depth. These depths correspond to the thickness of the dry crust and to the depth of the friction layer in borehole U07056 (Fig. 5.1). At the start, the exchange sites are in equilibrium with a salt sea water. Fresh water is added to the first cell. The water flows downwards in the clay and interacts with the exchange sites.

Three scenarios are used for the infiltrating water at 4 m depth:

1. Infiltrating water is rainwater, with a low content of all ions
2. Infiltrating water contains magnesium
3. Infiltrating water contains calcium

Fig. 5.2 Data in the PHREEQC input file

```
SOLUTION 1-250 Salt water
        units mmol/kgw
        pH          8
        temp        5
        Ca          5
        K           5
        Mg          27
        Na          243
        C           1
        Cl          283
        S           15
        -water 650
EXCHANGE 1-250 Unleached clay
        X  100
        -equilibrate 1
SOLUTION 0
        units       mg/L
        pH          7.6
        temp        5
        Mg          0.116       # or 5mmol/L
        Ca          0.153       # or 5mmol/L
        C           0           # or 5mmol/L
        Na          0.908
        K           0.120
        Cl           1.58
        S(6)        0.725
        -water  650
TRANSPORT
        -cells  22
        -lengths  1
        -time_step 2.2e9
        -shifts 99
        -diffusion_coefficient 1e-9
```

The modelled concentrations of sodium, potassium, calcium and magnesium are compared with the present day pore water composition, in order to see if ion exchange is sufficient to explain the observed pore water chemistry, or if other processes (e.g. weathering) are needed.

5.3.1 *Chemistry Setup in the Model*

The model used as few chemical processes as possible, in order to find the least complex solution. It contained only advection, diffusion and ion exchange. Minerals were not dissolved or precipitated, and the only role of the clay is to contribute ion exchange sites.

The initial pore water in the clay is a salt sea water. Fresh water from the melting glacier is mixed with seawater during deposition. To approximate these concentrations, the initial salt water is a 1:1 mixture of freshwater and present day seawater according to (Dickson and Goyet 1997; Fig. 5.2).

The number of exchange sites in the model is 100 eq/m^3, or about 90 meq/kg$_{dw}$. This is reasonable considering that the cation exchange capacity is 45 meq/kg$_{dw}$ for highly sorted sediments with a long distance to the bedrock in the C-horizon in County Västra Götaland (SLU 2011), and considering that exchangeable sodium on clay in Lilla Edet is 70 meq/kg_{dm} (SOU 1962).

Present day rainwater is used as fresh water for infiltration into cell 1 at 4 m depth (Fig. 5.2). The pH of the rainwater is 4.5, but processes in the dry crust increase the pH. Erroneous pH can give numerical instabilities, and the pH in the infiltrating water was set to 7.6, as measured at 3 m depth in the area (Löfroth et al. 2012a).

Arbitrary amounts of calcium and magnesium (5 mM) are added in scenarios 2 and 3 to observe the effect on the pore water concentration in the model. (No calcium or magnesium is added in scenario 1.) Carbonate is used as counter ion when calcium or magnesium is added.

5.3.2 Transport in the Model

The water flow calculations are rough approximations. However, the interest here is in the patterns of leaching and ion chemistry. The flow conditions and time frame need only be approximate in order to get a correct picture of the chemistry.

Advection was calculated from present day permeability and pore water pressure in the area. Diffusion was adjusted to fit the measurements of sodium.

Pore water pressure measurements carried out in situ about 10 m from borehole U07201Q show that the pore water pressure in the friction layer correspond to the water level in the Göta River (SGI 2011). Presumably the layer is in contact with the river. Assuming that the upper ground water level is 1 m below ground surface at borehole U07201Q, and that the water pressure in the friction layer corresponds to the water level in the river, it follows that the hydraulic gradient is 0.5.

The permeability in the model is 9×10^{-10} m^2/s. This is the average of the permeability measurements at 6, 8 and 12 m depth (SGI 2011). The resulting pore water velocity is 4.5×10^{-10} m/s.

Duration for the leaching is 7,000 years (Klingberg et al. 2006), and only numerical dispersion is included. The amount of water in each cell is 650 kg/m^3 (SGI 2011).

The model ignores the dry crust and the first metres of the clay, because many chemical reactions take place here and the purpose of the study is to test the necessity of weathering processes and mineral equilibria at greater depth in the clay. The first cell is at 4 m below the surface. The model extends to the sand layer at 22 m depth. The model results shown in this paper have a cell length of 1 m. Cell length 0.1 m has been tested for scenario 3, and shows the same results as for 1 m.

5.4 Results and Discussion

Three scenarios were modelled to investigate their effect on the pore water chemistry. In the first scenario, natural fresh water was infiltrated in the model. In the second and third scenarios, the infiltrating water contained magnesium and calcium ions respectively. These could be products of weathering or mineral dissolution in the dry crust.

5.4.1 Present Day Conditions

The modelled ion concentrations in the pore water were very low during fresh rainwater infiltration (i.e. scenario 1) (Fig. 5.3). A reason for this is that there are few ions in the infiltrating fresh water. At the start of the model calculations, sodium on the exchange sites is 50 mol/m^3. After 7,000 years, 34 mol/m^3 sodium remains on the exchange sites at 5 m depth, and 47 mol/m^3 remained at 22 m depth (Fig. 5.4). The sodium in the pore water is around 0.1 % of the sodium on the exchange sites.

In scenario 2, rainwater containing magnesium infiltrates the soil model. In this case, the sodium concentration in the pore water is more realistic than in scenario 1 (Fig. 5.3). The total amount of sodium in the modelled soil is lower than in the precipitation-scenario, because the magnesium in the infiltrating solution competes with sodium for the exchange sites. Therefore, more sodium enters the pore water. Sodium on the exchange sites is now only 0.2 mol/m^3 at 5 m depth after 7,000 years, and there is more magnesium than sodium on the exchange sites (Fig. 5.4).

The magnesium concentration in the pore water in scenario 2 increases continuously upwards in the clay profile. At the surface, the concentration is close to the concentration of the infiltrating solution. The magnesium concentration then decreases steadily downwards, as magnesium enters the exchange sites and is removed from the pore water.

Potassium in the pore water shows a peak concentration around 16 m depth (Fig. 5.3). Potassium has a higher affinity for the exchange sites than sodium, therefore sodium is released first. However, potassium has a lower affinity for exchange sites than magnesium. Therefore magnesium displaces potassium from the exchange sites. The result is a peak concentration of potassium in the pore water at a depth between the peak concentration of magnesium and the peak concentration of sodium, as shown in Fig. 5.3.

In scenario 3 the infiltrating water contains calcium. The effect on sodium and potassium is the same as in the second scenario (Fig. 5.3). Calcium competes with sodium and potassium in a way similar to magnesium. Calcium also competes with magnesium. There is a considerable amount of magnesium on the clay exchange sites owing to the salt water at the time of deposition. The magnesium from the exchange sites moves to the pore water, similar to potassium in the second scenario. Since calcium is now the ion in the infiltration, it increases steadily upwards in the profile (Fig. 5.3).

Scenarios 2 and 3 both approximate the measurements of the pore water. However, scenario 3 with calcium provides a better agreement with the measurements than scenario 2 with magnesium, since the measured magnesium concentrations decrease upwards in the profile (Fig. 5.3). Magnesium in the infiltrating solution leads to a higher concentration all the way to the top of the model.

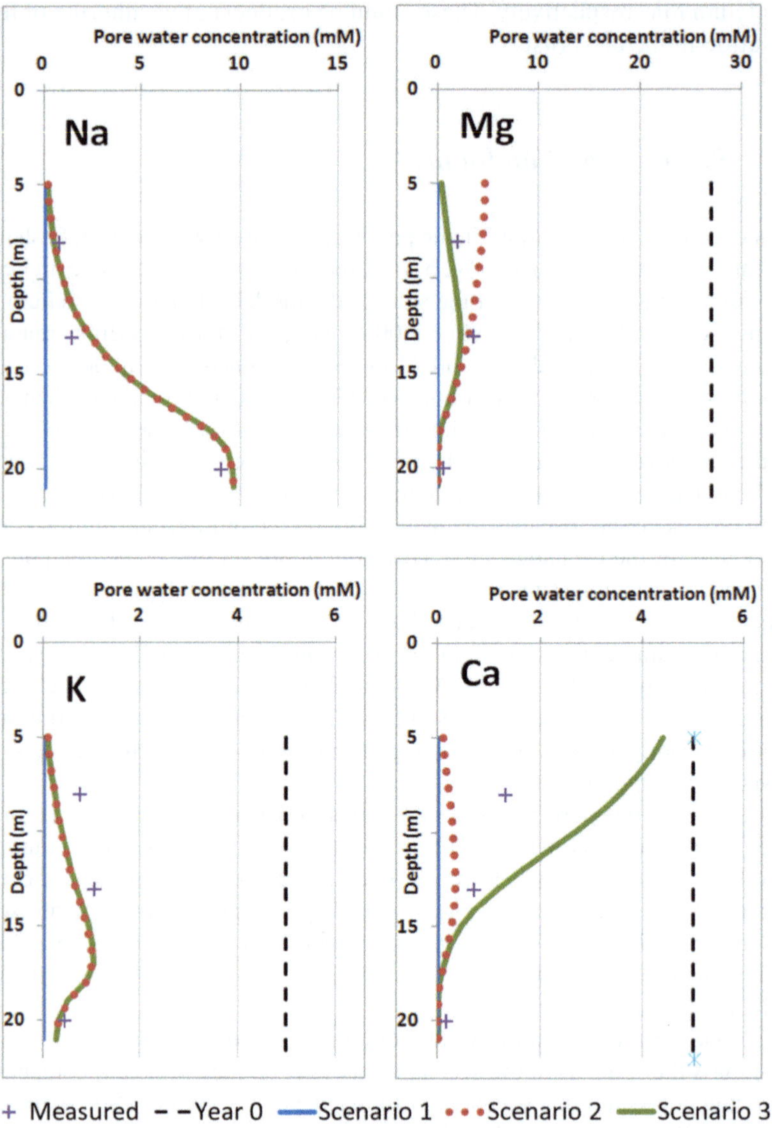

Fig. 5.3 Measured and modelled concentrations of elements in the *pore water* after 7,000 years, as a function of depth for the three scenarios. Sodium (Na), magnesium (Mg), calcium (Ca) and potassium (K) in mM. NB Sodium starting concentration is out of scale (243 mM)

5.4.2 Development Over Time for the Calcium Scenario

The concentrations of sodium and magnesium over time are shown in Fig. 5.5 in order to elucidate the processes that could control the concentration of magnesium in the pore water.

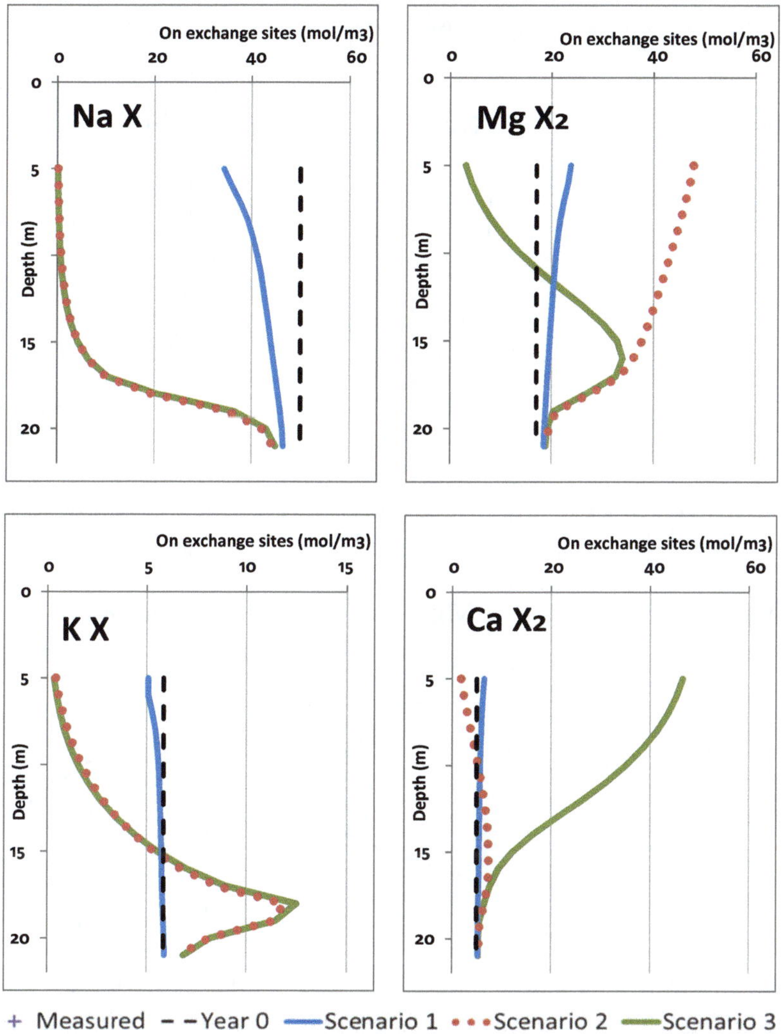

Fig. 5.4 Modelled amounts of ions on the *exchange sites* as a function of depth in the four scenarios. Sodium (Na X), magnesium (Mg X_2), potassium (K X), and calcium (Ca X_2) in mol/m^3

Sodium at time t = 0, 7,000 years ago, is constant at 50 mol/m^3 on the exchange sites at all depths. Inflowing water gradually leaches sodium from the clay, and after 7,000 years there is a leaching front around 18 m depth.

Magnesium on the other hand is 19 mol/m^3 on the exchange sites at all depths at time t = 0. Near the surface, magnesium leaves the exchange sites and is replaced with calcium. The magnesium from the exchange sites at the top of the soil profile re-enters exchange sites further down in the model. Magnesium replaces sodium on the exchange sites, so that sodium leaches and magnesium increases on the exchange sites. Thus there is a peak of magnesium on the exchange sites around 15 m depth

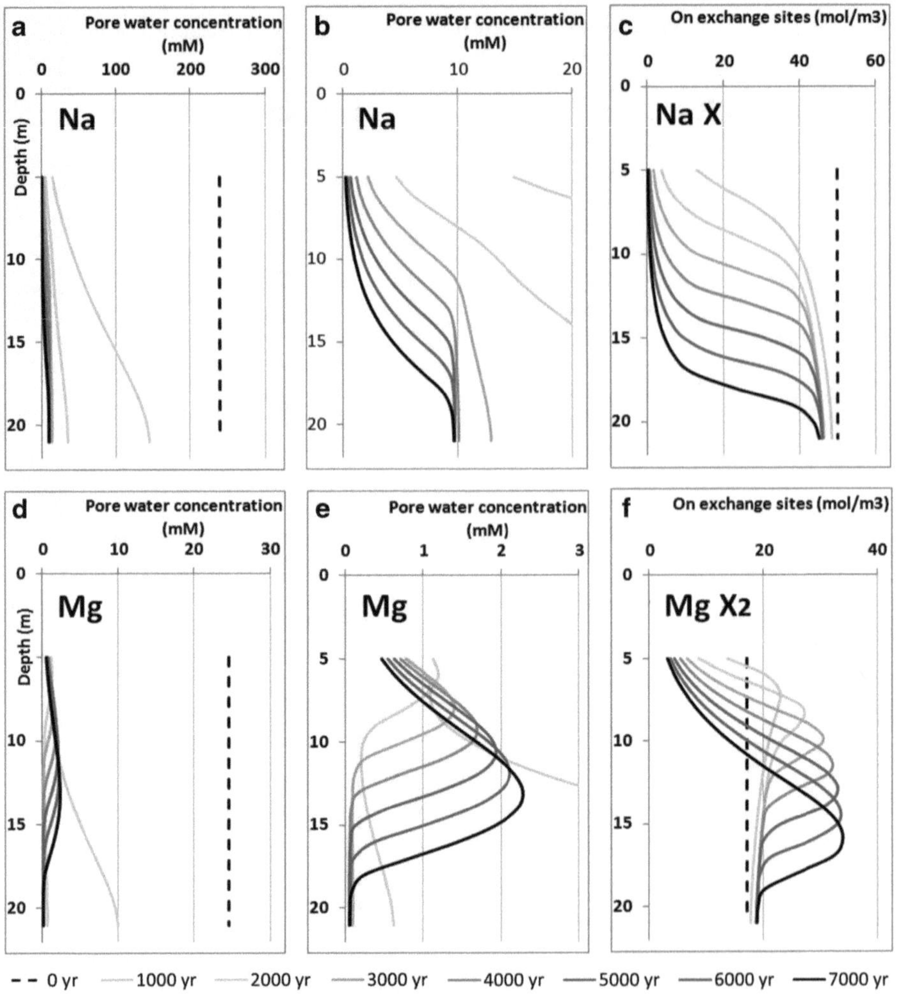

Fig. 5.5 Modelled ion concentration with time and as a function of depth (1,000 year intervals), with the calcium scenario. (**a, b**) Sodium in pore water at two different scales, (**c**) sodium on exchange sites. (**d, e**) Magnesium in pore water at two different scales, (**f**) magnesium on exchange sites. Concentrations in pore water in mM, exchange sites in mol/m^3

after 7,000 years. This agrees with the measurements of concentrations in the pore water shown in Table 5.1 and Fig. 5.3.

5.5 Conclusions

Ion exchange as a cause of natural restabilization of quick clay was modelled in this study using a 1D numerical model which included advection, diffusion and ion exchange processes. When compared to actual pore water chemistry measurements in borehole U07201Q, the best modelling results were obtained when considering a calcium-rich solution infiltrating at the top of the model.

The model of the clay profile does not require mineralisation or weathering below the dry crust in order to simulate the pore water composition. Instead magnesium comes from the seawater at the time of sediment deposition, and is displaced by calcium so that magnesium enters the pore water and moves through the soil. A cation, such as calcium, is necessary to force the magnesium out. The calcium source may be weathering for example from the dry crust, which is outside the model. The pore water concentrations below the dry crust can be explained by ion exchange together with advection and diffusion.

The results show that a calcium-rich infiltrating solution affects the soil beyond the spread of the calcium itself. The displacement of magnesium by calcium on the exchange sites is a likely source of the observed magnesium in the pore water. Calcium could originate naturally in the dry crust, where weathering processes are faster than deeper down in the clay profile. It is not necessary for magnesium minerals to dissolve deep down in the clays to cause increased magnesium concentrations here: the magnesium on the ion exchange sites is sufficient.

The present model is rather rough, and has so far only been applied to one borehole. Thus, the conclusions are tentative only. However, the results are important in order to better understand leaching and restabilisation processes in marine quick clay deposits. Improvement of the model should be performed in the future. This could be done with measurements of the amount and composition of the exchange sites, the age of the pore water and the diffusion coefficient for the clay. Whether the ion exchange is generally the dominating process is not shown here, but it is important enough to be part of any discussion of quick clay chemistry.

Acknowledgments Many thanks to Tonje Eide Helle, Jean-Sébastien L'Heureux and Ariane Locat who commented on the manuscript. We also gratefully acknowledge all those who contributed to the field work, laboratory work and interpretation of Löfroth et al. (2012a).

References

Appelo CAJ, Postma D (2005) Geochemistry, groundwater and pollution, 2nd edn. A. A. Balkema, Rotterdam

Dickson AG, Goyet C (eds) (1997) Handbook of methods for the analysis of the various parameters of the carbon dioxide system in sea water, Version 2.13. Carbon Dioxide Information Analysis Center, Oak Ridge National Laboratory

Klingberg F, Påsse T, Levander J (2006) Bottenförhållanden och geologisk utveckling i Göta älv, SGU Geological Survey of Sweden, Uppsala, Report K43 (in Swedish)

Löfroth H, Suer P, Dahlin T, Leroux V, Schälin D (2012a) Quick clay mapping by resistivity – surface resistivity, CPTU-R and chemistry to complement other geotechnical sounding and sampling, SGI Swedish Geotechnical Institute, Linköping. http://www.swedgeo.se

Löfroth H, Suer P, Dahlin T, Leroux V, Schälin D (2012b) Mapping of quick clay using sounding methods and resistivity in the Göta River valley. In ISC'4 (International conference on Geotechnical and Geophysical Site Characterisation). Brazil, ID236

Løken T (1968) Kvikkleiredannelse og kjemisk forvitring i norske leirer. (Formation of quick clay and chemical weathering in norwegian clays) Beretning over Norges geotekniske institutts virksomhet fra 1. januar 1966 til 31. desember 1967. NGI Norwegian Geotechnical Institute, Oslo, publication 75, pp 19–26 (In Norwegian)

Løken T (1970) Recent research at the Norwegian geotechnical institute concerning the influence of chemical additions on quick clay. Geologiska Föreningen i Stockholm Förhandlingar 92(2):133–147

Moum J, Sopp OI, Løken T (1968) Stabilization of undisturbed quick clay by salt wells. Väg- och vattenbyggaren 8:23–29

Parkhurst DL, Appelo CAJ (2011) PHREEQC for Windows – a hydrogeochemical transport model. http://pfw.antipodes.nl/

SGI (2011) Markteknisk undersökningsrapport, geoteknik. Delområde 7 (Intagan-Lilla Edet), Dnr 6-1001–0029. SGI Swedish Geotechnical Institute (in Swedish)

SLU (2011) MarkInfo, results from national survey of forest soils and vegetation (SK) and Swedish forest soil inventory (Markinventeringen), Swedish University of Agricultural Sciences. http://www-markinfo.slu.se/

SOU (1962) Rasriskerna i Göta älvdalen – Betänkande avgivet av Götaälvskommittén, Stockholm: Statens Offentliga Utredningar SOU 1962:48 (in Swedish)

Talme O (1968) Clay sensitivity and chemical stabilization. Dissertation, National Swedish Institute for Building Research. Byggforskningens rapport 56:1968

Chapter 6
Potassium Chloride as Ground Improvement in Quick Clay Areas – A Preliminary Study

Tonje E. Helle, Ingelin Gjengedal, Arnfinn Emdal, Per Aagaard, and Øyvind Høydal

Abstract Potassium chloride was added to low saline Norwegian post glacial clays to study its effect on strength parameters. The laboratory study was carried out on undisturbed sensitive clay samples from two locations in mid-Norway. The mechanical behaviour of sensitive clays is greatly influenced by their pore water ionic content. Adding salt changes the geotechnical properties of quick clay to such an extent that it appears as a total different clay. Salt migration is a time consuming process strongly dependent on diffusion through the soil. Deriving the effective diffusion coefficient from water content of 30–50 %, the clay will be de-sensitized over a length of 50 and 56 cm respectively after 1 year. Ground improvement with salt is done by installing salt wells. This study is part of a design project for such installations. The consumption of time depends on the distance between the wells, diffusion coefficient and maintenance of high concentration in the salt well.

Keywords Quick clay • Ground improvement • Potassium chloride • Diffusion

T.E. Helle (✉)
Department of Civil and Transport Engineering, Norwegian University of Science and Technology (NTNU), Trondheim, Norway

The Norwegian Public Road Administration, Trondheim, Norway
e-mail: tonje.eide.helle@ntnu.no

I. Gjengedal
Norwegian University of Science and Technology (NTNU), Trondheim, Norway

Norconsult, Molde, Norway

A. Emdal
Department of Civil and Transport Engineering, Norwegian University of Science and Technology (NTNU), Trondheim, Norway

P. Aagaard
Department of Geosciences, University of Oslo (UiO), Oslo, Norway

Ø. Høydal
Norwegian Geotechnical Institute (NGI), Oslo, Norway

J.-S. L'Heureux et al. (eds.), *Landslides in Sensitive Clays: From Geosciences to Risk Management*, Advances in Natural and Technological Hazards Research 36,
DOI 10.1007/978-94-007-7079-9_6, © Springer Science+Business Media Dordrecht 2014

6.1 Introduction

Safety measures to increase slope stability is challenging in quick clay areas. Counterfills may cause undesirable changes of the terrain with elevation of river bottoms, whereas chalk cement piles increase the pore water pressure and may reduce the stability in the slope during installation. Experiments adding salt in quick clays to increase their strength was carried out in the 60- and 70s. The results were promising, nevertheless using salt as ground improvement is not in use today.

Migration of salt in clay is a time consuming process, and is claimed to be costly due to ineffective and time consuming installation methods that were used in the 1970s. New installation methods may, however, be less time consuming and expensive. Nevertheless, the applicability of this method is not fully understood regarding time and necessary salt concentrations to increase the geotechnical strength parameters sufficiently. This paper presents preliminary results from a laboratory study on how potassium chloride affects the geochemical composition and geotechnical strength parameters over time for clays from two Norwegian quick clay sites (Fig. 6.1). The clay from Rissa presented in this paper is not

Fig. 6.1 Location of Tiller and Rissa site (close to Rein kirke)

sampled in the well-known quick clay slide area. Thus its properties deviate from earlier publications regarding the famous Rissa clay. The Tiller clay is sampled on NTNU's research site.

6.2 Background

Quick clay development depends on several factors i.e. mineralogy, sediment properties, pore water chemistry and consolidation (Rosenqvist 1946, 1953, 1955; Bjerrum 1955; Quigley 1980; Rankka et al. 2004; Torrance 1983, 2012). Both the total salinity and the composition of ions in the pore water will affect the bonds between the minerals. As many clay minerals normally have net negative surface charges, the electrostatic character of the pore water cations is of special importance. Sodium (Na^+) is abundant in the pore water in clays deposited in marine environment, whereas smaller amounts of magnesium (Mg^{2+}), calcium (Ca^{2+}) and potassium (K^+) are present. There are normally only traces of iron (Fe^{2+}) and aluminium (Al^{3+}). The geotechnical properties are significantly affected by the adsorbed ions on the mineral surfaces in addition to the pore water chemistry (Mitchell and Soga 2005; Løken 1968, 1970; Talme et al. 1966; Torrance 1983, 2012).

Changing the concentration of ions in the pore water may induce an exchange of ions in the adsorbed positions on the mineral surface. The replacing power of ions is in the following order (Mitchell and Soga 2005): $Na^+ < Li^+ < K^+ < Rb^+ < Cs^+ < Mg^{2+} < Ca^{2+} < Ba^{2+} < Cu^{2+} < Al^{3+} < Fe^{3+} < Th^{4+}$.

Ion exchange depends on valence, relative abundance of different ion types and ion size. Even though ions with higher valence have higher replacing power, introducing high concentrations of monovalent ions makes it possible to replace divalent ions at low concentrations (Appelo and Postma 2005; Løken 1970; Mitchell and Soga 2005).

Changing the ion composition or ion concentration in the pore water will influence which ions are in the adsorbed positions on the mineral surface. This will have a great impact on the physical properties of the soil (Appelo and Postma 2005; Mitchell and Soga 2005; Moum et al. 1968; Løken 1968; Torrance 1983, 2012).

According to Løken (1968) the cations ability to increase the shear strength follows in the order: $Na^+ < Fe^{2+} \leq Mg^{2+} \leq Ca^{2+} < Fe^{3+} < K^+ < Al^{3+}$.

The liquid and plasticity limit is also greatly affected by the cations, and the cations increase the Atterberg limits in the following order (Løken 1970): $Na^+ < Fe(OH)_3 = Fe^{2+} = Mg^{2+} = Ca^{2+} < Fe^{3+} < Al^{3+} = K^+ < Al(OH)_3$.

According to Moum et al. (1968), potassium chloride is easily dissolved in water and increase the remolded shear strength significantly compared to sodium chloride. Even though aluminium chloride gave an even larger increase of the remolded shear strength, it was not as easily dissolved in water. The laboratory experiments also showed that KCl had the greatest length of diffusion into the clay. Eggestad and Sem (1976) used KCl as ground improvement at a site in Oslo, Norway, observing an increase of the undisturbed shear strength of 200 % after 3 years.

Based on these findings, potassium chloride has been used in the present study to evaluate how salt affects the geotechnical parameters of leached (i.e. low-salinity) Norwegian marine clays.

6.3 Laboratory Investigations

Two laboratory setups were used to study how potassium chloride affects low-saline clays (Fig. 6.2). Setup (a) shows simplified diffusion cells made of Plexiglas. Dissolved potassium chloride (KCl) or KCl slurry was filled in the bottom reservoir, whereas the top reservoir was filled with de-aired distilled water. No filters or membranes were used to separate the clay specimen from the fluids. In addition a small salt well was made in the bottom half of a Sherbrooke block sample. The "mini salt well" was filled with KCl slurry and stored for 60 days. All samples were stored in a cold-storage chamber holding a temperature of 7 °C. At this temperature the solute had a concentration of 118 g K^+/l in contact with solid KCl salt.

All index testing was carried out in a laboratory holding a temperature of 7 °C. Fall cone shear strength, water content, Atterberg's limits, mineralogy and pore water chemistry were determined for the initial condition prior to KCl addition. These experiments were repeated after treating the clays with potassium chloride. The pore water chemistry was analysed using HR-ICP-MS, while the mineralogical composition was determined by XRD.

6.4 Results

6.4.1 Initial Conditions

The clay mineralogy of the clays from Tiller and Rissa is similar, dominated by illite and chlorite which is common for Norwegian clays (Rosenqvist 1955, 1975; Løken 1968).

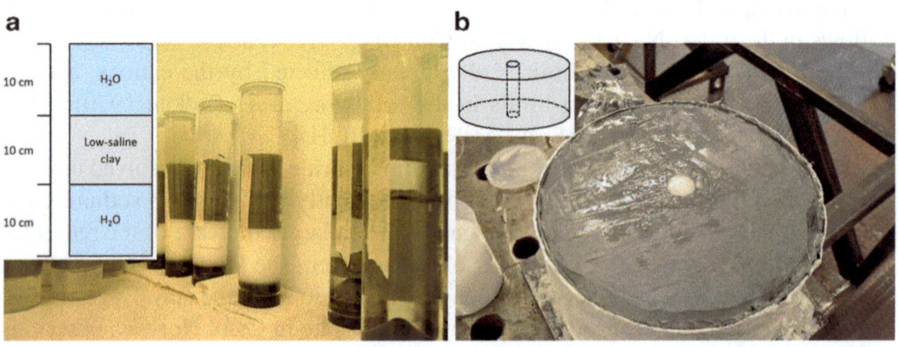

Fig. 6.2 The two laboratory setups used to study diffusion of potassium chloride in low-saline clays: (**a**) simplified diffusion cells (**b**) mini-salt well within a Sherbrooke block sample

Table 6.1 Ion composition in the pore water in clays from Tiller and Rissa

	Tiller (mEq/l)	Rissa (mEq/l)	Sea water[a] (mEq/l)
Na^+	9.48	32.97	485.00
K^+	0.65	0.75	10.60
Mg^{2+}	2.63	1.18	110.20
Ca^{2+}	0.58	0.44	21.40
Al^{3+}	[b]	0.04	
Fe^{3+}	[b]	0.02	
Cl^-	0.40	16.20	566.00
SO_4^{2-}	0.15	6.55	58.60
HCO_3^{-}[c]	12.78	12.64	2.40
Cl^- ‰ of sea water	0.7	28.6	

[a]Salinity of 35 ‰. Values from Appelo and Postma (2005)
[b]The amount of iron and aluminium was unlikely high probably due to occurrence of particles in the analysed pore water
[c]Tiller and Rissa may also contain CO_3^{2-} within these volumes

Table 6.2 Initial condition for clays from Tiller and Rissa

	Tiller	Rissa[a]
Water content, w (%)	33–42	37.5
Fall cone shear strength, s_u (kPa)	17.8–20.1	13.7–15.7
Remolded shear strength, s_r (kPa)	0.1–0.2	0.9–1.2
Sensitivity, St (–)	89–201	12.7–15.2
Liquid limit (Casagrande), w_L (%)	27.1–32.4	29.7
Plastic limit, w_P (%)	20.7–20.8	19.4
Total salt content (g/l)	0.62–0.85	2.00

[a]Results from Kornbrekke (2012)

It was not possible with the available equipment to measure nitrate in the pore water, and alkalinity was not determined. Bicarbonate is estimated by subtracting total negative charge from total positive charge given by the cations.

Comparing the natural sample from Tiller and Rissa with the ion composition of sea water (salinity 35 g/l), it appears that the concentration of chloride is much lower whereas the calculated occurrence of bi-carbonate has increased significantly. This condition appears to be caused by fresh water replacing original sea pore water, washing out the chlorides so that sodium appears as $NaHCO_3$ instead of NaCl. Tiller clay is heavily leached, whereas approximately 3 % of the chlorides remain in the clay from Rissa (Table 6.1).

The clay sample from Rissa is not considered quick according to the Norwegian definition (Table 6.2). Furthermore, the total salt content is lower in the Tiller clay than in the clay from Rissa.

6.4.2 *Salt Treated Clay*

Many of the diffusion cells in this study were in bad condition at the end of the preset time interval of diffusion. Therefore only results from a few of the diffusion

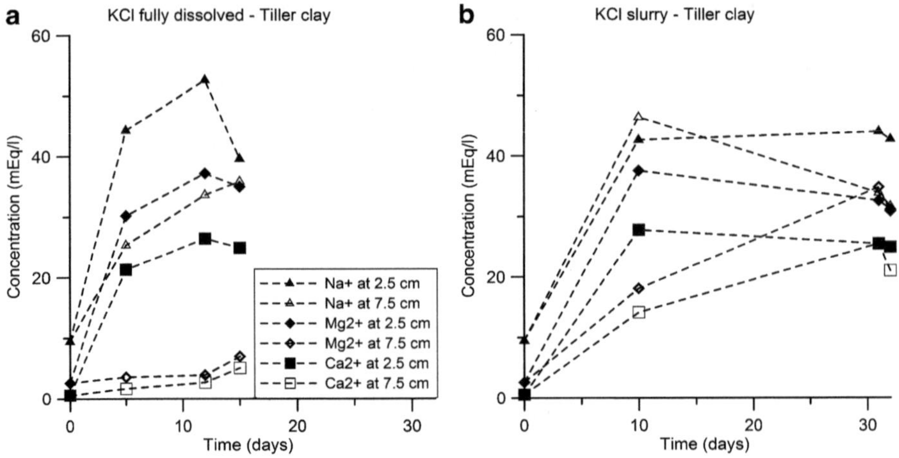

Fig. 6.3 Change in pore water composition over time in Tiller clay treated with (**a**) fully dissolved KCl and (**b**) KCl slurry

cells are included in this paper. The results are presented for a distance of 2.5 and 7.5 cm from the well. The geochemical data, however, are determined in water extracted from half the specimen in the diffusion cell, thus it gives an average of the content over a 5 cm long specimen.

An increase in sodium, calcium and magnesium is observed as well as the obvious increase of potassium and chloride (Fig. 6.3). The changes in aluminum, iron and sulphate are minor, thus these results are not presented herein.

The concentrations of the various ions seem to end up at approximately the same concentration regardless being treated with fully dissolved KCl or KCl slurry. The ion exchange reactions, however, seem to go quicker through the clay specimen treated with slurry.

Regardless of treating the clay with fully dissolved KCl or slurry, the increase of potassium and chloride is approximately the same close to the reservoir, whereas the potassium chloride front diffuses faster through the clay treated with slurry due to the maintenance of high concentration in the reservoir (Fig. 6.4).

There seem to be a limit for maximum attainable undisturbed shear strength (Fig. 6.5) as a result of increased salt content. The water content is very little affected. The liquid limit, however, is strongly influenced. The plastic limit increase as well. There seems to be an upper limit for both the plastic and liquid limits, beyond which an increase of potassium and chloride does not increase the limits further (Fig. 6.6).

All the salt in the mini salt well was diffused within the 60 days in the cold-storage chamber. A one sixth part of the block was divided into three specimens that were centrifuged to extract pore water for analyses. Thus the geochemical composition is not found at exact distances from the mini salt well. The amount of aluminum

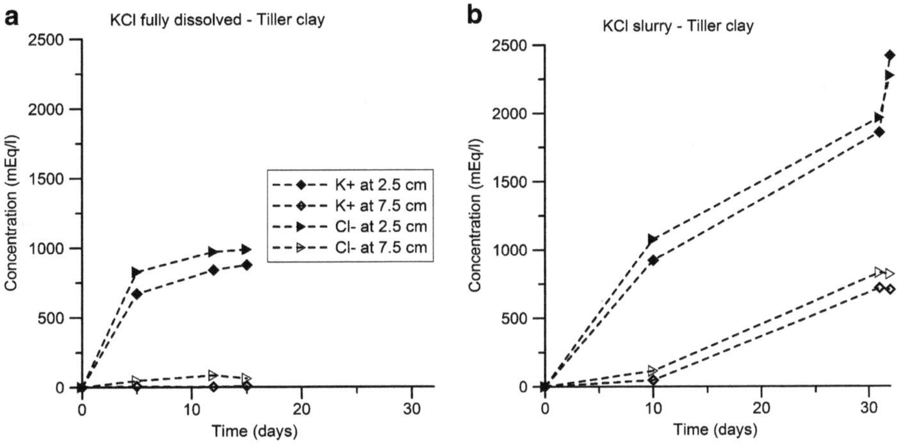

Fig. 6.4 Change of potassium and chloride in Tiller clay treated with (**a**) fully dissolved KCl and (**b**) KCl slurry

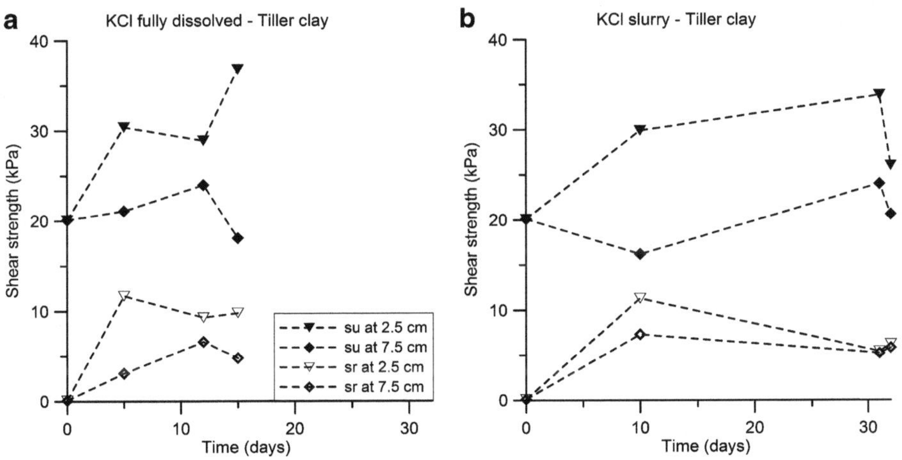

Fig. 6.5 Change in fall cone undisturbed shear strength and remolded shear strength for Tiller clay treated with (**a**) fully dissolved KCl and (**b**) KCl slurry

and iron are minor and does not change much, neither do sulphate thus the results are not presented herein. The potassium and chloride content increase throughout the block sample, the highest increase closest to the salt well (Fig. 6.7). In addition the concentration of sodium, calcium and magnesium increase as well.

The clay sample from Rissa used in this study is not quick, even so both the undisturbed and remolded shear strength increase throughout the sample. Interestingly the specimen closest to the salt well does not have the highest shear strength. The water

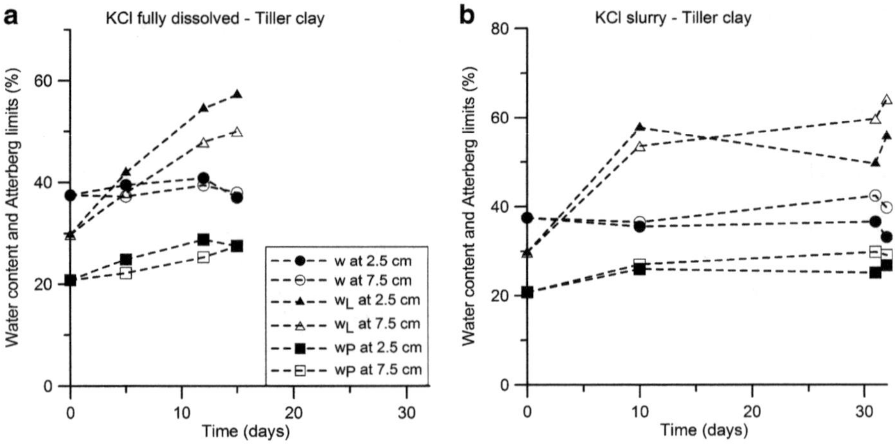

Fig. 6.6 Change in water content, liquid and plasticity limit in clay treated with (**a**) fully dissolved KCl and (**b**) KCl slurry

Fig. 6.7 Ion distribution with distance from the salt well. Point 4 shows the initial condition

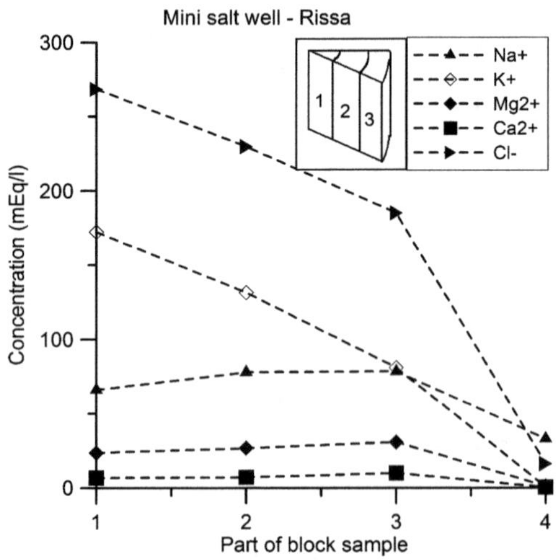

content is stable, while as both the liquid and plastic limit increase (Fig. 6.8). The increase of potassium and chloride and the geotechnical strength parameters would have been greater if the concentration in the salt well had been maintained.

6.4.3 *Diffusion of Potassium*

The diffusion coefficient in free water for potassium ion can be found in several textbooks. The effective diffusion coefficient in soils will vary depending on the

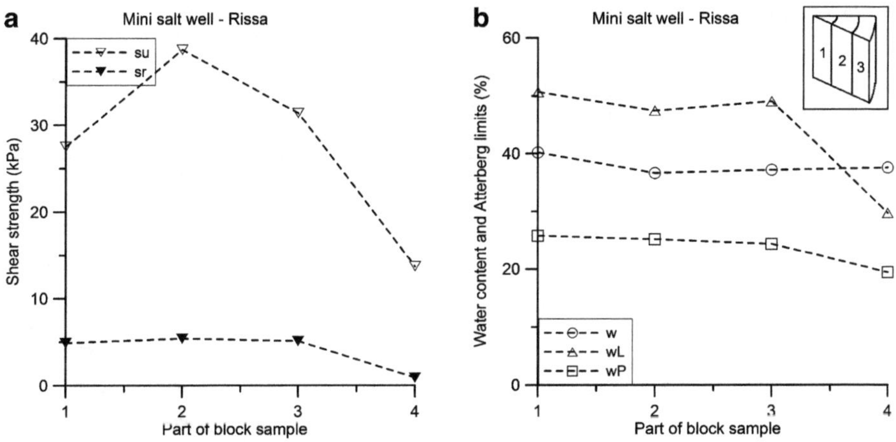

Fig. 6.8 Change in geotechnical parameters (**a**) fall cone shear strength and (**b**) water content, liquid and plasticity limits. Point 4 shows the initial condition

Table 6.3 Diffusion coefficient for K^+

	Appelo and Postma (2005)	Moum et al. (1968)[a]	Moum and Heiberg (1973)[a]	Bryhn (1981)[a]
Water content (%)		32.5	55	
Clay content (%)		36		42
D 7 °C (cm^2/s)		$6.2 \cdot 10^{-6}$	$3.8 \cdot 10^{-6}$	
D 22 °C (cm^2/s)			$6.3 \cdot 10^{-6}$	
D (cm^2/s)	$1.96 \cdot 10^{-5}$[b]			$4 \cdot 10^{-6}$[c]

[a]Determined on clay samples
[b]Diffusion coefficient in free water at 25 °C
[c]No information of temperature

water filled porosity, temperature, specific unit weight for the grain skeleton and the tortuosity. Several diffusion coefficients for potassium through clay has been determined (Table 6.3). Diffusion of cations and anions are coupled due to charge neutrality constraints. As chloride has a self-diffusion coefficient of $2.03 \cdot 10^{-5}$ cm^2/s, it will speed up potassium so that the diffusion in free water for KCl is $1.99 \cdot 10^{-5}$ cm^2/s (Mitchell and Soga 2005).

The effective diffusion coefficient (D_e) can be theoretically determined as the product of the diffusion coefficient in "free" water and the water filled porosity (Appelo and Postma 2005). Assuming 7 °C, 100 % saturation and specific unit weight of the soil 27 kN/m^3, the diffusion coefficients for normal range of water content (30–50 %) in Norwegian clays will be as presented in Table 6.4. Table 6.4 also presents theoretical derived effective diffusion coefficients for clays from Tiller and Rissa with specific unit weight of 27.5 and 27.8 kN/m^3 respectively.

The effective diffusion coefficients affects the time it takes to increase the geotechnical strength parameters sufficiently (Fig. 6.9). The same criterion as

Table 6.4 Theoretically derived effective diffusion coefficients for potassium

	Normal range low	Normal range high	Moum et al. (1968)	Tiller 1	Tiller 2	Rissa
Water content	30	50	32.5	38.9	42.1	37.5
D_e^a (cm²/s)	$5.17 \cdot 10^{-6}$	$6.64 \cdot 10^{-6}$	$5.40 \cdot 10^{-6}$	$5.97 \cdot 10^{-6}$	$6.20 \cdot 10^{-6}$	$6.36 \cdot 10^{-6}$

[a]Effective diffusion coefficient $D_e = D_f \cdot \varepsilon_w$, where ε_w is the water filled porosity (Appelo and Postma 2005)

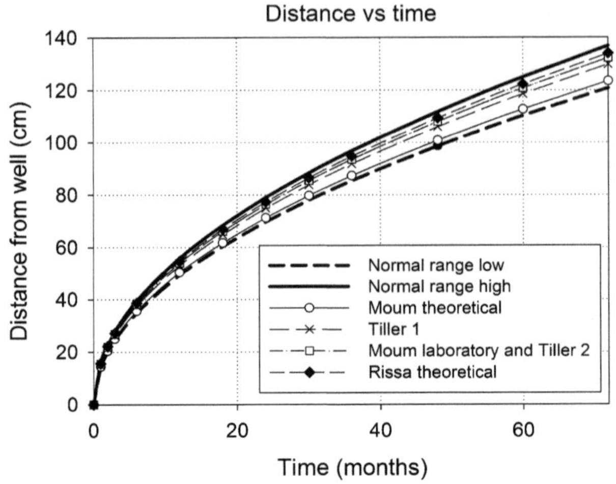

Fig. 6.9 Time and distance of effective front of diffusion of 0.5 g K+/l

Moum et al. (1968) for effective diffusion length is used in our calculations i.e. 0.5 g K^+/l. Using the effective diffusion coefficient derived from normal range of water content in Norwegian clays, it will take approximately 5–6 years to increase the content of potassium to such an extent that it increases the geotechnical strength parameters significantly in a distance of 1.2 m from the salt well.

6.5 Discussion

The laboratory setup was very simple, and the sources of error are numerous, amongst others the effect of storage which itself may increase the undisturbed and disturbed shear strength. No membrane was applied in the diffusion cells. This may have caused leaking of salt along the walls. With no support under the clay specimens, the clay was allowed to disperse in the water and salt reservoirs. Furthermore, no pressure was applied in the cells, consequently the clay specimens lost most of their in situ properties.

In addition to the obvious increase of potassium in the pore water, there was an increase of sodium, magnesium and calcium. This indicates an ion exchange on the mineral surface. Even though potassium has less ability to be adsorbed to the mineral surface than calcium and magnesium, the concentration of potassium is much higher thus it will be able to exchange sodium, calcium and magnesium and occupy the adsorbed positions on the mineral surface.

This paper presents results from 1D-calculation of diffusion of potassium chloride in clays assuming constant concentration in the well. However, the salt will migrate radially. Even so the salt concentration will be reduced over time, consequently the diffusion process will slow down and eventually come to an end if the well is not refilled after a certain period of time. The calculations are, however, based on diffusion coefficients for potassium alone. The diffusion front will be affected by contribution from all present ions and ion exchange reactions.

6.6 Conclusions

The mineralogy of the clays from Tiller and Rissa is quite similar, while the pore water chemistry is not, thus the initial geotechnical parameters seem to be governed by the pore water chemistry.

The laboratory study was carried out on undisturbed sensitive clay samples from two locations in mid-Norway. The preliminary results prove that the mechanical behaviour of sensitive clays is greatly influenced by their pore water salinity and ionic content. These preliminary results show that low-saline clay treated with potassium chloride changes its properties completely. Over time and with sufficient access to salt, the clay transforms from very sensitive, low plasticity quick clay to non-sensitive clay with medium plasticity.

Adding salt as ground improvement in sensitive clays is possible. Salt diffusion, however, is time consuming. Finding the in situ diffusion coefficient is necessary to establish the right distance between the wells and necessary salt concentration. Even so, finding an installation method ensuring low excess pore pressure, causing minor, or preferably no deformation in the ground and of the terrain surface is necessary.

Acknowledgments The authors would like to thank the Norwegian Geotechnical Society and the research program "Natural Hazards": Infrastructure for floods and slides (NIFS) initiated by the Norwegian Public Roads Administration, Norwegian Water Resources and Energy Directorate and Norwegian National Railways Administration for funding this project. Prof. Yudhbir is thanked for his valuable comments on the manuscript. We would also like to thank Elisabeth Gundersen, Vikas Thakur, Frode Oset, Tor Løken, Odd Gregersen, Eirik Traae and Steinar Nordal for good discussions, in addition to the great effort carried out by the field- and laboratory staff at NTNU Geotechnical Research Group Jan Jønland and Gunnar Winther.

References

Appelo C, Postma D (2005) Geochemistry, groundwater and pollution, 2nd edn. Balkema, Leiden

Bjerrum L (1955) Norske marine leirers geotekniske egenskaper, NGI publication no. 7. Norwegian Geotechnical Institute, Oslo

Bryhn O (1981) Stabilization of Norwegian quick clay with potassium chloride and lime. NGI report no. 52752–1. Norwegian Geotechnical Institute, Oslo

Eggestad A, Sem H (1976) Stability of excavations improved by salt diffusion from deep wells. Sechste Europaeische Konferenz Fuer Bodenmechanik Und Grundbau, vol 1

Kornbrekke H (2012) Skråningsstabilitet ved Rein kloster med utgangspunkt i resultater fra blokkprøver. Master thesis. Norwegian University of Science and Technology, Trondheim

Løken T (1968) Kvikkleiredannelse og kjemisk forvitring i norske leirer, NGI publication no. 75. Norwegian Geotechnical Institute, Oslo

Løken T (1970) Recent research at the Norwegian Geotechnical Institute concerning the influence of chemical additions on quick clay. Geologiska Föreningen i Stockholm Förhandlingar 92(2):133–147

Mitchell JK, Soga K (2005) Fundamentals of soil behavior, 3rd edn. Wiley, New York

Moum J, Heiberg S (1973) An experimental determination of the diffusion constant for high in situ salt concentrations in Norwegian marine clays. NGI internal report no. 50703–2. Norwegian Geotechnical Institute, Oslo

Moum J, Sopp OI, Løken T (1968) Stabilization of undisturbed quick clay by salt wells, NGI publication no. 81. Norwegian Geotechnical Institute, Oslo

Quigley RM (1980) Geology, mineralogy and geochemistry of Canadian soft soils: a geotechnical perspective. Can Geotech J 9:261–285

Rankka K, Andersson-Sköld Y, Hultén C, Larsson R, Leroux V, Dahlin T (2004) Quick clay in Sweden. SGI report no. 65. Swedish Geotechnical Institute, Linköping

Rosenqvist IT (1946) Om leirers kvikkaktighet. Medd.f. Vegdirektoren 4: 29–36

Rosenqvist IT (1953) Consideration on the sensitivity of Norwegian quick-clays. Geotechnique 3:195–200

Rosenqvist IT (1955) Investigations in the clay-electrolyte-water system, NGI publication no. 9. Norwegian Geotechnical Institute, Oslo

Rosenqvist IT (1975) Origin and mineralogy glacial and interglacial clays of Southern Norway. Clay Clay Miner 23:153–159

Talme O, Pajuste M, Wenner C-G (1966) Secondary changes in the strength of clay layers and the origin of sensitive clay. Rapport – Byggforskningsrådet, vol 46

Torrance JK (1983) Towards a general model of quick clay development. Sedimentology 30:547–555

Torrance JK (2012) Landslides in quick clay. In: Clague JJ, Stead D (eds) Landslides – types, mechanisms and modeling. Cambridge University Press, Cambridge

Part II
Landslide Characterization

Chapter 7
Inventory of Large Landslides in Sensitive Clay in the Province of Québec, Canada: Preliminary Analysis

Denis Demers, Denis Robitaille, Pascal Locat, and Janelle Potvin

Abstract The Ministère des Transports du Québec is currently conducting a comprehensive inventory of large historical landslides that have occurred in sensitive clays in Québec. One hundred and eight cases have been identified during the period from 1840 to 2012 through historical documents or aerial photos. Detailed data have been collected for numerous sites with respect to pre- and post-failure topography, soil properties, the position of the failure surfaces and the groundwater flow regime. In addition, piezocone soundings have been conducted at approximately 50 other undated large landslide sites in various regions in Québec. This article summarizes the preliminary analysis of these data.

Keywords Quick clay landslide • Earthflow • Spread • Historic landslides • Predictive method • Retrogression distance

7.1 Introduction

Large landslides occur frequently in the sensitive clays of Eastern Canada, and their retrogression distances can attain values of several 100 m (Evans 2001; Locat et al. 2011a). These landslides, which are referred to as large retrogressive landslides, occur in the clayey soils deposited in the postglacial seas of Québec and Ontario. The province of Québec is particularly exposed to this hazard, with 89 % of its population settled within the marine limits of the ancient Champlain, Laflamme, and Goldthwait seas (Fig. 7.1), which represent a total land area of approximately 76,000 km^2.

D. Demers (✉) • D. Robitaille • P. Locat • J. Potvin
Service de la Géotechnique et de la géologie (Geotechnique and Geology Branch), Ministère des Transports du Québec (MTQ), Québec City, QC, Canada
e-mail: Denis.Demers@mtq.gouv.qc.ca

J.-S. L'Heureux et al. (eds.), *Landslides in Sensitive Clays: From Geosciences to Risk Management*, Advances in Natural and Technological Hazards Research 36,
DOI 10.1007/978-94-007-7079-9_7, © Springer Science+Business Media Dordrecht 2014

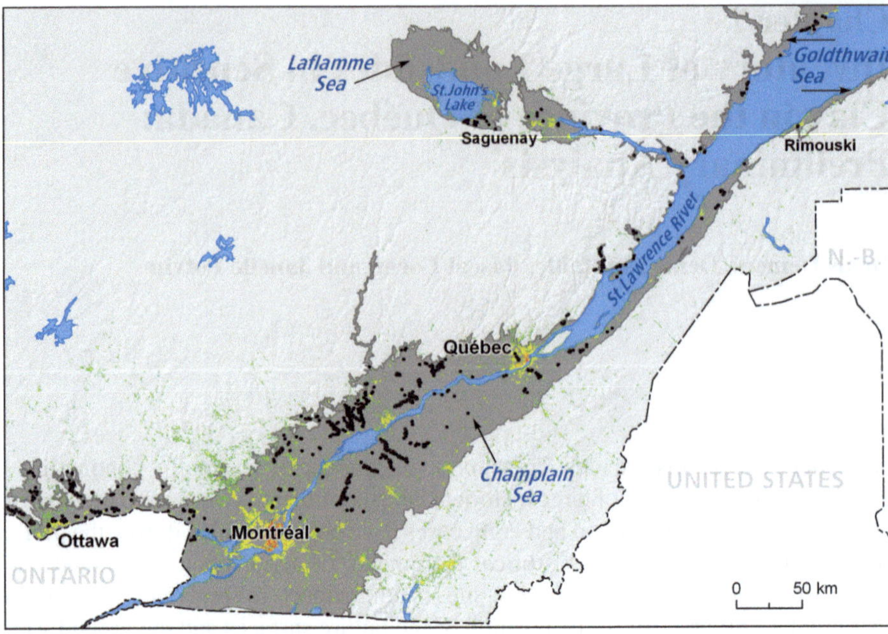

Fig. 7.1 Location of the marine limits (*dark gray*) and the scars left by large landslides (*black dots*) in southern Québec, along with graphical representation of the population density

This article was derived from an internal document produced by the Ministère des Transports du Québec (Demers et al. 2013), and is based on a preliminary analysis of geotechnical and geomorphological data from an inventory of large historical landslides in Québec since 1840 that currently underway. The purpose of this article is to create an initial general portrait of these historical landslides in order to more clearly identify the characteristics of flowslides as opposed to those of lateral spreads. Given that the "pre-failure" conditions associated with many of these historical cases can be reconstructed, the data were also used to test various existing methods for predicting retrogression distances.

7.2 Definitions and Investigative Methods

Based on the criteria used in cartography in Québec (Chap. 26 by Potvin et al., this volume), "large retrogressive landslides" are those with a retrogression distance that exceeds twice the height of the original slope (or more than 40 m for any slope of 20 m or higher). Within the context of its mandate to map landslide-prone areas in clayey soils (Demers et al. 2008), the Ministère des Transports du Québec has collected extensive data in various regions pertaining to both ancient undated landslide sites and historical cases.

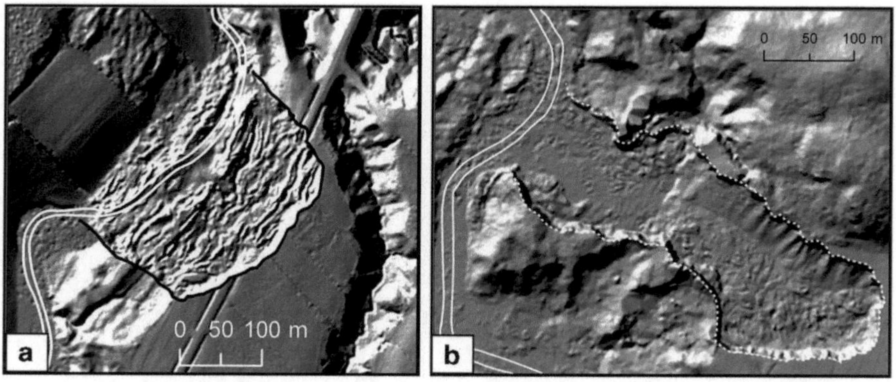

Fig. 7.2 Lidar hillshades of the spread at St-Jude in 2010 (**a**) and of the flowslide at Notre-Dame de la Salette in 2010 (**b**)

For the purposes of this compilation, 108 historical cases and 74 undated ancient scars of large retrogressive landslides have been documented. In the vast majority of cases, the dimensions of the scars, whether ancient or historical, and the heights of the slopes have been measured precisely, based on airborne light detection and ranging (lidar) for most of the area that was once covered by the three ancient postglacial seas that are depicted in Fig. 7.1. The data pertaining to the five Tyrell Sea sites, which are located in northern Québec (outside Fig. 7.1), were taken from cases published in the literature. In the vast majority of cases, a combined analysis of the airborne lidar images and the aerial photos also made it possible to identify the type of landslide (spread or flowslide, as defined below) based on the form of the debris materials within the scars (Fig. 7.2).

The position of the failure surface of the landslides was determined by means of piezocone soundings for 45 of the historical cases and 52 of the ancient scar sites. Documentation of the soil properties was carried out in 56 of the historical cases: grain-size analysis, water content and Atterberg limits, intact and remoulded undrained shear strength using Swedish cone and field vane testing, preconsolidation pressure. Piezometric data were also obtained for 40 of the historical sites.

In this document, the description of the sensitivity is based on the classification found in the 2006 "Canadian Foundation Engineering Manual". This Manual does not define the descriptor "quick", and therefore, the Norwegian definition (Chap. 28 by Kalsnes et al., this volume) is used (undrained remoulded shear strength of less than 0.5 kPa).

7.3 Historical and Spatial Aspects

With the exception of five cases in Ontario (four along the South Nation River and one in Hawkesbury), the other 103 occurred within the boundaries of the province of Québec. These large retrogressive landslides mainly occur on clayey plains

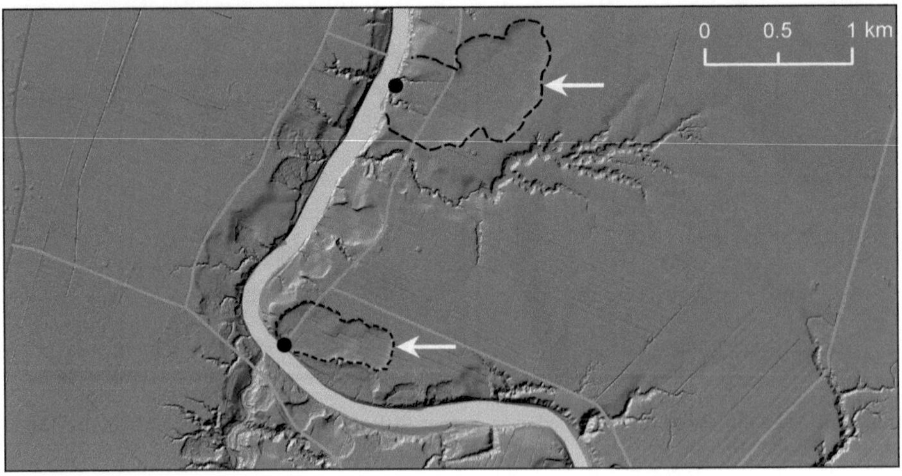

Fig. 7.3 Lidar hillshade showing an area of the Yamaska River that is prone to large retrogressive landslides. The two *arrows* indicate flowslide scars that propagated at an angle relative to their point of origin (*black dot*)

that generally have a surface with a very mild slope (gradient of less than 1 %). The slopes where landslides originate are mainly formed by the banks of watercourses (Fig. 7.3) that cut into the marine deposits during the isostatic rebound, or by the cliffs formed by the ancient marine terraces. On lands that are prone to large retrogressive landslides, the slope heights range from 9 to 108 m.

During the period from 1840 to 2012, there was an average of one large retrogressive landslide every 21 months (approximately 1.7 years). However, if only the 76 Québec cases with an area greater than one hectare are considered (Lebuis et al. 1983), the result is a mean interval of 27 months (approximately 2.3 years).

These averages must be considered with care. It is worth noting that the cases from 1840 to 1950 account for only 30 % of all of the cases in the inventory, despite this interval of time cover 64 % of the total period under consideration. Looking at only the period from 1950 to 2012, the mean intervals given above would instead work out to 11 months (0.9 year) and 15 months (1.3 years). This disproportionality can be explained by the vastness of the sparsely inhabited portions of the territory, such that there were few witnesses to events, and the increased frequency of aerial photography activities in Québec beginning in the 1950s. In fact, many cases that were never reported by anyone were identified through traces visible on aerial photos. Therefore, the averages indicated above must be considered to be minimum values.

Of the 108 historical cases that were inventoried, 79 have a known occurrence date, while the season of occurrence is known for 83 cases. Figure 7.4 shows a strong predominance of cases (total of 50 %) during the months of April to June. Relatively high values can also be noted for the months of March, October, November, and December. Only the very cold winter period from late December to mid-March is practically inactive, with only two historical cases recorded.

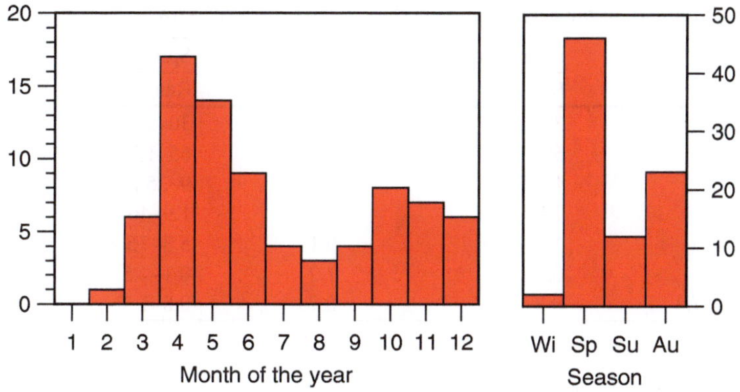

Fig. 7.4 Distribution of 79 historical landslides with known dates by month of the year and by season

The distribution of the historical cases in the inventory among the ancient postglacial seas is as follows: 59 cases in the Champlain Sea, 31 cases in the Goldthwait Sea, and 13 cases in the Laflamme Sea. Five other cases that occurred in the Tyrell Sea have also been inventoried.

The 108 historical large retrogressive landslides range in area from 0.3 to 447 ha, with 19 cases exceeding 10 ha in area. The largest historical case, which occurred in Saint-Alban in 1894, was of colossal dimensions, with a retrogression distance of 1.6 km and a width of 3.6 km (447 ha). This case ranks 9th among all scars (historical and ancient) inventoried by means of our lidar data for the entire province (Demers et al. 2013). The second largest historical event is that of Grandes-Bergeronnes in 1896, at 54 ha. The well-known case of Saint-Jean-Vianney in 1971, which resulted in 31 deaths, is the 5th largest historical event, with an area of 32 ha.

Finally, 90 % of the 108 historical cases are of natural origin, while 11 cases are the result of human activity: fill work at the slope crest or excavation at the base (5 cases), blasting activity (4 cases), concentration of water drainage (1 case), and pile driving (1 case). It is important to note that all of the cases of natural origin occurred along watercourses where erosion played a more or less active role.

7.4 Types of Large Retrogressive Landslides

Based on the Varnes classification (1978), the vast majority of large retrogressive landslides in Québec are flowslides and spreads. Of the 108 historical cases inventoried, 58 % would be classified as flowslides, 37 % would be classified as spreads, and the remaining 5 % would either be classified as another type or considered to be of indeterminate type. Tables 7.1 and 7.2 summarize the geometric characteristics and geotechnical properties of historical flowslides and spreads, which are discussed in the following sections.

Table 7.1 Summary of the geometric properties of the historical cases

	Flowslide		Spread	
	Range	Average	Range	Average
Retrogression (R)	38–1,340 m	225	30–560 m	145
Width (W)	50–3,500 m	250	65–1,725 m	370
Height (H)	10–73 m	27	9–35 m	18
R/H	1.9–47.4	9.5	1.9–46.7	8.0
W/H	1.4–47.9	9.1	5.3–76.2	20.7
R/W	0.3–3.1	1.2	0.1–2.5	0.5
Failure depth/H	0.3–1.0	0.7	0.8–1.3	1.0

Table 7.2 Summary of the geotechnical properties of the historical cases

Property	Flowslide	Spread
PI (average)	1–47 (%)	3–46 (%)
LI (min. value)	1.5–16.5	1.3–5.1
$S_{ur\text{-}Swedish\text{-}cone}$ (min. value)	0.08–0.80 (kPa)	0.08–1.30 (kPa)
St (from Swedish Cone)	10–1,890	12–1,500
OCR	1.2–11	1.1–7.9
%<2 μm	13–88 (%)	27–88 (%)

7.4.1 Characteristics of Flowslides

According to eyewitness reports related to a number of historical events, landslides of the flowslide type begin with an initial failure of the rotational landslide type, most often on the steep bank of a watercourse. These initial landslides, which are followed by a succession of rapid rotational failures under undrained conditions (Lefebvre 1996; Tavenas 1984), have sometimes been observed just a few hours before the onset of the flowslide (Saint-Alban 1894 and Saint-Thuribe 1898), and sometimes several days earlier (Kenogami 1924, Saint-Joachim-de-Tourelle 1963, Saint-Jean-Vianney 1971, Longue-Rive 2005).

The process of retrogression develops rearward as long as the sufficient quantities of materials are carried away, the soil conditions remain similar, and the height of the rear escarpment remains sufficient. During this process, part of the clayey deposit is transformed into a more or less liquid state at the moment when it slides downward. Therefore, the failure of each of the soil layers involved in a flowslide is not the result of the liquefaction of the clayey soil, but rather its cause.

The overlying debris materials, which do not liquefy, are carried away by the underlying liquefied clayey soils. This implies that the sensitive clay layer must be thick enough and appropriately located within the slope for the retrogression phenomenon to develop and continue. When they are very liquid, debris materials can flow along the watercourses for up to several kilometres from their starting point.

It frequently happens that a new series of failures occurs at a higher elevation during the process of retrogression, resulting in a stepped pattern of failure floors

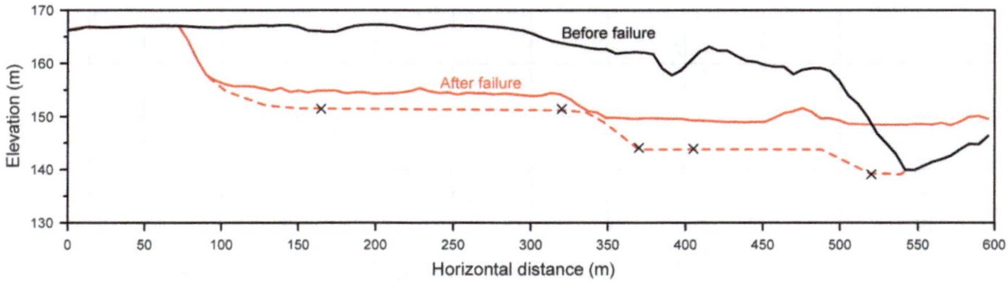

Fig. 7.5 2010 flowslide in Notre-Dame-de-la-Salette. The position of each failure surface levels as determined by piezocone sounding is indicated by an "×"

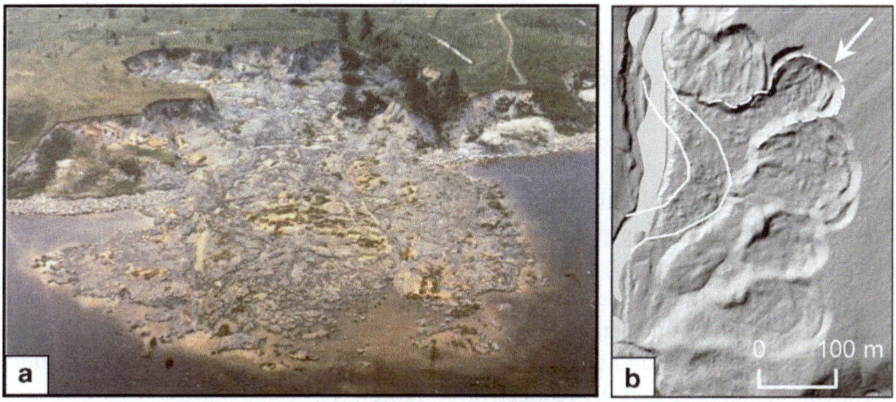

Fig. 7.6 (**a**) Typical example of a flowslide (Desbiens 1983) formed by successive rotational landslides propagating in different directions, resulting in a scalloped scar. (**b**) 1955 Isle-Verte flowslide, squeezed in among ancient scars

(Fig. 7.5). Accordingly, for the 19 historical flowslide scars for which piezocone sounding was conducted, the ratio between the depth of the floor and the slope height for 12 of the scars was between 0.28 and 0.85, which means that the base of the failure surfaces was much higher than the level of the watercourse in most cases. None of the cases subjected to sounding revealed a failure surface lower than the base of the slope.

With the exception of a few cases with multiple failure levels, the thickness of the debris materials above the floor of the scars of the historical cases is generally low, with a mean value of 0.18 times the height of the original slope. When a scar has a failure surface higher than the level of the watercourse, or several levels, the thickness of the debris on the top level is even lower, and sometimes even practically zero, allowing for a direct view of the surface of the landslide (Lefebvre et al. 1991), often with scattered striations caused by the flow of debris materials (Perret et al. 2011).

Flowslide scars can take a wide variety of shapes, ranging from cases with a narrow mouth (relatively rare) to shapes spreading in various directions (Figs. 7.3 and 7.6). It is not unusual for scars to extend in various directions relative to their

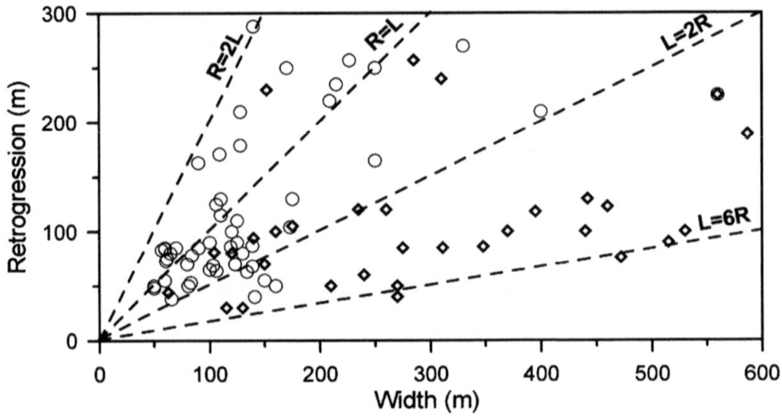

Fig. 7.7 Retrogression distance versus width of landslide scars for historical cases, by type (○: flowslide; ◇: spread)

original axis (Fig. 7.3). This characteristic means that flowslides can develop from relatively narrow starting points along a watercourse, and then spread behind the slope crest (Fig. 7.6b).

As a result of this process, flowslides often tend to have proportionately higher ratios of retrogression distance to width, as compared to spreads (Fig. 7.7). The ratios of retrogression and width to the slope height of flowslides are approximately identical, namely 9.5 and 9.1 respectively (Table 7.1).

The compilation of geotechnical data in Table 7.2 shows that the profiles of historical flowslides all had very low (0.8 kPa or less) remoulded shear strength values (S_{ur}) and high to very high (1.5–16.5) liquidity indices (LI). However, all of their other characteristics are similar to those of all postglacial marine clays in Québec (Leroueil et al. 1983).

7.4.2 *Characteristics of Spreads*

Many cases of spreads have been studied over the last 15 or so years in Québec (Grondin and Demers 1996; Demers et al. 2000; Ouehb 2007; Locat et al. 2008; Locat et al. 2011a, b; Fortin-Rhéaume 2013; Locat et al. 2013a, b). The classic cases generally have a ribbed appearance, with nearly intact or less disturbed strips of land (grabens) alternating with clay pinnacles (horsts). The horsts consist of intact clay whose strata indicate no movement other than a translational movement on a sub-horizontal plane. However, more disturbed soils can often be observed at the area of contact between the grabens and the horsts, which seem to have been extruded from deeper zones.

It often occurs that the land surfaces of grabens and any overlying structures are relatively well preserved (Fig. 7.8), which makes classic spreads less deadly than

Fig. 7.8 Spread in St-Luc-de-Vincennes in 1986. Note that the plant cover and the roadway remain practically intact on top of the grabens (Source: MTQ)

flowslides. The landslide debris materials spread out in the forward direction until they crash into the opposite bank, and sometimes even overrun it. In some cases, the bed of the watercourse is raised several metres and pushed onto the other bank. The landslide debris materials, which often show little evidence of remoulding, flow just a little bit upstream and downstream of the scar. The thickness of the debris in the historical spreads, as established by piezocone sounding in 16 cases, ranges between a ratio of 0.4 and 0.8 (mean of 0.6) of the original slope height.

Witnesses to spread-type historical landslides are quite rare in Québec. However, witnesses in Grandes-Bergeronnes in 1896 and in Notre-Dame-de-la-Salette in 1908 reported that an initial landslide had occurred between 4 and 18 h before the onset of the spread. In the cases of Poupore (1903) and St-Liguori (1989), witnesses who were directly involved in the event reported that it had lasted a few seconds, or a few minutes at the most.

The account given by the owner of the land directly involved in the case of the gigantic spread in Poupore (R = 560 m and L = 914 m) is quite instructive. Barlow (1905) reports that this individual was on his doorstep when he saw a very large fissure forming, from which there emerged an immense block of soil (a horst) that seemed to be slowly rising above him to a height of 7 m. In reality, the man and his dwelling were on a graben that was slowly sinking. This movement was so gradual and steady that when the movement stopped, a full water glass sitting on a window-sill remained undisturbed. This valuable account and the data collected in the recent studies of spreads clearly shows that landslides of this type arise through an entirely different mechanism than the one underlying flowslides.

Piezocone soundings have been conducted within the scars of 23 historical sites and in 9 cases of well-preserved ancient spreads. In all of these cases, the failure surface was flat and practically horizontal (or sloping very slightly toward the watercourse). Only the case of St-Jude (Locat et al. 2012) showed the presence of a second failure level higher up in the rear section of the scar. The ratio of failure surface depth to slope height generally varies between 0.84 and 1.25 (Table 7.1). The well-documented case of St-Jude in 2010 (Locat et al. 2011b, 2012) clearly shows that the failure surface developed some metres below the talweg of the river.

Other than cases where the lateral extension has been halted by adjacent ravines (e.g.: St-Ambroise-de-Kildare – 1974), the vast majority of spreads have a width that is considerable greater than their retrogression distance. Figure 7.7 shows that almost all cases where this ratio is between 2 and 6 are spreads.

The geotechnical data compiled in Table 7.2 show that the sites of historical spreads presented remoulded shear strength values that were much more variable than those for flowslides (between 0.08 and 1.3 kPa), along with comparatively lower liquidity indices (1.3–5.1). The cases of Yamaska in 1974 (Robert and Chagnon 1976) and St-Barnabé in 2005 (Locat et al. 2008) also show that spreads can occur in much less sensitive clays, where flowslide could not occur.

7.5 Estimation of Retrogression Distance

Estimation of the retrogression distance is essential in several aspects of risk management for large retrogressive landslides. Several approaches for estimating the potential retrogression distance have been proposed in the literature. Mitchell and Markell (1974), and later Mitchell (1978), proposed an empirical method based on using the stability number ($N_s = \gamma H/S_u$). Carson (1979) developed a method that is roughly comparable, but that also takes into account the sensitivity of the clay. Quinn et al. (2011) proposed a simple relationship based on the assumption that the complete failure surface develops before the onset of significant movement.

Figure 7.9 shows the stability number "N_s" as a function of retrogression distance for all of the sites where soundings have been conducted, along with the

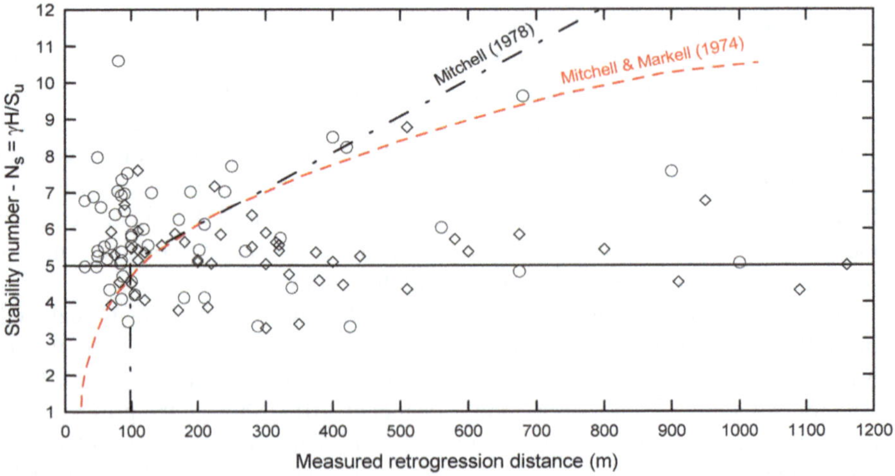

Fig. 7.9 Retrogression distance as a function of the stability number for historical cases (○) and ancient scars (◇)

relationships proposed by Mitchell and Markell (1974) and by Mitchell (1978). Our own measurements at many sites mentioned by these authors show significantly different values. Based on our own data, there does not seem to be a clear relationship between the retrogression distance and this parameter in the case of large retrogressive landslides in Québec. Our results also indicate that large retrogressions have been measured where the N_s values were less than 5, and as low as 3.3.

Carson's method (1979) requires numerous assumptions, and does not represent a significant improvement over the preceding approach. As for the method proposed by Quinn et al. (2011), the results obtained may differ from the actual retrogression distance by a factor of 10 in either direction.

For the purposes of cartography in Québec (Chap. 26 by Potvin et al., this volume), the regional statistical approach described by Lebuis et al. (1983) and Rissmann et al. (1985) is still in use.

7.6 Conclusion

The compilation of data pertaining to historical landslides has made it possible to better define the principal characteristics of large retrogressive landslides and their particularities. This exercise reveals that there are many factors that remain to be clarified and methods that must be improved, specifically including the approaches used to predict retrogression distances. It also does not seem to be possible at the present time to predict what type of landslide (flow or spread) may occur at a given site based on its current geotechnical properties. However, our inventory confirms that flowslides require the presence of very sensitive clay layers (quick clay), which is not the case of the spreads. Given these uncertainties, it is important to continue to take a cautious approach when tackling the question of large retrogressive landslides.

Acknowledgments The authors wish to thank Mélissa Raymond for her help in preparing figures and Mr. Guy Lefebvre for reviewing the manuscript. This paper is published with the permission of the Ministère des Transports du Québec.

References

Barlow AE (1905) A landslide on the Lievre River. The Ottawa Naturalist 18:181–191

Carson MA (1979) On the retrogression of landslides in sensitive muddy sediments: reply. Can Geotech J 16:431–444

Demers D, Robitaille D, Perret D (2000) The St.Boniface Lanslide of April 1996: a large retrogressive landslide in sensitive clay with little flow component. In: Landslides in research, theory and practice, Thomas Telford, London, pp 447–452

Demers D, Robitaille D, Potvin J, Bilodeau C, Dupuis C (2008) La gestion des glissements de terrain dans les sols argileux au Québec. In: Proceedings of the 4th Canadian conference on geohazards: from causes to management. Presse de l'université Laval, Québec, pp 519–526

Demers D, Robitaille D, Locat P, Potvin J, Paradis S, Therrien J (2013) Inventaire des grands glissements de terrain dans les argiles sensibles du Québec: analyse préliminaire. Ministère des Transports du Québec, Service Géotechnique et Géologie, Rapport MT13-01

Evans SG (2001) Landslides. In: Brooks GR (ed) A synthesis of geological hazards in Canada, Geological survey of Canada, bulletin 548. Geological Survey of Canada, Ottawa, pp 43–79

Fortin-Rhéaume A (2013) Étude de l'étalement latéral de 1988 et des autres glissements de terrain le long de la vallée à Brownsburg-Chatam, Québec. Mémoire de maîtrise, Université Laval

Grondin G, Demers D (1996) The 1989 Saint-Ligouri flakeslide: characterization and remedial works. In: Proceedings of the 7th international symposium on landslides, Trondheim, Norway, Balkema, Rotterdam, pp 743–748

Lebuis J, Robert J-M, Rissmann P (1983) Regional mapping of landslide hazard in Quebec. In: Proceedings of the symposium on slopes on soft clays. Swedish Geotechnical Institute report no. 17, Linköping, pp 205–262

Lefebvre G (1996) Soft sensitive clays. In: Turner AK, Schuster RL (eds) Landslides: investigation and mitigation, Special report 247, Transportation Research Board. National Academy Press, Washington, DC, pp 607–619

Lefebvre G, Rosenberg P, Paquette J, Lavallée J-G (1991) The September 5, 1987, landslide on the La Grande River, James Bay, Quebec, Canada. Can Geotech J 28:263–275

Leroueil S, Tavenas F, Le Bihan JP (1983) Caractéristiques des argiles de l'est du Canada. Can Geotech J 20:681–705

Locat A, Leroueil S, Bernander S, Demers D, Locat J, Ouehb L (2008) Study of a latral spread failure in an Eastern Canada clay deposit in relation with progressive failure: the Saint-Barnabé-Nord slide. In: Proceedings of the 4th Canadian conference on geohazards: from causes to management. Presse de l'université Laval, Québec, pp 89–93

Locat A, Leroueil S, Bernander S, Demers D, Jostad H (2011a) Progressive failures in Eastern Canadian and Scandinavian sensitive clays. Can Geotech J 48:1696–1712

Locat P, Fournier T, Robitaille D, Locat A (2011b) Glissement de terrain du 10 mai 2010, Saint-Jude, Montérégie – Rapport sur les caractéristiques et les causes. Ministère des Transports du Québec, Service Géotechnique et Géologie, Rapport MT11-01

Locat P, Demers D, Robitaille D, Fournier T, Noël F, Leroueil S, Locat A, Lefebvre G (2012) The Saint-Jude landslide of May 10, 2010, Québec, Canada. In: Proceedings of the 11th international and 2nd North American symposium on landslides, Taylor and Francis Group, London, pp 635–640

Locat A, Leroueil S, Fortin A, Demers D, Jostad HP (2013a) The 1994 landslide at Sainte-Monique-de-Nicolet, Quebec: geotechnical investigation and application of the progressive failure analysis. Paper submitted to Can Geotech J

Locat A, Leroueil S, Demers D, Robitaille D (2013b) L'étalement de 1986 à Saint-Luc-de-Vincennes, Québec. Paper submitted to the 66th annual Canadian geotechnical conference, Montréal

Mitchell RJ (1978) Earthflow Terrain Evaluation in Ontario. Min. of Transportation and Communications of Ontario, Research and Development Division, Report RR213, 29 p

Mitchell RJ, Markell A (1974) Flowsliding in sensitive soils. Can Geotech J 11:11–31

Ouehb L (2007) Analyse du glissement de Saint-Liguori (1989) dans l'optique d'une rupture progressive. Mémoire de maîtrise, Université Laval, Québec, 271p

Perret D, Mompin R, Bosse F, Demers D (2011) Stop 2-5B: the Binette Road earth flow induced by the June 23, 2010 Val-des-Bois earthquake. In: Deglacial history of the Chaplain Sea basin and implications for urbanization, Joint Annual Meeting GCMAC-SEG-SGA, Ottawa, ON, 25–27 May 2011, Field guide book, Geological Survey of Canada, Open File 6947, pp 72–74

Quinn PE, Diederichs MS, Rowe RK, Hutchinson DJ (2011) A new model for large landslides in sensitive clay using a fracture mechanics approach. Can Geotech J 48:1151–1162

Rissmann P, Allard JD, Lebuis J (1985) Zones exposées aux mouvements de terrain le long de la rivière Yamaska, entre Yamaska et Saint-Hyacinthe. Min. de l'Énergie et des Ressources du Québec, rapport DV83-04, 64 p
Robert JM, Chagnon JY (1976) Caractéristiques et correction d'un glissement e terrain dans les dépôts argileux de la mer de Champlain a St-Michel de Yamaska. In: Proceedings of the 29th Canadian geotechnical conference. Vancouver, British-Columbia, Canada, pp XI-1–XI-22
Tavenas F (1984) Landslides in Canadian sensitive clays – a state-of-the-art. In: Proceedings of the 4th international symposium on landslides, Toronto, vol 1, pp 141–153
Varnes DJ (1978) Slope movement types and processes. In: Schuster RL, Krizek RJ (eds) Landslides: analysis and control, special report 176, TRB. National Research Council, Washington, DC, pp 11–33

Chapter 8
Characterization of Post-failure Movements of Landslides in Soft Sensitive Clays

V. Thakur, S.A. Degago, F. Oset, R. Aabøe, B.K. Dolva, K. Aunaas, T. Nyheim, E. Lyche, O.A. Jensen, M.B. Sæter, A. Robsrud, M. Viklund, Daniel Nigussie, and J.-S. L'Heureux

Abstract Assessment of landslides in soft sensitive clays require, in addition to geometrical aspects, a combined and complete understanding of the engineering properties of the soil materials at their original as well as disintegrated state. In light of the geotechnical characterization of 33 Norwegian landslides, this paper attempts to discuss some of the material properties which influence the extent of post-failure movements of landslides in terms of retrogression and the run-out distance. The energy involved in remolding sensitive clays has been studied for the Norwegian sensitive clays to signify its importance with regards to the estimation of extent of landslides.

Keywords Sensitive clays • Landslides • Retrogression distance • Quickness • Remolding energy

---

V. Thakur (✉) • F. Oset • B.K. Dolva • K. Aunaas
Geotechnical and Landslide Division, Norwegian Public Roads Administration (NPRA), Oslo, Norway
e-mail: vikas.thakur@vegvesen.no

S.A. Degago • R. Aabøe
Norwegian Public Roads Administration (NPRA), Oslo, Trondheim, Norway

T. Nyheim
Landslides, Flood and River Management Division, Norwegian Water Resources and Energy Directorate (NVE), Oslo, Norway

E. Lyche • O.A. Jensen
Norwegian Water Resources and Energy Directorate (NVE), Trondheim, Norway

M.B. Sæter • A. Robsrud • M. Viklund
Norwegian National Rail Administration (NNRA), Oslo, Norway

D. Nigussie
Department of Civil and Transport Engineering, Norwegian University of Science and Technology (NTNU), Trondheim, Norway

J.-S. L'Heureux
Norwegian Geotechnical Institute (NGI), Trondheim, Norway
e-mail: jsl@ngi.no

J.-S. L'Heureux et al. (eds.), *Landslides in Sensitive Clays: From Geosciences to Risk Management*, Advances in Natural and Technological Hazards Research 36,
DOI 10.1007/978-94-007-7079-9_8, © Springer Science+Business Media Dordrecht 2014

8.1 Background

Soft sensitive clay slopes of Scandinavia, when provoked by manmade or natural causes, could fail and often result in a substantial post-failure ground movement due to their ability to flow when remolded. An overview of 33 Norwegian landslides in soft sensitive clays is given in Table 8.1. From these landslides, it is known that a seemingly stable area can be subjected to a major landslide after a small initial slide. Post-failure movement in these clays is occasionally fast moving which may involve massive soil volume in the order of millions of cubic meters. As a result, such landslides pose a threat to human lives and properties. Within the framework of landslide hazard assessment, understanding the extent of post-failure movements is a crucial element in decision making process for adopting appropriate design approaches and mitigations. Leroueil et al. (1996), D'Elia et al. (1998) Vaunat and Leroueil (2002) suggested four different stages to characterize the kinematics of slopes referred to as pre-failure, onset of failure, post-failure and reactivation stages.

The post-failure movements in terms of the retrogression of landslides and run-out of slide debris are inadequately understood because it would require, among others, a complete understanding the flow behavior of quick clays (e.g. Leroueil et al. 1996; Locat and Demers 1988; Thakur and Degago 2012), accurate modeling of the progressive failure and creep effects (e.g. Bjerrum 1955; Bishop 1967; Bjerrum and Kjærnsli 1957; Eigenbrod 1972; Aas 1981; Bernander 2000; Jostad and Andresen 2002; Locat et al. 2011; Quinn et al. 2011), the physical thickness of the localized shear failure zone (e.g. Jostad et al. 2006; Thakur 2007, 2011; Gylland 2012), adequate assessment tools for the landslide mobility and hazards (e.g. Mitchell and Markell 1974; Lebuis and Rissmann 1979; Locat and Demers 1988; Trak and Lacasse 1996; Leroueil 2001; Vaunat and Leroueil 2002; Hungr 2005; L'Heureux 2012a) and knowledge about the effect of sample disturbance on the material behavior (e.g. Lefebrve 1981; Tavenas and Leroueil 1981; Lacasse et al. 1985; Berre et al. 2007; Degago et al. 2011). However, owing to crucial nature of assessing the landslide hazards in sensitive clays and the need to identify material parameters that govern-post-failure movements, several simplified approaches have been proposed in literature (e.g. Mitchell and Markell 1974; Lebuis et al. 1983; Leroueil et al. 1983, 1996; Tavenas et al. 1983; Karlsrud et al. 1985; Locat et al. 2003; Thakur and Degago 2012, 2013).

Characterizing post-failure movement of flow slides ranges from prediction of the potential for its occurrence before failure (pre-failure stage) and estimation of the extent of retrogression after failure (post-failure stage). In this paper several parameters used in literature for prediction of flow slide potentials are briefly discussed and an approach based on a new test procedure is proposed. Additionally, the paper attempts to understand the parameters governing the post-failure movement using methods from literature as well as by proposing a method based on energy approaches. The study is supported by extensive Norwegian landslide data.

Table 8.1 Sensitive clay landslides reported in Norway[a]

Year	Landslide (*Ref.*[b])	Type	L_R [m]	L_F [m]	V [$10^5 \times m^3$]	c_{ur} [kPa]	S_t [–]	I_L [–]	I_P [%]
1940	Asrumvannet[1]	F				0.1	200	3.1	13
1626	Bakklandet[2]	FL	70	50		0.1	30	2	6
1988	Balsfjord[3,22]	F	400		8	1	30	3	6
1974	Båstad[4]	F	230	700	15	0.53	35	1.8	8
1953	Bekkelaget[5]	FL/F	145	20	1	0.11	150	2.4	11
1953[c]	Borgen[6]	RR	165		1.6	0.7	100	1.2	20
1928	Brå[7–9]	FL	197	300	5	0.24	75	2	
2012	Byneset[10,20]	FL	400	870	3.5	0.12	120	3.9	4.8
1955	Drammen[5]	RT	45		0.04	2.5	4	1.1	11
1625	Duedalen[8,9,11,21]	FL	410		5	0.07	209		
1996	Finneidfjord[12]	RR	150	850	10	0.4	60		
1980	Fredrikstad[13,14,15]	RR	45	22	1	<0.5	20	1	20
1959	Furre[16]	FL/F	300	90	30	0.1	115	2.1	11
1974	Gullaug[17]	F/FL	150		1.25	2	7.5		
1967	Hekseberg[18]	FL	700	300	2	0.25	100	2.4	4
2009	Kattmarka[19]	RR	300	350	3–5	0.24	63	2.9	8
1994	Kåbbel[20]	F	100	10	1	<0.5	>50	>1.2	20
1944	Lade[8,9,13,21]	FL	40	62	0.05	2.12	6.6	1	
2002	Leistad[22,15]	F	250	25		0.15	110	1.5	6
1989	Lersbakken[15,22]	F	65	75	0.75		38–62		
1954	Lodalen[23]	FL	40	10	0.1	17	3	0.8	17
2010	Lyngen[20]	F	153	411	2–3	0.14	51.4	2.1	
2000	Nedre Kåbbel[20]	F	120	10	1.8	<0.5	>50	>1.2	20
1978	Rissa[24]	RR&F	1,200		50–60	0.25	100	2	5
1995	Røesgrenda[25]	RR	100	50	0.02	0.1	186	>1.2	<10
1974	Sem[15,26]	FL	100	20	0.68	1.4	8–14		
1965	Selnes[27]	F	230	>400	1.4	0.35	100	2.3	7
1962	Skjelstadmarka[28]	F	600	2,800	20	0.83	80	1.1	10
1816	Tiller[8,10,22,23]	FL			55	0.1	90	2.7	4
2012	Torsnes[23]	RR	25		0.063	<0.5			22
1953[c]	Ullensaker[29,30]	RR	195	1,500	2	0.35	42	1.9	6.7
1893	Verdal[6,10,11,21]	FL	2,000	5,000	650	0.2	300	2.2	5
1959	Vibstad[31]	F	250	250	10	5	8	0.2	17

[a]L_R Retrogression distance measured from the toe of slope, L_F run-out distance measured from the toe of slope, *H* slope height, *V* slide volume, c_{ur} remolded shear strength along slip surface, S_t sensitivity, *w* water content, w_L liquid limit, I_p plasticity index, I_L liquidity index, *NA* Exact year data not available, *F* flow slide, *FL* flake slide, *RR* retrogressive slide, *RT* rotational slide

[b]References: [1]Mayerhof (1957), [2]Egeland and Flateland (1986), [3]Rygg and Oset (1996), [4]Gregersen and Løken (1979), [5]Eide and Bjerrum (1955), [6]Trak and Lacasse (1996), [7]Holmsen (1929), [8]Reite et al. (1999), [9]Trondheim Municipality Reports 1981, [10]Thakur (2012), [11]Furseth (2006), [12]Longva et al. (2003), [13]Holmsen and Holmsen (1946), [14]Karlsreud (1983), [15]Thakur et al. (2012), [16]Huchinson(1961), [17]Karlsrrud (1979), [18]Drury (1968), [19]Nordal et al. (2009), [20]NVE Reports 2012, [21]Natterøy(2011), [22]NPRA reports 1994, [23]Sevaldsen (1956), [24]Gregersen (1981), [25]Larsen (2002), [26]NGI (1974), [27]Kenney (1967), [28]Janbu (2005), [29]Bjerrum (1955), [30]Jørstad (1968), [31]Huchinson (1965)

[c]These two names represent the same landslide

8.2 Indicators for the Extent of Post-failure Movements

Initiation of large landslides, which begins with a small initial slide, can be attributed to natural or manmade causes such as erosion, creep, ground water variations, construction activities, loading and excavation. Once failure of slope is initiated, there are several factors that govern the extent of the post-failure movements of retrogressive flow slide. These factors can grossly be classified into topographical (geometrical) aspects of the slope and engineering characteristics of the soil in the slope. The topographical aspects relate to initial stability of the natural slope and possibility of the slide debris to flow after failure. The behavioral aspects of the soil that play a central role in assessing the potential for large landslides include remolded shear strength, sensitivity, liquidity index, rapidity number, stability number, quickness and rheology. These aspects are discussed briefly and individually in this section in light of the Norwegian landslide data listed in Table 8.1.

Favorable Topography: Large landslides are often results of initial failure under a favorable topographic situation such that the remolded materials have a possibility to flow out from the slide pit. Hence, topography is vital aspects for occurrence large landslides. In absences of a suitable numerical tool, the retrogression of landslides is empirically predicted by topographical aspects.

Locat et al. (2008) recommends a relation between retrogression distance (L_R) and run-out distance (L_F) as $L_F/L_R^{0.8}=4.4–8.8$ for the Canadian soils. L'Heureux (2012b) suggests L_F/L_R up to 10 for the Norwegian landslides. In Norway, a L_R of 15 times the slope height (H) is recommended by NVE (2010).

Stability Number (Nc): After an initial slide and favorable topography, stability of the back scarp is vital in the successive development of a landslide (Mitchell and Markell 1974; Lebuis et al. 1983; Tavenas et al. 1983; Karlsrud et al. 1985; Leroueil et al. 1996). The stability of back scarp is quantified using a parameter referred to as stability number (Nc) and is defined as $\gamma H / c_{ui}$ (Bjerrum 1955; Trak and Lacasse 1996). Leroueil et al. (1996) suggests that a significant post-failure movement (retrogression) of soil mass could happen if the stability of the area behind the initial slide zone has $Nc>4$ if $I_p \sim 10$ and $Nc>8$ if $I_p \sim 40$. Figure 8.1a presents Nc versus L_R observed for the Norwegian and Canadian landslides. Despite a large scatter in the data, $L_R>100$ m was mostly observed only when $Nc>4$. Majority of Norwegian soft sensitive clays have low I_p and therefore $Nc>4$ seems to give a good indicator of a possibility for occurrence of large landslides.

Remolded shear strength (c_{ur}): Mitchell and Markell (1974) suggest a direct relationship between c_{ur}, S_t and L_R and proposed that sensitive clays with $c_{ur}<1$ kPa could lead to retrogressive flow slides with an $L_R>100$ m. The c_{ur} and L_R as observed from Norwegian landslide data (Table 8.1) and Canadian landslide data (Lebuis et al. 1983) are presented in Fig. 8.1b. Based on both sets of data given in Fig. 8.1, $c_{ur}<1$ kPa may define the occurrence of large landslides.

Soil sensitivity (S_t) is often thought to be correlated with the extent of landslides. Lebuis et al. (1983) and the Norwegian landslides data presented in Table 8.1

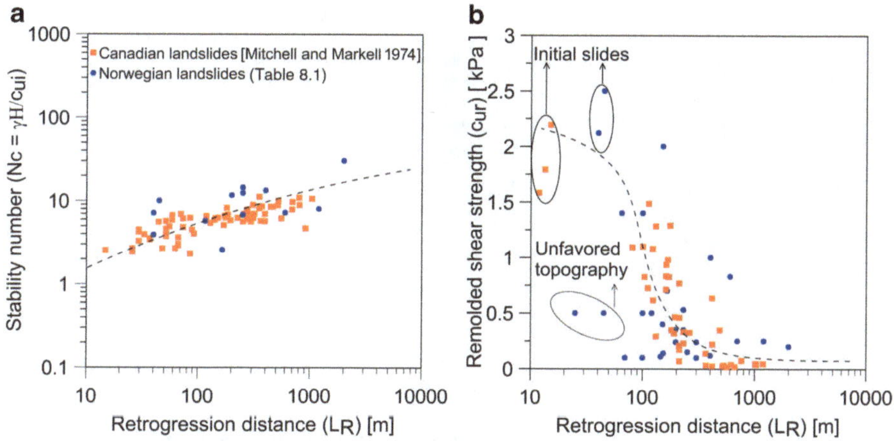

Fig. 8.1 (**a**) Stability number and (**b**) remolded shear strength versus retrogression distance

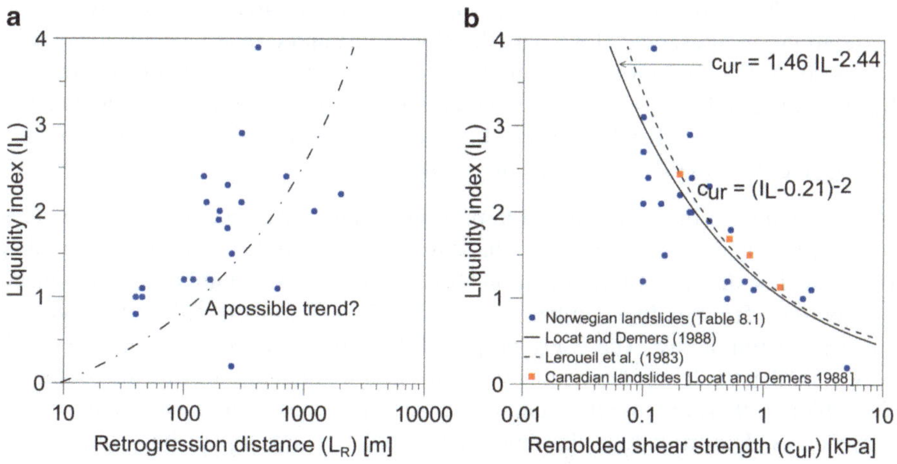

Fig. 8.2 Liquidity Index versus (**a**) retrogression distance and (**b**) remolded shear strength

suggests that large landslides may occur in sensitive clays having $S_t > 30$. However, S_t is a parameter derived from c_{ur} and c_{ui}. Therefore, the authors believe it is more logical to use c_{ur} rather than S_t itself to assess the landslide potential.

Liquidity index (I_L): Leroueil et al. (1983), Locat and Demers (1988) and Leroueil et al. (1996) suggest that sensitive clays with $I_L > 1.2$ are susceptible to flow slides. This finding is well supported by the landslide data presented in Table 8.1. It must, however, be noted that $I_L > 1.2$ is only possible when $c_{ur} < 1$ kPa. Hence, this criterion is equivalent to the criterion based on c_{ur} presented earlier. Leroueil et al. (1996) suggests that I_L is positively correlated with the extent of post-failure movements of slopes and this is supported by the Norwegian landslides and is shown in Fig. 8.2a.

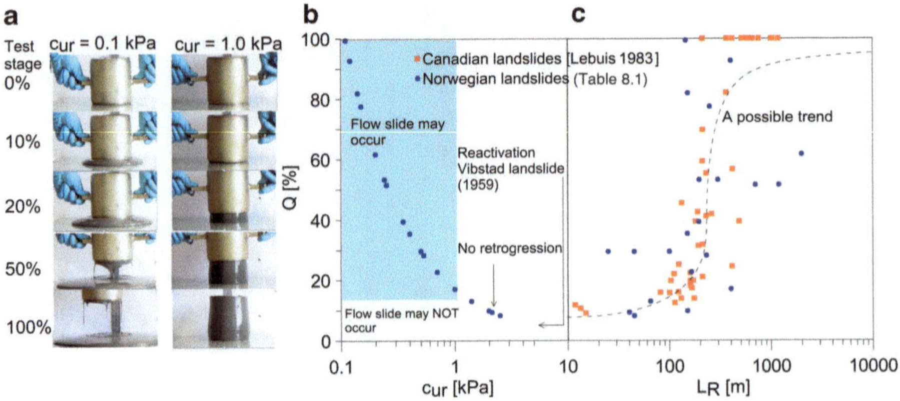

Fig. 8.3 The quickness concept developed by Thakur and Degago (2012) (**a**) to identify the potential for flow slides in soft sensitive clays (**b**) is correlated with the retrogression distance (**c**)

Leroueil et al. (1983) and Locat and Demers (1988) presents correlations between c_{ur} and I_L. The correlations are $c_{ur}=(I_L-0.21)^{-2}$ and $c_{ur}=1.46\,I_L^{-2.44}$ respectively. The correlations are presented along with the Norwegian and Canadian landslide data in Fig. 8.2b. Both correlations are seen to match well for the Canadian landslide data but a significant deviation from these equations is observed for the Norwegian landslide data when $c_{ur}<0.5$ kPa. Such deviation is also seen for the Canadian soils (Leroueil et al. 1983; Locat and Demers 1988).

Rapidity number: In 1974 Söderblom proposed a method to classify sensitive clays by determining the ease with which the clays can be remolded. The rapidity number is assigned on the basis of deformation observed after a fixed number of blows made on samples. The rapidity number ranges from 1 to 10, in the order of increasing easiness of remolding. The study proposed that highly sensitive clays, like quick clays, shall have a rapidity number of at least 8. In other words, according to Söderblom (1974) the rapidity number ≥ 8 is an indicator of large landslides in sensitive clays. Unfortunately, this approach could not develop to be a universally accepted practice as the observed remolding seems to be non-unique for different sensitive clays of a given a rapidity number.

Quickness (Q): Thakur and Degago (2012) suggest a new engineering parameter called quickness (Q) to assess the flow potential of sensitive clays. The quickness test is a simple procedure that is performed by filling an open ended cylinder with remolded sensitive clay, then slowly lifting the cylinder, and finally measuring the deformation (height and lateral spreading) as the material is subjected to flow due to its own weight, Fig. 8.3a. The ratio of the difference in height between the cylinder and the slumped material to the cylinder height defines Q. Quickness test gives a better visualization of the flow behavior of sensitive clays where small c_{ur} values have big implications in regards to understanding of the potential for retrogressive landslides. The quickness test was performed on more than 60 different samples extracted from three different locations in Norway (Thakur et al. 2012). Accordingly, the study showed that sensitive clays with $c_{ur} \approx 0.5$ kPa were not as fluid as they were originally assumed and sensitive clays with 0.5 kPa$< c_{ur} <$1.0 kPa were semisolid in nature. Thakur et al. (2012) shows that $Q = 15$ % corresponds to $c_{ur} = 1$ kPa. Based on conservative approach a correlation of Q with c_{ur} as $Q = 15c_{ur}^{-0.7}$ is suggested. Furthermore the tests showed that $Q = 15$ % gives the threshold limit below which the registered collapse of the remolded material was negligible. In fact, this finding is in line with Mitchell and Markell (1974), Lebuis et al. (1983) and Leroueil et al. (1996) where, based on the landslide data, they suggested that large landslides are less likely in sensitive clays having $c_{ur} > 1$ kPa. Figure 8.3b presents this graphically along with the Norwegian landslide data. The quickness tests supplements this suggestion by demonstrating the flow behavior of sensitive clays. Based on the lower bound equation for Q, a rough attempt is made to calculate Q for the Norwegian landslide (Table 8.1) and the Canadian landslide data presented by Lebuis et al. (1983), Fig. 8.3c. The calculate Q values that show that the extent of retrogression will be larger for sensitive clays having higher Q values i.e. highly flowable materials.

Rheological properties: Rheological properties of sensitive clays such as viscosity and yield strength are important input parameters for numerical modeling of run-out distance of debris flow e.g. Edgers and Karlsrud (1982), Norem et al. (1990), Locat (1997) and Hungr (2005). Locat and Demers (1988) and Locat and Lee (2005) have suggested several correlations between viscosity, liquidity index and yield strength for both fresh and marine water conditions.

8.3 Remolding Energy

Tavenas (1984), Karlsrud et al. (1985), Locat and Lee (2005), Leroueil et al. (1996), Hutchinson (2002), Locat et al. (2008) and Quinn et al. (2011) emphasized that to get an accurate run-out distance and the velocity of slide debris, one need to look at the energy balance and the volume of soil involved in landslides. Flon (1982), Yong and Tang (1983), Tavenas et al. (1983) were pioneers who extended the concept of remolding energy (RE) to the assessment of landslides in sensitive clays. They identified RE as a crucial aspect in the characterization of post-failure movements of landslides, however very limited amount of work has been carried out in this direction since Tavenas et al. (1983). In fact, no such study is reported for the Norwegian clays apart from some preliminary work by Thakur et al. (2012) and Thakur and Degago (2013). Keeping this in view, this paper makes an attempt to study the RE required for the Norwegian sensitive clays.

Remolding energy (RE) refers to the strain energy absorbed in remolding or disintegration of an intact soil material. A closer examination of the concept of RE provides an understanding of the overall mechanical behavior of sensitive clays during flow slides. RE of a material is quantified numerically by calculating the area under the shear stress–strain curve of the material. Thakur et al. (2012) simplified such stress–strain curve by considering a linear elastic hardening followed by a

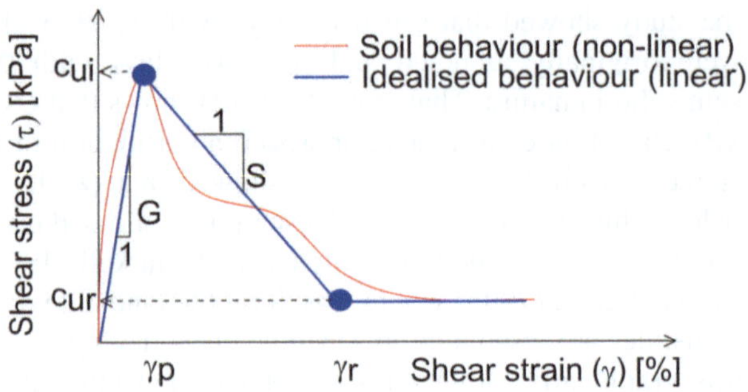

Fig. 8.4 Estimation of RE. The *dotted line* represents a real soil response and the *solid line* gives the idealization adopted for Eq. 8.1 (Thakur and Degago 2013)

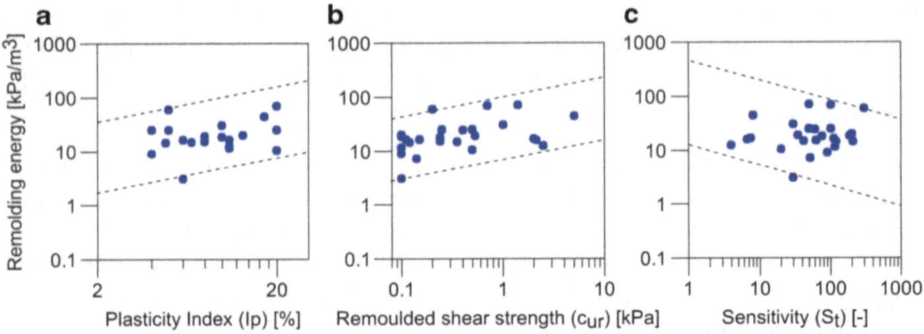

Fig. 8.5 Remolding energy of Norwegian sensitive clays, calculated using Eq. 8.1, versus (**a**) plasticity index, (**b**) remolded shear strength and (**c**) sensitivity

linear strain-softening behavior, Fig. 8.4, and proposed analytical equation for RE as given in Eq. 8.1.

$$RE = c_{ur} \times \gamma_r - \frac{c_{ur}^2}{2G} + \frac{1}{2}\left[\left(S_t - 1\right) \times c_{ur}\right]^2 \times \left[\frac{1}{G} + \frac{1}{S}\right] \tag{8.1}$$

The RE equation, Eq. 8.1, consists of stiffness and strength parameters; where, G and S are the secant shear and average softening modulus respectively; γ is the shear strain, where the subscripts *i* and *r* represent the peak and the residual strain levels, respectively. In order to illustrate the use of RE equation and get the feeling of its implications, the RE involved in Norwegian landslides listed in Table 8.1 is computed and presented in Fig. 8.5. In doing so, further simplifications were necessary due to lack of data demanded by the RE equation and this was done based on laboratory and field tests on sensitive clays by Berre et al. (2007) and Gylland et al. (2012). Accordingly, G is assumed to be $150c_{ui}$ whereas γ_p and γ_r were adopted as 0.005 and 3 respectively.

The RE obtained for the sensitive clays involved in Norwegian landslides with reference to a selected index property are shown in Fig. 8.5. The dotted lines are

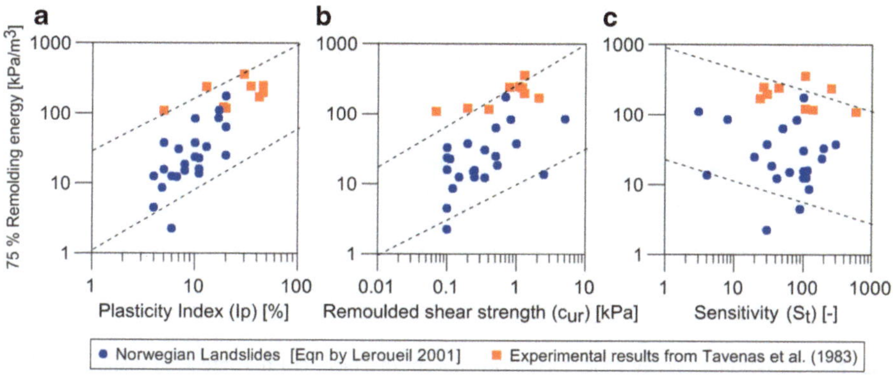

Fig. 8.6 Remolding energy versus (**a**) plasticity index, (**b**) remolded shear strength and (**c**) sensitivity

theoretical lines plotted based on Eq. 8.1 to support a possible theoretical bounds of trends between RE and the index properties and they are not necessarily meant to show the trend of the actual data. Accordingly, the RE seems to be positively correlated with I_p and c_{ur}, (Fig. 8.5a, b) and negatively correlated with S_t, (Fig. 8.5c). Despite these trends the results in Fig. 8.5 show large scatters and further investigation is necessary to establish a robust correlation.

Based on the laboratory experiments by Tavenas et al. (1983), Leroueil et al. (1996) proposed a pragmatic approach to calculate 75 % RE as $= 12.5c_{ui}I_p$ for the Canadian sensitive clays. This equation is adopted to calculate 75 % RE for the Norwegian clays and is presented along with experimental results of Tavenas et al. (1983) in Fig. 8.6 in relation to I_p, and c_{ur} and low S_t. The theoretical trend lines shown in Fig. 8.5 are also given in Fig. 8.6. Accordingly, the calculated 75 % RE per unit volume is up to 100 kPa for Norwegian sensitive clays whereas Tavenas et al. (1983) experimentally showed that for Canadian sensitive clays the required 75 % RE per unit volume ranges between 110 and 400 kPa. The experimental results by Tavenas et al. (1983) (Fig. 8.5) supports the observation in Fig. 8.4 that sensitive clays with higher I_p and c_{ur} and low S_t require greater RE. In general similar trends are observed in Figs. 8.5 and 8.6 except that the required RE for Canadian clays is higher due to their higher c_{ui} attributed by their high over-consolidation ratio.

In connection to this study, an interesting aspect to investigate is the role played by the RE in deciding the post-failure movements i.e. L_R and L_F. These movements depend on several factors including the thickness of the dry crust and sensitive clay layers, boundary conditions, and topographical aspects which may allow sensitive clays to "escape" from the slide scarp; however for sake of simplicity a rough study is performed. Accordingly, the RE in relation to initial geometry can be formulated by defining the remolding energy ratio, i.e. RE divided by the maximum available potential energy. The remolding energy ratio versus L_R, L_F and N_c is plotted in Fig. 8.7.

The remolding energy ratio indicates how large amount of potential energy will be consumed in the remolding of sensitive clays. Here, the potential energy per unit

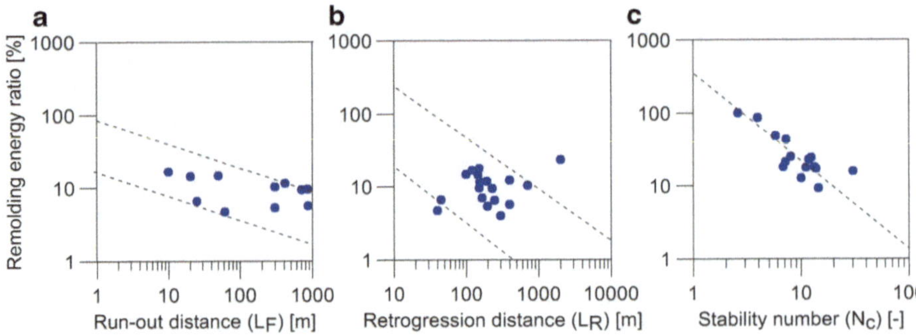

Fig. 8.7 Remolding energy ratio versus (**a**) run-out distance, (**b**) retrogression distance and (**c**) stability Number for Norwegian clays

volume was simply calculated using $2\rho gH/3$. Here ρ is the mass density of sensitive clay and g is the gravitational acceleration. Figure 8.7 indicates that as low as only 5 % of total potential energy would have been sufficient to remold Norwegian sensitive clays and the remaining energy would have been left for the possible movement of the sensitive clays that were subjected to L_R of L_F> 100 m. The fact that the lower remolding energy ratio was obtained for the Norwegian sensitive clays subjected to large post-failure movements. Figure 8.7c shows that Norwegian landslides that had higher stability numbers (N_c) had needed a lower remolding energy ratio. A similar study has been done by Locat et al. (2008) for the Canadian sensitive clays using a normalized energy index referred to as destructurisation index (I_D). The results in Locat et al. (2008) show somewhat similar level of scatter between the I_D and the L_R, as in Fig. 8.7a, b. Some degree of scatter in Fig. 8.7a, b could be attributed by the landslide types e.g. flow, flake and rotational slides, which are not differentiated in the figures. Accordingly, a well-defined trend between remolding energy ratio and L_R and L_F is not found for the Norwegian landslides and this is mainly believed to be due to the involved simplifications. A better correlation, in the form of the trend lines shown in Fig. 8.7a, b, was obtained among L_R, L_F and the RE when five similar landslides were studied (Thakur and Degago 2013). Therefore, it is authors believe that taking into account topographical aspects, the nature of regression and run-out process could yield a better correlation among L_R, L_F and the RE than shown in Fig. 8.7a, b. The results presented in this study are a preliminary attempt to assess post-failure movements of landslides. However, to make this promising approach robust and practical further research is necessary.

8.4 Conclusions

This paper presents a comprehensive overview over several parameters that may influence the extent of landslides in sensitive clays. The Norwegian landslide data supports the fact that large landslides are only possible when c_{ur} < 1.0 kPa or I_L > 1.2

or Q>15 % and N_c>4. These criteria are useful and can be used as indicators to assess the potential for occurrence of large landslides. However, the extent of a landslide with only an individual soil parameter may not be sufficient. This work advocates that a complete stress–strain behavior of soft sensitive clays must be accounted in the calculation of the post-failure movement of landslides. This paper suggests an analytical solution based on energy concepts. A simplified calculation approach shows that Norwegian soft sensitive clays may need as low as only 5 % of total potential energy available in a slope to remold the clays; whereas the remaining energy will be available for the post-failure movement of slide debris. This analytical observation demands further verification using laboratory testing.

Acknowledgments National research program "Natural hazards: Infrastructure for Floods and Slides (NIFS)", by the Norwegian Public Roads Authority, Norwegian Water Resources and Energy Directorate and Norwegian National Railways Administration is acknowledged for the support. The authors wish to acknowledge Professor Jacques Locat Laval University, for his valuable comments and the constructive feedback on this paper.

References

Aas G (1981) Stability of natural slopes in quick clays. In: Proceedings of the 10th ISSMGE, 3. Stockholm, Sweden, pp 333–338

Bernander S (2000) Progressive landslides in long natural slopes. Licentiate thesis, Luleå University

Berre T, Lunne T, Andersen KA et al (2007) Potential improvements of design parameters by taking block samples of soft marine Norwegian clays. Can Geotech J 44(6):698–716

Bishop AW (1967) Progressive failure – with special reference to the mechanism causing it. In: Proceeding of the geotechnical conference on Shear Strength Properties of Natural Soils and Rocks. Norwegian Geotechnical Institute, Oslo, pp 142–150

Bjerrum L (1955) Stability of natural slopes in quick clay. Géotechnique 5(1):101–119

Bjerrum L, Kjærnsli B (1957) Analysis of the stability of some Norwegian natural clay slopes. Géotechnique 7(1):1–16

D'Elia B, Picarelli L, Leroueil S et al (1998) Geotechnical characterization of slope movements in structurally complex clay soils and stiff jointed clays. Rivista Italiana di Geotecnica XXXIII:5–32

Degago SA, Nordal S, Grimstad G et al (2011) Analyses of Vasby test fill according to creep hypothesis A and B. In: Proceedings of the 13th international conference of IACMAG, vol 1, Melbourne, pp 307–312. Centre for Infrastructure Engineering and Safety, ISBN 978-0-9808244-1-4

Drury P (1968) The Hekseberg landslide, March 1967. NGI publication 75

Edgers L, Karlsrud K (1982) Soil flows generated by submarine slides. NGI publication 143

Egeland A, Flateland A (1986) Bakkalget landslide in Trondheim. Master thesis, NTNU

Eide O, Bjerrum L (1955) The slide at Bekkelaget. Géotechnique 5(1):88–100

Eigenbrod KD (1972) Progressive failure in oc clays and mudstones. PhD thesis, Univ. of Alberta

Flon P (1982) Énergie de remaniement et régression des coulées d'argiles. MSc thesis, Department of Civil Engineering, Laval University, Québec

Furseth A (2006) Skredulykker i Norge. Tun Forlag, Oslo

Gregersen O (1981) The quick clay landslide in Rissa, Norway. NGI publication 135

Gregersen O, Løken T (1979) The quick-clay slide at Båstad, Norway, 1974. Eng Geol 14:183–196

Gylland A, Jostad HP, Nordal S (2012) Failure geometry around a shear vane in sensitive clays. In: Proceedings of the 16th Nordic geotechnical meeting, vol 1, Copenhagen, pp 103–110, dgf-Bulletins (Danish Geotechnical Society) ISBN 78-87-89833-27-9

Holmsen G (1929) Lerfaldene ved Kokstad, Gretnes og Braa. NGU report 132
Holmsen G, Holmsen P (1946) Leirfall i årene 1940–1945. NGU report 167
Hungr O (2005) Classification and terminology. In: Jakob M, Hungr O (eds) Debris-flow hazards and related phenomena. Springer, Chichester, pp 9–24. ISBN 3-540-20726-0
Hutchinson DJ (1961) A landslide on a thin layer of quick clay at Furre, central Norway. Geotechnique 11(2):69–94
Hutchinson JN (1965) The landslide of February, 1959, at Vibstad Namdalen. NGI publication 61
Hutchinson JN (2002) Chalk flows from the coastal cliffs of northwest Europe. In: Evans SG, DeGraff JV (eds) Catastrophic landslides: effects, occurrence, and mechanisms, Reviews in engineering geology. Geological Society of America, Boulder, pp 257–302
Janbu N (2005) The 1962 quick clay slide in Skjelstadmarka, Norway. In: Proceedings of the ICFL, pp 195–203
Jørstad FA (1968) Clay slides in Norway. Nor Geol J 22:214–219
Jostad HP, Andresen L (2002) Bearing capacity analysis of anisotropic and strain-softening clays. In: Taylor, Francis Proceedings of the 8th international symposium NUMOG VIII, Rome, pp 469–474
Jostad HP, Andresen L, Thakur V (2006) Calculation of shear band thickness in sensitive clays. In: Proceedings of the 6th numerical methods in geotechnical engineering. Graz, Austria, pp 27–32
Karlsrud K (1979) Skredfare og planlegging. Lecture notes NIF-Course, Hardanger
Karlsrud K (1983) Analysis of a small slide in sensitive clay in Fredrikstad, Norway. In: Symposium on slopes on soft clays, Linkoping, Swedish Geotechnical Institute Report No. 17, pp 175–184
Karlsrud K, Aas G, Gregersen O (1985) Can we predict landslide hazards in soft sensitive clays? Summary of Norwegian practice and experience, NGI publication 158. Norwegian Geotechnical Institute bulletins, Oslo, Norway
Kenney TC (1967) Slide behaviour and shear resistance of a quick clay determined from a study of the landslide at Selnes, Norway. NGI publication 76
L'Heureux JS (2012a) A study of the retrogressive behaviour and mobility of Norwegian quick clay landslides. In: Proceedings of the 11th INASL, vol 1, Banff, pp 981–988. CRC Press
L'Heureux JS (2012b) Characterization of historical quick clay landslides and input parameters for Q-Bing. NGI report nr 20120753-02-R. Available at www.naturefare.no
Lacasse S, Berre T, Lefebvre T (1985) Block sampling of sensitive clays. In: ISSMGE conference. ASCE, San Francisco, USA, pp 887–892
Larsen JO (2002) Some aspects of physical weather related slope processes. PhD thesis, NTNU
Lebuis J, Rissmann P (1979) Les coulées argileuses dans le région de Québec et de Shawinigan. In: Argiles sensibles, pentes instables, mesures correctives et coulées des régions de Québec et Shawinigan, Geological Association of Canada Guidebook, pp 19–40
Lebuis J, Robert JM, Rissmann P (1983) Regional mapping of landslide hazard in Quebec. In: Proceedings of the symposium slopes on soft clays, SGI report 17, Swedish Geotechnical Institute, Linkoping, Sweden, pp 205–262
Lefebvre G (1981) Strength and slope stability in Canadian soft clay deposits. Can Geotech J 18:420–442
Leroueil S (2001) Natural slopes and cuts: mov. and fail. mechanisms. Géotechnique 51(3):197–243
Leroueil S, Tavenas F, Le Bihan JP (1983) Propriétés caracteristiqués des argiles de I'est du Canada. Can Geotech J 20:681–705
Leroueil S, Locat J, Vaunat J et al (1996) Geotechnical characterization of slope movements. In: Taylor, Francis, Senneset K (ed) Proceedings of the 7th international symposium on landslides, Trondheim, pp 53–74
Locat J, Demers D (1988) Viscosity, yield stress, remolded strength, and liquidity index relationships for sensitive clays. Can Geotech J 25:799–806
Locat J, Lee HJ (2005) Subaqueous debris flow. In: Jakob M, Hungr O (eds) Debris-flow hazards and related phenomena. Springer, Chichester, pp 203–246. ISBN 3-540-20726-0
Locat P, Leroueil S, Locat J et al (2003) Characterization of a submarine flow-slide at Pointe-du-Fort, Saguenay Fjord, Quebec, Canada. In: Lykousis V, Sakellariou DD, Locat J (eds)

Proceeding of the 1st symposium on submarine mass movements and their consequences. Springer, Dordrecht, pp 521–529
Locat P, Leroueil S, Locat J (2008) Remaniement et mobilité des débris de glissements de terrain dans les argiles sensible de l'est du Canada. In: Proceedings of the 4th Canadian conference on geohazards: from causes to management. Presse de l'Université Laval, Québec, pp 97–106
Locat A, Leroueil S, Bernander S et al (2011) Progressive failures in eastern Canadian and Scandinavian sensitive clays. Can Geotech J 48:1696–1712
Longva O, Janbu N, Blikra LH et al (2003) The 1996 Finneidfjord Slide: seafloor failure and slide dynamics. Submarine Mass Movements and their Consequences. In: Proceedings first international symposium. Kluwer Academic Publishers, Dordrecht, pp 531–538
Mayerhof GG (1957) The Mechanics of flow slides in cohesive soils. Geotechnique 7(1):41–49
Mitchell RJ, Markell AR (1974) Flow slides in sensitive soils. Can Geotech J 11(1):11–31
Natterøy A (2011) Skredkatalog om kvikkleire. Semester Project, NTNU
Nordal S, Alen C, Emdal A et al (2009) Landslide in Kattamrka in Namsos 13. March 2009. Transportation Ministry, 2009, Report. ISBN 978-82-92506-71-4
Norem H, Locat J, Schieldrop B (1990) An approach to the physics and the modeling of submarine flowslides. Mar Geotechnol 9:93–111
NPRA Reports (1994) Geotechnical reports Xd866A, Ud 937A, project 603330
NVE (2010) Construction on brittle clays. Guidelines by NVE, Norway
NVE Report (2012) Geotechnical reports 33–2012, 34–2012, 40–2012
Quinn PE, Diederichs MS, Rowe K et al (2011) A new model for large landslides in sensitive clay using a fracture mechanics approach. Can Geotech J 48(8):1151–1162
Reite AJ, Sveian H, Erichsen E (1999) Trondheim frå istid til nåtid – landskapshistorie og løsmasser. Gråsteinen 5, NGU
NGI Report (1974) Geotechnical investigation of the Sem slide. NGI report 71082
Rygg N, Oset F (1996) The Balsfjord landslide. Landslides. In: Taylor, Francis, Senneset K (ed) Proceedings of the 7th international symposium on landslides, Trondheim, pp 573–577
Sevaldson RA (1956) The slide in Lodalen, October 6th, 1954. Géotechnique 6(4):167–182
Söderblom R (1974) A new approach to the classification of quick clays. SGI reports 55, pp 1–17
Tavenas F, Leroueil S (1981) Creep and failure in slopes in clays. Can Geotech J 18(1):106–120
Tavenas F, Flon P, Lerouil S et al (1983) Remolding energy and risk of slide retrogression in sensitive clays. In: Proceedings of the symposium on slopes on soft clays, Linköping, Swedish Geotechnical Institute, pp 423–454
Thakur V (2007) Strain localization in sensitive soft clays. PhD thesis, NTNU
Thakur V (2011) Numerically observed shear bands in soft sensitive clays. Geomech Geoengin 6(2):131–146
Thakur V, Degago SA (2012) Quickness of sensitive clays. Géotechnique Lett 2(3):87–95
Thakur V, Degago SA (2013) Disintegration energy of sensitive clays. Géotechnique Lett 3:20–25. doi:10.1680/geolett.12.00062
Thakur V, Oset F, Aabøe R et al (2012) A critical appraisal of the definition of Brittle clays (Sprøbruddmateriale). In: Taylor, Francis Proceedings of the 16th Nordic geotechnical meeting, vol 1, Copenhagen, pp 451–462
Trak B, Lacasse S (1996) Soils susceptible to flow slides and associated mechanisms. In: Taylor, Francis Proceedings of the 7th international symposium on landslides, vol 1, Trondheim, pp 497–506
Trondheim Municipality Reports (1981) 0049, 0168, 1168–2, 1409
Vaunat J, Leroueil S (2002) Analysis of post-failure slope movements within the framework of hazard and risk analysis. Nat Hazards 26:83–102
Yong RN, Tang KY (1983) Soil remolding and sensitivity measurements. Geotech Test J 6(2):21–29

Chapter 9
Controls on the Dimensions of Landslides in Sensitive Clays

Marten Geertsema and Jean-Sébastien L'Heureux

Abstract The dimensions of landslides are subject to geometric, material and external controls. Geometric controls include: slopes and orientation of the ground surface and bedding planes, valley width and geometry, depth of failure, plus the presence and incision depth of bounding streams. Material controls include undrained and remoulded shear strength, sensitivity, stability number, remoulding index, rapidity, and the mobility of the depleted mass. Dense networks of trees can influence mobility. The position of layers prone to landsliding in the slope is important.

Keywords Sensitive clay • Flow • Spread • Geometric control • Material control • Stability number • Shear strength • Mobility • Travel angle • Retrogression

9.1 Introduction

Sensitive clays are most commonly associated with low elevation, flat-lying glacio-marine sediments in previously glaciated areas. They occur in the valley bottoms of uplifted fjords such as those in Scandinavia (Chap. 29 by Hansen et al., this volume) and British Columbia, or in the sediments of larger Holocene ice-contact seas such as the Champlain Sea which covered significant parts of southern Ontario and Quebec. Especially in mountainous areas, steep topography may relegate human settlements and infrastructure to these valley bottom settings.

M. Geertsema (✉)
Ministry of Forests, Lands, and Natural Resource Operations, Prince George, BC, Canada
e-mail: marten.geertsema@gov.bc.ca

J.-S. L'Heureux
Norwegian Geotechnical Institute (NGI), Trondheim, Norway
e-mail: jsl@ngi.no

J.-S. L'Heureux et al. (eds.), *Landslides in Sensitive Clays: From Geosciences to Risk Management*, Advances in Natural and Technological Hazards Research 36,
DOI 10.1007/978-94-007-7079-9_9, © Springer Science+Business Media Dordrecht 2014

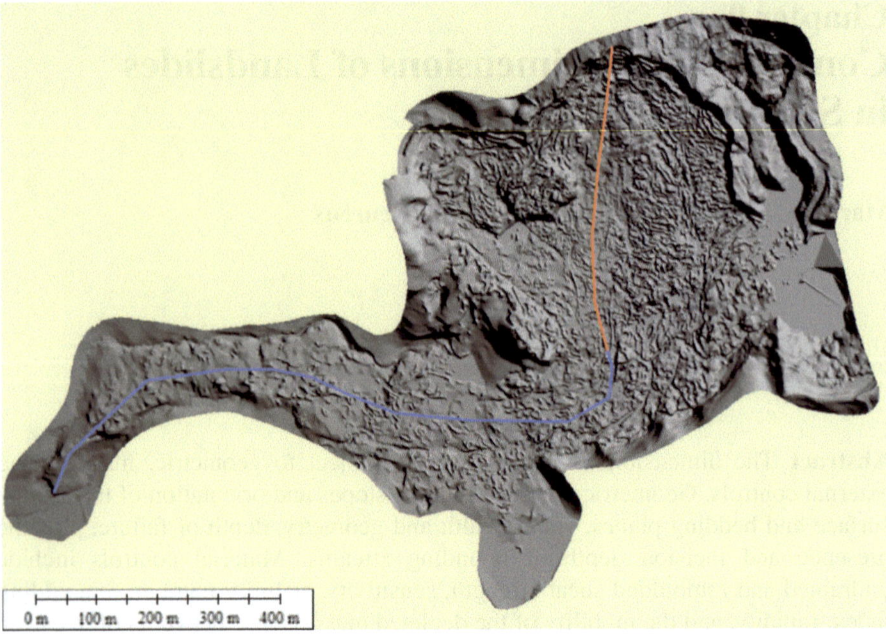

Fig. 9.1 Digital elevation model of the Mink Creek landslide near Terrace, British Columbia. Distance of retrogression (*orange line*), measured along the length of the zone of depletion (from the edge of the original slope to the crown) is some 500 m. Runout length (*blue line*), a measure of the central path of the zone of accumulation is 1,000 m, yielding a total travel distance of 1,500 m

Large, rapid landslides in flat-lying terrain are puzzling phenomena – usually unexpected and difficult to predict. The smaller end members include shallow movements on steep banks and rotational slides. The larger movements include flows and spreads (see Cruden and Varnes 1996 for terminology).

Terrain hazard maps, important for land-use zonation, require information about how far landslides might retrogress or encroach into valley sediments, and also about how far the displaced material might travel. We break landslide length into two components (c.f. L'Heureux 2012). Distance of retrogression (or penetration) is a measure of the length of the zone of depletion. The length of the zone of accumulation is a measure of the travel distance of displaced material beyond the original slope (Fig. 9.1). Taken together these values yield a total travel distance from which a travel angle can be calculated (Geertsema and Cruden 2008).

In this paper we discuss factors that control the dimensions of landslides. We review four criteria, the first two dealing with material properties, and the second two with geometry and topography. The first criterion is the presence of sensitive clay at depth. Changes in material type or decreases in sensitivity can limit landslide size. Secondly, the character of the displaced material influences its mobility, and hence its ability to move out of the zone of depletion. Thirdly, there has to be somewhere for the displaced landslide material to go to prevent buttressing of the back scarp. Finally, the inclination of the rupture surface influences potential landslide size. We draw heavily on reviews by Geertsema and Schwab (1997) and L'Heureux (2012).

9.2 Material Controls

The preconditions for the development of sensitive soils are summarized by Torrance (1983). Essentially deposition of low activity clays and silts as floccules into a saline waterbody, slow load increase, subaerial uplift, and leaching by freshwater to a salt content of less than 1 g/l, are required.

9.2.1 Shear Strength and Sensitivity

Sensitivity (S_t) – the ratio of undisturbed (s_u) to remoulded shear strength (s_r) is one of the underpinning requisites for the development of large rapid landslides in glaciomarine sediments. The material has to transform from a relatively strong or rigid state to a much weaker state in brittle fashion.

Landslide kinematics relate to material properties, and the type of movement may control their dimensions. Spreads for example, tend to retain a thicker cover of displaced material in their zones of depletion than do flows, indicating a lower mobility. Geertsema et al. (2006) show that spreads have higher s_u values than flows (Fig. 9.2). In Fig. 9.3 we plot landslide data from L'Heureux (2012) that show that flows have larger travel distances than spreads per unit landslide volume. It follows then that s_u is perhaps a crude predictor of relative landslide dimension.

The ability of the clay to be remoulded during a landslide controls the dimensions of landslide in sensitive clays. This process depends on the mechanical properties of the clay and on the available potential energy (i.e. the height of the slope). Because of this, the distance of retrogression has previously been related to Janbu's stability number Ns (Mitchell and Markell 1974; Trak and Lacasse 1996). Using this concept, these authors found that the stability number ($Ns = \gamma H/s_u$) needs to exceed the value 6 for earthflows to occur (where γ the bulk unit weight of the soil, H is the height of the slope and s_u is the average undrained shear strength). The distance of retrogression increased positively with the stability number. They suggest that the distance of retrogression is related to the height of the slope above a surface of rupture, to the stability number Ns, and to the liquidity index of the clay. Note that the Ns threshold is likely higher than 6 in Norway (Fig. 9.4).

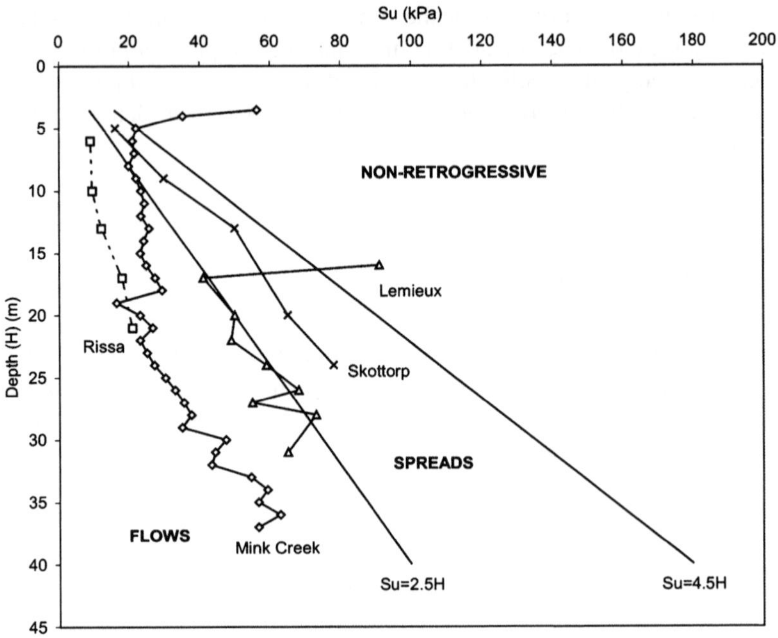

Fig. 9.2 Plot of undisturbed shear strength (s_u) with depth (Geertsema et al. 2006). Spreads plot between flows and non-retrogressive landslides, with flows having the lowest s_u values

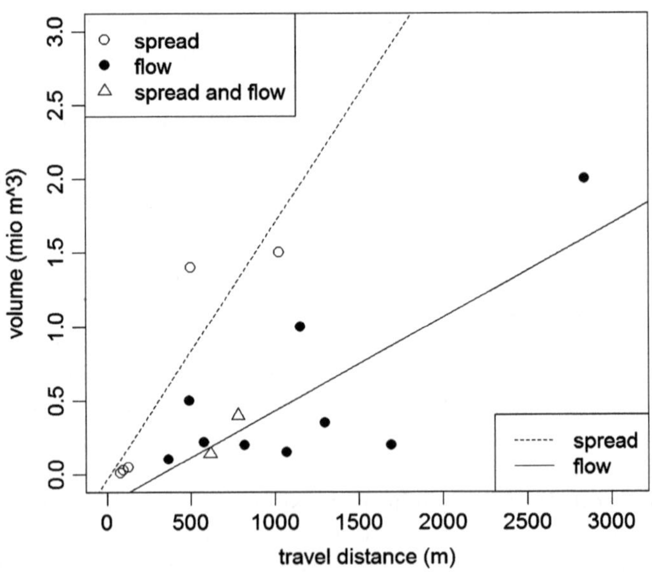

Fig. 9.3 Plot of travel distance (retrogression plus runout) against volume for flows and spreads. Here flows have greater travel distances per unit volume than spreads (Select data from Table 1 in L'Heureux (2012) for cases where both retrogression and runout was reported)

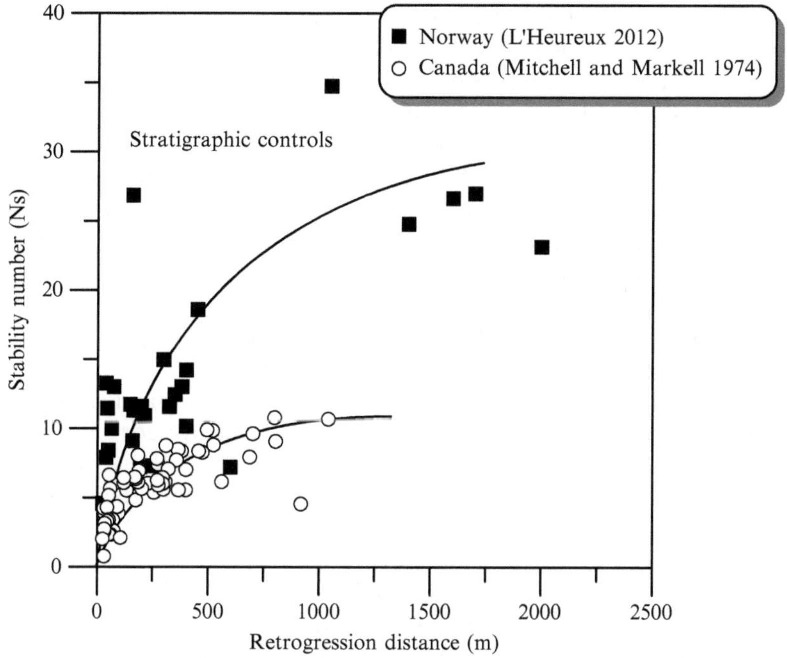

Fig. 9.4 Stability number (Ns) plotted against distance of retrogression (Note that Norwegian Ns values are much higher than those from eastern Canada)

Mitchell (1978) proposed that for sensitivities (determined by field vane) greater than 10 the criterion R = 100 (Ns–4) could be applied for regional planning and route selection in Canada, where R is the distance of retrogression. This criterion appears to fit well for the 1993 Lemieux slide on the South Nation River in Ontario (Evans and Brooks 1994).

Lebuis and Rissman (1979) and Lebuis et al. (1983) employ empirical correlations of remoulded shear strength, sensitivity, and plastic limit with distance of retrogression. They show that low remoulded strengths (less than 1 kPa as measured by fall cone), and sensitivities greater than 25 are necessary for retrogression to exceed 100 m. They also show that most large earthflows have associated plasticity indices below 20 and liquidity indexes above 1.2. Similar results are found for Norwegian earthflows as shown in (L'Heureux 2012; Chap. 8 by Thakur et al., this volume).

9.2.2 Remoulding Energy and Rapidity

Essentially for a landslide in sensitive clays to attain large dimensions, displaced material has to move rapidly out of the zone of depletion. The long run-out distances observed for landslides in sensitive clays have previously been related to the low remoulded shear strength (or viscosity) of the soil following failure. The mobility of most flows in such material is acquired at the time of failure as some energy is available for remoulding. Tavenas et al. (1983) use an energy dissipation approach to assess the risk of retrogression in Champlain Sea clays. They include the rapidity of the clay in their calculations, a concept introduced by Söderblom (1974, 1983), which represents the ease with which clay can be brought to its remoulded state (this concept has also been suggested by Carson 1977). They considered that rapidity actually represents the energy required to remould the clay. Tavenas et al. (1983) stress that for retrogression to occur, not only must the initial slope and the temporary backscarps become unstable, but the landslide debris must remould rapidly to flow away from a high temporary backscarp, keeping its toe unloaded. From these two conditions they proposed four criteria.

First, H/s_u must be greater than or equal to Ns/OCR, for initial slope failure to occur (where Ns is Janbu's stability factor and OCR is the overconsolidation ratio). Second, H/Su must be greater than or equal to Ns, where Ns is close to 4, for continued retrogression to occur. Third, they used criteria from Lebuis and Rissman (1979), where the remoulded shear strength must be less than 1 kPa, or the liquidity index greater than 1.2, for the slide mass to flow away from the backscarp if it becomes remoulded. Their fourth criterion deals with the ability of clay to become sufficiently remoulded during failure for retrogression to occur. They showed experimentally that the degree of remoulding (defined by a remoulding index Ir) varied differently as a function of the normalized remoulding energy (W_N) between clays. They further suggest that for retrogression to occur the Liquid Limit should be less than 40 % and Ir needs to exceed 70 % for $10W_N$. This criterion implies that increases in slope, stability number, and energy will decrease the strength of remoulded landslide debris and consequently increase the risk of retrogression (Tavenas 1984).

Locat (1992) has studied relationships between viscosity, yield strength, and mud flow mobility of sensitive clays. He showed how slope angle and critical flow depth varied with yield strength (essentially remoulded strength). Although he did not do so, this relationship could be applied to critical states where mud would either flow away from or support an unstable backscarp, and thus it could possibly be applied to prediction of the distance of retrogression, and may be particularly useful in cases where sloping slip surfaces are abundant.

Trees also can play a role in strengthening (increasing cohesion) of displaced material, limiting its mobility. Geertsema et al. (2006) made this observation at the Mink Creek landslide in British Columbia. Here displaced material formed a wall against an interlocking network of coniferous tree stems (Fig. 9.5).

The environment in which landslides occur may in some cases directly affect the mobility of the masses. L'Heureux (2012) showed that channelized landslide events or landslides entering larges water bodies (i.e. fjords and/or lakes) often show larger runout distances than those constrained on land. When entering rivers, lakes and/or fjords, the low permeability of the clays may ensure water entrapment below the landslide masses resulting in hydroplaning and longer run-out distances (Fig. 9.6). Water incorporation also helps remoulding the clay material.

Fig. 9.5 Interlocking networks of large trees add cohesion to the displaced mass (locally) at the Mink Creek landslide (Note how the displaced material has piled up against trees (*arrow*))

9.3 Geometric and Topographic Controls

9.3.1 Topographic Controls

Carson and Lajoie (1981) described various topographic factors that influenced the extent of retrogression. Bjerrum et al. (1969), Mitchell and Markell (1974), Carson (1979), Lebuis et al. (1983), Tavenas et al. (1983), Karlsrud et al. (1984), Vaunat et al. (1992), and Lefebvre (1996) give brief discussions of stratigraphic and topographic controls of retrogression. In addition to discontinuities in material type, such as till, bedrock, or less sensitive clay, topography also limits landslide dimensions. Topography that constricts displaced material, such as a narrow valley, may allow the displaced material to support the back scarp (Carson 1977; Carson and Lajoie 1981). Landslides entering wide valleys, on the other hand, or at subparallel angles to the downstream river axis in a river bend, do not have the same constrictions, and thus continued removal of displaced material from the backscarp allows retrogression to ensue (Fig. 9.7). In these cases, if the displaced material liquefies, excess retrogression may be expected (Carson and Lajoie 1981). Large retrogressions are often observed in the case where landslide enters a larger water basin such as a lake or a fjord (see also Chaps. 1 and 12 by L'Heureux et al., this volume). In such cases, the gently sloping fjord/lake bottom enables excessive landslide debris accommodation. This is well exemplified for the 1978 Rissa landslide in Fig. 9.6.

Another control of retrogression includes gully erosion. Landslides at South Nation River (Fig. 9.8) (Eden et al. 1971), Lemieux (Brooks et al. 1994), and Mink

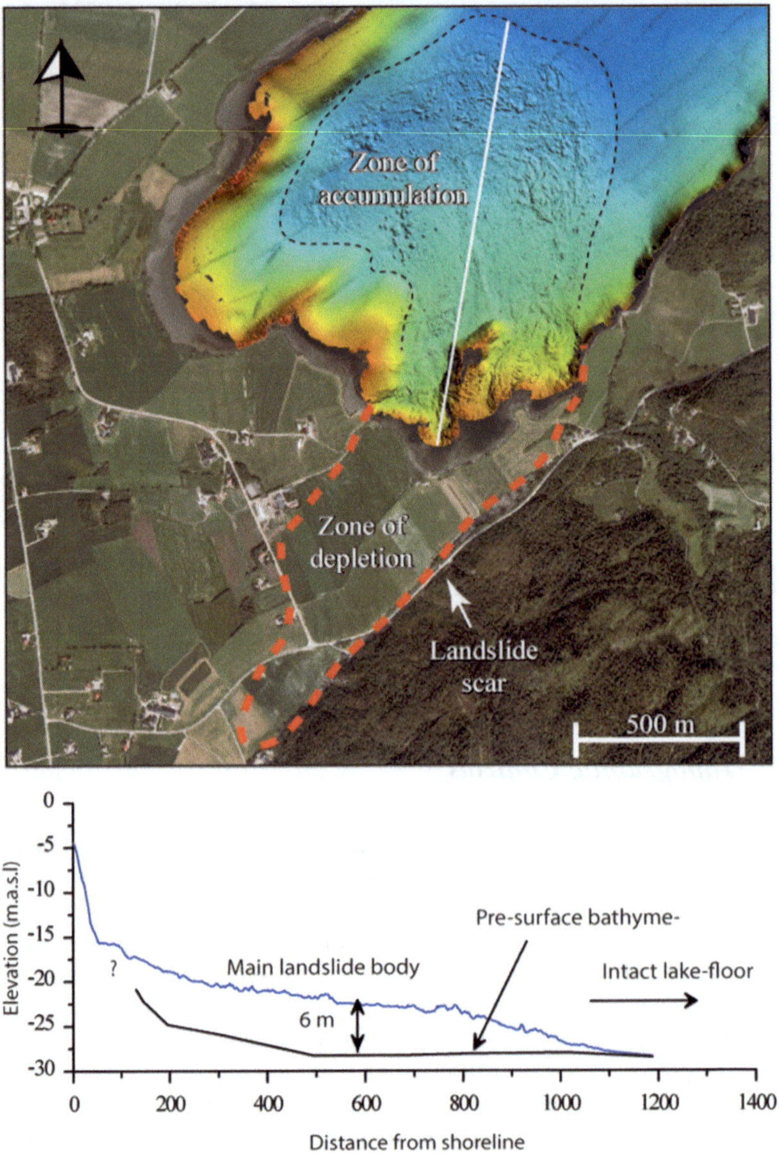

Fig. 9.6 Subaqueous morphology of the 1978 Rissa landslide in Lake Botn. The lower panel presents a slope profile prior and after the 1978 landslide. The thickness of the landslide debris was evaluated using seismic reflection data in Lake Botn (After L'Heureux et al. 2012a)

Creek (Geertsema et al. 2006) are all bounded by pre-failure gullies. The gullies pose barriers in three ways: (1) by providing a negative topography which reduces lateral earth pressure (and reduces Ns); (2) by developing a deeper strong weathered crust (in the interfluve areas between gullies); and (3) by lowering the water-table.

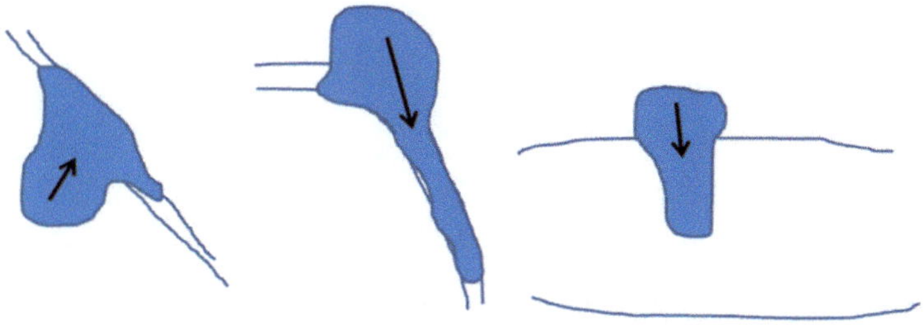

Fig. 9.7 Topographic constrains such as a narrow valley with its axis normal to landslide movement may limit the removal of displaced material and thus support a landslide back scarp. Movement at a river bend in the down-valley direction does not restrict displaced material, and neither does a landslide depositing into a large water body such as a lake

Fig. 9.8 The 1971 South Nation River landslide. This rapid spread was contained laterally, by pre-failure gullies (Canada Airphoto A30362: 152)

9.3.2 Inclination of the Rupture Surface

Carson (1977, 1979) and Carson and Lajoie (1981) include the slopes of both pre-failure surface and the surface of rupture in formulae designed to predict the distance of retrogression, R, with respect to the depth to the rupture surface, H. They

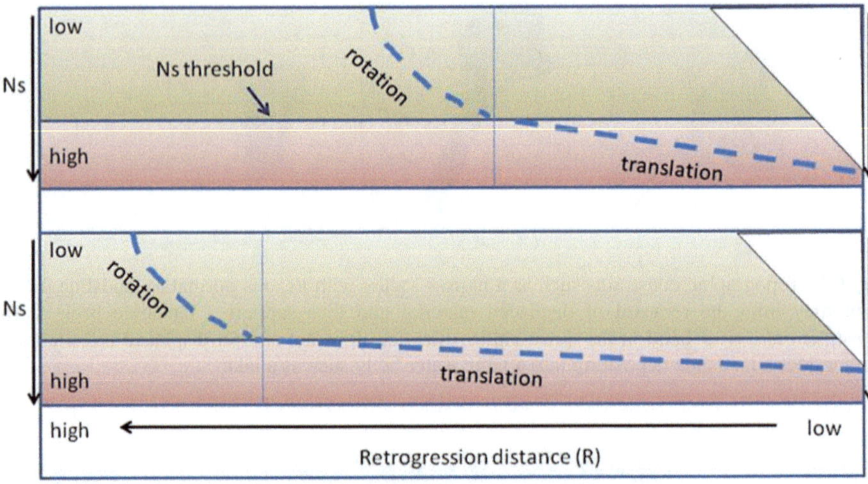

Fig. 9.9 Influence of the inclination of the rupture surface (often bedding) on distance of retrogression. Steeper inclinations intersect the Ns threshold over a shorter distance than do more gently sloped rupture surfaces. Once the threshold is reached movement transforms from translation to rotation. The intersection of the rotational surface with the ground surface is generally the crown of the landslide

show how low remoulded shear strengths influence whether or not retrogression will occur, and how sensitivity plays a role in continued retrogression. Indeed, Carson (1979, Fig. 13 page 441) plots R/H values for incremental S and Ns for different pre surface and rupture surface gradients. For some low gradient values R/H essentially becomes infinite. Here landslides will continue until they encounter a material change such as bedrock in the valley wall. Such was the case at the Lakelse landslides near Terrace, BC (Geertsema and Schwab 1997).

Similarly the orientation of the rupture surface, can either assist or resist the retrogressive process depending on whether it dips towards or away from the valley. While a bedding plane or surface of rupture sloping towards the valley bottom can facilitate removal of displaced material by adding a gravitational force to the slide mass, such a slope will also limit retrogression by decrease in backwall height during retreat (reduction of a threshold Ns (Fig. 9.9)) and can abort retrogression by inclining the rupture surface into more resistant material.

With inclined slip surfaces (once certain threshold conditions for retrogression are met) the geometry, the dip and the extent of the pre-failure surface are the dominant factors controlling the extent of landslide penetration. A further increase in sensitivity, liquidity index, or remoulding index, or a further decrease in remoulded shear strength, would cause no further increase of retrogression in such geometrically controlled flows. Flows along Norwegian fjords are often controlled by pre-existing, regional, weak layers (L'Heureux et al. 2012b, 2013). In such case the length of retrogression seems to be in part controlled by the changes in inclination for these layers.

9.4 Conclusions

Predicting the dimensions of landslides in sensitive sediments is complicated, by a number of factors including material properties, geometry and topography. Sensitivity, shear strength, remoulded shear strength, liquidity indices, degree of remoulding, and rapidity all control the mobility of the displaced material and its ability to leave the zone of depletion. The incorporation of thick networks of trees can reduce the mobility of displaced material, whereas run-out into waterbodies can increase mobility. Sometimes the main control of landslide dimension is another material type. Geometry and topography are also important. Narrow valleys can restrict the movement of displaced material from the zone of depletion, and provide toe support to a temporary back scarp, limiting retrogression. Similarly both the direction and degree of inclination of the rupture surface along bedding planes is important. Dip-slope direction favours movement along a bedding plane. If a rupture surface along bedding is steep threshold stability numbers will be achieved earlier, halting landslide penetration.

Acknowledgements We are grateful to Professor Dave Cruden for a constructive review that improved the manuscript.

References

Bjerrum L, Løken T, Heiberg S, Foster R (1969) A field study of factors responsible for quick clay slides. In: Proceedings of the 7th ICSMFE, vol 2, Mexico City, pp 531–540

Brooks GR, Aylsworth JM, Evans SG, Lawrence DE (1994) The Lemieux landslide of June 20, 1993, South Nation Valley, southeastern Ontario – a photographic record. Geological Survey of Canada, Misc. Report 56

Carson MA (1977) On the retrogression of landslides in sensitive muddy sediments. Can Geotech J 14:582–602

Carson MA (1979) On the retrogression of landslides in sensitive muddy sediments: reply. Can Geotech J 16:431–444

Carson MA, Lajoie G (1981) Some constraints on the severity of landslide penetration in sensitive deposits. Géographie physique et Quaternaire XXXV:301–316

Cruden DM, Varnes DJ (1996) Landslides types and processes. In: Turner AK, Schuster RL (eds) Landslides investigation and mitigation, Special report 247, Transportation, Research Board, National Research Council. National Academy Press, Washington, DC, pp 37–75

Eden WJ, Fletcher EB, Mitchell RJ (1971) South Nation River landslide, 16 May 1971. Can Geotech J 8:446–451

Evans SG, Brooks GR (1994) An earthflow in sensitive Champlain Sea sediments at Lemieux, Ontario, June 20, 1993, and its impact on the South Nation River. Can Geotech J 31:384–394

Geertsema M, Cruden DM (2008) Travels in the Canadian Cordillera. In: Proceedings of the 4th Canadian conference on geohazards. Quebec City, pp 383–390

Geertsema M, Schwab JW (1997) Retrogressive flowslides in the Terrace-Kitimat, British Columbia area: from early post-deglaciation to present – and implications for future slides. In: Proceedings of the 11th Vancouver geotechnical society symposium, pp 115–133

Geertsema M, Cruden DM, Schwab JW (2006) A large, rapid landslide in sensitive glaciomarine sediments at Mink Creek, northwestern British Columbia, Canada. Eng Geol 83:36–63

Karlsrud K, Aas G, Gregersen O (1984) Can we predict landslide hazards in soft sensitive clays? Summary of Norwegian practice and experiences. In: Proceedings of the 4th international symposium on landslides, vol 1, Toronto, 16–21 September 1984. University of Toronto Press, Toronto, pp 107–130

L'Heureux J-S (2012) A study of the retrogressive behaviour and mobility of Norwegian quick clay landslides. In: Eberhardt E, Froese C, Turner AK, Leroueil S (eds) Landslide and engineered slopes: protecting society through improved understanding. Taylor & Francis Group, London, pp 981–988

L'Heureux J-S, Eilertsen RS, Glimstad S, Issler D, Solberg I-L, Harbitz CB (2012a) The 1978 quick clay landslide at Rissa, mid-Norway: subaqueous morphology and tsunami simulations. In: Mosher D, Shipp C, Moscardelli L (eds) Submarine mass movements and their consequences, Advances in natural and technological hazards research, 31. Springer, Dordrecht, pp 507–516

L'Heureux J-S, Longva O, Steiner A, Hansen L, Vardy ME, Vanneste M, Haflidason H, Brendryen J, Kvalstad TJ, Forsberg CF, Chand S, Kopf A (2012b) Identification of weak layers and their role for the stability of slopes at Finneidfjord, northern Norway. In: Yamada Y et al (eds) Submarine mass movements and their consequences, vol 31, Advances in natural and technological hazards research. Springer, Dordrecht, pp 321–330

L'Heureux J-S, Longva O, Hansen L, Vanneste M (2013) The 1930 landslide in Orkdalsfjorden: morphology and failure mechanism. In: Proceedings of the 6th international symposium on submarine mass movements and their consequences. Kiel, September 2013

Lebuis J, Rissmann P (1979) Earthflows in the Quebec and Shawnigan areas. Ministère des Richesses naturelles, Québec, pp 18–38

Lebuis J, Robert J-M, Rissmann P (1983) Regional mapping of landslide hazard in Québec. In: Proceeings of the symposium on slopes on soft clays, Linkoping, Swedish Geotechnical Institute Report, no. 17, pp 205–262

Lefebvre G (1996) Soft sensitive clays. In: Turner AK, Shuster RL (eds) Landslides investigation and mitigation, Special report 247, TRB, National Research Council. National Academy Press, Washington, DC, pp 607–619

Locat J (1992) Viscosity, yield strength, and mudflow mobility for sensitive clays and other fine sediments. In: Proceedings of the 1st Canadian conference on geotechnique and natural hazards, Vancouver, pp 389–396

Mitchell RJ (1978) Earthflow terrain evaluation in Ontario. Ontario Ministry of Transportation and Communications, Toronto, 30 pp

Mitchell RJ, Markell AR (1974) Flowsliding in sensitive soils. Can Geotech J 11:11–31

Söderblom R (1974) A new approach to the classification of quick clays. Swedish Geotechnical Institute reprints and preliminary reports no. 55, pp 1–17

Söderblom R (1983) Studies of the rapidity number: can this number be used to determine the slide tendency of a clay? In: Bergren B, Lindgren J (eds) Symposium on slopes on soft clays. Swedish Geotechnical Institute. Report No. 17, pp 381–395

Tavenas F (1984) Landslides in Canadian sensitive clays – a state-of-the-art. In: Proceedings of the 4th international symposium on landslides, vol 1, Toronto, 16–21 September 1984. University of Toronto Press, Toronto, pp 141–153

Tavenas F, Flon P, Leroueil S, Lebuis J (1983) Remoulding energy and risk of slide retrogression in sensitive clays. In: Proceedings of the symposium on slopes on soft clays, Linköping. Swedish Geotechnical Institute, SGI Report No. 17, pp 423–454

Torrance JK (1983) Towards a general model of quick clay development. Sedimentology 30:547–555

Trak B, Lacasse S (1996) Soils susceptible to flow slides and associated mechanisms. In: Proceedings of the seventh international symposium on landslides, vol 1, Trondheim, Norway, Balkema, Rotterdam, pp 497–506

Vaunat J, Leroueil S, Tavenas F (1992) Hazard and risk analysis of slope instability. In Proceedings of 1st Canadian symposium on géotechnique and natural hazards (GEOHAZARD'S 92), Vancouver, pp 397–404

Chapter 10
Prehistoric Sensitive Clay Landslides and Paleoseismicity in the Ottawa Valley, Canada

Gregory R. Brooks

Abstract The ages of 39 large retrogressive landslides and three areas of disturbed terrain in the Ottawa Valley were compiled using 122 published and unpublished radiocarbon ages. The chronological dataset includes 15 confined-valley landslides, 23 scarp-side landslides, and four special case features (one massive landslide, three disturbed terrain areas). The ages of the features range from 'modern' to 8,000 ^{14}C cal BP. Distinct clusters of 10 and 11 coincidentally-aged landslides at ~1,000 and ~5,150 ^{14}C cal BP are the groups of landslides previously interpreted to have been triggered by paleoearthquakes. Scarp-side landslides with scars between 0.1 and 10 km^2 are the dominant failure morphology forming the two age clusters and constitute an important component of the interpreted paleoseismic evidence. Five of the confined-valley landslides are part of the ~1,000 ^{14}C cal BP cluster, but the other ten failures are of widely varying ages. One of the special case features, a massive landslide originating from a source area of ~20 km^2, falls within the ~1,000 ^{14}C cal BP cluster. Radiocarbon ages representing the age of the Treadwell and Wendover disturbed terrains, suggest that these areas are contemporary with the Lefaivre disturbed area at ~7,900 ^{14}C cal BP, but this is not an unequivocal interpretation. Notwithstanding sampling bias within the dataset, it is inferred from the high proportion of dated landslides falling within the two interpreted paleoearthquake clusters that there is a 'strong' paleoseismic signature within the temporal pattern of landsliding within the Ottawa Valley. Based on the Ottawa Valley dataset, scarp-side landslides with preserved debris fields and scars greater than 0.1 km^2, and landslides in general with scars greater than 1 km^2, are more promising targets than confined-valley landslides for paleoseismic studies.

Keywords Landslide • Sensitive sediments • Radiocarbon chronology • Paleoearthquake • Holocene • Champlain Sea • Canada

G.R. Brooks (✉)
Natural Resources Canada, Geological Survey of Canada,
601 Booth Street, Ottawa, ON, Canada K1A 0E8
e-mail: gbrooks@NRCan.gc.ca

J.-S. L'Heureux et al. (eds.), *Landslides in Sensitive Clays: From Geosciences to Risk Management*, Advances in Natural and Technological Hazards Research 36,
DOI 10.1007/978-94-007-7079-9_10, © Springer Science+Business Media Dordrecht 2014

10.1 Introduction

The occurrence of rapid landslides is generally considered to be climatically triggered, but this is not always the case, as temporally and spatially well-clustered prehistoric landslides have been interpreted to be triggered by paleoearthquakes (e.g., Crozier 1992; Borgatti and Soldati 2010). The study of prehistoric landslides is an established method for investigating regional paleoseismicity and can produce estimates of paleoearthquake magnitude and epicenter location, thus contributing to paleoseismic hazard assessment (see Jibson 2009). A key step using landslides in a paleoseismic study is establishing a landslide chronological record.

Large areas of the St. Lawrence Lowlands-Ottawa Valley region are underlain by fine-grained glaciomarine silty-clay and clayey-silt deposits that can be geotechnically sensitive and prone to rapid, large (>1 ha) earth flows and earth spreads (e.g., Chagnon 1968; Mitchell and Markell 1974). This style of failure can be triggered by seismic shaking, as occurred during the recent AD 2010 Val-des-Bois (M_w 5.0), and AD 1988 Saguenay (M_w 5.9) earthquakes (Lefebvre et al. 1992; Locat 2011).

Aylsworth et al. (2000) and Brooks (submitted) hypothesized that paleoearthquakes occurred in the Ottawa Valley at ~5,120 and ~1,020 ^{14}C cal BP, respectively, on the basis of clusters of coincidentally-aged, retrogressive, sensitive clay landslides, but with due consideration of geomorphic settings and other triggering mechanisms. The purpose of this paper is to place the interpreted paleoearthquake-triggered landslides into the context of the available sensitive clay landslide chronological record for the Ottawa Valley area (Fig. 10.1). Morphological consistencies

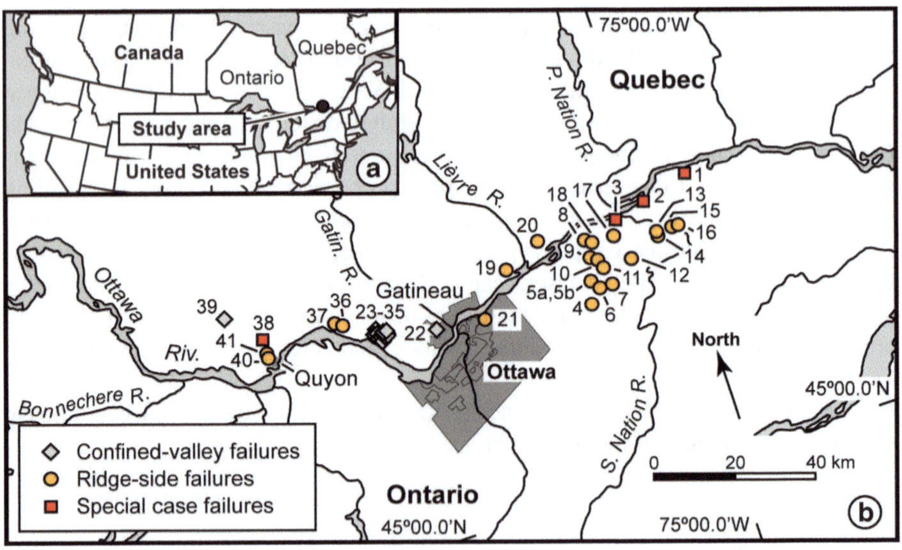

Fig. 10.1 (**a**) Map showing the general location of the Ottawa Valley study area in eastern Canada. (**b**) Map showing the locations of the 42 dated landslide and disturbed terrain sites within the Ottawa Valley with each classified in the confined-valley, scarp-side or special case groups. Symbol size does not reflect the feature scale. Site numbers are keyed to the landslides/features listed in Table 10.1

are identified from the two landslide clusters to allow better targeting of undated landslides that are more likely to have been triggered by paleoearthquakes. Also, the proportion of landslides in the chronological dataset that are attributed to a paleoearthquake trigger is discussed in terms of an apparent landslide paleoseismic signature. The conclusions are provisional, however, since they are based on an incomplete chronological record.

10.2 Study Area

The Ottawa Valley, south-eastern Ontario/south-western Quebec, Canada, is located within the Ottawa-Bonnechere graben system and is a physiographic extension of the St. Lawrence Lowland (Fig. 10.1a). Much of the valley was overlain by thick glaciomarine-marine and fluvial-deltaic deposits that aggraded within the Champlain Sea, which inundated the area between 13.9 and 11.5 ka ^{14}C cal BP (see Cummings et al. 2011). Following the marine recession, a proto-drainage network developed along which the ancestral Ottawa River eroded major terraces and several large (paleo)channels, and its tributaries incised narrow valleys (see Aylsworth et al. 2000). Approximately 250 landslide polygons appear on surficial geology maps in the Ottawa Valley that denote individual or closely-spaced, large, retrogressive landslides (see St. Onge 2009). The Ottawa Valley is located within the West Quebec Seismic Zone (WQSZ), which encompasses parts of western Quebec, eastern Ontario, and northern New York state (see Lamontagne 2010). Significant historic seismic events in the WQSZ include the AD 1732 Montreal (M 5.8), AD 1935 Temiscaming (M 6.1) and AD 1944 Cornwall (M 5.8) earthquakes.

10.3 Landslides in the Ottawa Valley

For the purposes of this paper, landslides and 'landslide-like' features are grouped into three general categories:

<u>Confined-valley landslides:</u> Many retrogressive landslides in the region are clustered along the sides of narrow river or creek valleys (Fig. 10.2a; Fransham et al. 1976). The confined stream courses locally erode the valley sides, which have had a role in triggering many of the failures (e.g., Torrance 2012). Once initiated, retrogression cuts rapidly back into a valley-side slope, forming a well-developed scar, most of which are less than 1 km^2. Well-developed clusters of landslides in confined-valley settings are present along reaches of the South Nation River, Lièvre River, Gatineau River, Petit Nation River, Breckenridge Creek and Bonnechere River valleys (see St. Onge 2009). Isolated or small groups of confined-valley failures are present along reaches of many streams in the region.

<u>Scarp-side landslides:</u> A number of landslides occur along the margins of paleochannels and terraces of the Ottawa River that are eroded into Champlain Sea deposits (see Fransham et al. 1976; Aylsworth et al. 2000). The failures retrogressed into the scarp slopes, producing debris that splayed onto the

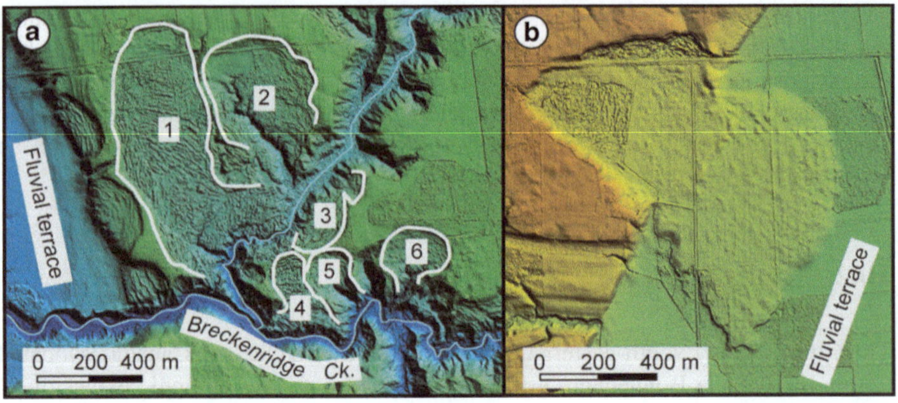

Fig. 10.2 (**a**) Lidar DEM showing six, closely-spaced confined-valley landslides along Breckenridge Valley. Note the three scarp-side landslide scars (to the left of scar 1) that open towards the fluvial terrace and have debris fields truncated at the edge of the scarp. (**b**) Lidar DEM of the Murphy Road scarp-side failure with a preserved debris field splayed on the adjacent fluvial terrace (Lidar DEMs, © Government of Quebec)

adjacent terrace or paleochannel surface. Preservation of the splayed debris ranges from uneroded (Fig. 10.2b) to partially eroded or completely truncated to the edge of the scarp slope (Fig. 10.2a), depending on whether the failure happened before or after the abandonment of the fluvial surface by the Ottawa River. Clusters of 'scarp-side failures' occur along the Bourget, Hammond and Plantagenet paleochannels, and along terraces on the north side of the Ottawa River (see St. Onge 2009).

Special case features (massive landslide and disturbed terrains): There are several large-scale landslides and landslide-like features in the region that warrant inclusion within a separate category than those above. One of these features is the Quyon Valley landslide (source area ~20 km^2), which resulted from a massive failure along both sides of the lower 11.5 km of Quyon Valley (landslide 38; Table 10.1; Fig. 10.1b; see Brooks submitted). There are also the 'disturbed terrains' at Lefaivre and Treadwell, covering ~56 and ~27 km^2, respectively, which include areas on both sides of the Ottawa River (features 1 and 2; Table 10.1; Fig. 10.1b; see Aylsworth et al. 2000; Aylsworth and Lawrence 2003). The origins of these two features, and the similar, but smaller disturbed area at Wendover (feature 3), are enigmatic and have been variously interpreted to be landslide deposits (Crawford 1961; Richard 1984), collapse features (Rodriguez et al. 1987), or "seismically disturbed" terrain (Aylsworth et al. 2000; Aylsworth and Lawrence 2003). The shallow subsurface deposits (up to ~50 m depth) within these areas are composed of variably deformed, folded, faulted and/or rotated deposits of glaciomarine silt and clay, and fluvial sand deposits, similar to sensitive clay landslide debris. For the purposes of discussion, these features are included in this paper.

Table 10.1 Summary radiocarbon chronological and morphological data for Ottawa Valley landslides and disturbed terrain features

Feature #	Landslide/feature location and name	^{14}C sample number	Selected ^{14}C age (BP), (# of dates avail.), 2σ calib. ^{14}C age range (cal BP)[b]	Landslide/feature type, scar area (km^2), dated material, age context, comment, data source
1	Lefaivre disturbed terrain	GSC-6470[a]	7,060±80, (2), 7,795–7,965	Sp. C., 56.4, wood, max. age[c,d]
2	Treadwell disturbed terrain	UCIAMS-71217	7,105±20, (3), 7,871–7,971	Sp. C., 27.8, macrofossil from base of peat, min. age[d]
3	Wendover disturbed terrain	UCIAMS-71211	7,140±20, (5), 7,937–8,004	Sp. C., 3.1.0, wood, min. age[c]
4	Mer Bleue paleoch., landsl. 1	Beta-127245	4,570±70, (1), 4,979–5,467	Scarp S., 1.6 , wood, min. age[c]
5a	Hammond paleoch., landsl. 2	Beta-90880	4,470±80, (1), 4,874–5,308	Scarp S., 0.8, wood, max. age[c]
5b	Hammond paleoch., landsl. 2	Beta-90881	3,050±70, (1), 3,007–3,438	Scarp S., 0.8, wood, max. age[c]
6	Hammond paleoch., landsl. 3	Beta-122473	4,590±40, (1), 5,055–5,460	Scarp S., 0.24, wood, min. age[c]
7	Mer Bleue paleoch., landsl. 4	Beta-122475	2,760±50, (1), 2,759–2,963	Scarp S., 0.05, wood, min. age[c]
8	Bourget paleoch.l, landsl. 5	Beta-127281	5,130±60, (1), 5,726–5,995	Scarp S., 0.6, wood, max. age[c]
9	Bourget paleoch., landsl. 6	Beta-127284	4,440±80, (1), 4,867–5,295	Scarp S., 1.38, wood, max. age[c]
10	Bourget paleoch., landsl. 7	Beta-127243	4,450±70, (2), 4,875–5,291	Scarp S., 2.1, wood, max. age[c]
11	Bourget paleoch., landsl. 8	Beta-122472	4,520±50, (1), 4,979–5,315	Scarp S., 3.1, organic materials, min. age[c]
12	Bourget paleoch, landsl. 9	Beta-127282	4,540±90, (1), 4,881–5,466	Scarp S., 2.3, wood, max. age[c]
13	Plantagenet paleoch, landsl. 10	Beta-127283	4,530±60, (1), 4,974–5,442	Scarp S., 0.17, wood, max. age[c]
14	Plantagenet paleoch, landsl. 11	Beta-122477	4,450±50, (2), 4,878–5,288	Scarp S., 0.07, seed, max. age[c]
15	Plantagenet paleoch, landsl. 12	Beta-122471	1,870±40, (1), 1,712–1,890	Scarp S., 1.0, organic materials, min. age[c]
16	Plantagenet paleoch, landsl. 13	Beta-127242	4,820±70, (1), 5,326–5,709	Scarp S., 0.38, wood, max. age[c]
17	Ottawa R. terrace, landsl. 14,	Beta-122474	4,470±50, (1), 4,891–5,302	Scarp S., 5.0, wood, min. age[c]
18	Bourget paleoch, landsl. at Rampage and Baseline roads	Beta-122476	Modern, (1), na	Scarp S., 2.5, wood, min.(?) age[d]
19	Angers landslide, Angers	GSC-6822[a]	6,140±70, (2), 6,945–7,160	Scarp S., 2.0, wood, max. age[e]
20	Masson landslide, Masson	GSC-1922[a]	4,620±80, (1), 5,087–5,469	Scarp S., 2.0, wood, max. age[f]
21	Beacon Hill landslide, Ottawa	GSC-550[a]	1,140±150, (1), 938–1,176	Scarp S., 0.2, wood, max. age[g]
22	Jomonville Street landsl., Hull	GSC-1741[a]	120±150, (1), –4 to 285	Conf., 1.0, wood, max.(?) age[f, h]
23	Breckenridge V., scar 1? landsl.	UCIAMS-88797	1,115±15, (3), 974–1,058	Conf., 3.0, wood, max. age[d, i]
24	Breckenridge V., scar 2 landsl.	UCIAMS-88802	3,390±15, (10), 3,581–3,690	Conf., 0.16, wood (same max. and min. ages[d]
25	Breckenridge V., scar 3? landsl.	GSC-6246[a]	6,980±80, (5), 7,702–7,929	Conf., 0.04, wood, max. age[d]

(continued)

Table 10.1 (continued)

Feature #	Landslide/feature location and name	^{14}C sample number	Selected ^{14}C age (BP), (# of dates avail.), 2σ calib. ^{14}C age range (cal BP)[b]	Landslide/feature type, scar area (km^2), dated material, age context, comment, data source
26	Breckenridge V., scar 4 landsl.	UCIAMS-88674	600 ± 15, (3), 545–650	Conf., 0.03, wood, max. age[d]
27	Breckenridge V., scar 9 landsl.	GSC-6449[a]	1.080 ± 70, (3), 932–1,056	Conf., 0.06, wood, max. age[i]
28	Breckenridge V., scar 11? landsl.	UCIAMS-88703	180 ± 20, (1), −2 to 286	Conf., 0.03, macro., min. age[d]
29	Breckenridge V., scar 12? landsl.	GSC-6318[a]	1,030 ± 70, (1), 803–1,053	Conf., 0.04, wood, max. age[i]
30	Breckenridge V., scar 13 landsl.	UCIAMS-88811	1,880 ± 70, (11), 1,721–1,890	Conf., 0.2, wood, max. age[d]
31	Breckenridge V., scar 15 landsl.	GSC-6481[a]	1,040 ± 60, (4), 918–1,051	Conf., 0.04, wood, max. age[i]
32	Breckenridge V., scar 17 landsl.	GSC-6433[a]	1,440 ± 50, (1), 1,299–1,375	Conf., 0.18, wood, max. age[d]
33	Breckenridge V., scar 18 landsl.	UCIAMS-88817	3,625 ± 15, (4), 3,890–3,980	Conf., 0.06, wood, min. age[d]
34	Breckenridge V., scar 19 landsl.	UCIAMS-88814	1,525 ± 15, (5), 1,355–1,509	Conf., 0.08, wood, min. age[d]
35	Breckenridge V., scar 22 landsl.	Beta-139135	310 ± 40, (1), 297–478	Conf. , 0.03, wood, min.(?) age[d]
36	Alary Road landslide, Luskville	UCIAMS-106657	1,145 ± 15, (3), 979–1,166	Scarp S., 0.19, wood, max. age[i]
37	Luskville landslide, Luskville	UCIAMS-122468	1,095 ± 20, (3), 959–1,056	Scarp S., 0.33, wood, max. age[i]
38	Quyon Valley landsl.	UCIAMS-106653	1,120 ± 15, (17), 976–1,060	Sp. C., 20.0, wood, max. age[i]
39	Upper Quyon River landsl.	UCIAMS-106651	970 ± 15, (3), 799–930	Conf., 0.54, wood, min. age[i]
40	Murphy Road landsl., Quyon	UCIAMS-106577	1,140 ± 15, (3), 978–1,077	Scarp S., 0.25, wood, max. age[i]
41	Verner landsl., Quyon	UCIAMS-106581	5,830 ± 20, (3), 6,563–6,727	Scarp S., 0.5, wood, min. age[c, j]

Sp. C. special case feature, *Scarp S.* scarp-side landslide, *Conf.* confined valley landslide (see text for explanation)

[a] GSC dates are reported to 2σ

[b] All of the radiocarbon ages were calibrated to calendar years (years before AD 1950) using Calib 6.1 (see Stuiver and Reimer 1993) and the calibration dataset of Reimer et al. (2009)

[c] Date and landslide name after Aylsworth et al. (2000)

[d] Unpublished GSC data

[e] McNeely (1989)

[f] Lowdon and Blake (1973)

[g] Lowdon et al. (1967)

[h] Scar area is a crude estimate based on map polygon

[i] Date and landslide name after Brooks (submitted)

[j] Debris field truncated to edge of scarp

10.4 Data Source

The presented landslide chronological dataset for the Ottawa Valley is compiled from published and unpublished radiocarbon ages collected by the Geological Survey of Canada (GSC) or dated in the GSC Radiocarbon Laboratory, as listed in Table 10.1 and shown on Fig. 10.1. Dated material has been obtained opportunistically, e.g., from construction excavations, but mostly through focused sampling in specific areas. The sampling has not been random and contains bias: scarp-side failures with intact (uneroded) debris fields are over-represented, whereas small (less than 0.1 km^2) confined-valley and scarp-side failures, as well as scarp-side failures with truncated debris fields are under-represented.

10.5 Results

Ages are available for 39 landslides and three disturbed terrain features, as listed in Table 10.1, based on 122 radiocarbon ages. The ages of 21 features are represented by single dates that are assumed to be reasonably representative of the timing of the landslide/disturbance. The other 21 features are represented by between 2 and 17 radiocarbon ages. For this group, the timing of the failure/disturbance listed in Table 10.1 are represented by either a representative radiocarbon age interpreted from a group of maximum and minimum dates, or from the youngest maximum or oldest minimum ages where there is only one such set of ages available (which is most of the cases). In Breckenridge Valley, the specific origin of four dated debris deposits is uncertain because of the close spatial clustering of landslide scars, hence the identified source scars are marked with question marks in Table 10.1 (landslides 23, 25, 28 and 29). There is no question about the dates being related to landslides, however, as the dated organic materials were collected from organic layers buried beneath landslide deposits.

As plotted in Fig. 10.3a, b, the landslide ages range from 'modern' to 8,000 ^{14} C cal BP. Most striking, is the clustering of landslides at ~1,000 and ~5,150 ^{14}C cal BP composed of 10 and 11 landslides, respectively. A weaker cluster of four features occurs at ~7,900 ^{14}C cal BP, as is discussed below. The other 18 landslides are reasonably well distributed temporally over the 8,000 year range (Fig. 10.3a).

In Fig. 10.3c, scar area is plotted versus feature age and the data points are classified by the groups summarized above. The ages of 15 of the 23 scarp-side landslides fall within the clusters at ~1,000 and ~5,150 ^{14}C cal BP. One of the remaining failures (landslide 16, Table 10.1) has minor overlap with the age ranges of several dates in the ~5,150 ^{14}C cal BP cluster, whereas the ages of the other seven are distributed sporadically over the 8,000 ^{14}C cal BP time range (Fig. 10.3b). Twelve of the fifteen confined-valley landslides are comparatively young in age, post-dating 2,000 ^{14}C cal BP (Fig. 10.3b, c). The ages of four confined-valley landslides coincide within the ~1,000 ^{14}C cal BP cluster and one falls within

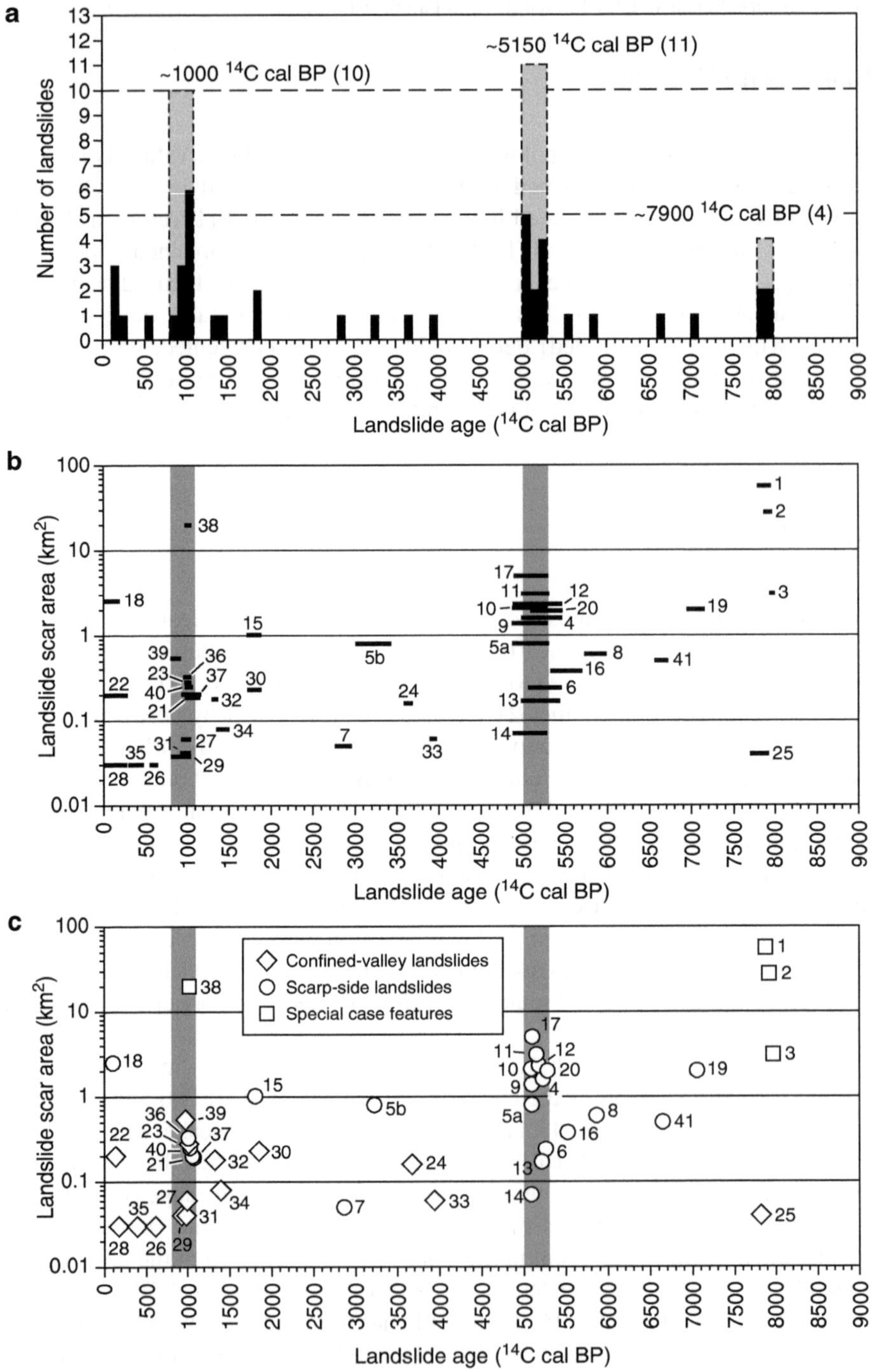

Fig. 10.3 (**a**) Plot showing the distribution of grouped landslides and disturbed terrain features, using the median of the calibrated age ranges. The groups at ~1,000 and ~5,150 ^{14}C cal BP are composed of the landslides interpreted by Brooks (submitted) and Aylsworth et al. (2000) as being evidence of paleoearthquakes. (**b**) Plot of the landslide scar/feature area versus calibrated age range at 2**σ** error range. The variation in age ranges between the dates in ~1,000 and ~5,150 ^{14}C cal BP clusters relates to the calibration of the radiocarbon ages. (**c**) Plot of scar area versus landslide age (median of calibrated age range) with the data points classified by the landslide groups

the ~7,900 ^{14}C cal BP cluster (Fig. 10.3). Of the four special case features, the three disturbed terrain areas (features 1, 2 and 3; Table 10.1) form the ~7,900 ^{14}C cal BP cluster. The fourth is the massive Quyon Valley landslide (landslide 38), which is one of the failures forming the ~1,000 ^{14}C cal BP cluster (Fig. 10.3b, c).

With regards to scar area, only three features, all in the special case groups (features 1 and 3 and landslide 38; Table 10.1; Fig. 10.3c), fall within the 10–100 km^2 range. The remaining landslides are reasonably well distributed with 11, 17 and 11 occurring within the 0.01–0.1, 0.1–1 and 1–10 km^2 ranges, respectively (Fig. 10.3c). There are obvious trends to the distribution of the morphological groups. Nine of the 11 failures within the 0.01–0.1 km^2 range are confined-valley landslides while the other two are scarp-side landslides. Five of the 11 failures coincide within the ~1,000 (three confined-valley landslides), ~5,150 (one scarp-side landslide) and ~7,900 (one confined-valley landslide) ^{14}C cal BP clusters. There are 17 failures within the 0.1–1 km^2 ranges, consisting of 11 scarp-side and 6 confined-valley landslides. Of these, two confined-valley and four scarp-side landslides, and three scarp-side landslides fall within the ~1,000 and ~5,150 ^{14}C cal BP clusters, respectively (Fig. 10.3c). The 1–10 km^2 range is dominated by scarp-side landslides (10), but it also contains one special case feature (feature 3; Table 10.1). Seven of these scarp-side landslides fall within the ~5,150 ^{14}C cal BP cluster, while the special case feature falls within the ~7,900 ^{14}C cal BP cluster. No landslides in the 1–10 km^2 range coincide with the ~1,000 ^{14}C cal BP cluster, but this cluster does contain a landslide in the 10–100 km^2 range (landslide 38).

10.6 Discussion

The clusters of 10 and 11 coincidentally-aged landslides at ~1,000 and ~5,150 ^{14}C cal BP are the groups of landslides interpreted by Brooks (submitted) and Aylsworth et al. (2000), respectively, to have been triggered by paleoearthquakes at ~1,020 and ~5,120 ^{14}C cal BP (~4,550 ^{14}C BP), respectively. Despite a common interpreted mechanism, the composition of landslide types and range of scar sizes within the two clusters differ. The failures composing the ~5,150 ^{14}C cal BP cluster all are scarp-side landslides with 7 of these falling within the 1–10 km^2 range of scar size (Fig. 10.3c). None of the failures forming the ~1,000 ^{14}C cal BP cluster are within this scar size range, but one (landslide 38) is larger and in the 10–100 km^2 range. Conversely, failures in both the ~1,000 (four scarp-side, two confined) and ~5,150 (three scarp-side) ^{14}C cal BP clusters fall in the smaller 0.1–1 km^2 range. Scarp-side landslides greater than 0.1 km^2 (14 of 21) show an obvious correlation with the two clusters, and thus constitute an important aspect of the interpreted paleoseismic evidence. The occurrence of 8 scarp-side landslides sporadically within the 8,000 year time span, either in isolation or paired with another event (e.g., landslides 5 and 30; Fig. 10.3b, c), indicates that not every dated scarp-side landslide can be assumed to represent a paleoearthquake. Nevertheless, the Ottawa Valley dataset indicates that scarp-side landslides with scars greater than 0.1 km^2 are promising failures to date for paleoseismic investigations. Although not shown in Fig. 10.3c, all of the dated scarp-side landslides in the Ottawa Valley dataset have intact debris fields, except landslide 41, indicating that the failures happened after the abandonment of the terrace/paleochannel by the ancestral Ottawa River. This characteristic is important geomorphic evidence supporting the paleoearthquake hypothesis, as the failures were not triggered by river erosion along the base of the scarp slope (see Aylsworth et al. 2000; Brooks submitted).

The five confined-valley failures within the ~1,000 ^{14}C cal BP cluster represent one-third of the 15 dated confined-valley landsides. None fall within the ~5,150 ^{14}C cal BP cluster, but this undoubtedly reflects the locations of the dated failures rather than meaning that no confined-valley landslides were generated by the interpreted ~5,150 ^{14}C cal BP paleoearthquake. The nearest dated confined-valley landslides are located in Breckenridge Valley, 50–75 km west of the Hammond, Bourget and Plantagnet paleochannels, where the ~5,150 ^{14}C cal BP cluster of failures are situated (Fig. 10.1b). It is possible that an undated confined-valley landslide(s) located closer to the paleochannels (and even one of the 13 undated confined-valley landslides within Breckenridge Valley) has an age coinciding with the ~5,150 ^{14}C cal BP cluster.

The variability of the dated landslides from Breckenridge Valley (landslides 23–35; Table 10.1) indicates that confined-valley landslides in a dense spatially-clustered setting can have substantially different ages. Such an age distribution probably reflects, in part, long-term landsliding triggered by active stream incision and lateral erosion within the confined valley setting. Because of this age variability, a large proportion of confined-valley landslides may need to be dated before any meaningful pattern of coincidentally-aged landslides becomes apparent. Variability in landslide ages has been observed within landslide chronologies from confined valleys elsewhere in the St. Lawrence Lowlands, for example, along the Yamaska River (9 dated landslides spanning 5,200 ^{14}C years; two pairs of coinciding ages; Rissman et al. 1985). On the other hand, there is a strong coincidence of landslide deposits containing young radiocarbon ages (less than 600 ^{14}C BP) along the Joffre River, Quebec (see Filion et al. 1991), although it is not clear how many landslides these dates actually represent. The Ottawa Valley dateset indicates that confined-valley settings can yield data on coincidentally-aged landslides, but a significant proportion of the dated failures may have widely scattered ages.

Also coincidental with the ~1,000 ^{14}C cal BP cluster is the massive Quyon Valley landslide (landslide 38), in the special case group, that is unique in scale for sensitive landslides within the Ottawa Valley. However, the ancient Saint Jean Vianney and the Colombier landslides, Quebec, are of similar scale and both are attributed to being triggered by the AD 1663 Charlevoix earthquake (see Locat 2011). In combination, these three massive events suggest that landslides in sensitive clay from scars of greater than 10 km^2 should also be a target for paleoseismic investigations, although landslides of this scale are obviously uncommon. Within the Gatineau River watershed, however, there are several large, undated landslides with scars in the 1–10 km^2 range. The features evidently are the product of large-scale failures occurring approximately synchronously on both sides of a tributary valley, which is a style analogous to the massive Quyon Valley landslide (see Brooks submitted). Such large failures represent obvious targets for paleoseismic studies, but, again, every dated landslide greater than 1 km^2 cannot be assumed to represent a paleoearthquake, as exemplified by the 1894 St. Albans landslide (6.5 km^2; Mitchell and Markell 1974), Quebec, which was not triggered by an earthquake (Locat 2011).

Three of the four features forming the ~7,900 ^{14}C cal BP cluster are the Lefaivre, Treadwell and Wendover areas (features 1, 2 and 3; Table 10.1; Fig. 10.3) that Aylsworth et al. (2000) and Aylsworth and Lawrence (2003) interpreted to be seismically-disturbed terrain, as mentioned above. A key aspect of their interpretation is that all three areas are the same age, although they lacked chronology for the Wendover and Treadwell areas. As listed in Table 10.1, radiocarbon ages are now available that suggest these two areas formed at ~7,900 ^{14}C cal BP (Fig. 10.3c). This chronological data, however, must be interpreted with caution because there is uncertainty about the relationship between dated organic materials and the terrain disturbance at the sampling sites. The radiocarbon ages were obtained from wood and macrofossils sub-sampled from the base of buried leaves/wood and peat deposits, interpreted to have accumulated in poorly-drained depressions on the surface of the features, which were later exposed in shallow excavations. It is not clear, however, if the start of leaves/wood or peat accumulation relates to the abandonment of the underlying fluvial surface by the Ottawa River as opposed to the timing of the disturbance of the underlying deposits. In the latter case, the radiocarbon ages confirm a coincidental clustering of ages of the three disturbed terrains, but in the former case, the ages of the Treadwell and Wendover areas probably differ from that of the Lefaivre area. The discovery of additional landslides with ages of ~7,900 ^{14}C cal BP in the Ottawa area would clarify the paleoearthquake hypothesis based on these features.

The 39 dated landslides represent a small proportion (~16 %) of the ~250 mapped landslide features within the Ottawa Valley. As mentioned above, 21 of the dated landslides form the ~1,000 and ~5,150 ^{14}C cal BP landslide clusters, which have been interpreted to represent two prehistoric paleoearthquakes. That ~54 % of the landslides are attributed to this mechanism implies there is a 'strong' paleoearthquake signature within the temporal pattern of landsliding within the Ottawa Valley. While the strength of the signature probably reflects sampling bias within the landslide dataset, it is difficult to believe that this signature is not the product of a significant causal linkage between paleoseismicity and the triggering of landslides within sensitive clay deposits. Such a link is exemplified by AD 1663 Charlevoix earthquake, the largest historical earthquake in the region, which is attributed to triggering numerous landslides in sensitive clay deposits within southern Quebec (see Locat 2011). If the interpretation of the apparent paleoseismic signature is correct, then landsliding within sensitive clay deposits represents a major secondary seismic hazard. It therefore should be anticipated that a significant earthquake in the region will trigger multiple landslides, including large failures along scarp slopes that are currently not subject to active toe erosion i.e., above terraces and paleochannels. Additional research into the chronology of prehistoric landslides within the St. Lawrence Lowlands region will help elucidate this relationship. It might also reveal the occurrence of unknown, significant paleoearthquakes in areas where the historical seismicity record has been relatively low, for example, between Montreal and Quebec City, and thus contribute to improving the understanding of the regional seismic hazard.

10.7 Conclusions

Based on the Ottawa Valley dataset, scarp-side failures with preserved debris fields and scars greater than 0.1 km^2 are promising sensitive clay landslides to date for paleoseismic studies. The age of confined-valley landslides are also useful to investigate, but a significant proportion of the failures may yield widely varying ages. In areas where both undated scarp-side and confined-valley features are present, it seems more efficient to focus on dating scarp-side landslides with preserved debris fields to recognize clusters of coincidentally-aged landslides.

Evidence from the Ottawa Valley and elsewhere in southern Quebec suggests that landslides in sensitive clay from scars of greater than 10 km^2 should also be a target for paleoseismic investigations, although failures of this scale are uncommon.

Radiocarbon ages are now available for the Treadwell and Wendover disturbed terrains, suggesting that these areas are contemporary with the Lefaivre disturbed area, but this is not an unequivocal interpretation.

The high proportion of failures (21 of 39 dated landslides) attributed to paleoearthquakes implies there is a 'strong' paleoearthquake signature within the temporal pattern of landsliding in the Ottawa Valley. Although probably partly a product of sampling bias, this signature likely reflects a significant causal linkage between paleoseismicity and the triggering of sensitive clay landslides.

Acknowledgments This research has benefited greatly from the landslide chronological work of now-retired GSC colleagues J. Aylsworth and T. Lawrence. S. Morton assisted in the compilation of the Ottawa Valley landslide chronology dataset. S. Wolfe, J. Hunter, D. Perret and M. Geertsema commented on earlier drafts of the paper. This research was supported by the Public Safety Geoscience Program, Earth Sciences Sector, Natural Resources Canada, and represents ESS Contribution 20120421.

References

Aylsworth JM, Lawrence DE (2003) Earthquake-induced landsliding east of Ottawa; a contribution to the Ottawa Valley Landslide Project. In: Proceedings of the 3rd Canadian conference on geotechnique and natural hazards, Edmonton, Alberta, pp 77–84, 9–10 June 2009

Aylsworth JM, Lawrence DE, Guertin J (2000) Did two massive earthquakes in the Holocene induce widespread landsliding and near-surface deformation in part of the Ottawa Valley, Canada? Geology 28:903–906

Borgatti L, Soldati M (2010) Landslides as a geomorphological proxy for climate change: a record from the Dolomites (northern Italy). Geomorphology 120:56–64

Brooks GR (submitted) A massive sensitive clay landslide, Quyon Valley, southwestern Quebec, Canada, and evidence for a paleoearthquake triggering mechanism. Quat Res

Chagnon JY (1968) Les coulées d'argile dans la Province de Quebec. Nat Can 95:1327–1343

Crawford CB (1961) Engineering studies of Leda clay. In: Legget RF (ed) Soils in Canada – geological, pedological, and engineering studies, Royal Society of Canada, special publication 3. University of Toronto Press, Toronto, pp 200–217

Crozier MJ (1992) Determination of paleoseismicity from landslides. In: Bell DH (ed) Landslides. Proceedings of the 6th international symposium, vol 2. Balkema, Rotterdam, pp 1173–1180

Cummings DI et al (2011) Sequence stratigraphy of a glaciated basin fill, with a focus on easker sedimentation. Geol Soc Am Bull 123:1478–1496

Filion L, Quinty F, Bégin C (1991) A chronology of landslide activity in the valley of Rivière du Gouffre, Charlevoix, Quebec. Can J Earth Sci 28:250–256

Fransham PB, Gadd NR, Carr PA (1976) Sensitive clay deposits and associated landslides in Ottawa Valley. Geol Surv Can Open File 352, Ottawa, 9 p

Jibson RW (2009) Using landslides for paleoseismic analysis. In: McCalpin JP (ed) Paleoseismology, International geophysics 95. Academic, San Diego, pp 565–601

Lamontagne M (2010) Historical earthquake damage in the Ottawa-Gatineau region, Canada. Seismol Res Lett 81:129–139

Lefebvre G, Leboeuf D, Hornych P (1992) Slope failure associated with the 1988 Saguenay earthquake, Quebec, Canada. Can Geotech J 29:117–130

Locat J (2011) La localisation et la magnitude du séisme du 5 février 1663 (Charlevoix) revues à l'aide des mouvements de terrain. Can Geotech J 48:1266–1286

Lowdon JA, Blake Jr W (1973) Geological Survey of Canada radiocarbon dates XIII. Geol Surv Can Paper 73–7, Ottawa

Lowdon JA, Fyles JG, Blake Jr W (1967) Geological Survey of Canada radiocarbon dates VI. Geol Surv Can Paper 67-2B, Ottawa

McNeely R (1989) Geological Survey of Canada radiocarbon dates XXVIII. Geol Surv Can Paper 88–7, Ottawa

Mitchell RJ, Markell AR (1974) Flowsliding in sensitive soils. Can Geotech J 11:11–31

Reimer PJ et al (2009) IntCal09 and Marine09 radiocarbon age calibration curves, 0–50,000 years cal BP. Radiocarbon 51:1111–1150

Richard SH (1984) Surficial geology, Lachute-Arundel, Québec-Ontario. Geol Surv Can Map 1577A, Ottawa

Rissman P, Allard JD, Lebuis J (1985) Zones exposées aux mouvements de terrain le long de la rivière Yamaska, entre Yamaska et Saint-Hyacinthe. Quebec Ministère de L'Ènergie et des Ressources DV 83–04

Rodriguez CG, Gadd NR, Richard NR (1987) Excursion B: invertebrate paleontology and lithostratigraphy of Champlain Sea sediments. In: Fulton RJ (ed) Quaternary of the Ottawa region and guides for day excursions. International Union for Quaternary Research, XII international congress, pp 25–36, July 31–August 9, 1987

St.-Onge DA (2009) Surficial geology, lower Ottawa Valley, Ontario-Quebec. Geol Surv Can Map 2140A, Ottawa

Stuiver M, Reimer PJ (1993) Extended ^{14}C data base and revised Calib 3.0 ^{14}C age calibration program. Radiocarbon 35:215–230

Torrance JK (2012) Landslides in quick clay. In: Clague JJ, Stead D (eds) Landslides: types, mechanisms and modeling. Cambridge University Press, Cambridge, pp 83–94

Chapter 11
Characterization and Post-failure Analysis of the 1980 Landslide in Sensitive Clays at Havre-St-Pierre, Québec, Canada

Pascal Locat, Serge Leroueil, Jacques Locat, and Denis Demers

Abstract Early in the morning of April 3rd, 1980, approximately 1.15×10^6 m^3 of soils from the coastal bluffs spread out over the tidal flat along the North Shore of the Jacques-Cartier Strait in the Gulf of St. Lawrence. This landslide had a width of 410 m and had retrogressed back up to 110 m from the slope crest, cutting through Highway 138, 9 km west of the town of Havre-St-Pierre. In this area, the coastal bluffs are approximately 23 m high and consist, from the bottom to the top, of a thick marine sensitive clay unit from the Goldthwait Sea, overlain by 3 m of sand and 3 m of peat on top. The debris morphology has a "thumbprint-like pattern" characterized by unbroken blocks of intact clay forming elongated ridges, surrounded by a mixture of remoulded clay, sand and peat, aligned parallel to the bluff and arched in the direction of the flow. The absence of obstacles and the non-channeled character of the flow path of the Havre-St-Pierre landslide provide a good opportunity to characterize and analyze the post-failure stage of this large landslide in sensitive clays.

Keywords Havre-St-Pierre landslide • Quick clays • Sensitive clays • Spread • Landslide mobility • Numerical simulation • Bing model • Post-failure analysis

P. Locat (✉) • D. Demers
Service de la Géotechnique et de la géologie (Geotechnique and Geology Branch), Ministère des Transports du Québec (MTQ), Québec City, QC, Canada
e-mail: pascal.locat@mtq.gouv.qc.ca

S. Leroueil
Department of Civil Engineering and Water Engineering, Laval University, Québec City, QC, Canada

J. Locat
Laboratoire d'études sur les risques naturels, Department of Geology and Geological Engineering, Laval University, Québec City, QC, Canada
e-mail: Jacques.locat@ggl.ulaval.ca

J.-S. L'Heureux et al. (eds.), *Landslides in Sensitive Clays: From Geosciences to Risk Management*, Advances in Natural and Technological Hazards Research 36,
DOI 10.1007/978-94-007-7079-9_11, © Springer Science+Business Media Dordrecht 2014

11.1 Introduction

Leroueil et al. (1996) proposed that slope movements should be characterized by means of a three-dimensional matrix whose three principal axes are: the type of movement, the stage of the movement and the type of material. In the case of slope movements in sensitive clays, the movement can be divided into four stages: (1) pre-failure; (2) failure; (3) post-failure; and (4) reactivation. The research studies carried out on landslides in sensitive clays of eastern Canada in the past focused mainly on the pre-failure stage (Tavenas and Leroueil 1981; Demers et al. 1999; Leroueil 2001), the failure stage (Lefebvre 1981, 1986; Leroueil 2001; Locat et al. 2011), and the retrogression stage in the post-failure phase (Mollard and Hughes 1973; Mitchell and Markell 1974; Carson 1977; Carson and Lajoie 1981; Lebuis et al. 1983; Tavenas et al. 1983; Chap. 7 by Demers et al., this volume). However, aspects related to mobility in the post-failure phase in these materials have been the subject of very few studies (Locat 1992; Locat et al. 2003). According to Locat et al. (2008), the mobility of the debris materials in a landslide in sensitive clays is a function of the potential energy available, the energy needed to remould the clay mass after the failure, the flow capacity of the soil (i.e. its liquidity index) and the environment in which the landslide occurs (inclination of the flow surface, whether or not the debris materials were confined or washed out by a river stream, etc.).

The Havre-Saint-Pierre landslide was studied for three reasons: (1) the topographic conditions before and after the failure are known; (2) the debris materials spread freely over flat terrain without encountering any obstacle; and (3) geotechnical data are available. This landslide has already been the subject of brief descriptions in the literature (Bergeron 1982; Carson and Lajoie 1981). This landslide was also the subject of a terrain description by Rissman and Lajoie (1980) and a geotechnical study conducted by the Ministère des Transports du Québec (MTQ) (Paré 1980), but this information was never published in detailed form. In addition, a field work and laboratory testing program were carried out in 2002 by Université Laval. The present article offers a detailed description of the morphology, the geology and the geotechnical properties of the clayey materials, along with a detailed analysis of the post-failure phase of this landslide.

11.2 Review of the 1980 Landslide at Havre-Saint-Pierre

Shortly before 6:00 a.m. on Wednesday, April 3, 1980, a large retrogressive landslide occurred and spread on the tidal flat of Pakuauashau Bay, on the north shore of the Jacques-Cartier Strait, approximately 9 km west of the municipality of Havre-Saint-Pierre and approximately 1,000 km northeast of Québec City, Canada (Fig. 11.1a). The landslide carried away a section of Highway 138 measuring approximately 350 m (Figs. 11.1a and 11.2), thus cutting off the only road linking the village of Havre-St-Pierre to the rest of Québec. No loss of life was recorded, but

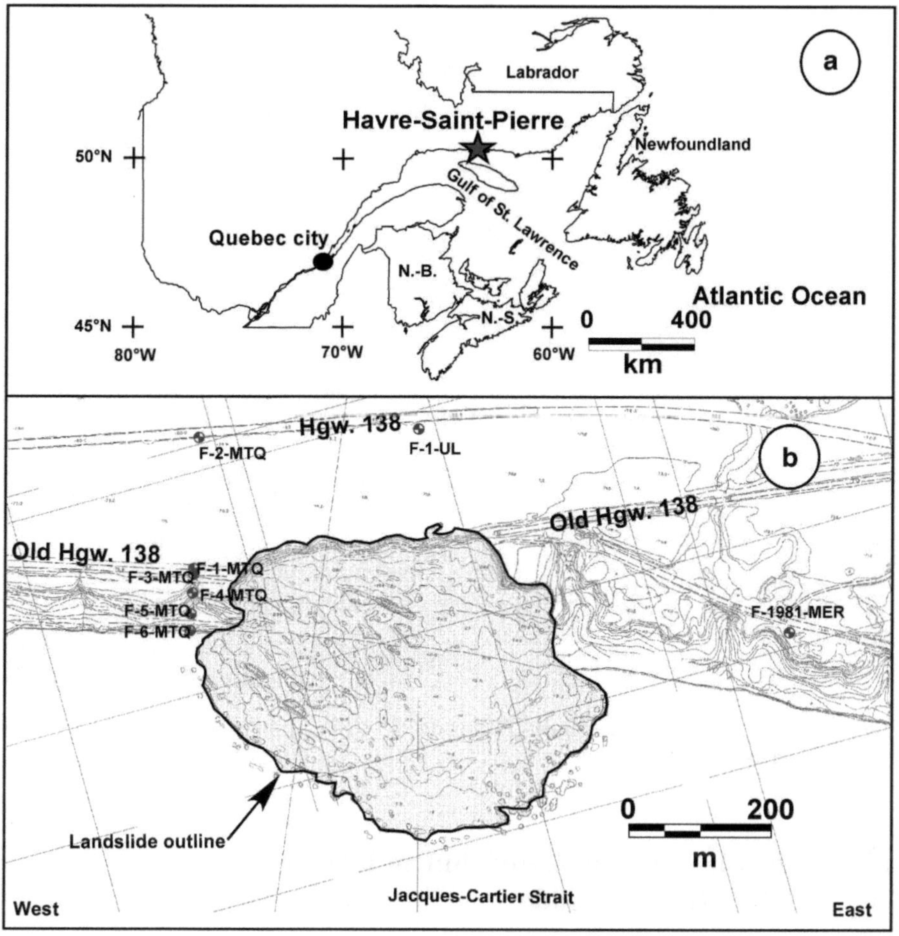

Fig. 11.1 General location map (**a**) and close-up map (**b**) of the 1980 landslide at Havre-Saint-Pierre. The exact coordinates of the landslide are 50°17′24″N and 63°43′12″W

a Sûreté du Québec police vehicle fell into the crater at the western end of the landslide, causing minor injuries to the police officers who were in it. According to the account provided by the police officers, who were travelling in the eastbound direction (toward Havre-Saint-Pierre), they had encountered a car coming from the opposite direction a few minutes before falling into the scar, which indicates that the landslide occurred very rapidly. This landslide is characterized by a width that is greater than its retrogression distance and by a debris morphology that displays a succession of ridges aligned perpendicular to the direction of the flow (Fig. 11.2). These morphological features are typical of landslides of the " spread" type (Cruden and Varnes 1996).

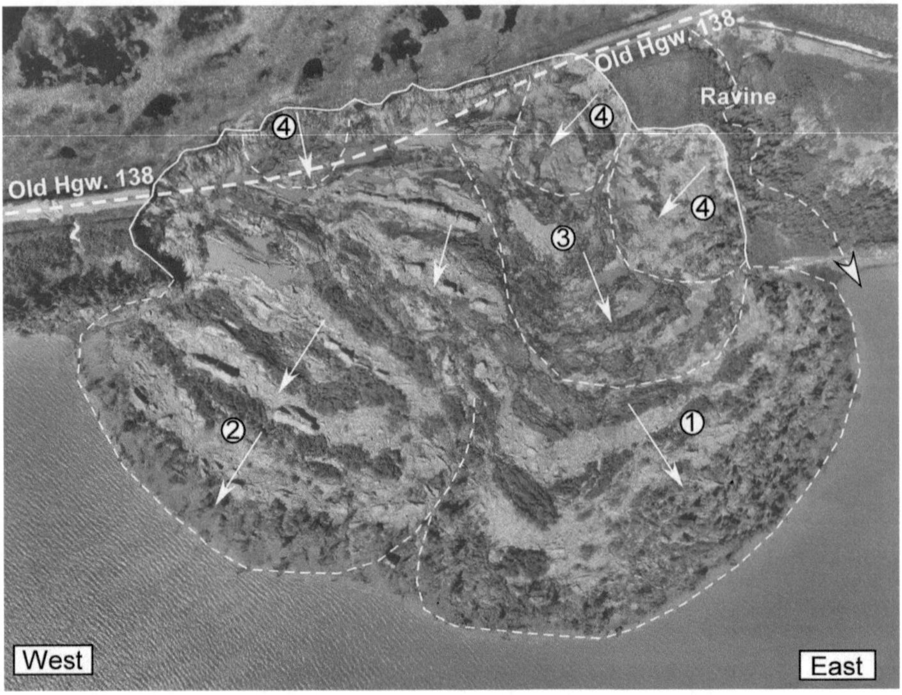

Fig. 11.2 Section of aerial photograph (Photo AP8035 n°87, Aero-Photo (1961) inc.) and morphological analysis of the landslide

11.3 Morphology and Geotechnical Properties Before the Landslide

The original slope was approximately 23 m high. The aerial photographs from 1975 (Q75408- 9 and 10) show that its base was subject to active erosion. A ravine, which was located near the centre of the coastal slope and was carried away by the landslide, collected and carried off the surface water from a section of Highway 138, which ran approximately 40 m behind the slope crest. Analysis of the digital terrain model (DTM) obtained by photogrammetry based on the 1975 aerial photographs shows that the mean slope inclination in the area corresponding to the landslide ranged between 20° and 35°, with some segments of the slope measuring as much as 40°.

Figure 11.1b presents the location of the soundings conducted at various times near the site of the landslide. The geotechnical profile shown in Fig. 11.3 was prepared based on the borehole that was carried out by Université Laval in 2002, approximately 200 m behind the landslide (F-1-UL on Fig. 11.1b), along the existing path of Highway 138. The profiles based on boreholes F-2-MTQ and F-1981-MER (Fig. 11.1b) are not shown here, because they present the same geotechnical characteristics as shown in Fig. 11.3, indicating a relative uniformity of the clay deposit

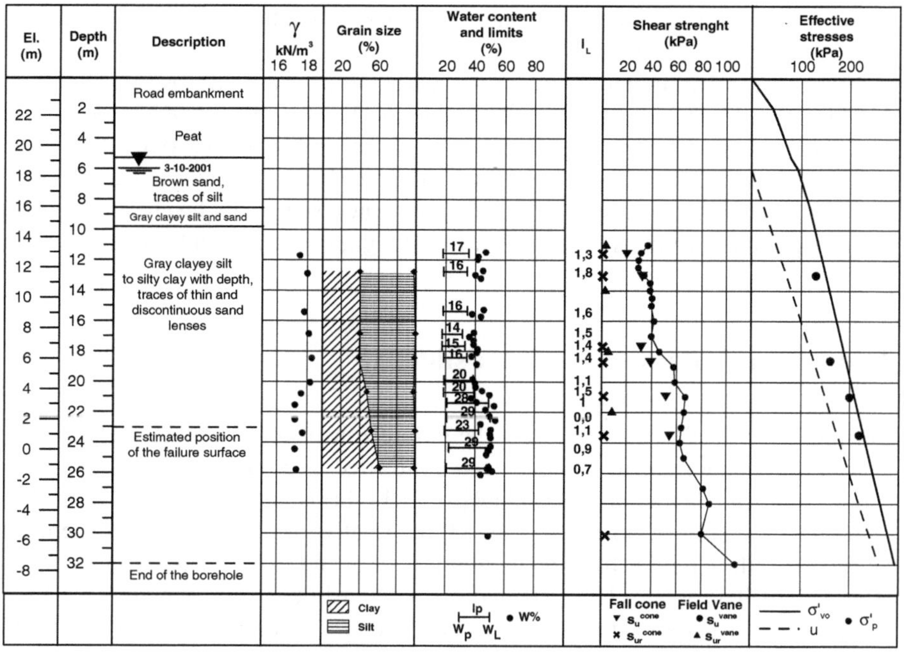

Fig. 11.3 Geotechnical profile of borehole F-1-UL

between these boreholes. From the bottom to the top (Fig. 11.3), the Quaternary deposits consist of sensitive clays from the Goldthwait Sea, overlain by approximately 3–5 m of sand deposits and topped by a relatively impermeable iron pan (ferruginous crust). On this layer, peaty materials grow to a thickness of 3–5 m, depending on the area. Based on the geotechnical properties obtained (Fig. 11.3), the clayey deposit can be subdivided into 2 units.

The upper unit, at depths of 8.5–21 m, consists of a clayey silt with a unit weight of approximately 18 kN/m^3, a water content of approximately 40 %, a mean plasticity index of 17, and a mean liquidity index of 1.4.

The lower unit, at depths of 21–32 m, thus essentially below the sea level, consists of a silty clay with a unit weight of 17.2 kN/m^3, a mean water content of 50 %, a mean plasticity index of 28, and a mean liquidity index of 0.9.

The salinity of the interstitial water increases with depth, ranging from approximately 1 g/l at 12 m depth in the upper unit to approximately 8 g/l at 26 m depth in the lower unit.

The undrained shear strength, measured with the field vane, is approximately 35 kPa in the upper portion of the clay deposit at a depth of 10 m, increasing gradually with depth to approximately 100 kPa at the bottom of the borehole, at a depth of 32 m. The undrained shear strength, measured in laboratory with the 100 g/30° Swedish fall cone, is approximately 25 kPa in the upper portion of the deposit (Fig. 11.3), increasing gradually with depth at the same rate as the undrained shear strength measured with the field vane. The higher strength, determined by the fall cone is 55 kPa at a depth of 23.5 m (Fig. 11.3).

From 13 to 23 m in depth, the preconsolidation pressure increases from 120 to 210 kPa (Fig. 11.3), indicating that the soil in this location has an OCR close to 1.0. The remoulded undrained shear strength of the soil averages 0.8 kPa for the upper unit and 2.2 kPa for the lower unit. The sensitivity, as determined using the Swedish fall cone, ranges from 26 to 46. The highest sensitivity levels are observed in the upper unit, which is consistent with the highest liquidity index computed.

11.4 Morphology of the Landslide

The mean width (W_m, Fig. 11.4a) of the failure zone is approximately 410 m. The eastward boundary seems to have been controlled by a large ravine (Fig. 11.2). The width of the landslide opening (W_o, Fig. 11.4a) is approximately 430 m. The slope retrogression distance (R, Fig. 11.4b) relative to the position of the crest before the failure ranges from 80 m at the western end to 110 m at the eastern end (Fig. 11.1b). The area of land affected by the landslide is approximately 6.1 ha.

The debris materials spread out over the tidal flat, which was ice-free at the time of the landslide, according to the photographs taken by Rissman and Lajoie (1980) a few hours after the event. In addition, the portion of the tidal flat across which the debris materials spread must have been exposed, because the tide was almost out at around 6:00 a.m., given that low tide was at 7:35 a.m., according to the Canadian tide tables for the day of the landslide. The central ravine, which is identified on the 1975 aerial photographs, was carried away by the landslide. The configuration of the frontal portion of the debris gives the impression of two lobes (Fig. 11.2) separated by the central ravine, hereinafter referred to as the East and West lobes. The aerial photos also show that additional sections of land broke off from the escarpments of the landslide scar in subsequent stages (marked 3 and 4 on Fig. 11.2), especially for the East side.

Overall, the debris materials, including those inside the scar, have a maximum width (W_d Fig. 11.4a) of approximately 580 m and a total area of approximately 17.1 ha. From the downstream extremity of the debris up to the main escarpment, the spread zone exhibits alternating light (clay pinnacles) and dark (peat) segments aligned perpendicular to the direction of the flow. On aerial photograph (Fig. 11.2), this variation in shading and texture gives the impression of a "fingerprint" ("thumbprint-like pattern", Mollard and Hugues 1973). Assuming that the failure surface developed through the base of the slope, the thickness of the earth mass moved by the landslide is approximately 23 m. The total volume of the intact mass involved in the landslide would be approximately 1.15×10^6 m^3.

Over the West lobe, five rectilinear ridges (horsts) can be distinguished, rising more than 6 m above the average debris level in some places. According to Rissman and Lajoie (1980), these ridges generally exhibited quite horizontal structures characterized by fine sand beds that were a few millimetres thick. This observation indicates that these ridges experienced practically no rotational movement, and

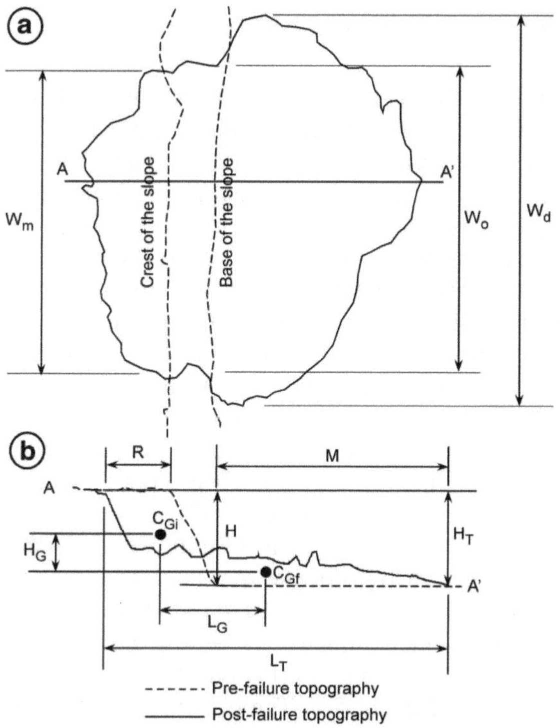

Parameter	East lobe	West lobe
H_T (m)	24.5	24
L_T (m)	465	360
H_G (m)	7,2	6.6
L_G (m)	138,3	92.6
M (m)	290	210
R (m)	110	80
$\tan^{-1} H_T/L_T$(°)	3.0	3.8

Parameter	Landslide
W_m (m)	410
W_o (m)	430
W_d (m)	580
H (m)	~23
S_f (ha)	6.1
S_d (ha)	17.1
V (m^3)	1 150 000

Fig. 11.4 Geometric characteristics of the landslide. S_f, S_d, and V represent the area of the failure zone, the total debris area, and the debris volume, respectively

moved horizontally, suggesting that the failure surface is sub-horizontal. For the East lobe, the ridges seem to have collapsed, and the land strips take on a crescent shape. A zoom on the 1980 aerial photograph (Fig. 11.2) shows pieces of intact clay of all sizes floating in a mud matrix. These observations indicate that only some of the clayey soils were remoulded during the landslide, and that the degree of remoulding differs between the East and West lobes.

Differences on the degree of remoulding are reflected in the geometric parameters for the two lobes (Fig. 11.4b). For the East lobe, the total length (L_T) is 465 m, measured in the direction of the flow, and the total vertical drop (H_T) is 24.5 m. The vertical (H_G) and horizontal (L_G) displacements of the centre of gravity are 7.2 and 138.3 m, respectively. The thickness measured at the centre of gravity of the debris is approximately 5.8 m. The maximum debris travel distance (M on Fig. 11.4) is approximately 290 m. For the West side, the total length (L_T) is 360 m, measured in the direction of the flow, and the total vertical drop (H_T) is 24 m. The vertical (H_G) and horizontal (L_G) displacements of the centre of gravity are 6.6 and 92.6 m, respectively. The thickness measured at the centre of gravity of the debris is approximately 7.6 m. The maximum debris travel distance (M) is approximately 210 m.

11.5 Numerical Modelling of Debris Spreading

In the absence of post-failure models for spreads, the numerical model BING, developed by Imran et al. (2001), will be used hereafter mostly to analyse the overall mobility of the debris of the Havre-St-Pierre landslide, and to evaluate the average mobilised strength during the post failure phase, assuming a homogeneous mass with an average remoulded strength.

This model allows 1D numerical simulation of a mud or debris flow. Starting from an initial parabolic shape, BING allows the debris mass to deform and flow until it stops, following the rheological model and topography specified by the user (see Imran et al. 2001 for a parametric analysis). For this study, the results of the laboratory viscosity tests suggest that the material can be modelled based on a Bingham-type behaviour. This behaviour is assumed in BING when assigning the value 1 to the exponent n in the Herschel-Bulkley model. Laboratory tests have shown a ratio in the order of approximately 1–1,000 between plastic viscosity and yield stress, as observed for fine mud by Locat (1997). This ratio was taken for the numerical modeling. Densities of 1,800 and 1 kg/m^3 were assumed for the clay and the ambient fluid (air). In light of the fact that there are two debris lobes with slightly different soil volumes, two BING simulations were carried out, differing only in the assumed dimensions of the failure zone (Fig. 11.5). These dimensions were 150 m in length for the East lobe (Fig. 11.5a) and 130 m in length for the West lobe (Fig. 11.5b). The elevation of the mass remained fixed at 24 m in both cases. In order to reproduce the travel distances (M on Fig. 11.4) of the East and West lobes, yield stresses (τ_c) of 5 kPa (Fig. 11.5a)

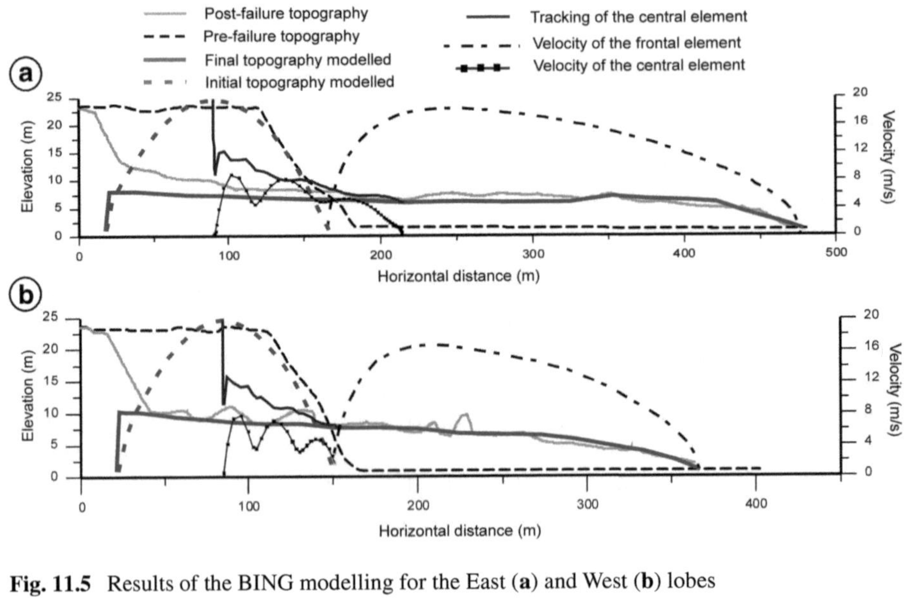

Fig. 11.5 Results of the BING modelling for the East (**a**) and West (**b**) lobes

and 7.5 kPa (Fig. 11.5b) respectively, were determined by a parametric study using BING.

Figure 11.5a, b show cross sections of the post-failure topography and debris profiles obtained with BING for the East and West lobes, respectively. These results indicate that it is possible to use BING to reproduce the topographic profiles of the debris in terms of distance and thickness. In addition, the median element in BING underwent a displacement of approximately 80 m, which is similar to the field observations of the displacements of the road segments that were originally located approximately at the centre of the mass involved in the landslide.

The maximum velocities reached by the frontal elements of the debris (Fig. 11.5a, b) are approximately 18 m/s for the East lobe and approximately 17 m/s for the West lobe. The mean frontal velocities for both cases are approximately 10.5 m/s. The maximum velocity reached by the median elements is approximately 7.5 m/s, which can be considered to be closer to the mean value for the debris mass as a whole. The mean velocity of the median elements is approximately 5.5 m/s.

11.6 Discussion of the Remoulding and Mobility of the Debris

According to Locat et al. (2008), the remoulding of sensitive clay in a landslide is a function of the potential energy available, the undrained shear strength and the plasticity of the material. In addition, based on the observations of Chap. 7 by Demers et al., this volume, it seems that the failure mode affects the degree of remoulding of sensitive clays with comparable geotechnical characteristics. For flowslides, the average thickness of the debris remaining in the scar is equivalent to 18 % of the initial height of the slope, and they can flow downstream and upstream of a river for distances of up to several kilometres from the scar. For spreads, the average thickness of the debris remaining in the scar is equivalent to 60 % of the initial height of the slope, and there is generally very little downstream and upstream debris flow outside the scar.

In Fig. 11.5, the debris morphology indicates that for both lobes, the remoulding was more significant in the front half than in the rear half. In fact, the frontal thickness of the debris is approximately 5 m, which is equivalent to approximately 22 % of the initial height of the mass that slid. Within the scar, from 20 to 160 m on the horizontal axis on Fig. 11.5, the thickness of the debris is approximately 8 m for the East lobe and approximately 10 m for the West lobe, corresponding to approximately 33 and 42 % of the initial height of the soil mass that slid, respectively.

For a spread landslide, it is generally assumed that the horsts and grabens move along a sub-horizontal failure surface, and that the remoulding is mostly observed to be more significant at the level of the grabens than at the level of the horst. In the case of the Havre-Saint-Pierre landslide, horsts and grabens are observable within the West lobe. However, no horst has been clearly observed in the East lobe, even though alternating soil layers that slid (Fig. 11.2) can be detected. The remoulding seems to have been relatively more uniform and significant in the eastern part of the landslide.

Locat and Demers (1988) show that the relationship between the yield stress for sensitive clays that was obtained in the laboratory and their remoulded shear strength (s_{ur}) is practically 1–1. In comparison, the yield stresses back calculated with BING to fit to the Havre-Saint-Pierre landslide morphology are approximately equivalent to ten times the mean s_{ur} of 0.7 kPa (Fig. 11.3) obtained for the clay involved in the landslide. This represents somehow a mean undrained shear strength mobilized in the post-failure phase. On the other hand, it is clear from the presence of intact blocks, as observed on the aerial photographs, that the clay was not completely remoulded. In addition, the first 8 m of soils that were carried away by the landslide consist of sand and peat, which could not be modelled in BING. These materials became partly mixed with the remoulded clay during the movement, likely modifying its overall rheological properties. In the present work, adjusting the yield stress as a function of the travel distance made it possible to indirectly account for the aspects related to the partial remoulding of the soil. In addition, it is interesting to note that adjusting the yield stress as a function of the travel distance in BING results in a mean thickness of the final spread that very closely approximates what was observed in the field.

In addition, the fact that the yield stresses values were adjusted as a function of the travel distance made it possible to obtain realistic mean velocities. In fact, the frontal and median mean velocities of 10.5 and 5.5 m/s modelled by BING for the Havre Saint-Pierre case are of the same order of magnitude as those measured for the cases of Rissa in Norway and Saint-Jean-Vianney in Québec. For these two landslide cases, respective mean velocities of 11 m/s (L'Heureux 2012) and 7.2 m/s (Tavenas et al. 1971) have been reported. However, these two landslides were not of the spread type. The debris materials in these two landslides were strongly remoulded, and were largely evacuated from their scars. In addition, the sensitivity of the clay in these two landslides (Potvin et al. 2001; L'Heureux 2012) was higher than that at Havre-Saint-Pierre, and the debris materials at Saint-Jean-Vianney were channelled into a ravine.

This analysis shows the limitations of the BING model, especially the fact that it cannot model the formation of horsts and grabens at the beginning of the post-failure phase. To this end, it would be worthwhile to use an approach similar to the one that was developed by Gauer et al. (2005) for the case of the Storegga slide.

11.7 Conclusion

This article presents the Havre-Saint-Pierre landslide, which occurred April 3, 1980, within a coastal bluff of approximately 23 m height in sensitive clay. This spread landslide has a mean width of 410 m and a volume of approximately 1.15×10^6 m^3. The landslide retrogressed by approximately 100 m from the slope crest. The debris materials spread out over sub-horizontal terrain that was free of obstacles. The maximum spreading distance of the debris was 290 m. The clayey soil involved in the landslide consists of a clayey silt with a unit weight of approximately 18 kN/m^3, a water content of approximately 40 %, a mean plasticity index of 17, and a mean liquidity index of 1.4 (mean s_{ur} of 0.8).

This landslide represents an interesting case for studying the spreading distances of debris materials for landslides in sensitive clays. In fact, the topographic conditions before and after the failure are known, the debris materials were able to spread freely, and detailed geotechnical data are available.

Numerical modelling of the spread of the debris in the Havre-Saint-Pierre landslide using the BING model shows that it is possible to reproduce the final shape of the topographic profile of the debris, provided that the yield stress used in this software is increased to approximately ten times the value determined in the laboratory, which is one way of compensating for the fact that BING treats the soil as being completely remoulded and retaining its properties during the event. This aspect may be explained by the fact that a portion of the mass of the clayey debris materials underwent little remoulding, and by the presence of the upper layers of sand and peat, which constituted approximately 25 % of the height of the slope. However, BING cannot allow for consideration of multiple soil layers with different rheological properties, and it does not allow for taking into account the failure mode and the parameters related to remoulding of the clay. Although these aspects are ignored in BING, it was still possible to obtain a mean velocity of the same order of magnitude as in the cases of landslides in sensitive clays for which mean velocities were estimated in the field.

Acknowledgments The authors would like to thank the Natural Sciences and Engineering Research Council of Canada (NSERC) for its financial support and the Ministère des Transports du Québec for access to its archives. The authors would also like to thank Geneviève Cauchon-Voyer for the revision of the manuscript and her helpful comments.

References

Bergeron R (1982) Introduction historique au phénomène des coulées argileuses québécoises. Ministère de l'Environnement, Gouvernement du Québec, Québec, 24 pp

Carson MA (1977) On the retrogression of landslides in sensitive muddy sediments. Can Geotech J 14:582–602

Carson MA, Lajoie G (1981) Some constraints on the severity of landslide penetration in sensitive deposits. Rev Géo Phys et Quat 35:301–316

Cruden DM, Varnes DJ (1996) Landslide types and processes. In: Turner AK, Shuster RL (eds) Landslides investigation and mitigation, special report 247, Transportation Research Board, Washington, DC, pp 36–75

Demers D, Leroueil S, D'Astous J (1999) Investigation of a landslide in Maskinongé, Québec. Can Geotech J 36:1001–1014

Gauer P, Kvalstad TJ, Forsberg CF, Bryn P, Berg K (2005) The last phase of the Storegga Slide: simulation of retrogressive slide dynamics and comparison with slide-scar morphology. Mar Pet Geol 22:171–178

Imran J, Harff P, Parker G (2001) A numerical model of submarine debris flows with graphical user interface. Comput Geosci 274:717–729

L'Heureux J-S (2012) Characterization of historical quick clay landslides and input parameters for Q-Bing, Norwegian Geotechnical Institute, Oslo, Norway. NGI Report 20120753-02-R. Also available at www.naturafare.no

Lebuis J, Robert J-M, Rissmann P (1983) Regional mapping of landslide hazard in Quebec. Paper presented at the symposium on slopes on soft clays, Swedish Geotechnical Institute Report No 17, Linköping, pp 205–262

Lefebvre G (1981) Fourth Canadian geotechnical colloquium: strength and slope stability in Canadian soft clay deposits. Can Geotech J 18:420–442

Lefebvre G (1986) Slope instability and valley formation in Canadian soft clay deposits. Can Geotech J 23:261–270

Leroueil S (2001) 39th Rankine Lecture: natural slopes and cuts: movement and failure mechanisms. Géotechnique 51:197–243

Leroueil S, Vaunat J, Picarelli L, Locat J, Faure R, Lee H (1996) A geotechnical characterization of slope movements. Paper presented at the 7th international symposium on landslides, Trondheim, pp 53–74

Locat J (1992) Viscosity, yield strength and mudflow mobility for sensitive clays and other fine sediments. Paper presented at the 1st Canadian conference on geotechnique and natural hazards, Vancouver, pp 389–396

Locat J (1997) Normalized rheological behaviour of fine muds and their flow properties in a pseudoplastic regime. Paper presented at the 1st ASCE international conference on debris-flow hazards mitigation: mechanics, prediction and assessment, San Francisco, 7–9 Aug 1997, pp 260–269

Locat J, Demers D (1988) Viscosity, yield stress, remolded strength, and liquidity index relationships for sensitive clays. Can Geotech J 25:799–806

Locat J, Leroueil S, Locat P (2003) On the mobility of quick clays: the cases of the St. Jean-Vianney flowslides of 1663 and 1971. Paper presented at 2nd symposium on rapid mass movements, Naples

Locat P, Leroueil S, Locat J (2008) Remaniement et mobilité des débris de glissements de terrain dans les argiles sensibles de l'Est du Canada. Paper presented at the 4th Canadian conference on geohazards: from causes to management, Presse de l'Université Laval, Québec

Locat A, Leroueil S, Bernander S, Demers D, Jostad HP, Ouehb L (2011) Progressive failures in Eastern Canadian and Scandinavian sensitive clays. Can Geotech J 48:1696–1712

Mitchell RJ, Markell AR (1974) Flowsliding in sensitive soils. Can Geotech J 11:11–31

Mollard JD, Hughes GT (1973) Earthflows in the Grondines and Trois Rivières Areas, Québec: discussion. Can J Earth Sci 10:324–328

Paré D (1980) Étude géotechnique, glissement de Havre-Saint-Pierre. Dossier 01381180(17)001, Ministère des Transports du Québec, Québec

Potvin J, Pellerin F, Demers D, Robitaille D, La Rochelle P, Chagnon J-Y (2001) Revue et investigation supplémentaire du site du glissement de Saint-Jean-Vianney. Paper presented at the 54th Canadian geotechnical conference, Calgary, 2, pp 792–800

Rissman P, Lajoie G (1980) Rapport Sommaire de visite. Ministère de l'Énergie et des Ressources, Québec, pp 3

Tavenas F, Leroueil S (1981) Creep and failure in slopes in clays. Can Geotech J 18:106–120

Tavenas F, Chagnon J-Y, La Rochelle P (1971) The Saint-Jean-Vianney landslide: observations and eyewitness accounts. Can Geotech J 8:463–478

Tavenas F, Flon P, Leroueil S, Lebuis J (1983) Remolding energy and risk of slide retrogression in sensitive clays. Paper presented at the symposium on slopes on soft clays, Swedish Geotechnical Institute Report No 17, Linköping, pp 423–457

Chapter 12
The Evolution of Material Properties Within an *In Situ* Shear Zone in Sensitive Clay

Jean-Sébastien L'Heureux, Ragnar Moholdt, Vidar Gjelsvik, and Einar Lyche

Abstract In this paper, we present the geotechnical properties and microstructure of a clay from Mosjøen, Northern Norway, where a landslide occurred in June 2011. The evolution of the clay properties at the depth of failure (i.e. shear zone) are studied based on an integrated set of samples, laboratory tests, CPTU data and piezometer readings. Results from SEM analyses show a general lowering of the macro porosity of the clay due to shearing during the landslide. As a consequence, the lower permeability within this zone slowed down the consolidation process when compared to the surrounding intact material. Despite of this, as pore pressure decreased from a situation with nearly zero effective stress to a situation close to the hydrostatic pore pressure, the shear resistance of the clay slowly increased; from its remoulded strength and towards values slightly higher than typically found for normally consolidated clays. It was also found that the peak strength envelope of the destructured clay was above the envelope of the intact material. The reasons could be attributed to the microstructure of the remoulded material and to its silty nature.

Keywords Sensitive clay • Microstructure • Landslide • Reconsolidation • Shear strength

12.1 Introduction

The shear resistance of clay plays a major role in the quantitative evaluation of slope stability. Its determination constitutes a complex problem as this parameter is not unique for a given clay material, but evolves with the stress and strain path

J.-S. L'Heureux (✉) • R. Moholdt • V. Gjelsvik
Norwegian Geotechnical Institute (NGI), Trondheim, Norway
e-mail: jsl@ngi.no

E. Lyche
Norwegian Water Resources and Energy Directorate (NVE), Trondheim, Norway

J.-S. L'Heureux et al. (eds.), *Landslides in Sensitive Clays: From Geosciences to Risk Management*, Advances in Natural and Technological Hazards Research 36,
DOI 10.1007/978-94-007-7079-9_12, © Springer Science+Business Media Dordrecht 2014

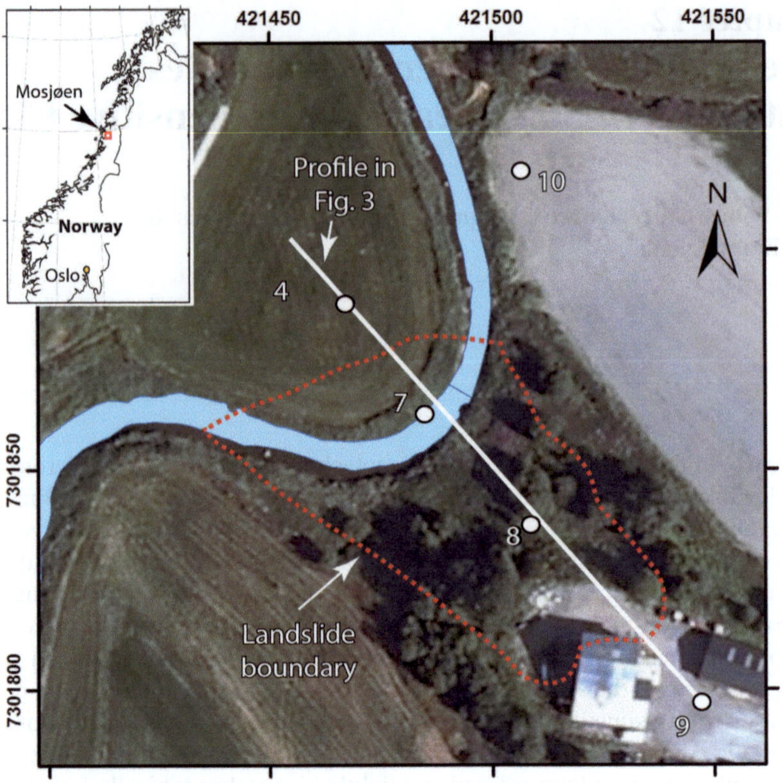

Fig. 12.1 Location of the study area near Mosjøen, Northern Norway (UTM-33N) and aerial photography prior to June 2011. The boundary of the landslide is shown in *red* while the *white circles* indicate the test site locations

to which the soil has been subjected. For sensitive clay, it is generally accepted that part of this complexity results from its microstructure (e.g. Mitchell and Soga 2005). Alteration or collapse of the clay structure is generally invoked to account for the changes in the mechanical properties. Albeit a few laboratory studies comparing the behaviours of the same soil in the intact state, remoulded state, reconstituted and reconsolidated, there are few direct field studies discussing the effect of destructuration and reconsolidation in relation to the geotechnical behaviour of sensitive clays.

This paper presents field and laboratory results collected at three occasions between 2011 and 2012 in and around the shear zone of a landslide which occurred in June 2011 near Mosjøen in Northern Norway (Fig. 12.1). The aim of the study is to evaluate the effect of reconsolidation on the geotechnical properties of the clay within the shear zone. In parallel, the microstructure of the clay has been studied by means of scanning electron microscopy (SEM) in order to evaluate the effect of destructuration.

Fig. 12.2 Landslide seen from the northwest (Photo R. Moholdt)

12.2 Site Description

The study area is located near the river Døla, 3 km east of Mosjøen, in Northern Norway (Fig. 12.1). Deposits in the area consist mainly of fjord and marine sediments which emerged from the sea following a relative fall of sea-level during the Holocene (Follestad 1989). The landslide occurred on June 19th 2011 between the river Døla and the larger farm house presented in Fig. 12.1. Parts of the landslide debris flowed into the river. Here, the total length and width of the landslide masses were c. 100 and 60 m, respectively (Fig. 12.1). To the southeast, the 10 m high and 30 m long headwall stopped just in front of a farm house (Fig. 12.2). The absence of sensitive clay deposits at this location, and the presence of coarse anthropogenic fills, prevented further retrogression towards the farm houses. The compression and heave observed along the foot of the landslide and the subsidence near the headwall (i.e. depletion zone) suggests that movement was dominantly monolithic and rotational. The landslide was likely triggered progressively due to river erosion at the foot of the slope. Placement of a fill at the head of the slope over several years is also believed to have played an important role in destabilizing the slope.

Engineering work was performed on the slope in October and November 2011 for protection purposes. The work included establishment of a new river bed at c. 40 m in front of the landslide foot and levelling of some parts of the landslide area (Fig. 12.3). The levelling included some placement of fills and was done to gain access to the site with a drill rig. A series of vertical drains were thereafter installed with c.t.c. distance of 1.5 m between locations 4 and 8 (Fig. 12.1) along the entire

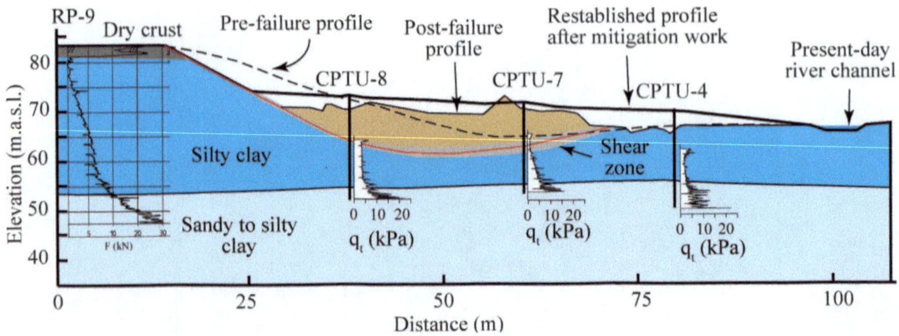

Fig. 12.3 Presentation of the different slope profiles (pre-failure, post-failure and after mitigation work) with results from CPTUs and rotary pressure soundings (RP) performed in June 2011. The failure surface is drawn in *red* with the interpreted shear zone in light *gray*

Table 12.1 Fill heights and depth of vertical drains at the borehole locations in the landslide area

Borehole location	Terrain level after the landslide/top of vertical drain (m.a.s.l.)	Terrain level after placement of fill (m.a.s.l.)	Fill thickness (m)	Depth of vertical drain (m)	Bottom level of vertical drains (m.a.s.l.)
4	+66.8	+69.6	2.8	0–13	+53.8
7	+69.7	+71.6	1.9	0–13	+56.7
8	+70.2	+73.3	3.1	0–17	+53.2

width of the landslide. The drains are 13 m long in the lower landslide area and 17 m long in the upper part. Finally, a berm and some fill were constructed to stabilize the debris and the headwall. An overview of fill thicknesses and depth of vertical drains is given in Table 12.1 for the different borehole locations.

12.3 Field and Laboratory Investigations

The site investigation programme carried out in the study area includes a total of four 54 and 75 mm fixed piston sample series, ten piezocone tests (CPTU) and the installation of seven piezometers. The *in situ* tests and sample extractions were carried out at three occasions (i.e. June 2011, Oct. 2011, 2012) at a reference locality outside the landslide area, and at a few localities within the landslide (Fig. 12.1a). A summary of *in situ* tests and sampling carried out at the different test locations is given in Table 12.2. Due to length restriction only the most relevant data are presented in this paper. We refer to Moholdt (2013) for a full overview of *in situ* test results.

In the laboratory, routine tests for determination of water content, grain size analyses (falling drop method), salt content, bulk density, and consistency limits

Table 12.2 Summary of *in situ* tests and sampling carried out at the different test locations (see Fig. 12.1 for location of test sites)

Test location	June 2011	October 2011	October 2012
4	PZ, RP, CPTU	–	CPTU
7	RP, CPTU	PZ, PC, CPTU	PC, CPTU
8	RP, CPTU	CPTU	CPTU
9	RP	–	–
10	–	–	PZ, RP, PC, CPTU

PZ piezometer, *RP* rotary pressure sounding, *PC* piston sampling, *CPTU* cone penetration test with pore pressure measurements

were carried out on a large number of samples. A series of four triaxial CAUC tests and four oedometer (CRS) tests were performed on samples collected in October 2012. We refer to Moholdt (2013) for more information on these tests.

A total of 6 samples from both boreholes 7 and 10 were analyzed using scanning electron microscopy (SEM) at the University of Oslo. The samples were prepared by first freezing them in liquid nitrogen. The surfaces to be observed were obtained by cutting while the sample was still frozen. Cutting while frozen normally ensures that the surface is as planar as possible and that the porosity is well exposed to the viewer. The samples were thereafter freeze-dried before SEM analysis. Magnifications used for SEM analyses varied from 20 to a maximum of 20,000. During inspection, micrographs were taken at increasing magnifications. Characteristics of the micro-structure (orientation, alignments, grain sizes, shapes), types of particles present and features of special interest were noted.

12.4 Results and Interpretations

12.4.1 Soil Conditions and Failure Plane

The stratigraphy in the study area consists of a thin crust of dry clay (1–2 m thick) reposing on top of a silty clay deposit (Fig. 12.3). This marine deposit is up to 30 m thick in borehole 9 and thins down to about 15 m in borehole 4. Several thin seams of silt and fine sand were found in the silty clay sequence. The sandy intervals increase in the boreholes and the *in situ* tests from about 55 m.a.s.l. This shift in sediment composition likely represents the boundary between marine and glacio-marine sediment.

Samples from borehole 10 (i.e. reference site) show a clay content varying from 20 to 26 % in the marine deposit. The water content varies from 26 to 33 % and the plastic index is in the range 9–14 %. The salt content is lower than 1 g/l in all samples tested while the sensitivity (i.e. the ratio of undrained shear strength to remoulded shear strength) was measured up to 50. Oedometer test results show that the clays are overconsolidated at the reference site with an OCR of 5.6.

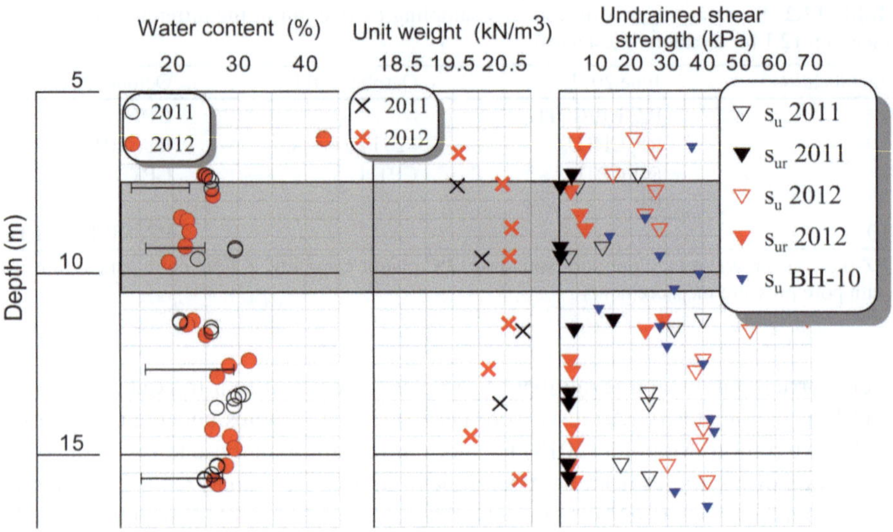

Fig. 12.4 Water content, unit weight and undrained shear strength (fall cone) at test location seven measured on samples retrieved in October 2011 and 2012. The undrained shear strength for borehole 10 (reference site) are also shown. The *grey* interval represents the location of shear zone

A profile through the main axis of the landslide is shown in Fig. 12.3. In the compression zone, borehole 7 shows soft and sensitive clay in the interval 61–64 m.a.s.l. (Figs. 12.3 and 12.4). Undrained shear strength values are low and close to the remoulded strength of the soil in some samples (Fig. 12.4). The same interval, interpreted as the failure zone, also shows low values of corrected tip resistance (q_t) on CPTU-7 (i.e. less than 250 kPa). However, the slide plane and the surrounding failure zone are not so discernible in the middle of the landslide at borehole location 8. Here, results from rotary pressure sounding shows that the top 6 m of soil consists of sand and gravel (landslide debris). The slide plane is assumed to occur immediately below these coarser deposits, likely at the top of the soft clay layer observed on CPTU-8. Unfortunately, no sampling was performed at this location.

12.4.2 Microstructural Changes

The microfabric of the clay was investigated through a series of SEM analyses. A few micrographs were selected to illustrate the main signatures related to the clay found outside the landslide area (borehole 10) and within the failure zone (borehole 7). Representative SEM micrographs are shown in Fig. 12.5. As expected, sediment deposition in the marine environment led to an open and flocculated particle assemblage governed by edge-to-face and edge-to-edge particle associations (Fig. 12.5a–d). Clay connectors are observed between silt particles on the

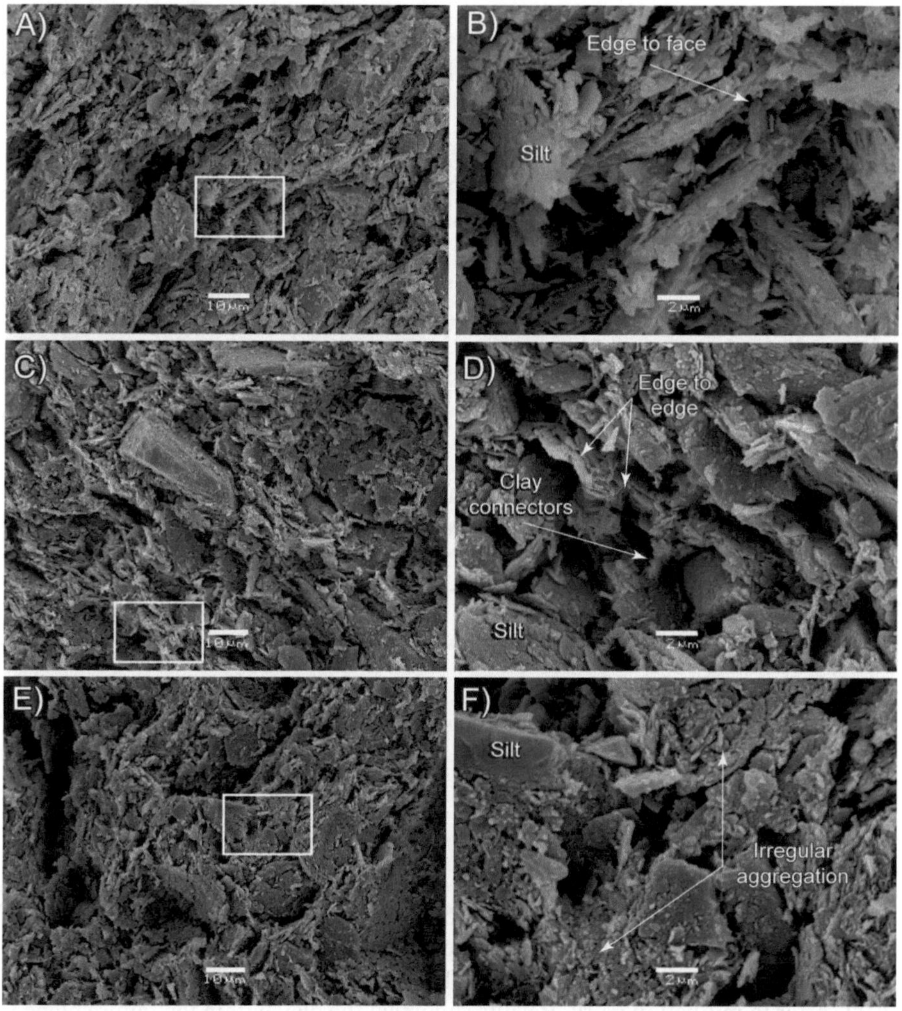

Fig. 12.5 SEM examples of the clay soil within and outside the Døla landslide. (**a, b**) From borehole 10 at a depth of 6.57 m, (**c, d**) from borehole 10 at a depth of 6.82 m, and (**e, f**) from borehole 7 at a depth of 7.47 m. The *small squares* on the lower magnification micrographs show the location of the larger magnification. All samples are oriented normal to bedding planes

micrographs (Fig. 12.5d). For the samples collected within the shear zone, micrographs show that most of the clay connectors have been crushed during the failure process.

This led to a more irregular aggregation of clay particles around the silt grains and generally to a lower porosity (Fig. 12.5f). It seems, however, that only the largest macro pores were destroyed during the landslide process since several smaller pores with clay bridges were also found in the samples from the shear zone.

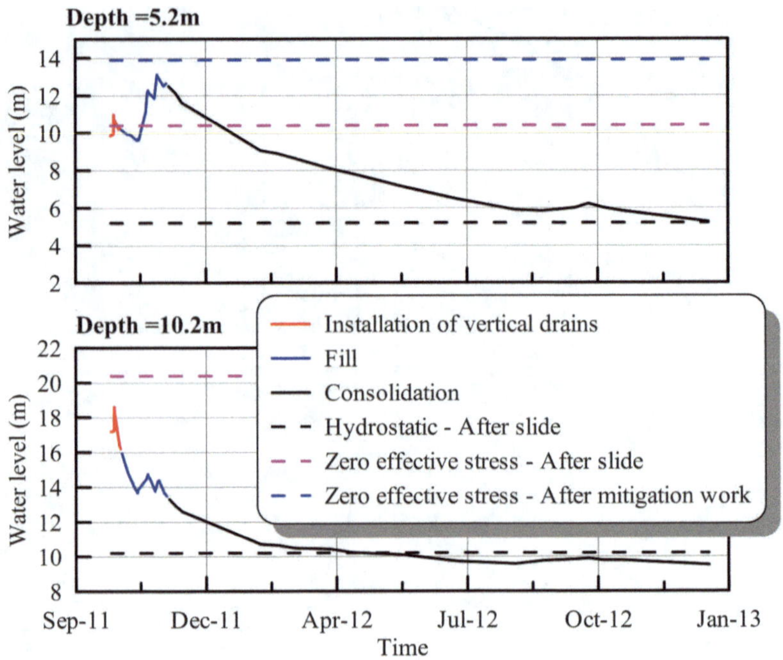

Fig. 12.6 Pore-pressure evolution following the June 19th landslide at Døla. The results are given for depth of 5.2 and 10.2 m at location 7 (see Fig. 12.1 for location)

12.4.3 Pore Pressure Evolution

Piezometers were installed at a depth of 5.2 and 10.2 m at location 7. The depth of the upper sensor lies within the shear zone, while the lower sensor is situated immediately below the remoulded clay zone. The first readings were recorded in October 2011, prior to mitigation work. Results show that the clay within the failure plane is still in a liquefied state 4 months after the landslide since the pore water pressure at a depth of 5.2 m is close to the total stresses in the soil (i.e. 105 kPa; Fig. 12.6). High excess pore pressures are also registered at a depth of 10.2 m (Fig. 12.6). The piezometer readings show that the pore pressure increased slightly following the installation of vertical drains in October 2011. Again, due to mitigation work and to the placement of a fill, the pore pressure increased at location 7 in November 2011. The latter is noticeable especially for the sensor at 5.2 m. In contrast, the response was minimal for the sensor at 10.2 m and for the piezometer at location 4.

Pore pressure evolution further shows that the vertical drains were effective at 10.2 m depth at location 7 and at test location 4. Here the excess pore pressure dissipated rapidly and hydrostatic conditions were achieved 3–4 months after the installations of the drains (Fig. 12.6). Within the assumed shear zone, dissipation of excess pore pressure was slower and it took over 1 year after the installation of the

Table 12.3 Estimated geotechnical parameters from oedometer (CRS) tests and triaxial (CAUC)=tests on samples from collected in 2012 at borehole localities 7 and 10

				Oedometer test results (CRS)		Triaxial test results (CAUC)			
Borehole	Depth (m)	Elevation (m.a.s.l.)	$\Delta e/e_0$	OCR	Permeability (m/s)	φ (a=0)	φ (a=10)	s_{uA}/p_0'	α*
7	8.5	63.1	0.08	1	–	36.5	32.1	0.51	0.51
7	9.6	62.0	0.07	1	$1.97 \cdot 10^{-10}$	36.5	32.5	0.50	0.50
7	11.6	60.0	0.08	1	$5.11 \cdot 10^{-10}$	–	–	–	–
10	6.5	60.5	0.07	–	$7.32 \cdot 10^{-10}$	36.6	29.4	0.74	–
10	8.5	58.5	0.02	5.6	$9.39 \cdot 10^{-10}$	34.8	30.0	0.82	0.29

$\Delta e/e_0$ change in void ratio during the consolidation phase, s_{uA} active undrained shear strength, p_0' in situ effective stress, φ friction angle, a attraction
$*\alpha = s_{uA}/(p_0' \cdot OCR^{0.6})$

drains before reaching hydrostatic conditions (Fig. 12.6). The reason for the large differences in the time taken to reach hydrostatic conditions between the different sensors is attributed to the microstructural changes in the clay within the shear zone as discussed above. The reduced porosity in this zone directly affect the permeability of the material and hence the time for consolidation. In support to this, oedometer test results show lower permeability values for the clay within the shear zone (Table 12.3).

12.4.4 Changes Over Time in the Failure Zone

Figure 12.4 shows results of water content, unit weight and undrained shear strength from samples collected at location 7 in October 2011 and in October 2012. During the reconsolidation process, the water content in the failure zone decreased from about 28 to 22 %. Similarly, an increase in unit weight (about 0.5 kN/m^3) can be observed. Measurements of water content and unit weight are more or less constant for samples below 10 m depth (Fig. 12.4). Furthermore, results from laboratory investigations show a general increase in undrained shear strength in the failure zone with time. The undrained shear strength in October 2012 was measured to c. 22 kPa compared to values of about 5 kPa in October 2011 (Fig. 12.4). The undrained shear strength values in October 2012 are close to and slightly higher than those measured at the reference location outside the landslide (i.e. BH-10; Fig. 12.4).

A summary of results from oedometer tests and triaxial tests carried out on samples borehole 7 and 10 is presented in Table 12.3. The oedometer data show that the reconsolidated clay within the shear zone no longer exhibits its former overconsolidation. OCR values are close to unity in samples from borehole 7. During the triaxial tests, all samples showed an initial contractive behaviour and then dilatancy towards the failure line. Although the number of tests is not extensive, results show that the friction angle is somewhat higher for the reconsolidated samples from the

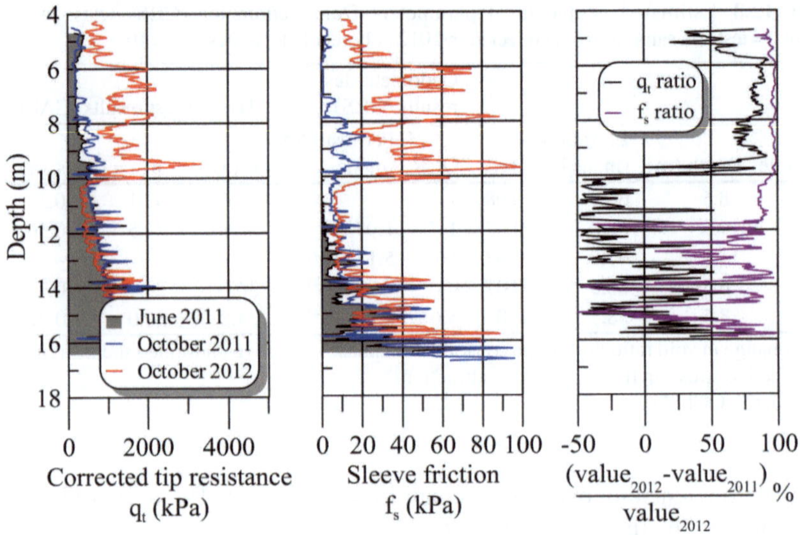

Fig. 12.7 Comparison of CPTU results between June 2011 and October 2012 at location 7 (see Fig. 12.1)

shear zone (Table 12.3). Similarly, and when compared in terms of effective stress and OCR, the normalized shear strength (i.e. α* in Table 12.3) is found to be slightly higher in the reconsolidated clay material.

Changes in material properties can also be evaluated from CPTU test results. CPTU tests were carried out at three different occasions at locations 7 and 8 in the landslide area and a two occasions at location 4 (Table 12.2). Results over time for the CPTUs carried out at location 8 show little variation in measured tip resistance and/or sleeve friction around the failure plane. As mentioned above, the depth of the failure plane is difficult to interpret at this location and the remoulded shear zone is most likely thin. Similarly, the tests carried out at location 4 gave very similar results in June 2011 and in October 2012. The most important changes over time are observed from the CPTUs performed at location 7. Here a distinct increase in tip resistance and sleeve friction can be observed in the remoulded zone (depth interval 5–10 m; Fig. 12.7). In the shear zone, the corrected tip resistance increased from about 300 to 1,250 kPa, while the average sleeve friction increased from only c. 1 to 37 kPa. The largest increase in tip resistance and sleeve friction occurred between October 2011 and 2012, which corresponds to the consolidation period (Fig. 12.6). With the exception of the lower q_t values measured in October 2012, the observed changes in q_t and f_s are minimal below the depth of 10 m. This suggests that the clay from this level and downwards was not much affected by the landslide and the reconsolidation process. The lower values of q_t in October 2012 are difficult to explain and these might be inferred to e.g. zeroing effect of the CPTU probe or saturation problems. Other differences in q_t and f_s between 2011 and 2012 can also be associated to soil heterogeneities such as sand/silt layers. Unfortunately, the measured excess pore pressure u_2 for CPTU-7 in October 2012 is not reliable and does not permit comparison with previous sounding results.

12.5 Discussion and Conclusions

For landslides in sensitive clay material, results presented in this study show that the excess pore pressure dissipation and consolidation is slower within the shear zone than in the surrounding intact material. The main reason is attributed to the microstructural changes within the shear zone associated to the failure process. Crushing of the particle assemblage leads to a lower macro porosity and lower permeability. Similarly, results from SEM analyses seem to show that only the largest macro pores collapsed during the landslide. The fact that intact macro pores and parts of the fragile particle assemblage still exists in the passive landslide zone may indicate that the remoulding process was not completed during the landslide. This is similar to observations made by Delage and Lefebvre (1984) on clays from eastern Canada where it was shown that intact macro pores may subsist at pressures exceeding the preconsolidation pressure.

For the Døla landslide, a great number of vertical drains were used to facilitate reconsolidation of the remoulded clay material around the failure plane. During the reconsolidation process the water content of the clay diminished and the unit weight increased. Consequently, the shear resistance of the clay slowly increased to values slightly higher than those of the intact material (i.e. BH-10). This is similar to findings obtain from studies on e.g. friction piles and on laboratory tests (e.g. Roy and Lemieux 1985). The design of protection measures at the Døla landslide anticipated an increase in soil strength as a result of pore pressure dissipation in time. Long term stability calculations assumed hydrostatic pore pressure and a relationship between shear strength and effective overburden stress which is typically found for normally consolidated clay ($s_{ua}/p_0' = 0.28$). Results from triaxial tests show that this assumption was conservative as the s_{ua}/p_0' was found to be slightly higher than anticipated (i.e. $s_{ua}/p_0' = 0.35$).

Results from triaxial tests showed that the peak strength envelope of the destructured and reconsolidated clay is above the envelope of the intact clay. This contradicts observations made on clays from eastern Canada (e.g. Leroueil et al. 1979). The reasons for this are not fully understood, but could be due to the differences in microstructure between the Canadian and Norwegian clays. Shearing in clays from eastern Canada is controlled by the aggregate nature of the clays which are very stiff and angular. After breaking down of the bonds between particles and aggregates, shearing occurs through the aggregates and could be controlled by the particles themselves as the aggregates become smoother and closer to each other (Saihi et al. 2002). On the other hand, shearing of the destructured clay from Døla seems to be controlled by its silty nature (clay content less than 26 %), following breaking of the bonds.

It is often difficult for an engineer to correctly characterize the long-term behaviour of a remoulded clay and its effect of the stability of a slope when designing slope protection measures. Field and laboratory observations presented in this study are consistent with results found in the literature and show that it is conservative to use shear strength and effective stress relationships typical for normally consolidated clays. Note, however, that the time to reach full pore pressure dissipation may vary greatly from site to site and will depend mostly on e.g. the soil properties, stratigraphy, geometry of the slope, thickness of the remoulded material (i.e. shear zone) and the amount and the type of vertical drains. After reconsolidation, further strength increase is also anticipated due to the combined influence of ageing and electro-chemical processes. This is believed to be a long-term effect and should be investigated in the future.

Acknowledgments This project was financed by the Norwegian Research Council through the strategic project (SP1) at NGI and by the National research program "Natural hazards: Infrastructure, Floods and Slides (NIFS)". We also thank Dr. Anders Gylland for is constructive review.

References

Delage P, Lefebvre G (1984) Study of the structure of a sensitive Champlain clay and its evolution during consolidation. Can Geotech J 21:31–35

Follestad BA (1989) Fustavannet 1926 IV, Quaternary geology map. 1:50.000. Geological Survey of Norway (NGU). (In Norwegian)

Leroueil S, Tavenas F, Brucy F, La Rochelle P, Roy M et al (1979) Behaviour of destructured natural clays. J Geotech Eng Div ASCE 105(GT6):759–778

Mitchell JK, Soga K (2005) Fundamentals of soil behavior. Wiley, New York

Moholdt R (2013) Skred ved Døla i Vefns. Undersøkelse av materialegenskaper. NGI report 20120166-01-R. (In Norwegian)

Roy M, Lemieux M (1985) Long-term behaviour of reconsolidated clay around a driven pile. Can Geotech J 23:23–29

Saihi F, Leroueil S, La Rochelle P, French I et al (2002) Behavior of the stiff and sensitive Saint-Jean-Vianney clay in intact, destructured, and remoulded conditions. Can Geotech J 39:1075–1087

Part III
Integrated Geotechnical and Geophysical Site Investigations

Chapter 13
The Use of Geophysics for Sensitive Clay Investigations

Shane Donohue, Michael Long, Jean-Sébastien L'Heureux, Inger-Lise Solberg, Guillaume Sauvin, Magnus Rømoen, Thomas Kalscheuer, Mehrdad Bastani, Lena Persson, Isabelle Lecomte, and Peter O'Connor

Abstract Marine clay deposits in coastal, post-submarine areas of Scandinavia and North America may be subjected to quick clay landslides and hence significant efforts are being taken to map their occurrence and extent. Recently, considerable efforts by a number of researchers have been made to investigate areas of sensitive

S. Donohue (✉)
School of Planning, Architecture and Civil Engineering, Queen's University Belfast, Belfast, Northern Ireland, UK
e-mail: s.donohue@qub.ac.uk

M. Long
School of Civil, Structural and Environmental Engineering, University College Dublin, Dublin, Ireland

J.-S. L'Heureux • M. Rømoen
Norwegian Geotechnical Institute (NGI), Oslo, Norway
e-mail: jsl@ngi.no

I.-L. Solberg
Geological Survey of Norway (NGU), Trondheim, Norway
e-mail: inger-lise.solberg@ngu.no

G. Sauvin • I. Lecomte
International Centre for Geohazards (ICG), Oslo, Norway

Department of Geosciences, University of Oslo (UiO), P.O. Box 1072, Blindern 0316 Oslo, Norway

NORSAR, P.O. Box 53, N-2027 Kjeller, Norway
e-mail: Guillaume.Sauvin@norsar.no; Isabelle.Lecomte@norsar.no

T. Kalscheuer
ETH Zürich, Zürich, Switzerland

M. Bastani • L. Persson
Geological Survey of Sweden (SGU), Uppsala, Sweden

P. O'Connor
Apex Geoservices, Gorey, Ireland

J.-S. L'Heureux et al. (eds.), *Landslides in Sensitive Clays: From Geosciences to Risk Management*, Advances in Natural and Technological Hazards Research 36,
DOI 10.1007/978-94-007-7079-9_13, © Springer Science+Business Media Dordrecht 2014

clay using a range of geophysical techniques. Although the majority of this work has focussed on measurements of electrical resistivity, other electromagnetic and seismic geophysical techniques have also received attention in the literature. The purpose of this paper is to review recent research concerning the effectiveness of a number of geophysical techniques for investigating sensitive clays. In addition to discussing a number of case studies, this review will also consider recent work showing the correlation of geophysical measurements, and in particular electrical resistivity, with a range of relevant engineering properties.

Keywords Sensitive clay • Geotechnical properties • Geophysics, resistivity • Seismic • Electromagnetics

13.1 Introduction

A large amount of marine clay in the Northern Hemisphere that was deposited during the Pleistocene epoch currently lies above sea level, as a result of isostatic uplift following deglaciation. The pore water chemistry of these materials may have been altered as a consequence of the change from a marine to a freshwater environment. Salt, which originally contributed to the bonding between the clay particles may, therefore, have been leached from these materials by ground water and percolating surface water. If sufficient leaching of salt from the soil pore water occurred, a highly sensitive or "quick" material may develop.

Some of the most densely inhabited regions of Scandinavia, are located in potential quick clay areas. Over recent years, considerable efforts have been made with respect to mapping of quick clay formations using combined geotechnical and geophysical methods. Although it was recognised that some intrusive geotechnical investigations will always be necessary, the objective of these studies was to develop techniques to maximize the use of non-intrusive geophysical surveys. Although most recent research efforts on this topic have taken place in Scandinavia, quick clays continue to pose a hazard in other countries such as Canada (Geertsema and Torrance 2005) and Japan (Torrance and Ohtsubo 1995).

The purpose of this paper is to review recent research concerning the effectiveness of a range of geophysical techniques for investigating sensitive clays. Geophysical measurements to be discussed will include resistivity, electromagnetic and seismic methods. In addition to discussing a number of case studies, illustrating the use of these techniques, this paper will also consider recent work showing the correlation of geophysical measurements, and in particular electrical resistivity, with a range of relevant engineering properties.

13.2 Geo-electrical

Recently there has been a considerable amount of work published on the use of geo-electrical measurements for mapping quick clay. These field studies have compared measured resistivity values with salt content (Solberg et al. 2008, 2012a; Long et al.

2012), as well as remoulded shear strength and sensitivity (Rankka et al. 2004; Dahlin et al. 2005; Lundström et al. 2009; Donohue et al. 2012; Long et al. 2012; Löfroth et al. 2012). Unleached marine clay, which maintains a large concentration of ions in its pore water, has been shown to have very low values of resistivity, generally less than 10 Ωm. On the other hand, for quick clay, where significant leaching of salt has occurred, the resistivity value is expected to be higher than that of the unleached clay. Experience has shown that an effective first-order interpretation of Electrical Resistivity Tomography (ERT) profiles can be performed by using the following classification of resistivity values (Solberg et al. 2012a): Unleached clay deposits: 1–10 Ωm; Leached clay deposits, possibly quick: 10–100 Ωm; Dry crust clay deposits and coarse sediments: >100 Ωm. Also, quick clay may become non-quick with further leaching as more stabilising ions begin to dominate the pore water (Solberg et al. 2008). In this case the concentration of ions may be the same as quick clay and as such the resistivity will remain in the 10–100 Ωm range. The values quoted above are based on Norwegian marine clays. According to Dahlin et al. (2001), quick clays in Western Sweden usually have a resistivity higher than about 7 Ωm. Recently, Lundström et al. (2009) reported that Swedish quick clays generally exhibit lower resistivity values (as low as 5 Ωm) and attributed this to the higher clay content generally present in Swedish clays.

After this first evaluation, a refined site-specific classification can be made, taking into account local mineralogical, geological and hydrogeological conditions, etc. (see further descriptions on method, applications and classification in Solberg et al. this volume).

13.2.1 Electrical Resistivity Tomography

The use of Electrical Resistivity Tomography (ERT – also called 2D resistivity) as a tool for identifying potential quick clay has expanded during the last decade due to advances in the measurement technique and the data acquisition and processing software. The development has also been driven by the relatively high cost of traditional drilling and sampling techniques. In an area without previous investigations, ERT gives an overview of the subsurface as a basis for determining the optimum locations for further investigation (Solberg et al. 2008). The method is a valuable supplement to drilling and once calibrated it can separate intact marine clay deposits (high salt content – low resistivity) from quick clay (low salt content – higher resistivity), in addition to indicating coarser material and bedrock.

An example of ERT measurement from Melhus (mid Norway) is presented in Fig. 13.1. The study area is located in the scar of a relatively large quick-clay landslide, dated to ca. 1665 (Rokoengen 1998). Thick marine deposits are surrounded by bedrock hills, and a small glaciofluvial deposit (Reite and Sørensen 1980). The ERT measurements are based on the Lund-system developed by Dahlin (1993). The measuring equipment was an ABEM Terrameter SAS 4000 (ABEM 1999), using a current of about 200 mA. The multiple gradient electrode array was used with steel

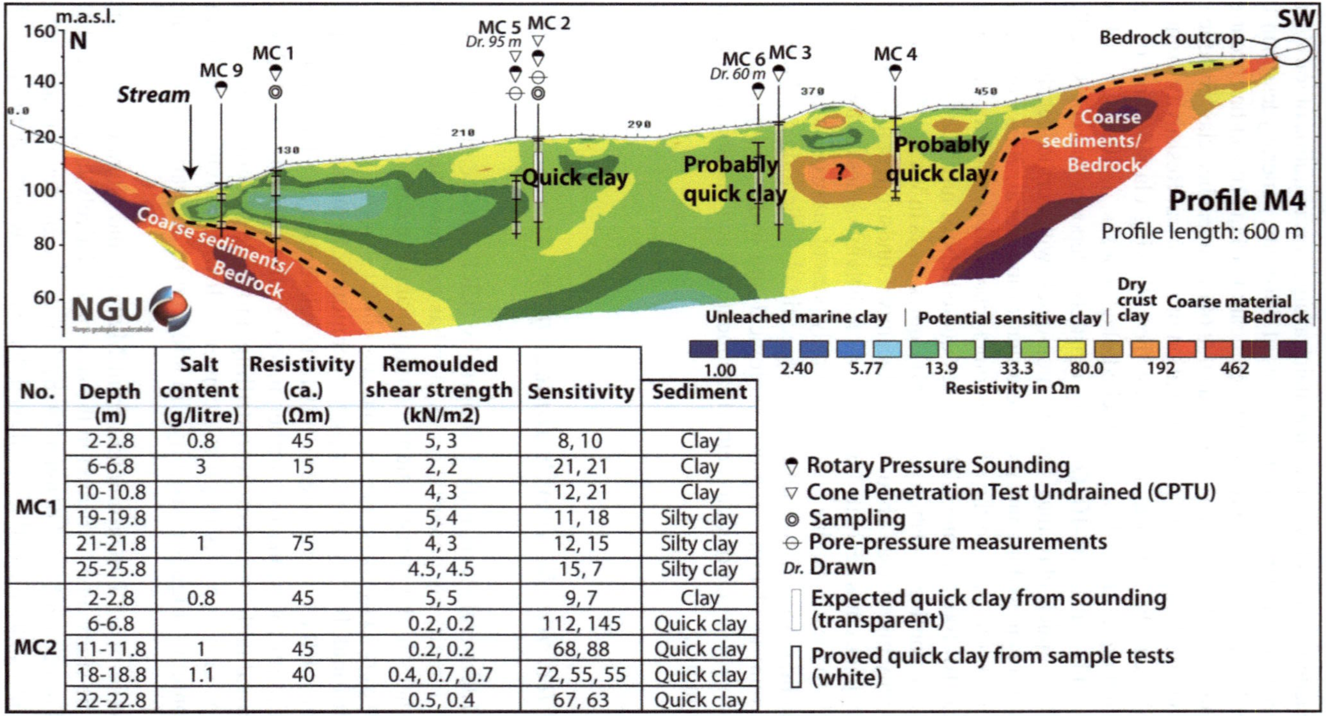

No.	Depth (m)	Salt content (g/litre)	Resistivity (ca.) (Ωm)	Remoulded shear strength (kN/m2)	Sensitivity	Sediment
MC1	2-2.8	0.8	45	5, 3	8, 10	Clay
	6-6.8	3	15	2, 2	21, 21	Clay
	10-10.8			4, 3	12, 21	Clay
	19-19.8			5, 4	11, 18	Silty clay
	21-21.8	1	75	4, 3	12, 15	Silty clay
	25-25.8			4.5, 4.5	15, 7	Silty clay
MC2	2-2.8	0.8	45	5, 5	9, 7	Clay
	6-6.8			0.2, 0.2	112, 145	Quick clay
	11-11.8	1	45	0.2, 0.2	68, 88	Quick clay
	18-18.8	1.1	40	0.4, 0.7, 0.7	72, 55, 55	Quick clay
	22-22.8			0.5, 0.4	67, 63	Quick clay

Fig. 13.1 ERT profile from Melhus, Mid Norway (Solberg and Dalsegg 2012), compared to geotechnical data (Sandven and Vik 2011). There is good agreement between the datasets

electrode separations of 5 m. Raw data from the resistivity measurements give the apparent resistivity (ρa) of the subsurface. This represents a weighted mean of all the resistivity values that fall within the soil volume of influence. To obtain the specific resistivity distribution (ρ in Ωm), the data are inverted, in the present case using the least-squares method (Loke 2010).

The ERT profile from Melhus shows high resistivities (>100 Ωm) in both ends indicating bedrock, possibly overlain by coarse sediments (Fig. 13.1). The intervening deposits are interpreted as probable leached clay (Solberg and Dalsegg 2012).

Geotechnical investigations were carried out in seven locations after the resistivity profiling (Sandven and Vik 2011). Sounding indicated quick clay in several places and the table included in Fig. 13.1 shows material properties from sampling. All the samples in borehole MC2 show quick clay, with corresponding resistivities of about 40–45 Ωm. In borehole MC1 quick clay is not detected, and here the clay is less leached than the other parts of the profile (lower resistivities). Several of the soundings along the profile indicate medium sensitive or quick clay. Also salt content measurements have been carried out, showing that the clay is highly leached, and the resistivity values corresponds very well.

Even though the area has previously been exposed to a landslide, the remaining material is still quick, or has been further leached and subsequently become quick. There is also very little or no dry crust clay. The resistivity profile helps the interpolation between drillings, and also shows that there are no barriers against landsliding (e.g. unleached clay, buried bedrock peaks) in the centre of the area.

13.2.2 Resistivity Cone (RCPT/RCPTU)

A number of authors have recently reported the results of resistivity cone measurements in sensitive fine grained materials (e.g. Rømoen et al. 2010; Solberg et al. 2012a; Löfroth et al. 2012). There is some evidence that RCPTUs give slightly lower resistivity values than inverted ERT data in highly sensitive clays (Schälin and Tornborg 2009). However, Sauvin et al. (2011) and Löfroth et al. (2012) found good agreement between ERT and RCPTU data for quick clays sites in Norway and Sweden, respectively. Dahlin et al. (2004) also found good correlation between ERT and RCPTU data, but their study did not include quick clay.

A recent example of some RCPTU data acquired in central Oslo (Norway) is shown in Fig. 13.2. As part of this investigation, three RCPTU tests were performed at locations which had been previously sampled. Of these three boreholes one (1–35) displayed a decrease in remoulded shear strength (s_{ur}) below 11 m and consisted of quick clay below 17 m depth.

The upper 8–10 m of tests 1–23 and 1–35 exhibit relatively high resistivities, when compared to test 1–4, possibly due to weathering and to a higher amount of sand and silt found at these locations. From around 11 m depth, however, a clear difference is observed in the resistivity measured at 1–23 and at 1–35. While the resistivity measured at 1–4 and 1–23 remain below 5 Ωm, the resistivity gradually

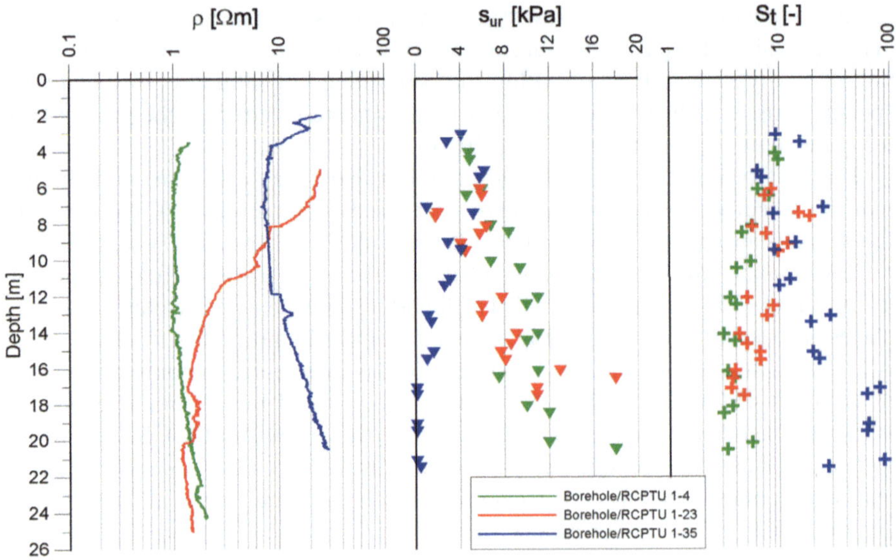

Fig. 13.2 Three resistivity profiles acquired using RCPTU, along with corresponding remoulded strength (s_{ur}) and sensitivity data

increases from 8 to 12 Ωm in 1–35. A corresponding gradual reduction in remoulded strength and increase in sensitivity is observed below 11 m and the material may eventually be considered "quick" from about 16 m depth.

13.3 Electromagnetics

Although the use of geoelectrical measurements has received some attention in the literature, as described above, most of this work has focussed on ERT, with Solberg et al. (2008) also making use of Induced Polarisation (IP) for differentiating between bedrock and sediments and seismic refraction. Electromagnetic conductivity mapping has, in contrast only received limited attention for investigating quick clay. Calvert and Hyde (2002) used a Geonics EM-34 ground conductivity meter for mapping the variation in electrical conductivities on unstable slopes that are underlain by sensitive marine sediments in the Ottawa valley, Canada. Donohue et al. (2012) used a Geonics EM-31 ground conductivity meter in their quick clay investigations at a site in Southern Norway and detected a contrast in conductivity, relating to areas of leached (i.e. quick) and unleached marine clay. They noted, however, as the depth is limited to 6 m, the EM-31 can only give an indication of the clay 'outcropping' just below the dry-crust.

Recently, other electromagnetic techniques, such as Radiomagnetotelluric (RMT) and Controlled Source Audio Magnetotelluric (CSAMT) have been used to map quick clay (Kalscheuer et al. 2013).

13.3.1 Radiomagnetotelluric (RMT) and Controlled Source Audio Magnetotelluric (CSAMT) Methods

The radiomagnetotelluric method (e.g. Pedersen et al. 2005) is a passive electromagnetic method that employs the signals from remote radio transmitters in the VLF and LF frequency bands between 10 and 300 kHz. The sources of the RMT fields are fixed distant transmitters used for communication as well as long-wavelength (LW) radio transmitters. These transmitters are normally vertical electric dipoles that are located all over the world.

The CSAMT method (e.g. Bastani et al. 2009) employs grounded cables or closed loops of wire as aerials to actively transmit signals at a number of fixed frequencies. The typical frequency range employed in this method is 1Hz to 10 kHz. It is advantageous to use horizontal magnetic dipoles, as they are safer, easier to set up and their range is sufficient to cover several hundred meters.

An example of some inverted data from these techniques is presented in the next section for a site in southern Norway (Kalscheuer et al. 2013). For this example, an EnviroMT instrument, was used to conduct a combined RMT, CSAMT survey (called Controlled Source/Radio Magnetotelluric CSRMT), which enabled measurements in the frequency range 1–250 kHz to be acquired and an electrical resistivity model of the ground be generated (Bastani 2001). The apparent resistivity is proportional to the ratio between the horizontal electric and magnetic field components. The 15–250 kHz frequency range is covered by the RMT method and 1–12.5 kHz by the CSAMT method. The lower frequency signals penetrate deeper.

13.3.2 Joint Inversion of ERT and CSRMT Data

An extensive geophysical and geotechnical field program has been carried out in Smørgrav (Southern Norway) in order to investigate the quick clay distribution. RCPT tests performed at two positions along the profile (RCPT 524 and 525) identified unleached and quick clay to have resistivity ranges of 1–10 Ωm and 10–80 Ωm, respectively (Helle et al. 2009; Donohue et al. 2012). The ERT data were collected with a multiple gradient array with electrode spacing of 5 m. RMT and CSAMT data were collected with a station spacing of 10 m.

Kalscheuer et al. (2013) describe the joint inversion theory in detail and inverted the ERT and CSRMT data to 2D models of electric resistivity. In Fig. 13.3, the final single inversion models of the ERT data (Fig. 13.3a) and the CSRMT data (Fig. 13.3b) are compared to the joint inversion model of the ERT and CSRMT data (Fig. 13.3c). A geological interpretation of the joint inversion model is given in Fig. 13.3c. The shallow resistive structure underneath the south-eastern half of the profile (3,000–4,000 Ωm) is interpreted as limestone. As to be expected from other joint inversion examples (Kalscheuer et al. 2010), the data involved in the joint inversion constrained this structure much better than any single data set alone.

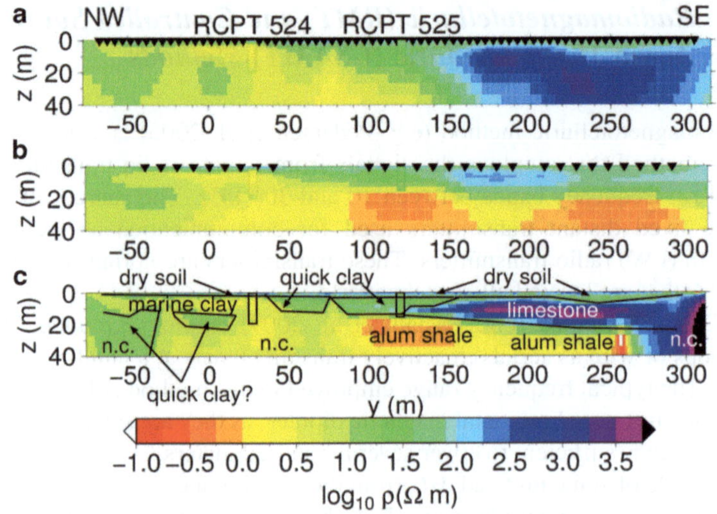

Fig. 13.3 Inversion models of (**a**) ERT data, (**b**) CSRMT data, and (**c**) joint ERT and CSRMT data with geological interpretation (After Kalscheuer et al. 2013). *Inverted triangles* indicate ERT electrodes in (**a**) and RMT stations in (**b**). The resistivity logs of RCPT 524 and 525 are bounded by *black lines*. Model regions which are not constrained by the data are marked with "n.c."

The conductive structure at a depth of more than 20 m and with a resistivity of about 0.5 Ωm is predominantly constrained by the CSRMT data, and interpreted to be alum shale. Due to the overlying resistive limestone, ERT currents are impeded from entering the conductor.

In agreement with the RCPT logs (Fig. 13.3) and rotary pressure soundings in the vicinity of the profile, shallow structure at RCPT 524 with resistivities of 1–10 Ωm is interpreted as unleached marine clay, whereas more resistive structures of 10–80 Ωm at RCPT 525 and at $y = 0$ m along the profile and $z = 12$ m to 20 m below ground surface are considered to be quick clay.

13.4 Seismic

13.4.1 Seismic Refraction

This method uses the first arrivals in the seismic signals to derive a velocity model and assumes that the velocity increases with depth. In engineering geology, this method is used to determine the depth to bedrock. For sensitive clay investigation, seismic refraction has proved to be applicable for locating the undisturbed bedrock below the landslide prone area and therefore help evaluating the thickness of the sensitive clay deposit and the preferential leaching paths (Donohue et al. 2012; Solberg et al. 2012a, b; Chap. 18 by Sauvin et al. this volume; Sauvin et al. 2013).

As part of the seismic refraction method, seismic tomography consists of inverting first-arrival travel-times to derive a velocity field. This method requires more travel-time and is more field demanding compare to classical refraction, but allows retrieving lateral velocity variations.

13.4.2 Surface Wave Analysis

The analysis of surfaces waves is now increasingly used to derive shear wave velocity (V_s) and thus the small strain shear modulus (G_{max}) in subsurface investigations of clay terrain (e.g. Long and Donohue 2007; Donohue and Long 2008; Sauvin et al. 2013). Note that G_{max} is directly related to V_s by:

$$G_{\max} = \rho V_s^2 \tag{13.1}$$

where ρ = density. If a sufficient recording time is considered during acquisition surface waves may be recorded together with P-wave refraction data. Only a few 1D analyses have been performed on sensitive clay. Nevertheless, recent studies (e.g. Long and Donohue 2007, 2010) show that V_s profiles derived from surface wave analysis are comparable to those obtained from correlation with cone penetration test (CPT) or seismic-CPT and correlate well with index properties for these soils. This is discussed further in Sect. 5 of this paper.

Donohue et al. (2012) found that shear wave velocities measured in the quick clay area of the Smørgrav test site appeared to be slightly less than those measured for unleached clay. In order to investigate this further all the available Multichannel Analysis of Surface Waves (MASW) data for Norwegian marine clays are plotted on Fig. 13.4. It is interesting to observe that the V_s profiles for the quick clay sites are all very similar and increase from about 100 m/s at the surface to 200 m/s at 10 m depth. The data for the non-quick sites are more scattered. Particularly low V_s measurements were recorded at Onsøy, where the soil is particularly soft, and in the organic clay at Farriseidet. Perhaps not surprisingly, it seems that MASW measurements are unable to distinguish whether or not clay is quick alone.

13.4.3 High Resolution Seismic Reflection

High resolution seismic reflection data is commonly used in the marine realm and also close to shoreline for landslide investigation in sensitive clay terrain (e.g. L'Heureux et al. 2012; Vardy et al. 2012). Seismic reflection data acquisition using a P-wave source (e.g., TOPAS or boomer) gives very high resolution allowing for detailed interpretation, direct correlation with CPTU and boreholes, and mapping of landslide-prone layers. However, seismic reflection is not commonly used onshore for mapping sensitive clay deposits. The reason is attributed to the low resolution of this method due to high velocity layers at the ground level.

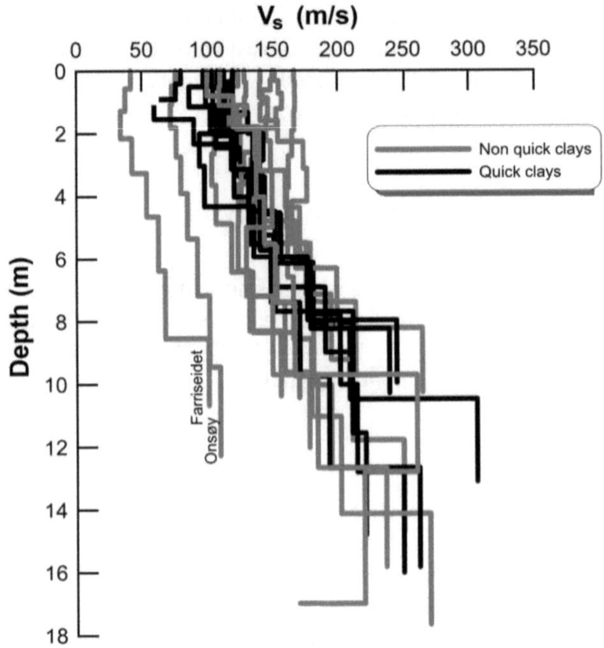

Fig. 13.4 All available MASW data for Norwegian marine clays

To avoid this problem, high resolution seismic reflection data collected onshore using landstreamers and shear-wave vibrator sources has recently been used and the results are promising in clay terrain investigation (Polom et al. 2011; Malehmir et al. 2013a, b; Hansen et al. 2013). In addition to providing stratigraphic information and revealing the sedimentary history (Fig. 13.5), shear wave velocities collected in the reflection mode yields useful quantitative information for geotechnical applications (e.g. L'Heureux et al. 2013).

One of the main drawbacks of the seismic reflection method using a shear-wave source is that it is time consuming and costly especially in rough topography compared to other geophysical methods. Moreover, good signal to noise ratio and recording of high frequencies are mandatory in order to successfully perform shallow seismic reflection with the necessary resolution.

13.5 Relationship Between Geophysical and Geotechnical Properties

13.5.1 Index Parameters

Using a database of 15 sites in Norway, Long et al. (2012) compared inverted ERT resistivity values to basic geotechnical properties obtained from measurements on samples from adjacent boreholes. This work suggested that there is:

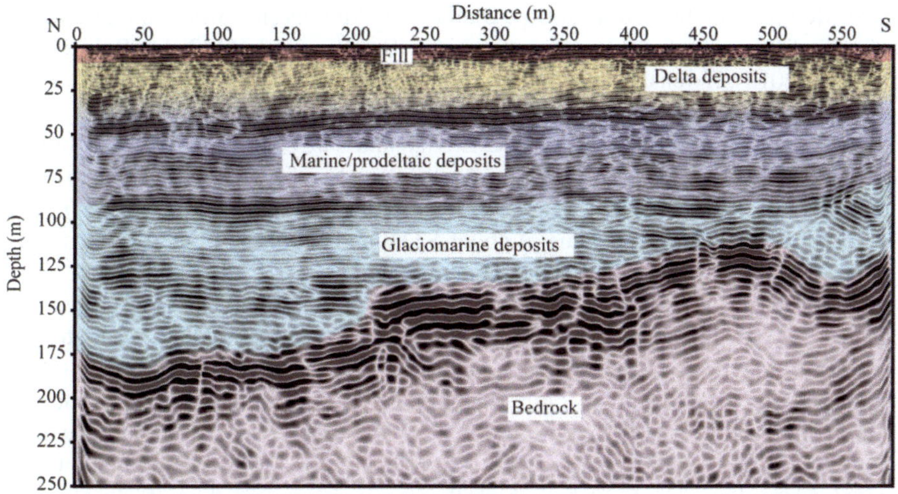

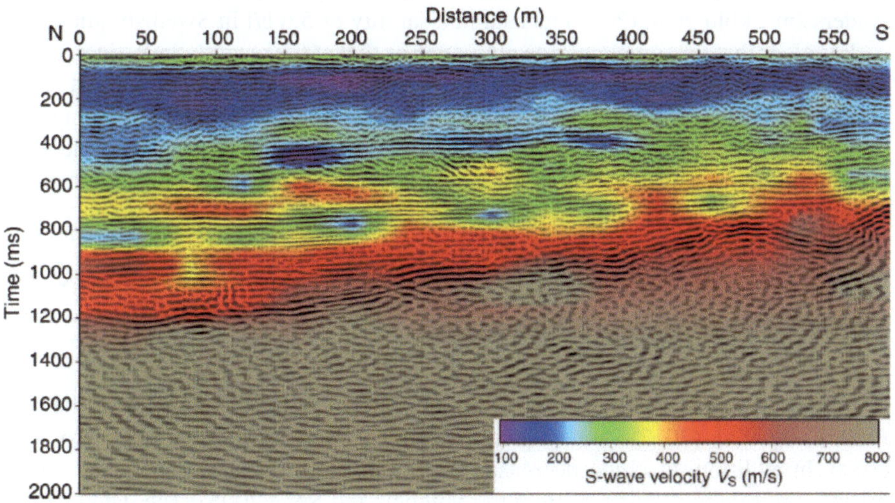

Fig. 13.5 SH-wave seismic reflection profile from Trondheim harbour (After Polom et al. 2011). (**a**) Interpreted seismic profile with units and subunits. (**b**) Shear-wave velocities (See also Hansen et al. 2013 for further details)

- a strong relationship between resistivity and salt content of the pore fluid,
- a reasonable correlation between resistivity and both clay content and plasticity index. Resistivity decreases as these two parameters increase,
- a weak tendency for a decrease in resistivity with increasing bulk density,
- no clear relationship between resistivity and water content.

Some of the most important findings from this work are illustrated here. In addition to the original database, data from three additional Norwegian sites are also included to confirm that the previously established trends. Two of these sites are near Melhus, and the third is at Esp. Data for the Melhus sites were obtained as part of an investigation for a large road project (Sandven and Vik 2011; Solberg and Dalsegg 2012). Extensive investigations have been carried out at the Esp site, where a slide took place in January 2012 (Solberg et al. 2012b; Thakur 2012).

As expected, the link between resistivity and salt content of the pore fluid is strong (Fig. 13.6a). The new data from Melhus and Esp fits well with that previously presented by Long et al. (2012). Resistivity decreases rapidly with increasing salt content and reaches a low value of about 5 Ωm and becomes more or less constant once the salt content exceeds approximately 8 g/l.

In the past, authors such as Bjerrum (1954) and Rosenqvist (1955) have suggested that clay becomes quick (i.e. sensitivity $S_t > 30$ and remoulded shear strength $s_{ur} < 0.5$ kPa) when the salt content is less than 5 g/l. Subsequently, Torrance (1974) suggested the limit should be 2 g/l. The plot of sensitivity versus salinity of the pore fluid (Fig. 13.6b), shows that although all of the quick clay data points have salt content less than 5 g/l, there are also a significant number of data points with salt content less than 2 or 5 g/l for which the sensitivity is less than 30. In addition, Andersson-Sköld et al. (2005) measured a salinity of 5.6 g/l in Swedish quick clay.

It is clear that non quick marine clay, such as that at the recently investigated sites in Melhus and Esp, may also contain very low salt content due to continued leaching or weathering, confirming that quick clay formation/clay leaching is an ongoing dynamic process. This indicates that although salt content of the pore fluid is an important factor, sensitivity of marine clay is also influenced by other factors (c.f. Mitchell 1993).

13.5.2 Shear Strength Parameters

As remoulded shear strength (s_{ur}) is directly related to the salt content of the pore fluid, Long et al. (2012) showed a strong link between resistivity and s_{ur} (from fall cone). In addition, leaching decreases the liquid limit of marine clays and consequently the remoulded shear strength (Mitchell and Soga 2005). The data of Long et al. (2012) are reproduced on Fig. 13.7a, with additional data from four sites, the three sites mentioned earlier and a fourth site at Hvittingfoss, Southern Norway. This site has been used for research purposes into integrated geotechnical and geophysical testing of quick clays (Chap. 18 by Sauvin et al. this volume; Sauvin et al. 2013).

On Fig. 13.7a, the focus is on those values where s_{ur} is less than 0.5 kPa, which is the threshold for quick clay, according to NGF (1982). There is a clear trend of increasing resistivity with decreasing s_{ur}. This is consistent with the fact that s_{ur} will decrease with increasing intensity of leaching. Although most of the data discussed here is within the resistivity range (i.e. 10–100 Ωm) defined by Solberg et al. (2012a), some of the sites with relatively high silt content exhibit significantly higher resistivity values up to 150 Ωm.

Resistivity values are plotted against sensitivity on Fig. 13.7b. There is a reasonably good linear relationship between the two properties. The increase in scatter of the data

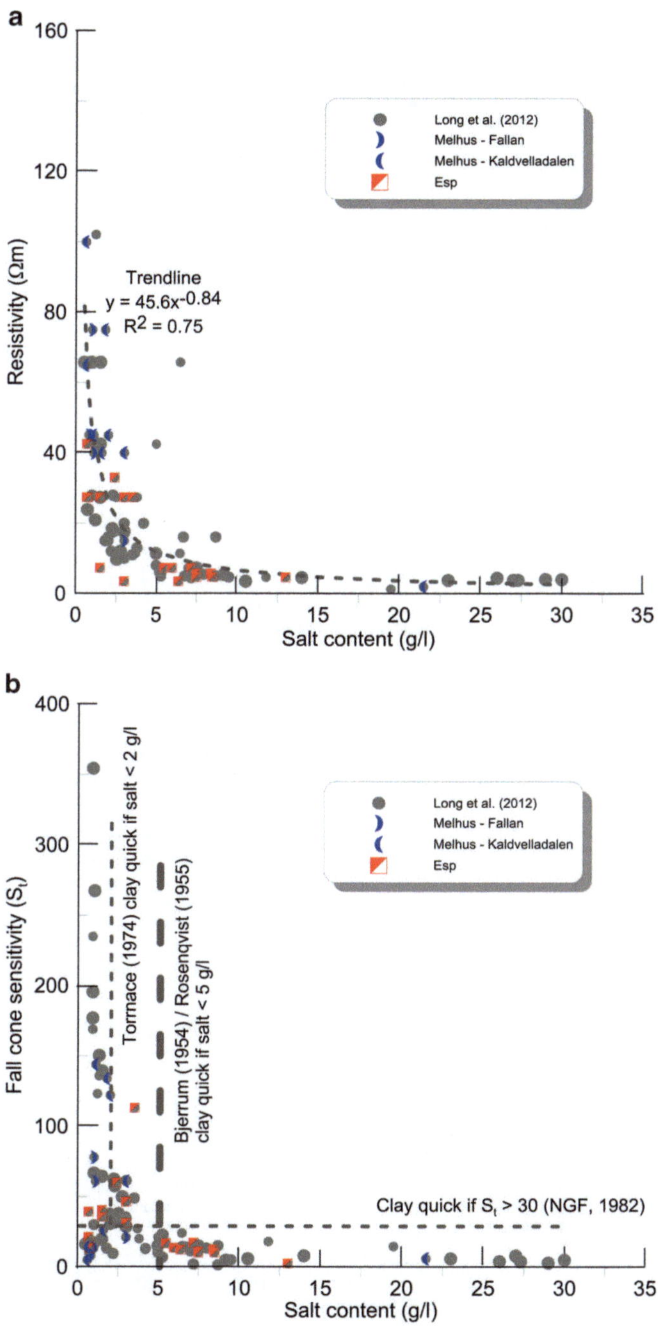

Fig. 13.6 Relationship between (**a**) resistivity and salt content and (**b**) sensitivity and salt content

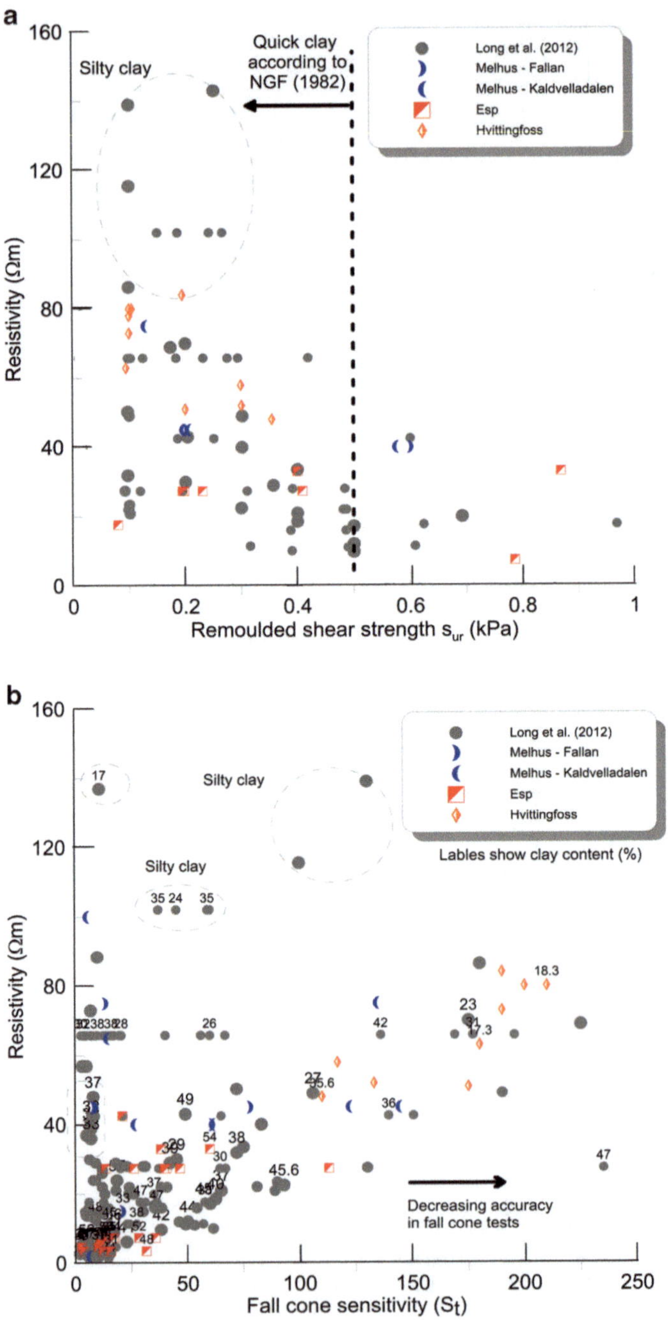

Fig. 13.7 Relationship between (**a**) remoulded shear strength (up to 1 kPa) and resistivity, (**b**) sensitivity and resistivity

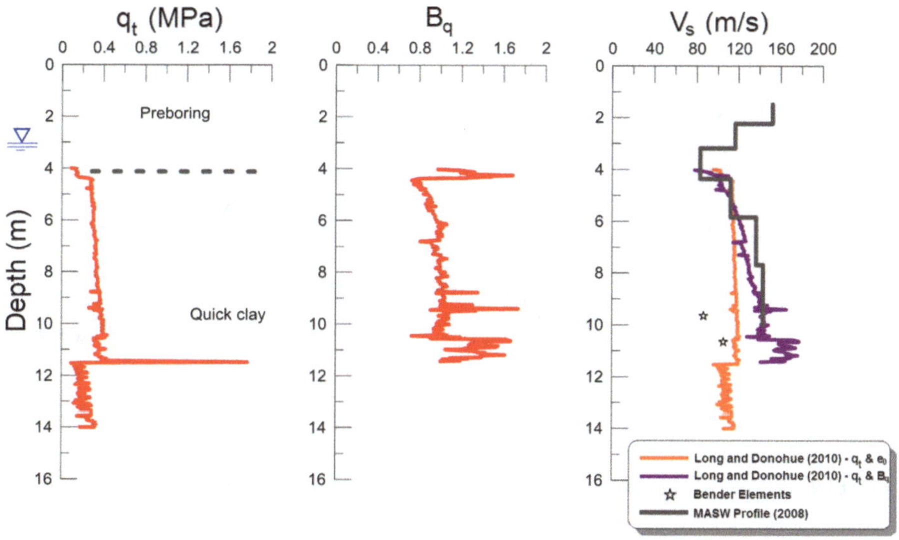

Fig. 13.8 Comparison of measured and predicted V_s data – Fredrikstad Site – Line 1

with increasing S_t is due to the decreasing accuracy of the fall cone measurements. Scatter could also be due to ERT inversion ambiguity. The high values of resistivity, at some of the sites, are attributed to the silty nature of the material.

In addition to knowledge of remoulded shear strength and sensitivity it is very important that intact undrained shear strength (s_u) is obtained for landslide stability assessment. A well-established practice is to estimate s_u from CPTU data. Long and Donohue (2010) developed relationships between V_s and corrected CPT cone resistance (q_t) specifically for Norwegian marine clays:

$$V_s = 65.00 q_t^{0.150} e_0^{-0.714} \tag{13.2}$$

$$V_s = 1.961 q_t^{0.579} \left(1 + B_q\right)^{1.202} \tag{13.3}$$

Equation 13.2 relies on e_0 (in-situ void ratio) as input, a parameter not always readily available. Equation 13.3 requires the CPTU pore water pressure parameter (B_q) as input but has the advantage of not needing laboratory data. Use of these formulae was subsequently tested on a commercial project for a site near Fredrikstad in Norway (data provided by Mr Roy Nalbant and Prof. Rolf Sandven of Multiconsult). MASW data and parallel CPTU data were available at several locations. V_s predictions from CPTU data using the formulae of Long and Donohue (2010) are shown on Fig. 13.8 for Line 1 (w ≈ 55 %, I_p ≈ 12 %, S_t ≈ 72). Both formulae give similar results and reasonable fits to the measurements. Equation 13.3 (using q_t and B_q) arguably performs better here as it captures the increase in V_s with depth. The laboratory V_s values (from bender elements) are less than the in-situ measurements, probably due to sampling disturbance (Donohue and Long 2010).

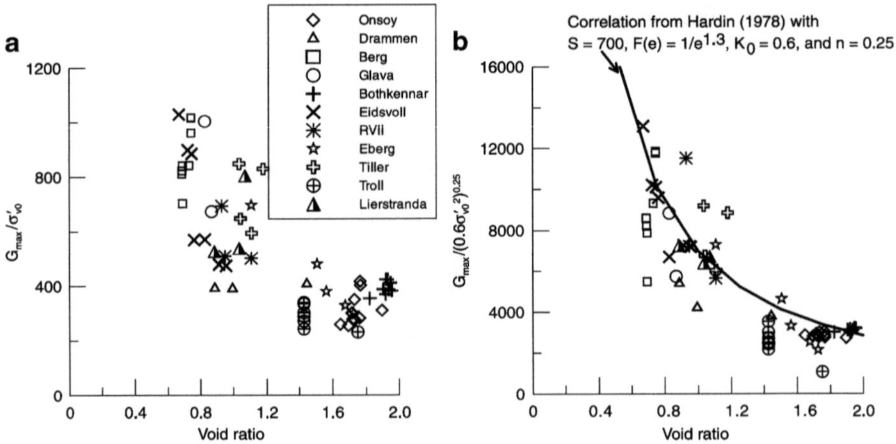

Fig. 13.9 (**a**) and (**b**) Normalised small strain shear modulus (From Long and Donohue 2010)

13.5.3 Small Strain Stiffness

Long and Donohue (2010) related G_{max} to natural water content (w) or in-situ void ratio (e_0) for 11 Norwegian marine clay sites. Only data from high quality Sherbrooke block samples were used, to minimise errors due to sample disturbance. Normalised values (G_{max}/σ_{v0}') are plotted on Fig. 13.9a. It can be seen that G_{max}/σ_{v0}' decreases with increasing e, as expected. On Fig. 13.9b the data have been normalised as suggested by Hardin (1978) and Hight and Leroueil (2003):

$$G_{max} = S \cdot F(e)\left(\sigma'_v \sigma'_h\right)^n p_a^{(1-2n)} \tag{13.4}$$

where S is a dimensionless "structure" parameter, F(e) the void ratio function, σ_v' and σ_h' the vertical and horizontal effective stresses, n a parameter indicating the influence of stress and p_a the atmospheric pressure. The fit is good confirming that G_{max} for Norwegian clays are consistent with other published experimental data.

13.6 Discussion and Conclusions

This paper has considered the use of a number of geophysical techniques for investigations in sensitive and quick clay terrain, and for estimation of soil parameters. Based on the author's experience, data available in the literature as well as in this paper, the applicability of geophysical methods for determination of a number of soil parameters is tentatively presented in Table 13.1.

Electrical Resistivity Tomography (ERT) appears to be the most popular technique for sensitive clay investigations, owing to the relatively strong relationship

Table 13.1 Applicability and usefulness of geophysical methods in clay terrain

	Soil parameters and deposit information									
Method/technique	Soil type	Structure/ stratigraphy	w	I_p	S	G_{max}	s_u	s_{ur}	σ'_p	OCR
P-wave seismic (reflection, refraction)	B[a]/C	A/B	–	–	–	–	–	–	–	–
S-wave seismic (MASW, SASW, reflection, S-CPTU)	A/B	A/B	A/B[b]	B/C	–	A	B/C[c]	–	A/B[d]	B/C[d]
Electromagnetic (e.g. EM-31, EM-34, CSRMT)	A/B	A/C	B/C[e]	–	B/C	–	–	–	–	–
Geo-electrical (ERT, R-CPTU)	A	A/C	C	C	A	–	C	C	–	–

Applicability: A high, B moderate, C low, – none

Soil parameter definitions: w water content, I_p plasticity index, S salinity, G_{max} small strain shear modulus, s_u undrained shear strength, s_{ur} remolded undrained shear strength, σ'_p preconsolidation stress, OCR overconsolidation ratio

[a]Using empirical relationship between the acoustic quality factor (Q) and geometric grain size (e.g. Pinson et al. 2008; Vardy et al. 2012)

[b]Through empirical relationships with the normalized small strain shear stiffness (g_{max})

[c]Through empirical relationships with G_0

[d]Through correlations with Vs (L'Heureux et al. 2013)

[e]Through empirical correlations (e.g. Topp et al. 1980)

between resistivity and a number of relevant geotechnical properties. The addition of instrumentation to conventional geotechnical tools enabling resistivity measurements to be simultaneously acquired, also illustrates the importance that this geophysical measurement has assumed for sensitive clay investigations.

The data discussed above shows that resistivity is strongly influenced by the salt content of the pore fluid and also by the clay content and plasticity of the material. A strong link between resistivity and s_{ur} was also established.

Recent work carried out using refraction, high resolution onshore reflection and surface wave seismic methods have demonstrated the potential use of these methods for mapping landslide prone sensitive clay. V_s profiles derived from surface wave analysis are shown to be comparable to those obtained from correlation with cone penetration test (CPT) and correlate well with index properties for these soils.

In contrast with geo-electrical measurements, electromagnetic conductivity mapping has only received limited attention for investigating sensitive clays. This paper has discussed the potential application of CSRMT for mapping sub-surface conductivity, particularly when jointly inverted with ERT data.

Acknowledgement The authors would like to thank Prof. Torleif Dahlin for his interesting and positive review of this paper.

References

ABEM (1999) ABEM Terrameter SAS 4000/SAS 1000. Instruction manual. ABEM Printed Matter 93101. ABEM, Sundbyberg

Andersson-Sköld Y, Torrance JK, Lind B, Odén K, Stevens RL, Rankka K (2005) Quick clay – a case study of chemical perspective in southwest Sweden. Eng Geol 82:107–118

Bastani M (2001) EnviroMT – a new controlled source/radio magnetotelluric system. PhD thesis, Acta Universitatis Upsaliensis, Uppsala Dissertations from the Faculty of Science and Technology 32

Bastani M, Malehmir A, Ismail N, Pedersen LB, Hedjazi F (2009) Delineating hydrothermal stockwork copper deposits using controlled-source and radio-magnetotelluric methods: a case study from northeast Iran. Geophysics 74(5):B167–B181

Bjerrum L (1954) Geotechnical properties of Norwegian marine clays. Géotechnique 4(1):49–69

Calvert HT, Hyde CSB (2002) Assessing landslide hazard in the Ottawa Valley using electrical and electromagnetic methods. In: Proceedings of the Symposium on the Application of Geophysics to Engineering and Environmental Problems (SAGEEP), Las Vegas. Environmental and Engineering Geophysical Society, Wheat Ridge, 10–14 February 2002

Dahlin T (1993) On the automation of 2D resistivity surveying for engineering and environmental applications. Doctoral thesis, Department of Engineering Geology, Lund Institute of Technology, Lund University, Lund

Dahlin T, Larsson R, Leroux V, Svensson M, Wisén R (2001) Geofysik I släntstabilitetsutredningar. Report Nr. 62. Swedish Geotechnical Institute, Linköping

Dahlin T, Palm M, Garin H (2004) Combined resistivity imaging and RCPT for geotechnical preinvestigation. In: Proceedings of the 14th NGM, Nordic geotechnical meeting, Ystad, pp C81–C88

Dahlin T, Larsson R, Leroux V, Larsson R, Rankka K (2005) Resistivity imaging for mapping of quick clays for landslide risk assessment. In: Proceedings of the 11th annual meeting EAGE – environmental and engineering geophysics, Palermo, 4–7 September 2005, A046

Donohue S, Long M (2008) An assessment of the MASW technique incorporating discrete particle modelling. J Environ Eng Geophys 13(2):57–68

Donohue S, Long M (2010) Assessment of sample quality in soft clay using shear wave velocity and suction measurements. Géotechnique 60(11):883–889

Donohue S, Long M, O'Connor P, Eide-Helle T, Pffaffhuber AA, Rømoen M (2012) Geophysical mapping of quick clay: a case study from Smørgrav, Norway. Near Surf Geophys 10(3):207–219

Geertsema M, Torrance JK (2005) Quick clay from the Mink Creek landslide near Terrace, British Columbia: Geotechnical properties, mineralogy, and geochemistry. Can Geotech J 42(3):907–918. doi:10.1139/t05-028

Hansen L, L'Heureux JS, Sauvin G, Polom U, Lecomte I, Vaneste M, Longva O, Krawczyk C (2013) The effect of mass-wasting on the stratigraphic architecture of a fjord-valley fill: correlation of onshore, shear wave seismics and marine seismic data at Trondheim, Norway. Sedimentary Geol 289:1–18

Hardin BO (1978) The nature of stress – strain behaviour for soils. In: Proceedings of the ASCE speciality conference on earthquake engineering and soil dynamics, Pasadena, pp 3–90

Helle TE, Pfaffhuber AA, Rømoen M, Forsberg CF (2009) SIP12 – correlation between horizontal and vertical resistivity measurements. NGI Report 20081135-1. Norwegian Geotechnical Institute, Oslo, Norway

Hight DW, Leroueil S (2003) Characterisation of soils for engineering purposes. In: Tan TS, Phoon KK, Hight DW, Leroueil S (eds) Proceedings of the international workshop on characterisation and engineering properties of natural soils. Balkema, Rotterdam/Singapore, pp 255–360

Kalscheuer T, García M, Meqbel N, Pedersen LB (2010) Non-linear model error and resolution properties from two-dimensional single and joint inversions of direct current resistivity and radiomagnetotelluric data. Geophys J Int 182:1174–1188

Kalscheuer T, Bastani M, Donohue S, Persson L, Pfaffhuber AA, Reiser F, Ren ZY (2013) Delineation of a quick clay structure at Smørgrav, Norway, with electromagnetic methods under geotechnical constraints. J Appl Geophys 92:121–136. doi:10.1016/j.jappgeo.2013.02.006

L'Heureux JS, Longva O, Steiner A, Hansen L, Vardy ME, Vanneste M, Haflidason H, Brendryen J, Kvalstad TJ, Forsberg CF, Chand S, Kopf A (2012) Identification of weak layers and their role for the stability of slopes at Finneidfjord, northern Norway. In: Proceedings of the 5th international symposium on submarine mass movements and their consequences, Kyoto, October 2011. In: Submarine mass movements and their consequences, advances in natural and technological hazards research 29, Springer, Dordrecht

L'Heureux JS, Long M, Vanneste M, Sauvin G, Hansen L, Polom U, Lecomte I, Dehls J, Janbu N (2013) On the prediction of settlement from high-resolution shear-wave reflection seismic data: the Trondheim harbour case study, mid Norway. Engineering Geology (Accepted for publication)

Löfroth H, Suer P, Schalin D, Dahlin T, Leroux V (2012) Mapping of quick clay using sounding methods and resistivity in the Göta River valley. Geotechnical and Geophysical Site Characterisation 4. In: Couthino RC, Mayne PW (eds) Proceedings of the 4th international conference on site characterisation (ISC-4) 2, pp 1001–1008, CRC Press, Taylor & Francis Group, Balkema

Loke MH (2010) Res2DInv ver. 3.59.102 Geoelectrical imaging 2D and 3D. Instruction manual. Geotomo Software. www.geoelectrical.com

Long M, Donohue S (2007) In situ shear wave velocity from multichannel analysis of surface waves (MASW) tests at eight Norwegian research sites. Can Geotech J 44:533–544

Long M, Donohue S (2010) Characterization of Norwegian marine clay with combined shear wave velocity and piezocone cone penetration test (CPTU) data. Can Geotech J 47:709–718

Long M, Donohue S, L'Heureux JS, Solberg IL, Rønning JS, Limacher R, O'Connor P, Sauvin G, Rømoen M, Lecomte I (2012) Relationship between electrical resistivity and basic geotechnical parameters for marine clays. Can Geotech J 49(10):1158–1168

Lundström K, Larsson R, Dahlin T (2009) Mapping of quick clay formations using geotechnical and geophysical methods. Landslides 6(2009):1–15

Malehmir A, Bastani M, Krawczyk CM, Gurk M, Ismail N, Polom U, Persson L (2013a) Geophysical assessment and geotechnical investigation of quick-clay landslides—a Swedish case study. Near Surf Geophys 11:341–350. doi:10.3997/1873-0604.2013010

Malehmir A, Saleem MU, Bastani M (2013b) High-resolution reflection seismic investigations of quick-clay and associated formations at a landslide scar in Southwest Sweden. J Appl Geophys 92:84–102

Mitchell JK (1993) Fundamentals of soil behaviour. Wiley, New York

Mitchell JK, Soga K (2005) Fundamentals of soil behaviour. Wiley, Hoboken

NGF (1982) Veilding for symboler og definisjoner i goeteknikk – presentasjon av geotekniske undersøkelser. Norwegian Geotechnical Society (Norsk Geoteknisk Forening) (In Norwegian), Oslo

Pedersen LB, Bastani M, Dynesius L (2005) Groundwater exploration using combined controlled-source and radiomagnetotelluric techniques. Geophysics 70:G8–G15

Pinson LJW, Henstock TJ, Dix JK, Bull JM (2008) Estimating quality factor and mean grain size of sediments from high-resolution marine seismic data. Geophysics 73:G19–G28

Polom U, Hansen L, Sauvin G, L'Heureux JS, Lecomte I, Krawczyk CM, Vanneste M, Longva O (2011) High-resolution SH-wave seismic reflection for characterization of onshire ground conditions in the Trondheim harbour, central Norway. In: Richard D. Miller, John H. Bradford, Klaus Holliger (eds) Advances in near-surface seismology and ground-penetrating radar, Chap. 18, pp 297–312, SEG book

Rankka K, Anderssen-Skold Y, Hulten C, Larsson R, Leroux V, Dahlin T (2004) Quick clay in Sweden. Report 65. Swedish Geotechnical Institute, Linkoping

Reite A, Sørensen E (1980) Støren. Kvartærgeologisk kart 1621 II, M 1:50 000. NGU map

Rokoengen K (1998) Naturkatastrofer i Gauldalen. Department of Geology and Mineral Resources Engineering NTNU Report 33, Trondheim, 50 pp

Rømoen M, Pfaffhuber AA, Karlsrud K, Helle TE (2010) Resistivity on marine sediments retrieved from RCPTU-soundings: a Norwegian case study. In: International symposium on cone penetration testing, 2, CPT'10, Huntington Beach, Proceedings, vol. 2

Rosenqvist IT (1955) Investigations in the clay-electrolyte-water system. Norwegian Geotechnical Institute Publication 9: 36, Oslo

Sandven R, Vik A (2011) Ny E6 Haga-Skjerdingstad. Utredning for kommuneplan. Datarapport grunnundersøkelser. Beskrivelse av grunnforhold. Multiconsult Report 414622-1, Juli 5 2011

Sauvin G, Bazin S, Vanneste M, Lecomte I, Pfaffhuber AA (2011) Towards joint inversion/interpretation for landslide prone areas in Norway – integrating geophysics and geotechnique. In: Proceedings of near surface 2011 – 17th European meeting of environmental and engineering geophysics, Leicester. Paper E11

Sauvin G, Lecomte I, Bazin S, L'Heureux JS, Vanneste M (2013) Towards geophysical and geotechnical integration for quick clay mapping in Norway. Near Surf Geophys 11, (in press). doi: 10.3997/1873-0604.2012064

Schälin D, Tornborg J (2009) Evaluation of CPT-R and resistivity measurements in quick clay area. MSc thesis, Department of Civil and Environmental Engineering, Division of GeoEngineering, Chalmers University of Technology, Göteborg

Solberg IL, Dalsegg E (2012) Resistivitetsmålinger for løsmassekartlegging i Kaldvelladalen og ved Fallan i Melhus kommune, Sør-Trøndelag. Data og tolkninger. NGU Report 2012.013

Solberg IL, Rønning JS, Dalsegg E, Hansen L, Rokoengen K, Sandven R (2008) Resistivity measurements as a tool for outlining quick clay extents and valley fill stratigraphy: feasibility study from Buvika, Central Norway. Can Geotech J 45:210–225

Solberg IL, Hansen L, Rønning JS, Haugen ED, Dalsegg E, Tønnesen JF (2012a) Combined geophysical and geotechnical approach to ground investigations and hazard zonation of a quick clay area, mid Norway. Bull Eng Geol Environ 71(1):119–133. doi:10.1007/s10064-011-0363-x

Solberg IL, Dalsegg E, L'Heureux JS, Rønning JS, (2012b) Resistivitetsmålinger for løsmassekartlegging ved skredgrop på Byneset, Sør-Trøndelag. Geological Survey of Norway (NGU) Report 2012.004 (In Norwegian), Trondheim

Solberg IL, Hansen L, Rønning JS, Dalsegg E (this volume) Experiences with 2D resistivity measurements for quick-clay mapping in Norway. In: Proceedings of the 1st international workshop on landslides in sensitive clays (IWLSC), Quebec, 28–30 October 2013

Thakur V (2012) Datarapport for Kvikkleireskred ved Esp i Byneset i Januar 2012 Datarapport, NIFS Report 34/2012 (In Norwegian), Trondheim

Topp GC, Davis JL, Annan AP (1980) Electromagnetic determination of soil water content: measurements in coaxial transmission lines. Water Resour Res 16(1):574–588

Torrance JK (1974) A laboratory investigation of the effect of leaching on the compressibility and shear strength of Norwegian marine clays. Géotechnique 24(2):155–173

Torrance JK, Ohtsubo M (1995) Ariake Bay quick clays: a comparison with the general model. Soils Found 35(1):11–19. doi:10.3208/sandf1972.35.11

Vardy ME, L'Heureux JS, Vanneste M, Longva O, Steiner A, Forsberg CF, Haflidason H, Brendryen J (2012) Multidisciplinary investigation of a shallow near-shore landslide, Finneidfjord, Norway. Near Surf Geophys 10(4):267–277

Chapter 14
Applications of 2D Resistivity Measurements for Quick-Clay Mapping in Mid Norway

Inger-Lise Solberg, Louise Hansen, Jan Steinar Rønning, and Einar Dalsegg

Abstract During the last 10 years, several Norwegian projects have explored the possibilities of using 2D resistivity measurements for quick-clay mapping in combination with traditional geotechnical methods. Experience has shown that an effective first-order interpretation of 2D resistivity profiles can be performed by using the following classification: unleached clay deposits (1–10 Ωm); leached clay deposits, possibly quick (10–100 Ωm), and dry crust clay deposits and coarse sediments (>100 Ωm). However, resistivity values are influenced by local conditions and there is an overlap between the classes. The 2D resistivity method can prioritize areas for further investigation using other geophysical methods or drilling to refine the interpretation of the subsurface. Through several case-studies, the 2D resistivity method has proven useful for detecting potential layers of quick clay, for outlining the extent of these layers and their positions in slopes (also near the shoreline), and for engineering applications such as construction planning. To this end, the method has shown to be applicable for adjusting the extent of hazard zones to improve stability evaluations. Another important application is landslide investigations, to identify barriers that may deter further landslide propagation.

Keywords 2D resistivity • Marine deposits • Quick clay • Landslide

14.1 Introduction

Following the last ice age, several hundred meter thick glaciomarine and marine deposits accumulated in Norwegian fjords. Due to glacio-isostatic rebound, the sediments are now exposed on land with a ground surface elevation of

---

I.-L. Solberg (✉) • L. Hansen • J.S. Rønning • E. Dalsegg
Geological Survey of Norway (NGU), Trondheim, Norway
e-mail: inger-lise.solberg@ngu.no

J.-S. L'Heureux et al. (eds.), *Landslides in Sensitive Clays: From Geosciences to Risk Management*, Advances in Natural and Technological Hazards Research 36,
DOI 10.1007/978-94-007-7079-9_14, © Springer Science+Business Media Dordrecht 2014

up to 220 m a.s.l. Fine-grained marine deposits of clay and silt have an open grain structure with saline pore-water. Subsequent leaching by fresh groundwater alters the chemical composition of the pore-water, and "quick clay", which is highly sensitive clay, can develop (Rosenqvist 1953). Landslides that involve remoulded and liquefied quick clay can radically alter the landscape, as well as damage buildings and infrastructure. Numerous landslides in Norway have involved quick clay: see examples in L'Heureux and Solberg (2012). An understanding of ground conditions, including the distribution of quick clay, is important for the evaluation of landslide hazards, and results can be used for land-use planning, construction and mitigation measures. Numerous and expensive investigations are often necessary when outlining ground conditions with traditional drilling techniques and sampling. 2D resistivity measurements are a useful supplement to drilling as they can give a continuous and relatively detailed two-dimensional (2D) picture of the subsurface within a short time and allow for targeted drilling investigations to be planned less expensively. The method can also potentially distinguish unleached marine clay deposits from quick clay, in addition to identifying coarser material and bedrock.

The use of 2D resistivity measurements as a tool for identifying quick clay has expanded during the last 10 years in Norway, Sweden and Canada (e.g. Calvert and Hyde 2002; Solberg et al. 2008, 2012d; Lundström et al. 2009; Donohue et al. 2012; Long et al. 2012). After several years of testing, the method has proven to be useful for several purposes. Advancements in field equipment and software for data processing, such as the inversion method (Loke 2010), have simplified the investigations and produced more reliable results.

The purpose of this paper is to synthesize the key experiences from studies carried out over the last decade in Mid Norway, and to highlight the broad range of applications of the 2D resistivity method for investigations related to quick clay.

14.2 Methods

14.2.1 2D Resistivity Measurements

The 2D resistivity measurements carried out by the Geological Survey of Norway (NGU) are based on the Lund-system developed by Dahlin (1993). Gradient and/or Wenner electrode arrays were used. The measuring equipment was an ABEM Terrameter SAS 4000, using a current of 100 or 200 mA, and in more recent investigations, ABEM Terrameter LS. Steel electrode separations were 2, 5 and/or 10 m.

Raw resistivity data give the apparent resistivity (ρ_a) of the subsurface. This represents a weighted mean of all the resistivity values that fall within the soil volume of influence. To obtain the specific resistivity (ρ) in Ωm from different parts of the subsurface, the data are inverted to produce a model where the apparent resistivity pseudo-section matches the measured pseudo-section. Recorded resistivities were inverted using Res2DInv software (Loke 2010), with a least-squares method.

Where profiles intersect, a quasi-3D picture of the subsurface can be established, and mismatch between profiles can provide valuable information on 3D geological variations.

14.2.2 Geotechnical Methods

Different geotechnical sounding methods provide valuable indications on quick-clay occurrence and sediment stratification (e.g. Rotary Pressure Sounding (RPS) and Cone Penetration Test Undrained (CPTU)). Soundings which include resistivity measurements (e.g. RCPTU) support the 2D resistivity data (e.g. Rømoen et al. 2010; Solberg et al. 2012d). However, laboratory testing of core samples gives detailed information on material properties, like shear strength, and is currently the most reliable method assessing clay sensitivity.

In the studies presented here, sensitivity (S_t) is defined as the undisturbed shear strength divided by the disturbed (or remoulded) shear strength (s_r). When s_r is less than 0.5 kN/m^2, the clay is characterised as "quick". Sensitive clay is divided into the following (NGF 2011): low sensitivity ($S_t<8$); medium high sensitivity ($8<S_t<30$), and high sensitivity (quick clay) ($S_t>30$). When the pore-water salt content is reduced below 5 g/l, the marine clay can become sensitive and show quick-clay behaviour (Bjerrum 1954).

14.2.3 Classification of Sediments from Resistivity Values

The classification of clay properties and materials of larger grain sizes based on resistivity values is mainly empirical, deduced from a comparison of geophysical and geotechnical data from different studies in Scandinavia and Canada (see discussions in Solberg et al. 2008, 2012d, and Long et al. 2012). An effective first-order interpretation of 2D resistivity profiles can be performed by using the following classification (Fig. 14.1): unleached clay deposits (1–10 Ωm); leached clay deposits, possibly quick (10–100 Ωm), and dry crust clay deposits and coarse sediments (>100 Ωm).

2D resistivity data are influenced by a relatively large volume of soil around the measured profile. When interpreting a 2D resistivity profile it is important to be aware that the higher resolution data are in the top and centre of the profile. Problems in inversion can be caused by suppression and equivalence (Reynolds 2011), and there will always be gradual transitions between high and low resistivity values in a profile. In addition, the geology and material properties themselves will often show gradual transitions, not sharp boundaries. Therefore, depth estimates must be made with caution. An atlas providing numerous synthetic models for different geological conditions may help in the interpretation of resistivity profiles (Reiser et al. 2010).

Resistivity (Ωm)	Main characterisation	Description
1-10	Unleached marine clay deposits	Saline pore-water, stable clay structure. Good conductivity. Electronic conductive minerals like graphite, sulphides and some oxides, may also give low resistivity values.
10-100	Leached clay deposits	Low total electrolyte content. Still good conductivity, but poorer than for the unleached marine clay. Silt, fine-grained till, and leached, non-quick clay may fall into the same interval.
> 100	Dry crust clay deposits, coarse sediments, (bedrock)	Dry clay crust, remoulded, dry clay from quick-clay landslides, and coarser materials (sand, gravel) will have higher resistivity values than marine clay. Most bedrock types will have values of several thousand Ωm.

Fig. 14.1 First-order classification of sediments from resistivity values (Modified from Solberg et al. 2011, 2012d). Note that there are gradual transitions between the classes and there may be local variations related to factors such as pore-water chemistry, saturation, grain size distribution, mineral composition etc

Although the 2D resistivity method is able to identify the resistivities associated with quick clays (10–100 Ωm), both leached, non-quick clay, and silty, non-sensitive material can also fall into the same range. Local conditions may influence the resistivity values of different sediment types and properties, and quick-clay will occasionally fall outside the 10–100 Ωm interval. Some of our studies in Mid Norway show quick clay with resistivities higher than 100 Ωm, but have not verified quick clay below 10 Ωm. A few studies from Southern Norway and Sweden have examples of quick-clay resistivities lower than 10 Ωm (e.g. Lundström et al. 2009; Rømoen et al. 2010).

14.3 Applications of 2D Resistivity Measurements in Quick-Clay Areas

2D resistivity measurements for quick-clay mapping have been carried out in several study areas in Mid Norway during the last 10 years. Examples of resistivity data compared to geotechnical data are presented first, followed by a summary of various applications for the 2D resistivity method. Only highlights from selected case studies are presented below. For more details and full documentation of each case, see the inserted references.

14.3.1 Combined 2D Resistivity Measurements and Geotechnical Investigations for Quick-Clay Mapping

Comparisons of data from 2D resistivity measurements and geotechnical soundings are often in good agreement, but for more detailed information lab-tested samples

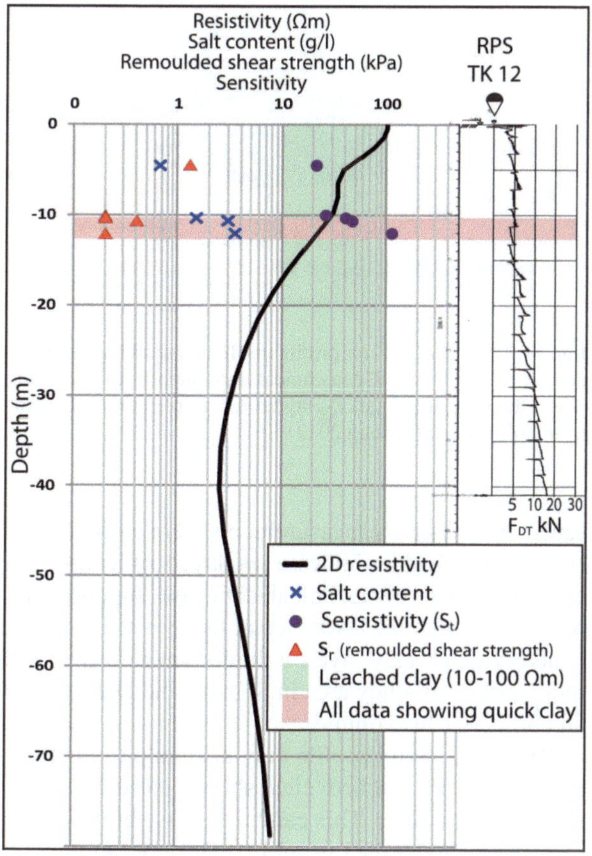

Fig. 14.2 2D resistivity data from Byneset compared to geotechnical data from laboratory tested samples (remoulded shear strength (s_r), sensitivity (S_t) pore-water salt content) and rotary pressure sounding (*RPS*) profile. See Fig. 14.4 for location. Sample results from 10 to 12 m depth show quick clay in an area where the resistivity is about 25 **Ω**m. Note that the resistivity data from 2D measurements are an average of the surroundings, and in deeper parts the resolution is relatively low. Data from Solberg et al. (2012d); NVE (2012); Trondheim kommune (2012)

are needed, as shown in Fig. 14.2. This is example is from Byneset close to Trondheim (see more from this site in Sect. 3.2). The resistivity data indicate leached clay down to about 18 m depth, with unleached low-resistivity clay below. This is in accordance with an RPS profile which shows low drill resistance down to 15 m depth (indicating highly sensitive or quick clay from 7 to 15 m depth). Samples from some of the depths were tested, showing medium sensitive clay around 4 m depth, and quick clay in the 10–12 m depth range. The pore-water salt contents are below 5 g/l for all the samples. In the transition between leached to unleached clay (~20 m), the drill resistance increases very little, indicating some sensitive material. However, the sensitivity is not verified by samples in this zone.

The extent of quick clay between geotechnical drilling profiles is always drawn conservatively when calculating slope stability (Solberg et al. 2012d). 2D resistivity data can provide crucial information for the interpreter when interpolating/correlating quick-clay pockets between geotechnical profiles. This is clearly illustrated in Fig. 14.3, which gives an example from Melhus. Here, the extent of quick clay was interpreted on the basis of five drillings, the result shown by the transparent layer.

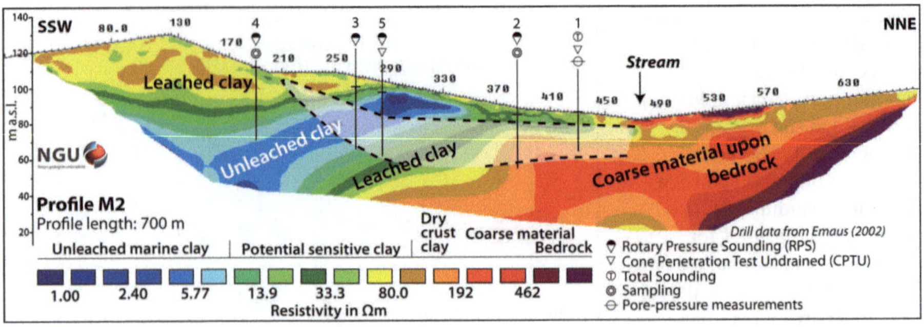

Fig. 14.3 A 2D resistivity profile in Melhus showing two layers of leached clay intersected by an unleached clay layer. Geotechnical data support this layering, but if geotechnical data is used alone, the most likely quick-clay layer interpretation is shown by the transparent layer (Modified from Solberg and Dalsegg 2012)

Later, 2D resistivity measurements were carried out along the same alignment, and the resistivity values indicate two layers of leached clay intersected by an unleached clay layer. The geotechnical data also support this layering. The model of two potential quick-clay layers separated by a stable unleached clay layer, instead of one coherent quick-clay layer, is important information for stability assessments. See more about this study area in Solberg and Dalsegg (2012) and Chap. 16 by Sandven and Solberg (this volume).

A 2D resistivity profile may also help to identify the degree of leaching when a sounding in leached clay shows increased resistance with depth and thus, increasing stability. This can be the case both where the clay has not been leached to a sufficient degree (around 10 Ωm), or has passed the quick-clay stage (around 100 Ωm) e.g. due to infiltration of stabilising ions present in the groundwater (Solberg et al. 2012d).

14.3.2 *Mapping of Barriers Against Landslide Development*

2D resistivity profiles give a continuous picture of the potentially unstable material over a large area. This is important information with regard to evaluation of propagation direction and size of a potential landslide. However, mapping of geological barriers to landslides can be just as important (e.g. unleached clay, coarse deposits and buried bedrock protuberances). This is illustrated through a case study where 2D resistivity measurements were applied, as a part of the method development, a few days after a landslide event in Byneset January 1 2012, near Trondheim (Mid Norway). The potential for further landslide retrogression needed to be assessed before evacuees would be permitted to return to their homes. One of the two 2D resistivity profiles measured near the landslide scar is shown in Fig. 14.4. 2D resistivity data confirmed that a small bedrock outcrop was part of a buried bedrock

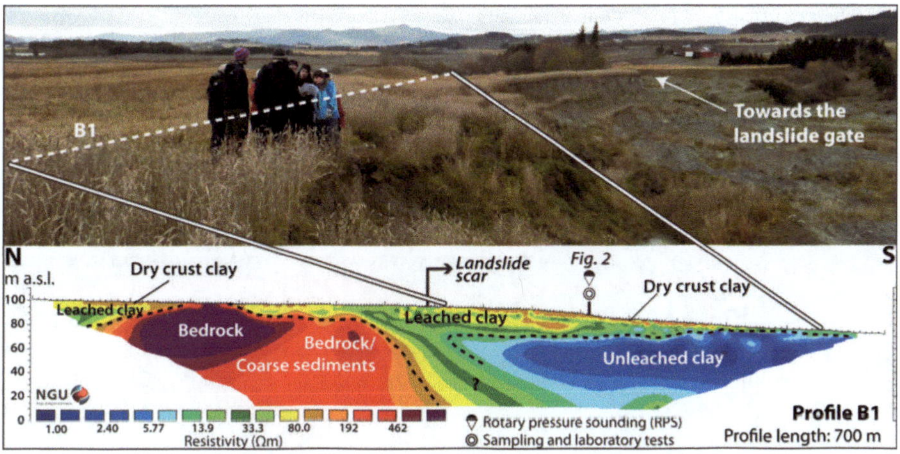

Fig. 14.4 Byneset landslide scar (01.01.2012), with location of a 2D resistivity profile measured few days after the event. The interpretation results from analysis of 2D resistivity data and geotechnical data. The presence of bedrock just northeast of the landslide scar was important for controlling the landslide retrogression (Photo: IL Solberg, data from Solberg et al. 2012c)

protuberance, which would limit the landslide development in a north-easterly direction (Solberg et al. 2012c). The profile also shows that the leached clay layer (partly quick as verified by sample tests and sounding, see one example in Fig. 14.2), thins out towards the backscarp of the landslide, near the bedrock. As a consequence, the height of the backscarp would be expected to decrease during retrogression, as would the potential energy of the slope, and landsliding will cease. If the quick-clay layer was thicker near the bedrock, the landslide would probably have retrogressed further (L'Heureux and Solberg 2012).

Subaqueous failure near a shoreline may propagate onshore with severe consequences if quick clay is present (e.g. L'Heureux 2009). In Rissa (Mid Norway), traces of slope failures are detected by seabed mapping (Fig. 14.5), and 2D resistivity measurements were carried out to link seabed information with data on onshore conditions (Solberg et al. 2012b). In this example, the probability for landslidepropagation was considered relatively low, at least for one of the landslides where the resistivity indicated coarse deposits/dry clay crust and bedrock behind the landslide backscarp.

14.3.3 Hazard Zonation

The limits of a hazard zone for potential quick-clay landslides are often drawn from the results of a few geotechnical investigations, terrain studies and inspection, and on slope stability calculations. Here, 2D resistivity measurements have shown to be useful in providing a better overview of potential quick-clay extents and subsurface landslide barriers, as shown by several cases (e.g. Solberg et al. 2012a, b, c, d).

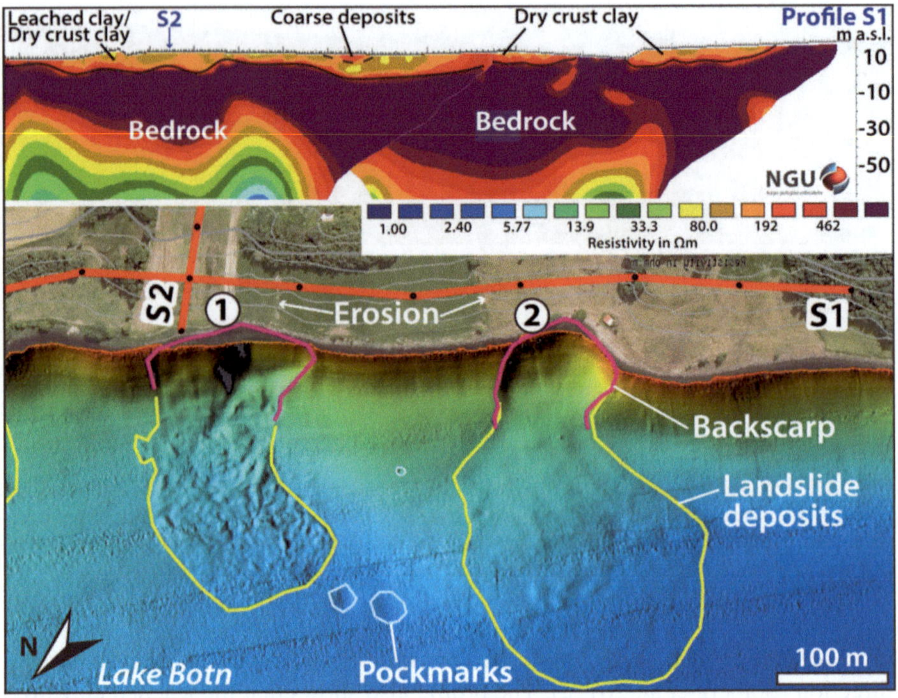

Fig. 14.5 Overlapping 2D resistivity profiles (S1) collected along the shore of Lake Botn in Rissa. Seabed mapping shows traces of slope failure close to the beach. Profile S1 shows a 5–10 m top layer of leached clay over bedrock behind landslide scar 1. An intersecting profile (S2, not shown) indicates bedrock in the shoreline which will limit further retrogression of landslide 1 towards the east. The material behind landslide scar 2 is coarse deposits/dry clay crust and bedrock. In the deepest parts of resistivity profile S1, the measurements are influenced by brackish water, showing too low resistivity values (Modified from Solberg et al. 2012b)

For hazard zonation, it is mainly the deposits below the land surface down to some level below a stream/river which are of interest. It is unlikely that quick clay found deeper could be affected by erosion or be disturbed by human intervention (Solberg et al. 2012d). This should be considered when deciding the on the acquisition parameters for resistivity profiling.

In Rissa, a hazard zone was mapped in 1989, based on limited amounts of geotechnical data (Gregersen and Korbøl 1989). In recent years, a dense network of data has been collected in the area, including 2D resistivity measurements (17 profiles in about 1 km^2), other geophysical surveys, and numerous geotechnical investigations (e.g. Sauvin et al. in press; Solberg et al. 2012a). Figure 14.6 shows bedrock depths and thicknesses of leached clay interpreted from these combined data sets. There are two partly exposed parallel bedrock ridges (NE-SW), which may act as a landslide barrier in the area. The hazard zone in Rissa could be limited in north due to bedrock and very thin sediment cover. However, between the ridges, the zone could be expanded due to the presence of thick leached clay.

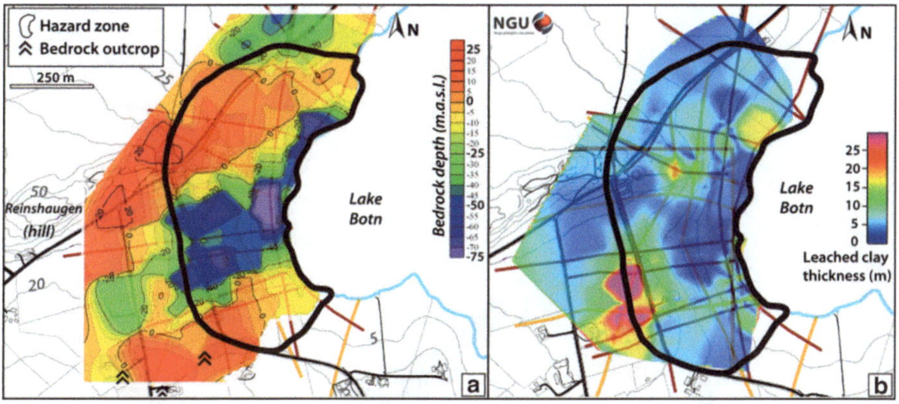

Fig. 14.6 An existing hazard zone in Rissa drawn upon depth to bedrock (**a**) and leached clay thickness (**b**) interpreted from 2D resistivity data and geotechnical data. Lake Botn is brackish water and located just above sea level. Data show that the limits of the hazard zone could be changed. Terrain contour line intervals are 5 m (Modified from Solberg et al. 2012a)

14.3.4 Engineering Purposes

In an area without previous or few investigations, 2D resistivity measurements will make a good basis for further exploration and help planning of optimal locations for geotechnical drilling and ground truthing. Since the 2D resistivity method gives continuous profiles, the need for costly and time-consuming geotechnical investigations can be reduced. Some areas may be difficult to reach by a conventional drilling rig, due to e.g. steep slopes or densely vegetated areas. Here, 2D resistivity data may be the only way to get information about the ground conditions.

For road planning and construction, it is crucial to have information about potential challenging ground conditions. An example of this is shown in Chap. 13 by Donohue et al. (this volume), where a 2D resistivity profile was measured as a part of the planning of a new highway. Even though this area has been exposed to a large landslide, the 2D resistivity data indicate that leached clay is still present in the scar, and the following sounding and sampling confirm large occurrences of quick clay. There is also very little/no dry clay crust which indicates that leached or quick clay is expected close to the land surface. The results from the surveys show that there will likely be stability challenges in the area which must be considered in the further road planning.

14.3.5 Landscape Development and Groundwater Drainage Pattern

Steep bedrock hills and the buried bedrock topography will influence leaching and quick-clay development. A high bedrock relief often implies artesian conditions,

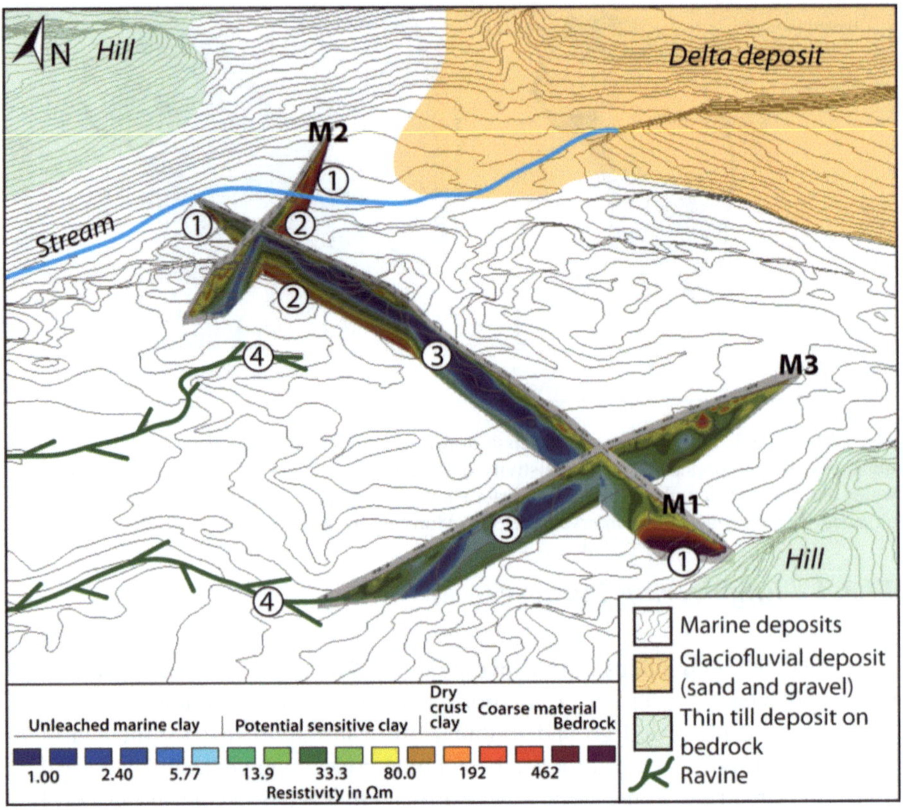

Fig. 14.7 "3D visualization" of 2D resistivity profiles in Melhus. Centrally there is a relatively coherent layer of unleached clay (*blue*). Leaching of clay (*green-yellow*) has occurred both downwards (close to the surface and in ravine slope) and upwards (due to shallow bedrock and coarse materials). See text for description of (1–4). Details of M2 is shown in Fig. 14.3 (Modified from Solberg and Dalsegg 2012; see also Chap. 16 by Sandven and Solberg this volume)

which must be considered in stability calculations. The presence of coarse layers within the clay will increase local groundwater movement. These layers may be too thin for resolution on a resistivity profile (depending on survey parameters), but in such cases e.g. large coarse-grained deposits nearby may indicate coarse layers within the clay. Soundings by CPTU can verify whether these layers are present.

Figure 14.7 presents three intersecting profiles, partly surrounded by bedrock, as shown in the terrain model and in profiles M1 and M2 (see number 1 on figure). To the northeast there is a large glacier marginal deposit (delta: orange), which is now a thick aquifer overlying bedrock, partly interbedded with the marine clay (as shown in sounding profiles). The geology can explain the distribution of leached and unleached clay: Facilitated groundwater flow near fractured bedrock and around coarse-grained layers/deposits have resulted in leaching of deeply buried clay as indicated by the resistivity values in the lower part of the 2D resistivity profiles (2). Leaching has also taken place in the clay near the land surface and in slopes, likely due to percolating meteoric water, as indicated by the resistivity values in the top of the 2D resistivity profiles. As a result, a more or less coherent layer of unleached clay remains in the centre of the profiles (3). Some of the soundings are deep enough to indicate medium sensitive or quick clay beneath the unleached clay layer. There are traces of many landslides and ravines in the area. The head of the ravines (4) is located in leached clay as indicated by resistivity values in profile M3. The ravine development may be efficient here due to groundwater seepage combined with the presence of leached clay, which can be soft and easily erodible.

14.4 Concluding Remarks

Experience gained over the last decade has shown that 2D resistivity measurements can be used to indicate unleached clay, potentially quick clay, coarse sediments, and bedrock. Comparison with geotechnical data shows that a layer of leached clay often is thicker than the actual quick-clay pocket. This is natural, since not all leached clay is quick, and since silty material may also occur in the same resistivity range. The measured profiles can also be used to locate areas warranting further investigations.

A good overview of ground conditions is important for understanding the geology, landscape development, and groundwater drainage patterns. For the correlation of drill data, stability assessments, and hazard zonation, it is important to delineate the extent of the quick clay and its position in the slope. For this, 2D resistivity profiling provides valuable information and maximises the data which can be obtained from subsequent intrusive investigations that are supported by laboratory tests. The studies also show that the method can be used to detect stable materials that may limit the potential size (hazard zonation) and development of a landslide.

Acknowledgments The authors thank reviewer Heather Crow at Geological Survey of Canada for valuable comments on the manuscript. Thanks also to Anne Liinamaa-Dehls for improving the English language.

References

Bjerrum L (1954) Geotechnical properties of Norwegian marine clays. Geotechnique 4:49–69

Calvert HT, Hyde CSB (2002) Assessing landslide hazard in the Ottawa Valley using electrical and electromagnetic methods. In: Symposium on the application of geophysics to environmental and engineering problems (SAGEEP). Environmental and Engineering Geophysical Society, Wheat Ridge

Dahlin T (1993) On the automation of 2D resistivity surveying for engineering and environmental applications. Doctoral thesis, Department of Engineering Geology, Lund Institute of Technology, Lund University

Donohue S, Long M, O'Connor P, Helle TE, Pfaffhuber AA, Rømoen M (2012) Multi-method geophysical mapping of quick clay. Near Surf Geophys 10:207–219

Gregersen O, Korbøl B (1989) Kartlegging av områder med potensiell fare for kvikkleireskred. Kartblad Rissa M 1:50 000. NGI Report 86054 (in Norwegian)

L'Heureux JS (2009) A multidisciplinary study of shoreline landslides: from geological development to geohazard assessment in the bay of Trondheim, Mid-Norway. PhD thesis, NTNU

L'Heureux JS, Solberg IL (2012) Utstrekning og utløpsdistanse for kvikkleireskred basert på katalog over skredhendelser i Norge. NGU Report 2012.040 (in Norwegian). Available at www.ngu.no

Loke MH (2010) Res2DInv ver. 3.59.102 Geoelectrical imaging 2D and 3D. Instruction Manual. Geotomo Software, www.geoelectrical.com

Long M, Donohue S, L'Heureux JS, Solberg IL, Rønning JS, Limacher R, O'Connor P, Sauvin G, Rømoen M, Lecomte I (2012) Relationship between electrical resistivity and basic geotechnical parameters for marine clays. Can Geotech J 49:1158–1168

Lundström K, Larsson R, Dahlin T (2009) Mapping of quick clay formations using geotechnical and geophysical methods. Landslides 6:1–15

NGF (Norwegian Geotechnical Society) (2011) Symboler og definisjoner i geoteknikk. NGF Melding 2, rev.1, 2011 (in Norwegian)

NVE (2012) Datarapport for Kvikkleireskred ved Esp i Byneset i januar 2012. Naturfareprosjektet: Delprosjekt Kvikkleire. NIFS, NVE Report 34/2012 (in Norwegian). Available at www.naturfare.no

Reiser F, Dahlin T, Rønning JS, Solberg IL (2010) Resistivity modelling for clay layer characterisation, possibilities and limitations. NGU Report 2010.047. Available at www.ngu.no

Reynolds JM (2011) An introduction to applied and environmental geophysics, 2nd edn. Wiley-Blackwell, Chichester

Rømoen M, Pfaffhuber AA, Karlsrud K, Helle TE (2010) Resistivity on marine sediments retrieved from RCPTU-soundings: a Norwegian case study. In: Proceedings of 2nd international symposium on cone penetration testing, Huntington Beach, 9–11 May 2010

Rosenqvist IT (1953) Considerations on the sensitivity of Norwegian clays. Geotechnique 3:195–200

Sauvin G, Lecomte I, Bazin S, L'Heuereux JS, Vanneste M, Solberg IL, Dalsegg E (in press) Towards geophysical and geotechnical integration for quick-clay mapping in Norway. Accepted for Near Surf Geophys

Solberg IL, Dalsegg E (2012) Resistivitetsmålinger for løsmassekartlegging i Kaldvelladalen og ved Fallan i Melhus kommune, Sør-Trøndelag. Data og tolkninger. NGU Report 2012.013 (in Norwegian). Available at www.ngu.no

Solberg IL, Rønning JS, Dalsegg E, Hansen L, Rokoengen K, Sandven R (2008) Resistivity measurements as a tool for outlining quick clay extents and valley fill stratigraphy: feasibility study from Buvika, Central Norway. Can Geotech J 45:210–225

Solberg IL, Hansen L, Rønning JS, Dalsegg E (2011) Veileder for bruk av resistivitetsmålinger i potensielle kvikkleireområder. Versjon 1.0. Geological Survey of Norway Report 2010.048 (in Norwegian). Available at www.ngu.no

Solberg IL, Dalsegg E, L'Heureux JS (2012a) Resistivitetsmålinger for løsmassekartlegging ved Rein kirke i Rissa, Sør-Trøndelag. Data og tolkninger. NGU Report 2012.018 (in Norwegian). Available at www.ngu.no

Solberg IL, Dalsegg E, L'Heureux JS, Rønning JS (2012b) Resistivitetsmålinger for løsmassekartlegging ved Rein kirke i Rissa, Sør-Trøndelag. Data og tolkninger. NGU Report 2012.018 (in Norwegian). Available at www.ngu.no

Solberg IL, Dalsegg E, L'Heureux JS, Rønning JS (2012c) Resistivitetsmålinger for løsmassekartlegging ved skredgrop på Byneset, Sør-Trøndelag. NGU Report 2012.004 (in Norwegian). Available at www.ngu.no

Solberg IL, Hansen L, Rønning JS, Haugen ED, Dalsegg E, Tønnesen JF (2012d) Combined geophysical and geotechnical approach for ground investigations and hazard zonation of a quick-clay area, Mid Norway. Bull Eng Geol Environ 71(1):119–133. doi:10.1007/s10064-011-0363-x

Trondheim kommune (2012) Esp – Byneset, kvikkleireskred. Grunnundersøkelser, datarapport. Trondheim kommune Report R.1527, 09.18.2012 (in Norwegian)

Chapter 15
An Integrated Approach to Quick-Clay Mapping Based on Resistivity Measurements and Geotechnical Investigations

Andreas Aspmo Pfaffhuber, Sara Bazin, and Tonje E. Helle

Abstract Quick clay is highly sensitive, marine clay with an unstable mineral structure due to post-glacial heaving and subsequent leaching of saline pore fluids by surface- and groundwater. Quick-clay layers pose a serious geo-hazard in Scandinavia and North America and need to be delineated in detail. Geophysical methods, especially resistivity methods, have been tested for quick-clay mapping at several sites across Norway. By scrutinizing results from Electric Resistivity Tomography (ERT) and integrating them with geotechnical borehole data including Resistivity Cone Penetrometer Testing (RCPT), we confirm the value of an integrated study for quick-clay hazard zonation. ERT is an ideal tool to interpolate limited borehole results and thus to provide a more cost efficient and detailed result than with borehole data alone. Our resistivity data from ERT, RCPT and lab measurements are generally consistent and appear isotropic. Geochemical analysis confirms that changes in resistivity are directly related to changes in clay salt content and secondarily related to sensitivity. It's a challenge to resolve small contrast in resistivity (to distinguish unleached from leached clay), in close vicinity to drastic changes in earth resistivity, for example the transition from clay to bedrock. To cope with this we improve results by means of constrained ERT inversion approaches based on drilldata and/or seismic refraction bedrock data. Though ERT is no silver bullet solution to detailed quick-clay mapping, it can provide a significant contribution to improve the risk assessment at comparably lower costs than extensive drilling campaigns. Remaining methodological ambiguities need to be handled by integration with further data.

Keywords Quick clay • ERT • RCPT • Geophysical inversion • Geochemistry

A.A. Pfaffhuber (✉) • S. Bazin
Norwegian Geotechnical Institute (NGI), Oslo, Norway
e-mail: aap@ngi.no; Sara.Bazin@ngi.no

T.E. Helle
Department of Civil and Transport Engineering, Norwegian University of Science and Technology (NTNU), Trondheim, Norway

The Norwegian Public Roads Administration, Trondheim, Norway

J.-S. L'Heureux et al. (eds.), *Landslides in Sensitive Clays: From Geosciences to Risk Management*, Advances in Natural and Technological Hazards Research 36,
DOI 10.1007/978-94-007-7079-9_15, © Springer Science+Business Media Dordrecht 2014

15.1 Introduction

Quick clay has developed over the last ~9,000 years by continuous leaching of marine clay. During leaching, pore-water salinity decreases, thus bulk electric resistivity increases (e.g. Løken 1968; Söderblom 1969). Hence, it should be feasible to distinguish resistive quick clay from conductive marine clay using Electric Resistivity Tomography (ERT). Due to the many factors affecting the resistivity measurements, ERT cannot be used as a standalone method to identify quick-clay deposits. Published resistivity data from various quick-clay sites (e.g. Rankka et al. 2004; Solberg et al. 2008, 2012; Lundström et al. 2009; Rømoen et al. 2010; Löfroth et al. 2011; Sauvin et al. 2013) indicate that the resistivity threshold for quick clay is site dependent. However, by combining both resistivity data and conventional geotechnical sounding data one should be able to detect low-saline/quick-clay deposits.

NGI's experience since 2009 from both commercial and research projects involving resistivity data and quick clay from eight different sites confirms the assumption that clay resistivity ultimately indicates salt content and thus indirectly the remoulded shear strength and sensitivity. Here we will show that ERT models fit with Resistivity Cone Penetrometer Testing (RCPT) logs as well as laboratory resistivity measurements within the methods' accuracy. Geochemical analysis links resistivity with salt content and remoulded shear strength obtained from sample tests.

The major limitation of ERT is the state of practice in geophysical inversion, finding the most fitting resistivity model to the measured data. The small resistivity difference between unleached marine clay and potential quick clay may be "washed out" by the potentially strong contrast between clay and bedrock. An approach to overcome this limitation is constrained inversion, where a known bedrock interface is used to increase the model resolution of the clay layer. In this study we illustrate the potential improvements by virtue of resistivity models constrained by bedrock topography. The resistivity of the clay layer may also be imaged more correctly by constraining the inversion model by RCPT data along the profile.

15.2 Field and Laboratory Work

In the following, two field sites belonging to two adjacent quick-clay hazard zones are described and discussed (Fig. 15.1) while the general conclusions are also supported by a third site in Central Norway that cannot be introduced in detail within the length of this paper.

15.2.1 Site Description Vålen and Smørgrav

The two main sites are located along the eastern banks of the Vestfossen River, approximately 65 km from Oslo, Norway. The area is covered with marine deposits; bedrock geology varies from north to south with alum shale in the north and limestone

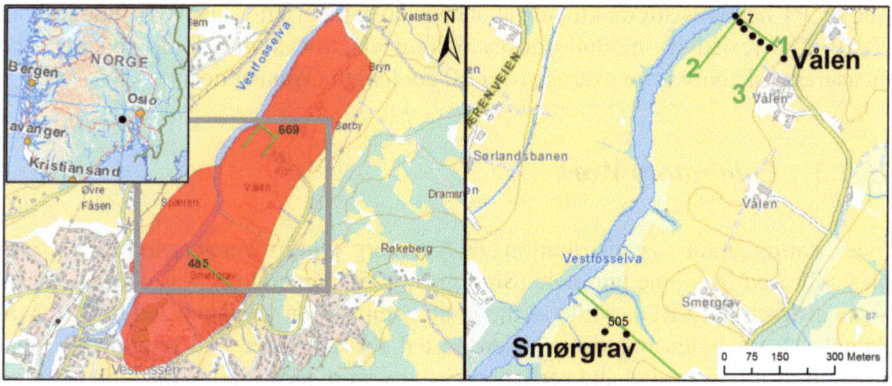

Fig. 15.1 Location of Vestfossen quick clay area and the two study sites, Vålen in the north and Smørgrav in the south. *Red* indicates high quick clay hazard level (Source www.skrednett.no). *Green lines* indicate ERT profiles considered in this study. Boreholes discussed are shown as *black dots*

in the south-east (from the Geological Survey of Norway, www.ngu.no). The southern site, Smørgrav, has been used for multi-method research as described by Pfaffhuber et al. (2010) and Donohue et al. (2012) including electromagnetic mapping, surface wave- and refraction seismic, ERT and controlled source radiomagnetotellurics (Kalscheuer et al. 2013).

The northern site, Vålen, is subject to ongoing monitoring of movements and pore pressure. Geotechnical data from 15 boreholes indicate two distinct quick-clay layers (a new interpretation updating earlier results from laboratory tests on samples from one borehole that were not sensitive enough to be classified as quick clay, Rømoen et al. 2010). Water content measured on samples from Vålen was above the liquid limit, also an indicator of quick clay. The quick-clay layers were finally identified based on these samples and rotary pressure soundings.

Common to both sites is that there is evidence for two isolated quick clay pockets and no continuous layer. The continuity and lateral extend of quick-clay bodies may have implication on the calculation of the safety factor. Standard quick-clay hazard assessment, often based on one drilling in one slope, fails to identify the lateral extent of potential quick-clay layers.

As the focus of this paper is integrating a range of data from various sources and disciplines, the used methods shall be briefly introduced especially emphasizing on the geophysical methods.

15.2.2 Geotechnical Site Investigations

Rotary pressure sounding (RPS) and RCPT (www.envi.se) were carried out to the depth of assumed bedrock. Piston sampling was carried out every meter from 1 to 20 m depth for selected boreholes. RPS gives a good indication on the occurrence of quick or sensitive clay as the resistance becomes constant or decrease with depth.

The RCPT has a resistivity module attached above the friction sleeve (Rømoen et al. 2010). The resistivity moduli consist of four electrodes with 5 cm spacing, and measures resistivity is an average over 20 cm length (Wenner array).

15.2.3 Laboratory Work

Index testing, grain-size distribution and Atterberg's limits were carried out on samples collected from one borehole (Smørgrav, 505). Mineralogy was determined by x-ray diffraction (XRD), while as the ionic composition was analyzed by ICP MS. Samples from Vålen borehole 7 were tested for undrained and remoulded shear strength. Both horizontal and vertical resistivity were measured in the laboratory on samples collected from Vålen and Smørgrav.

15.2.4 Electrical Resistivity Tomography (ERT)

ERT is a well-established method in near-surface geophysics combining vertical electrical sounding and profiling with modern, multi-electrode systems. These devices automatically perform measurements using numerous combinations of current and potential electrodes hooked up to multithread cables. The system used here, a Swedish Terrameter LS (ABEM 2010), switches up to 64 electrodes and measures potentials at 10 electrodes in parallel. To make use of this, all data presented here, where acquired in the gradient array configuration.

For standard ERT interpretation state-of-practice software RES2DINV was used (Loke 2010). As ERT measurements in principle can be explained by a multitude of resistivity models, certain constrains have to be applied to find a solution to the inverse problem. The "Occam" principle tries to find the simplest model that can explain the measured data (Constable et al. 1987). RES2DINV implements this by virtue of a smooth model with gradual changes of resistivity over a model with a fixed, rectangular mesh. The inversion is consequently fast and robust but will "smear out" sharp interfaces.

To obtain the more accurate constrained inversions, academic inversion software BERT was used (Günther et al. 2006). BERT is based on a triangular, unstructured mesh, making the forward model calculations more accurate than with the state of practice. Furthermore, known sharp interfaces (for example clay/bedrock) can be forced to result in a sharp change in resistivity.

15.3 Results

We have investigated three main aspects (1) the relationship between electrical resistivity and geotechnical sensitivity of marine clay, (2) the match of resistivity – depth profiles extracted from RCPT soundings and ERT results and finally (3) the potential improvement of ERT models by virtue of geometrically constrained inversion.

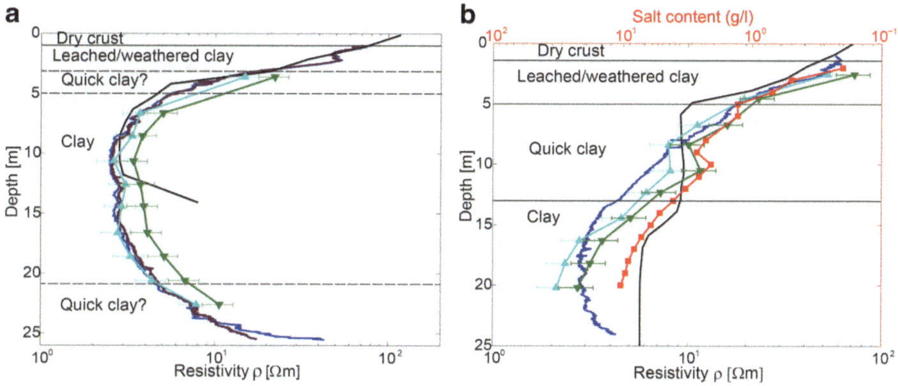

Fig. 15.2 Resistivities with depth. In situ (RCPT logs are shown *dark blue*), lab (vertical in *green* and horizontal in *light blue*), and ERT resistivities (*black*) for Vålen (**a** borehole 7) and Smørgrav (**b** borehole 505). The pore-water salt content was also measured in Smørgrav (*red*)

15.3.1 Resistivity and Sensitivity

RCPT was performed in several boreholes but only the most representative one is presented for each of the two sites. Resistivity values measured in situ and in the lab on collected samples are compared in Fig. 15.2. For Vålen the two types of resistivity values agree well (Fig. 15.2a).

The resistivities measured horizontally are in average 38 % lower than the ones measured vertically, bearing in mind a 20 % error estimate due to uncertainties in the sample size and contact resistance. The match between the in situ resistivities and those measured on Smørgrav samples is also good considering the measurement accuracy (Fig. 15.2b). The resistivities measured horizontally are in average 29 % lower than the ones measured vertically. This is slightly above the error estimate and we consider anisotropy neglectable. Figure 15.2b further compares measured resistivity with salt content data from the samples confirming the expected correlation between resistivity and salt content.

Further lab analyses was performed on those samples, indicating that the clay content increases with depth, from approximately 40 to 60 % below 17 m depth. The clay fraction mineralogy is dominated by illite with a significant amount of chlorite. The total salt content is 1.2–4.1 g/l in the quick clay, increasing with depth (Fig. 15.3a). The remoulded shear strength varied from 0.3 to 5 kPa in the borehole, with a sensitivity of 33–77 in the quick-clay section. The natural water content was well above the liquid limit through the quick-clay layer (Fig. 15.3b).

15.3.2 Resistivity from ERT and RCPT

Three ERT profiles (160–200 m long) were acquired in Vålen (Fig. 15.1). The electrode spacing is 2 m with some exceptions at the profile ends. Borehole 7 is located ~10 m away from profile 1. One 370 m long ERT profile was acquired at the

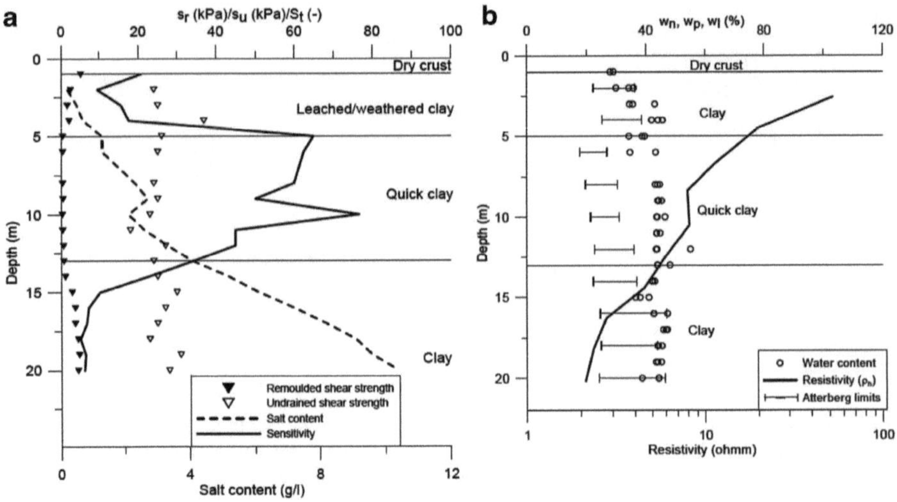

Fig. 15.3 Lab analyses from Smørgrav, borehole 505. (**a**) Undrained and remoulded shear strength, salt content and sensitivity. (**b**) Water content, Atterberg limits and laboratory horizontal resistivity

Smørgrav site (Fig. 15.1), with electrode spacing of 5 m. This profile is a new dataset, co aligned with two shorter profiles (160 m) discussed by Donohue et al. (2012). Kalscheuer et al. (2013) use both ERT datasets to jointly invert them with CSRMT data from the site.

Vålen profile 1 indicates a 1–5 m thick surface layer of dry crust (>50 Ω.m) with clay below (Fig. 15.4a). Interpretation from both ERT and geotechnical data indicate the presence of two distinct quick-clay layers in the shallow ESE part and in the deeper WNW part of the profile (10–50 Ω.m).This is confirmed by two perpendicular ERT profiles (not shown). On the top of the hill, a quick-clay layer extends from dry crust to bedrock, it thins out along the slope, and reappears at depth near the river. A 10 m thick pocket of unleached marine clay lies between these two layers of quick clay, as suggested by RPS resistance curves (Fig. 15.4b). The shale bedrock has a resistivity range quite similar to marine clay and thus doesn't stand out in the ERT model.

In Smørgrav, the ERT profile shows dry crust and quick-clay with similar resistivity ranges as in Vålen, yet the different bedrock geology is evident (Fig. 15.5). While shale in the center of the Smørgrav profile compares to the Vålen basement, a limestone body in the East is indicated with resistivity >200 Ω.m. A high voltage utility cable crossing the area generates a conductive anomaly in the ERT data. The high resistivity values (blue) in the SE part of the ERT profile indicate the presence of limestone (Fig. 15.5a).

The comparison between RCPTU and ERT in Smørgrav shows that the values of the in situ resistivities and of the inverted resistivities do not perfectly match while the overall trends agree (Fig. 15.2b). For such quantitative comparisons the different

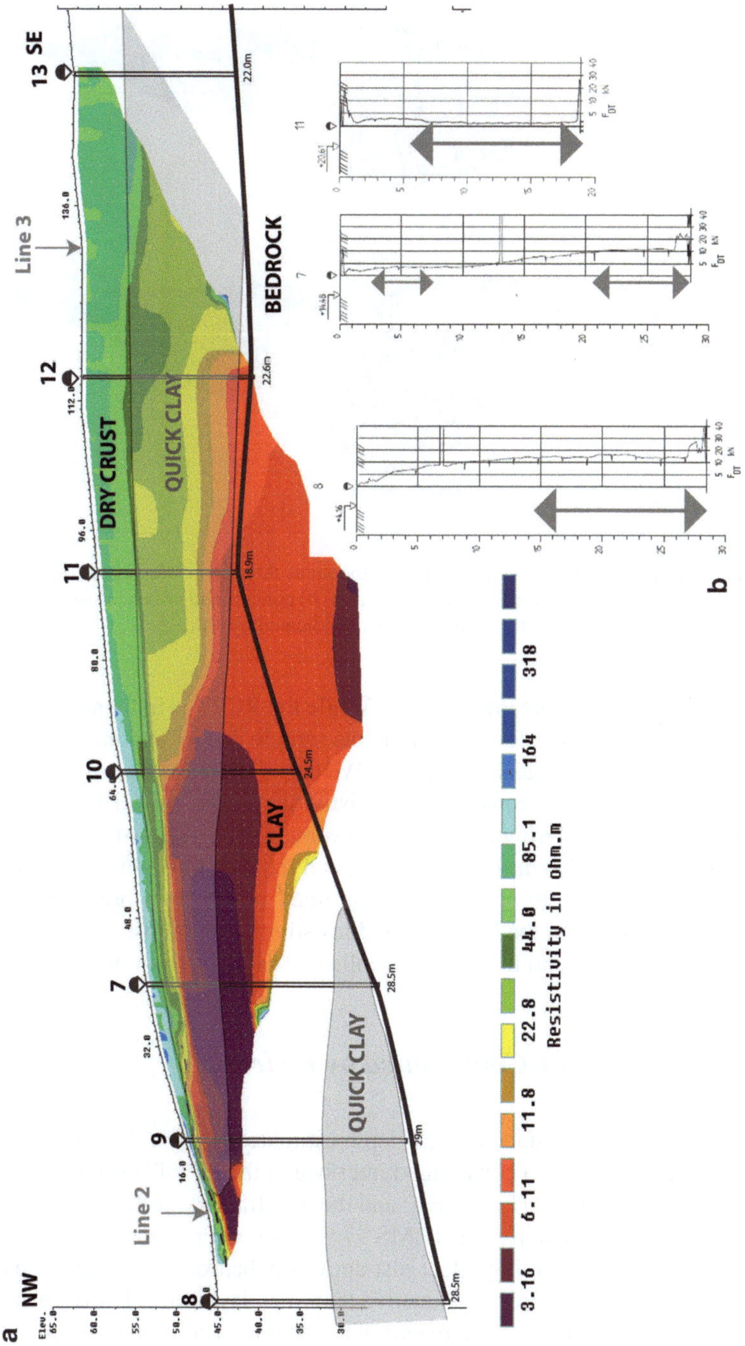

Fig. 15.4 Vålen. **(a)** ERT profile 1 with robust inversion constraint (no vertical exaggeration). Seven geotechnical sounding profiles indicate two quick-clay layers (*delineated in grey*) and the bedrock topography. **(b)** RPS profiles 8, 7 and 11 with interpreted quick-clay intervals

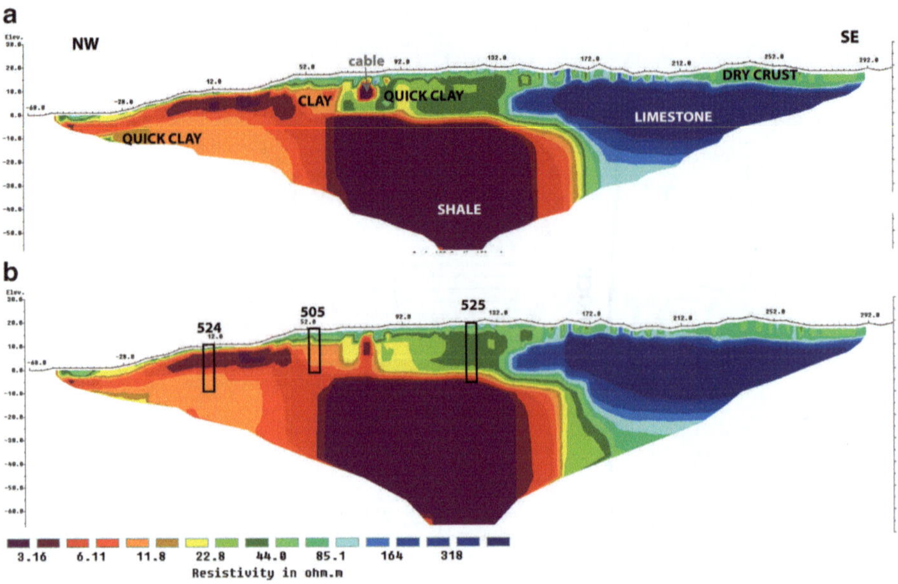

Fig. 15.5 Smørgrav ERT profile (length 370 m, depth extent 80 m from elevation 20 to –60 m). Resistivities are inverted (**a**) based on ERT data alone and (**b**) constrained with a priori information (block resistivities are fixed around three RCPT logs as indicated by *black boxes*)

sampling scales need to be take into account; While the RCPT is influenced by some tens of cm around the probe, ERT measurements sample deeper and not only downwards. Thus 3D effects can bias ERT results. We tested the influence of RCPT logs on ERT inversion results by constraining resistivities in the model blocks in the vicinity of three boreholes during RES2DINV inversion. The unconstrained and constrained models are visually similar (Fig. 15.5), and the data fit is almost unchanged (RMS = 1.9 versus 2.1). In this case, the constrained model does not improve the already excellent fit between the ERT-inverted resistivities and those measured both in situ with the RCPT tool and in the lab on collected samples from borehole 505.

15.3.3 Clay Resistivity Constrained by Bedrock

Several RPS soundings provided bedrock depth estimates along Vålen profile 1. We integrated this information as a constrained interface in the BERT inversion scheme. Both the unconstrained model (Fig. 15.6a) and the constrained model (Fig. 15.6b) fit well with the measured resistivity data (RMS ~7 %). We observe that the constrained inversion cancelled the smoothing effect just above the bedrock in the upper part of the slope, where the shale is more conductive than the clay above. However it does not have a significant effect in the centre part of the profile, where shale and clay have similar resistivities.

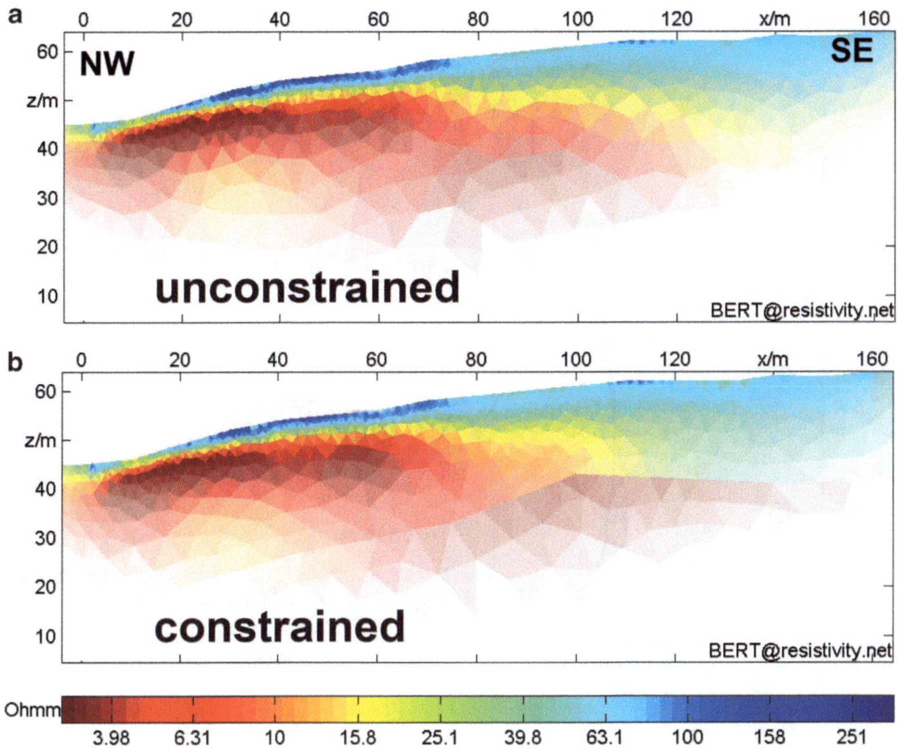

Fig. 15.6 Vålen profile 1. BERT resistivity section without (*top*) and with (*bottom*) a priori interface constraint such as bedrock depth. The bedrock topography is estimated from RPS profiles (see Fig. 15.4a)

ERT, RCPT and RPS were performed at another site, in central Norway, where the geometry and the extent of quick clay needed to be evaluated for slope stability assessment. A vertical penetration resistance curve indicates the presence of quick clay below 10 m depth (Fig. 15.7a). Due to the resistive bedrock at this site, in situ and inverted resistivities from RES2DINV disagree in the vicinity of bedrock (Fig. 15.7b). The sharp interface between clay and gneiss is modeled as a gradient. This implies that we cannot distinguish between the two alternatives for the lower part of the sediment unit: quick-clay upon bedrock, or clay/other sediments between quick clay and bedrock. Bedrock is also in this study constrained in the BERT model by estimated depth from sounding data. This gives resistivity values at the bottom of the clay unit of ~40 Ωm in yellow instead of the overestimated ~100 Ωm in turquoise (Fig. 15.7c, d). This is in coherence with the RPS profile indicating quick clay upon bedrock (Fig. 15.7a). The ERT profile helps the extrapolation from the discrete 1D geotechnical sounding, and thereby enhances the computation of the safety factor.

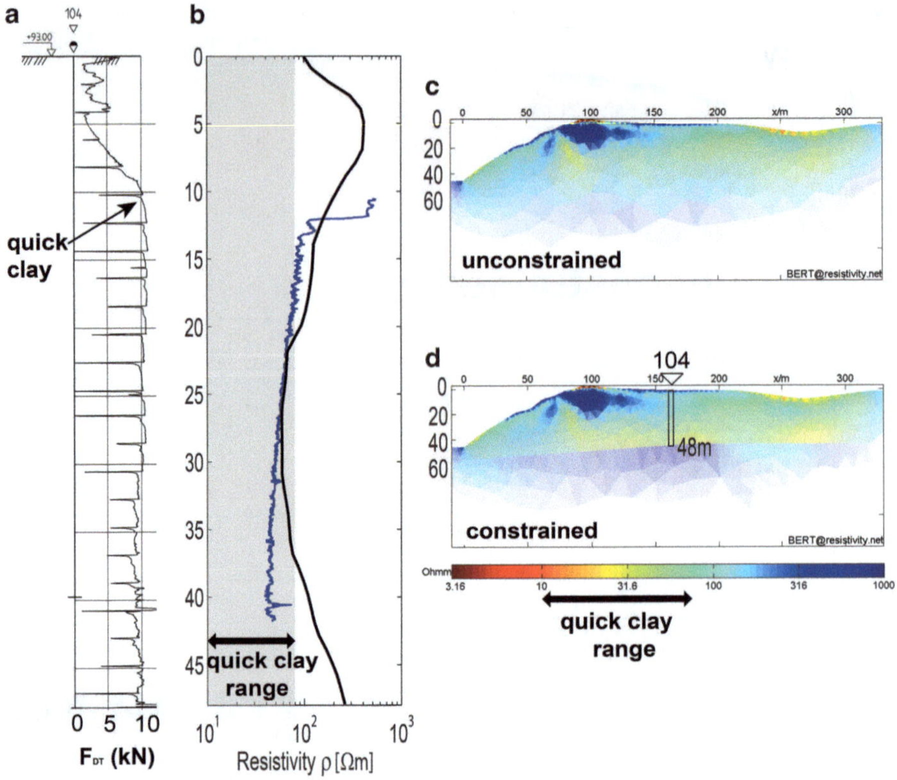

Fig. 15.7 Integrated study from central Norway (**a**) Borehole 104: the vertical penetration resistance curve below 10 m depth is interpreted as quick clay. (**b**) In situ RCPT (*blue*) and ERT inverted resistivities (*black*) in borehole 104. ERT profile without (**c**) and with (**d**) a priori interface constraint

15.4 Discussions and Conclusions

The comparison of sample test data from boreholes and resistivity data shows the relation between pore-water salt content and resistivity values, as also confirmed in other studies (e.g. Solberg et al. 2008, 2012; Long et al. 2012). Low salt content and consequently increased resistivity does not necessarily prove quick clay, however. Chemical weathering and further leaching might transform the quick clay to a non-quick stage, where the salt content remains low. Furthermore adding organic or inorganic dispersants in low-activity soils may decrease the liquid limit and the remoulded shear strength (Torrance 1974, 1983). Increased clay resistivity can thus not be solely used to identify quick clay.

Developing of quick clay depends on many factors and the combination of these; i.e. grain-size distribution, sediment properties, mineralogy, pore-water chemistry, and consolidation. A change in one of these factors will lead to different properties of the clay and in the measured resistivity. Thus a set interval for whereas quick clay can be interpreted is not achievable. The authors therefore recommend that a site-specific resistivity interval is necessary to interpret quick clay from ERT profiles (10–80 Ω.m for the presented sites). RCPT will be useful in this matter. These conclusions are based on several Norwegian quick-clay sites in addition to the two cases described here. Once a site-specific resistivity range for quick clay is established by comparison of resistivity and geotechnical data, borehole information can be extrapolated over larger areas than usually achievable with drillings.

Remaining methodological limitations in depth resolution, especially close to abrupt interfaces such as bedrock, can be improved by constrained inversion. Here, additional data such as drillings or refraction seismic data may be used. Development of dedicated geophysical inversion tools to resolve the subtle difference between clay and quick clay would be desirable.

Acknowledgments Funding for most of the presented results was provided by the research council of Norway through NGI research programs SIP12 and SP1, and the International Centre for Geohazards (ICG). We acknowledge Guillaume Sauvin, Marianne Lanzky Kolstrup, Matthew Lato for their help on the field. The authors also thank Samuel Cornée for his early contribution during an internship. We thank Magnus Rømoen, Gunvor Bårdvik, Alf Lund and many more NGI colleagues for providing and explaining RPS and CPT data, constructive discussion and many geotechnical insights. We thank Thomas Günther for providing BERT software and Reviewer Inger-Lise Solberg for valuable comments on the manuscript.

References

ABEM (2010) Terrameter LS, Instruction Manual, ABEM100709, p 63.. http://abem.se/files/upload/UserGuideTerrameterLS2020100709.pdf

Constable SC, Parker RL, Constable CG (1987) Occam's inversion: a practical algorithm for generating smooth models from electromagnetic sounding data. Geophysics 52(3):289–300

Donohue S, Long M, O'Connor P, Helle TE, Pfaffhuber AA, Rømoen M (2012) Multi-method geophysical mapping of quick clay. Near Surf Geophys 10(3):207–219

Günther T, Rücker C, Spitzer K (2006) Three-dimensional modelling and inversion of dc resistivity data incorporating topography – II. Inversion. Geophys J Int 166:506–517. doi:10.1111/j.1365-246X.2006.03011.x

Kalscheuer T, Bastani M, Donohue S, Perrsson L, Pfaffhuber AA, Reiser F, Ren Z (2013) Delineation of a quick clay zone at Smørgrav, Norway, with electromagnetic methods under geotechnical constraints. J Appl Geophys 92:121–136

Löfroth H, Suer P, Dahlin T, Leroux V, Schälin D (2011) Quick clay mapping by resistivity – surface resistivity, CPTU-R and chemistry to complement other geotechnical sounding and sampling, Swedish Geotechnical Institute, Report GÄU 30

Loke MH (2010) Res2DInv ver 3.59.102. Geoelectrical imaging 2D and 3D. Instruction manual. Geotomo Software. http://www.geotomosoft.com/

Løken T (1968) Kvikkleiredannelse og kjemisk forvitring i norske leirer. Norwegian Geotechnical Institute, Oslo. Publication 75, pp 19–26 (In Norwegian)

Long M, Donohue S, L'Heureux JS, Solberg IL, Rønning JS, Limacher R, O'Connor P, Sauvin G, Rømoen M, Lecomte I (2012) Relationship between electrical resistivity and basic geotechnical parameters for marine clays. Can Geotech J 49:1158–1168

Lundström K, Larsson R, Dahlin T (2009) Mapping of quick clay formations using geotechnical and geophysical methods. Landslides 6(1):1–15

Pfaffhuber AA, Bastani M, Cornée S, Rømoen M, Donohue S, Helle TE, Long M, O'Connor P, Persson L (2010) Multi-method high resolution geophysical and geotechnical quick clay mapping. Extended Abstract, Near Surface

Rankka K, Andersson-Sköld Y, Hultén C, Larsson R, Leroux V, Dahlin T (2004) Quick clay in Sweden, Swedish Geotechnical Institute, Linköping Report 65. www.swedgeo.se

Rømoen M, Pfaffhuber AA, Karlsrud K, Helle TE (2010) Resistivity on marine sediments retrieved from RCPTU-soundings: a Norwegian case study. International symposium on cone penetration testing, 2, CPT'10, Huntington Beach, CA. Proceedings 2, pp 289–296

Sauvin G, Lecomte I, Bazin S, L'Heureux JS, Vanneste M, Solberg IL, Dalsegg E (2013) Towards geophysical and geotechnical integration for quick-clay mapping in Norway. Near Surf Geophys 11 (in press). doi: 10.3997/1873-0604.2012064

Söderblom R (1969) Salt in Swedish clays and its importance for quick clay formation: results from some field and laboratory studies. Statens Geotekniske Institut, Stockholm, p 53

Solberg IL, Rønning JS, Dalsegg E, Hansen L, Rokoengen K, Sandven R (2008) Resistivity measurements as a tool for outlining quick clay extents and valley fill stratigraphy: feasibility study from Buvika, Central Norway. Can Geotech J 45:210–225

Solberg IL, Hansen L, Rønning JS, Haugen E, Dalsegg E, Tønnesen JF (2012) Combined geophysical and geotechnical approach to ground investigations and hazard zonation of a quick clay area, Mid Norway. Bull Eng Geol Environ 71:119–133

Torrance JK (1974) A laboratory investigation of the effect of leaching on the compressibility and shear strength of Norwegian marine clays. Geotechnique 24(2):155–173

Torrance JK (1983) Towards a general model of quick clay development. Sedimentology 30:547–555

Chapter 16
Geophysical and Geotechnical Investigations for a Major Highway in a Quick-Clay Area

Rolf Sandven and Inger-Lise Solberg

Abstract The paper presents results from site investigations for a major public highway project close to Trondheim in Mid Norway. The planned road is located in an area with thick deposits of sensitive and quick clays, and is expected to represent both local and global stability issues. The site investigations included 2D resistivity measurements and conventional geotechnical borings, such as rotary pressure soundings, total soundings, cone penetration tests (CPTU), undisturbed piston sampling and pore-pressure measurements. Extensive laboratory investigations were also carried out, including salinity measurements. The 2D resistivity measurements were carried out to detect the distribution of leached clay, and the results were used for planning of the subsequent geotechnical investigations. This is one of relatively few Norwegian studies using 2D resistivity measurements in evaluation of ground conditions in an industry project. There was very good agreement between the data sets with respect to presence of sensitive or quick clays. The investigation hence clearly points out that the combination of resistivity measurements and geotechnical borings has the potential of becoming a powerful site investigation strategy, particularly in mapping of large areas or long-stretched road corridors.

Keywords 2D resistivity • Geotechnical investigations • Marine deposits • Quick clay

R. Sandven (✉)
Multiconsult, Trondheim, Norway
e-mail: rolf.sandven@multiconsult.no

I.-L. Solberg
Geological Survey of Norway (NGU), Trondheim, Norway
e-mail: inger-lise.solberg@ngu.no

J.-S. L'Heureux et al. (eds.), *Landslides in Sensitive Clays: From Geosciences to Risk Management*, Advances in Natural and Technological Hazards Research 36,
DOI 10.1007/978-94-007-7079-9_16, © Springer Science+Business Media Dordrecht 2014

16.1 Introduction and Background

Large areas in Norway are prone to landslides in sensitive or quick marine clays. Landslides can be triggered naturally, but human activity is often an important factor. Slope failures in marine deposits occur regularly, however with varying sizes and consequences. Fortunately, landslide disasters with remoulding of large volumes of quick clay occur very rarely.

In Norway, presence of quick clay is usually detected by geotechnical soundings and undisturbed sampling in traditional site-investigation schemes. Laboratory testing on retrieved samples is hence considered to be the most reliable detection method for existence of quick clay. However, this method is expensive and time-consuming, and gives information only at scattered sampling locations within the investigated area. Cone penetration tests (CPTU) are also used commonly, but the method has shown some shortcomings in addressing the presence of quick clay.

In Canada and Sweden, 2D resistivity measurements have been carried out during the last decade for mapping of marine deposits and for landslide hazard evaluations (e.g. Dahlin et al. 2001; Calvert and Hyde 2002; Lundström et al. 2009; Löfroth et al. 2012). Several projects in Norway have in the same period studied the ability of the method to map potential quick-clay deposits (e.g. Solberg et al. 2008, 2012, this volume; Aasland 2010; Donohue et al. 2012). These investigations have shown that 2D resistivity profiling gives a good overview of the ground conditions, including the presence of leached clay. However, it is also concluded that it is essential to supplement the results with other surveys, using penetrating methods.

Advancements in field equipment and software for data processing, such as the inversion method, have simplified the investigations and produced more reliable results. Detailed mapping of the extent of quick clay is of great interest for planning and protection purposes, as the location of quick clay within slopes has a major impact on the landslide risk. Results from resistivity measurements moreover provide an excellent basis for selecting representative locations for more detailed geotechnical investigations. The 2D resistivity measurements may hence optimize the site investigation.

The main purpose of this article is to present the resistivity method and its ability to outline the extent of quick clay in a road planning project. Results from 2D resistivity measurements have been compared to a large amount of geological and geotechnical data. In addition, the results of the study will be used in order to interpret the general geological and stratigraphical distribution of sediments in the studied area, which is important for formation of quick clay and for the interpretation of quick-clay extent.

16.2 Setting and Project Description

The investigated area is located in Kaldvelladalen in the municipality of Melhus, about 25 km south of Trondheim in Mid Norway (Fig. 16.1). The area is cultivated and sparsely developed, and is surrounded by the hills Skardåsen and Aunåsen.

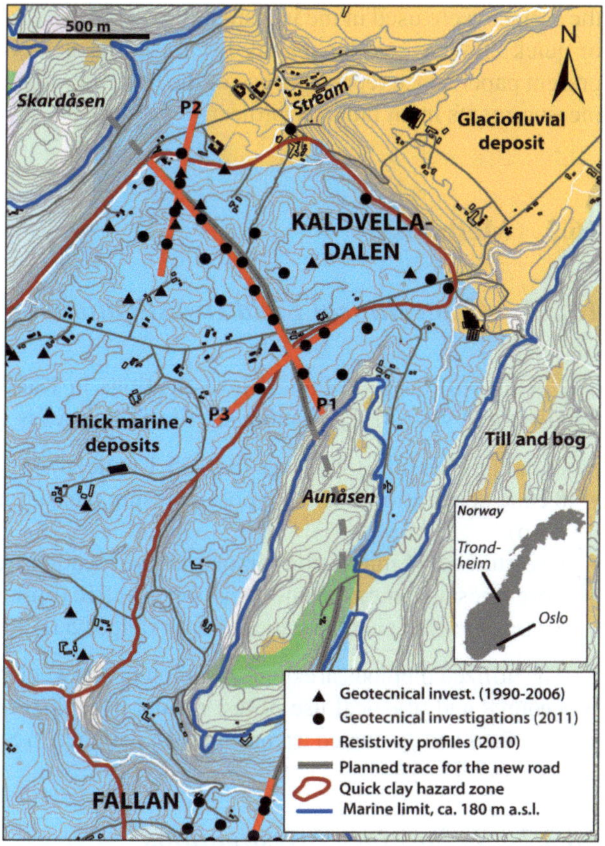

Fig. 16.1 Study area in Kaldvelladalen, Mid Norway, showing Quaternary sediments, 2D resistivity profiles, locations for geotechnical investigations, outline of quick-clay hazard zone and location of the marine limit. Terrain contour line intervals are 5 m

The terrain is undulated with elevations of 99–145 m a.s.l., and there are several traces of ravine erosion and landsliding. The study area is located within Engan quick-clay hazard zone (Gregersen et al. 1990). According to the Quaternary geological map, the area has thick deposits of marine clay (Reite and Sørensen 1980). Large amounts of glaciofluvial sediments were deposited towards the study area from a glacier located north-east in the valley.

The National Public Road Administration (NPRA), the consulting company Multiconsult AS and the Geological Survey of Norway (NGU) have carried out site investigations for an alternative road line of the European highway E6. Ground investigations were carried out in two separated areas along the planned road line: Fallan and Kaldvelladalen, with focus on the latter location in this paper. A profile from Fallan is shown in Donohue et al. (this volume).

The main scope of the investigations was to map the ground conditions and evaluate the geotechnical challenges for the new road, given the expected presence of large deposits of quick clay in the area. This was important for stability evaluations and the possibility of constructing remedial measures to improve local stability. Both conventional geotechnical investigations and 2D resistivity measurements on the surface were used in the investigation. 2D resistivity measurements for mapping of quick clay have so far mainly been used in method development projects. The present paper hence represents one of the first Norwegian studies using 2D resistivity measurements in an industry project.

16.3 Methods

16.3.1 Geotechnical Field Investigations

The details of the geotechnical investigations in Kaldvelladalen are described by Sandven and Vik (2011). A summary of the surveys is given below:

- 25 rotary pressure soundings (8.9–55.1 m deep)
- 8 cone penetration tests with pore-pressure measurements (CPTU) (30.0–40.0 m deep)
- Undisturbed 54 mm piston samples at five locations (depth intervals: 2.0–21.8 m)
- Pore-pressure measurements using electrical piezometers at four locations, 2 piezometers at each location (depth ranges 5.0–20.0 m).

The utilized methods are quite common in Norwegian site investigations. The test procedures and practical use of test results in evaluation of quick clay are recently summarized and evaluated according to experience in Sandven et al. (2012).

16.3.2 Laboratory Investigations

The samples were investigated in the geotechnical laboratory with respect to classification and identification of the encountered soils, together with determination of the mechanical parameters by oedometer and triaxial tests. The following index tests were carried out, using procedures described in Norwegian standards: Water content; density; fall cone test (including remoulded conditions (FCT)); unconfined compression test (UCT); grain size distribution including suspension analyses and Atterberg limits. Details from the laboratory investigations are given in Sandven and Vik (2011).

The natural salt content in marine, unleached clay is usually 35 g/l. When clay with salt pore-water is leached by fresh groundwater due to isostatic rebound, with former seabed becoming dry land, the salt content in the pore-water is reduced. Marine clay where the salt content in the pore-water is reduced below 5 g/l, can become sensitive and show quick behavior (Bjerrum 1954). The pore-water salinity in samples from Kaldvelladalen was determined using conductivity tests on pore-water sampled by vacuum sucking.

The most reliable information on sensitive material is obtained when clay from geotechnical core samples is tested in the laboratory. When the remoulded shear

Resistivity (Ωm)	Main characterisation	Description
1-10	Unleached marine clay deposits	Saline pore-water, stable clay structure. Good conductivity. Electronic conductive minerals like graphite, sulphides and some oxides, may also give low resistivity values.
10-100	Leached clay deposits	Low total electrolyte content. Still good conductivity, but poorer than for the unleached marine clay. Silt, fine-grained till, and leached, non-quick clay may fall into the same interval.
> 100	Dry crust clay deposits, coarse sediments, (bedrock)	Dry clay crust, remoulded, dry clay from quick-clay landslides, and coarser materials (sand, gravel) will have higher resistivity values than marine clay. Most bedrock types will have values of several thousand Ωm.

Fig. 16.2 First-order classification of sediments from resistivity values (Modified from Solberg et al. 2011, 2012)

strength (s_r) is less than 0.5 kPa, the clay is defined as quick clay. The sensitivity (S_t) is the undisturbed shear strength (s_u) divided by the remoulded shear strength (s_r). Sensitive clays are defined to have sensitivity $S_t > 15$ and remoulded shear strength $s_r < 2.0$ kPa. Together with the quick clays, these clays are grouped as brittle materials, showing significant strain softening at failure.

16.3.3 2D Resistivity Measurements

The 2D resistivity method is well suited for outlining pockets and layers of leached clay since this material has a slightly higher electric resistivity than intact unleached clay. This is due to a higher salt content and thereby higher ion concentration in the latter.

In this project, 2D resistivity measurements were carried out based on the Lund-system developed by Dahlin (1993). Four active cables were used with a Gradient electrode array, using electrode separations of 5 m. The grounding connections were very good. The measuring equipment was an ABEM Terrameter SAS 4000, using a current of 200 mA (Solberg and Dalsegg 2012).

Raw data from the resistivity measurements give the apparent resistivity (ρ_a) of the subsurface. This represents a weighted mean of all the resistivity values that fall within the soil volume of influence. To obtain the specific resistivity (ρ) in Ωm from different parts of the subsurface, the data are inverted to produce a model where the apparent resistivity pseudo-section matches the measured pseudo-section. Recorded resistivities were inverted by the computer program Res2DInv, using the least-squares method (Loke 2010).

The classification of clay properties based on resistivity values is mainly empirical, deduced from a comparison of geophysical and geotechnical data from different studies in Scandinavia and Canada (e.g. Solberg et al. 2008, 2012; Long et al. 2012). A first-order classification of sediments from resistivity values is shown in Fig. 16.2. Note that there are gradual transitions between the classes and there may be local variations related to factors such as pore-water chemistry, saturation, grain size distribution, mineral composition etc. The method is hence only indicative for quick-clay detection, since both leached, non-quick clay and silty, non-sensitive material can give resistivity values in the same range (10–100 Ωm). Quick clay will occasionally be detected outside the 10–100 Ωm interval.

2D resistivity data are influenced by a relatively large volume of soil around the measured profile, and there is always a trade-off between depth penetration and resolution. Suppression and equivalence will cause problems in the inversion (Reynolds 2011), and there are gradual transitions between high and low resistivity values in a profile. In addition, the geology and material properties themselves will often show gradual transitions, not sharp boundaries. Therefore, depth estimates must be made with caution. An atlas providing numerous synthetic models for different geological conditions may help in the interpretation of resistivity profiles (Reiser et al. 2010).

16.4 Results and Interpretations

16.4.1 Ground Conditions from Geotechnical Investigations

The field and laboratory investigations show that the ground in Kaldvelladalen consists of a thin top crust of organic material of 0.5–2.0 m thickness. Below this layer there are thick deposits of marine sediments, with soft and partly sensitive/quick clays covering large areas. The clay is apparently homogenous with some layers and pockets of silt.

The laboratory classification of undisturbed samples shows that quick-clay layers may be encountered at depths ranging from 7.0 to 22.0 m. In the borings, 18 of a total of 25 rotary pressure sounding profiles were containing quick or very sensitive clays, which agreed well with the laboratory results. However, quick clay is not found at all boring locations, and the quick-clay layers are probably not continuous over the whole area. A summary of the results from the laboratory tests is given in Table 16.1.

Table 16.1 Summary of results from laboratory tests for the Kaldvelladalen site (Sandven and Vik 2011)

	w (%)	ρ (g/cm^3)	w_p (%)	s_u (kPa) (FC and UCT)	S_t (–)	ε_f (%) (UCT)	Comment
Non-sensitive clay	28–40	1.90–1.92	5–20	40–100	2–23	5–10	Some sample disturbance, but satisfactory sample quality
Sensitive clay	30–40	1.90–1.92	3–8	25–40	54–350	Varying	Somewhat poorer sample quality

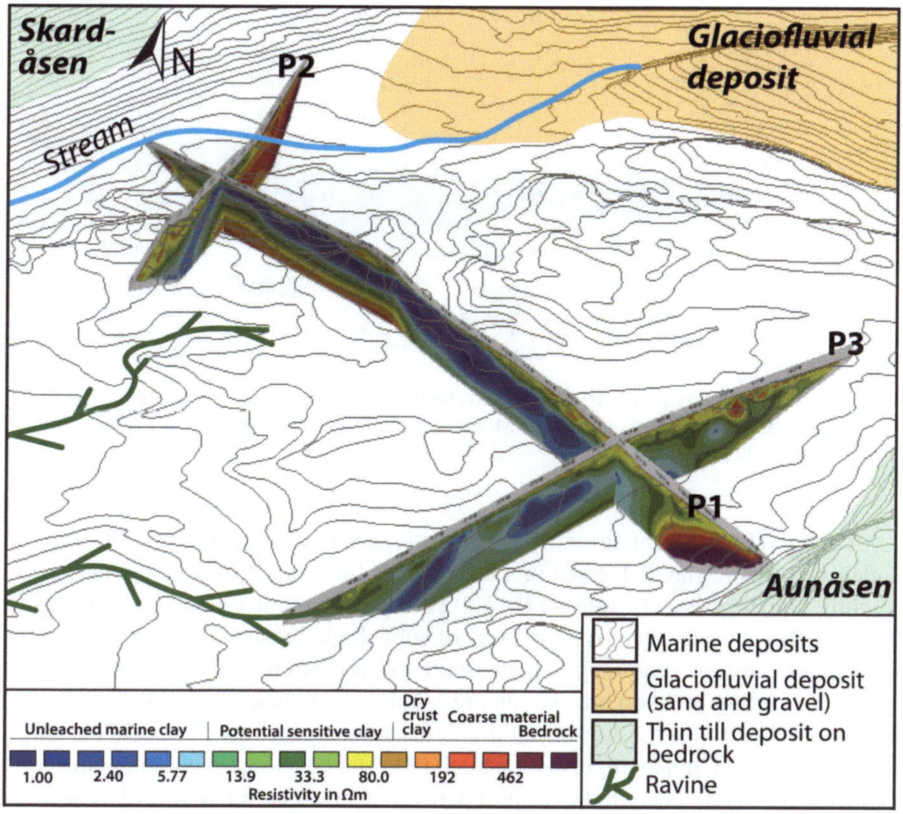

Fig. 16.3 "3D visualization" of 2D resistivity profiles in Kaldvelladalen. The profiles are displayed in ArcScene for comparison, and the scaling of the different profiles is equally set (Modified from Solberg and Dalsegg 2012)

Pore-pressure measurements indicate that the groundwater table (GWT) is located about 0.5–1.0 m below the terrain level. It is likely that the GWT shows some seasonal variations and will be at its highest close to streams and ravines. At some locations the pore-pressure distribution is lower than hydrostatic, with indication of an excess pore pressure by depth at other places.

16.4.2 *Interpretation of 2D Resistivity Profiles*

Three intersecting resistivity profiles were measured in the study area (P1-P3) (Figs. 16.1 and 16.3), and the profiles overlap very well in the intersections. The acquired 2D resistivity profiles are interpreted in terms of quick-clay extension, stratigraphy, bedrock and groundwater drainage patterns.

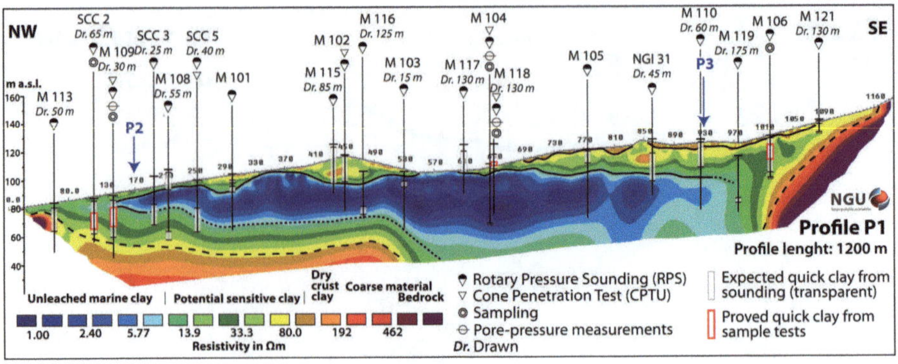

Fig. 16.4 2D resistivity profile P1 from Kaldvelladalen with interpreted geotechnical data (NGI: Gregersen et al. 1990; SCC: Emaus 2002; M: Sandven and Vik 2011; figure modified from Solberg and Dalsegg 2012)

A relatively coherent layer of low resistivities (<10 Ωm, blue) interpreted as unleached, stable clay is located in the area. This layer is surrounded by material of slightly higher resistivity values (10–100 Ωm, green/yellow/brown), which may be leached clay. Locally there is a top layer of some higher resistivity value interpreted as dry clay crust (brown/orange). In the south-eastern part of the area, high resistivity values exceeding 300 Ωm are encountered. This may indicate bedrock since profile P1 here is close to the hill Aunåsen. Similar resistivity values are present in the north-western part, which may represent bedrock (P1) and coarse-grained sediments lying upon bedrock (P2). Also in the deepest part of the north-eastern half of P1, bedrock and/or coarse-grained sediments may have influenced the measurements.

16.4.3 Comparison of Geotechnical Data and 2D Resistivity Data

The acquired 2D resistivity profiles are compared with data from geotechnical soundings and laboratory test samples. In profile P1, quite successful interpretation and comparison to sounding and laboratory test data was obtained (Fig. 16.4). The resistivity data reflects the presence of a continuous layer of leached clay in the upper part of the profile, and reveals that the thickness of this layer may vary significantly. The resistivity profiles also include a thick, continuous stratum of apparently unleached, non-sensitive clay in the central part of the studied area. However, some sounding profiles indicate quick clay within this zone, but this is mainly in the transition parts of the 2D resistivity profiles. Thin/small quick-clay layers/pockets may be present within the unleached clay, but have to be of a certain size in order to appear on the 2D resistivity profile (e.g. Reiser et al. 2010).

In profile P2, excellent agreement is obtained between the resistivity profiles and data from sounding and laboratory tests (Fig. 16.5). It is interesting to note how well

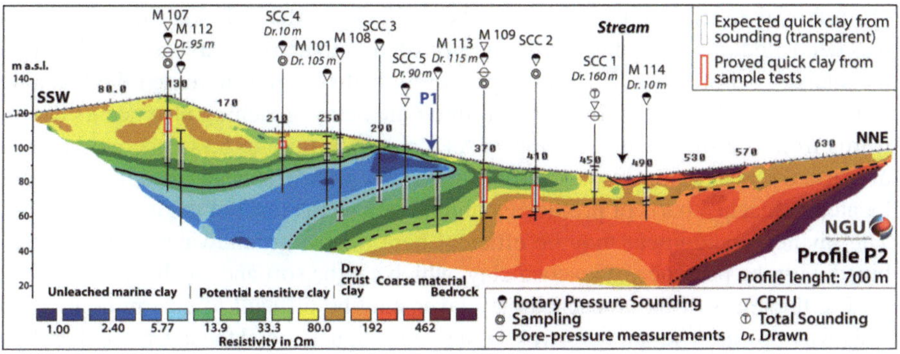

Fig. 16.5 2D resistivity profile P2 from Kaldvelladalen with interpreted geotechnical data (SCC: Emaus 2002; M: Sandven and Vik 2011). The 2D resistivity profile shows two layers of leached clay intersected by an unleached clay layer. Geotechnical data support this layering, but if the geotechnical data is used alone, the most likely interpretation is a coherent quick-clay layer in the slope (Modified from Solberg and Dalsegg 2012)

the resistivity measurements reveal the location of the wedge-shaped zone of unleached clay peaking up from larger depths in the south-western part of the profile. Based on geotechnical data alone, the interpretation between the borings would most likely be a coherent layer of quick clay down the slope. The model from the 2D resistivity profile with two layers of leached clay intersected by an unleached clay layer is also supported by the geotechnical data. This demonstrates how well geophysical and geotechnical information can be combined to obtain a good understanding of the ground conditions in such varying deposit types, giving important input for the stability assessments.

In the north-eastern part of resistivity profile P2, in the transition zone between leached clay and the coarse deposit, the measured resistivity values are up to 200 Ωm in the leached clay. Sample tests show that parts of these deposits are quick clay. These high resistivities are probably due to a high content of silt/coarse layers within the clay.

The resistivity values are closely connected to the ion content in the pore-water of the clay. In this study, pore-water salt content was measured in 17 samples. Most of the samples have low salt content (<3 g/l). The sensitivities of these samples vary from 5 to 190, and average resistivity values here vary from 15 to 100 Ωm. The one sample with high salt content (21.5 g/l in M 104, Fig. 16.4), has resistivity 2 Ωm. There is generally very good agreement between resistivity values, salt content and sensitivity.

16.5 Discussion

In Kaldvelladalen, the different data sets give a good overview of the ground conditions. The marine clay area is partly surrounded by bedrock, shown in the terrain and in the outer part of resistivity profiles P1 and P2. To the north-east the large glaciofluvial deltaic deposit acts as a thick aquifer upon the bedrock. Coarse material was deposited into the valley, partly interbedded with marine clay, as also shown in the sounding profiles. This, and the proximity bedrock, has increased the leaching process in the clay and partly turned it into quick clay. Leaching also occurs downwards from the surface and in slopes. As a result, a more or less coherent layer of unleached clay remains in the centre of the study area. The head of some ravines is located in leached parts of profiles P2 and P3, above the level of the unleached clay. The ravine development may be efficient here due to groundwater seepage combined with the presence of leached clay, which can be soft and easily erodible.

The thickness of an unleached clay layer may be difficult to determine in a 2D resistivity profile. The salt pore-water is a good conductor, and the layer will therefore seem thicker than it actually is (Reiser et al. 2010). When data are inverted, the subsurface is separated into blocks, and the thickness of the first layer of blocks is set at 0.5 times the electrode spacing. The thickness of each subsequent geophysical layer is normally increased by 10 %. This means that the resolution is better if the electrode spacing is small, and in the upper part of the profiles. In addition, the lower part of the profiles undergoes a more unrestrained inversion due to lack of data. In order to retrieve valuable information from a certain depth, it is therefore important to measure deeper than the actual depth of interest. Since the transition between high and low resistivities always is gradual, it may be difficult to determine the material type and property between the bedrock and the unleached clay. Some of the soundings in Kaldvelladalen are however deep enough to help the interpretation in such cases.

2D resistivity measurements give an outline of a large area within a relatively short time, compared to traditional geotechnical investigations. As shown in this case, in an area with few or no previous investigations, the resistivity method will give an overview that constitutes a basis for further exploration and define optimal locations for geotechnical drillings. Even if resistivity measurements may represent a valuable contribution to geotechnical investigations, there are some shortcomings and sources of error in the processing and interpretation of the data. Disagreements between intersecting profiles occasionally occur, and this is mainly related to different electrode spacing in the profiles, or it reflects geological variations in the three-dimensional soil volume (Solberg et al. 2008, 2012). In this study, however, the profiles overlap generally well in the fence diagram.

2D resistivity data give crucial information regarding interpolation/correlation between geotechnical profiles. The extent of quick clay between geotechnical drilling profiles is always drawn conservatively when calculating slope stability (Solberg et al. 2012). 2D resistivity profiles can thereby provide more details on the geometry of a quick-clay pocket between boreholes. This is clearly illustrated in profile P2 in Kaldvelladalen.

One of the most important questions in the evaluation of quick-clay areas is the presence of possible barriers that can prevent large quick-clay disasters to develop. The presence of the bedrock, thick strata of sand, gravel or non-sensitive clay may act as local subsurface barriers, and may limit the extension of a quick-clay landslide once it has been triggered (e.g. Solberg et al. this volume). A barrier must however be large enough to remain stable, even if the support disappears during a landslide. In this study, barriers are present by the tongue of saline clay peaking up in the middle of the slope in profile P2. The presence of non-sensitive clay could prevent further development of a landslide starting at the base of the slope and retrogressing upwards in the slope until it meets non-sensitive material.

The ground conditions in the study area are expected to be poor since the soft and sensitive clay layers are located close to the surface with small overburden stress. This may cause great challenges for both local and global stability, settlement of permanent and temporary road structures and bearing capacity in the construction phase. It might be possible to remedy the situation by construction of counter fills and/or use of light fill materials to reduce the weight on the subsurface. One interesting but expensive stabilization measure would be to install lime-cement columns in a continuous pattern at the most critical road sections.

16.6 Concluding Remarks

The study shows that the resistivity method is useful for detecting sections of possible quick clay when investigating deposits of marine clay. The agreement between geophysical and geotechnical data is in general very good, and the results support data from other studies in Norway, Sweden and Canada. Most of the data show quick clay within the 10–100 Ωm interval, with exceptions on interpretations from a few soundings. No samples verified quick clay below 10 Ωm, whereas one sample showed quick clay with resistivity of 200 Ωm. Resistivity measurements should hence be combined with geotechnical methods, such as sounding and sampling, to verify the local soil composition and properties. Nevertheless, resistivity measurements will be very important in an early stage of investigation, selecting areas for further investigations and for refining the interpretation of the ground conditions by continuous profiling. The latter feature is particularly important when discussing problems associated with mapping of quick-clay deposits.

Acknowledgments Dr. Professor Tim Länsivaara at University of Tampere, Finland, is acknowledged for his constructive remarks in reviewing this paper. Svein E. Hove at the National Public Roads Administration is greatly acknowledged for allowing the use of geophysical and geotechnical data in this paper. This is extended to the colleagues at Geological Survey of Norway for allowing the use of the data report from the geophysical surveys.

References

Aasland R (2010) Kartlegging av kvikkleire med 2D resistivitet og RCPT i Rissa. Master thesis, Norwegian University of Science and Technology (NTNU)

Bjerrum L (1954) Geotechnical properties of Norwegian marine clays. Géotechnique 22(1):27–52

Calvert HT, Hyde CSB (2002) Assessing landslide hazard in the Ottawa valley using electrical and electromagnetical methods. In: Proceedings of the symposium on the application of geophysics to engineering and environmental problems (SAGEEP), Las Vegas

Dahlin T (1993) On the automation of 2D resistivity surveying for engineering and environmental applications. PhD dissertation, Lund University, Lund, Sweden

Dahlin T, Larsson R, Leroux V, Svensson M, Wisén R (2001) Geophysics in slope stability evaluations. Report no 62 Swedish Geotechnical Institute, Linköping

Donohue S, Long M, O'Connor P, Helle TE, Pfaffhuber AA, Rømoen M (2012) Multi-method geophysical mapping of quick clay. Near Surf Geophys 10:207–219

Donohue S, O'Connor P, Long M, L'Heureux JS, Solberg IL, Lecomte I, Sauvin G, Bastani M, Persson L, Rømoen M (this volume) A review of geophysical techniques used in sensitive clays. In: Proceedings of 1st international workshop on landslides in sensitive clays (IWLSC), Quebec28–30 October 2013

Emaus K (2002) Nedre Langeland/Hoven Kaldvelladalen, Melhus. Grunnundersøkelse – Da-tarapport. Scandiaconsult Report 620298–01 (in Norwegian)

Gregersen O, Tuft P, Løken T (1990) Kartlegging av områder med potensiell fare for kvikk-leireskred, kartblad 1621 III Støren, M 1:50 000. NGI Report 81075–2 (in Norwegian)

Löfroth H, Suer P, Schälin D, Dahlin T, Leroux V (2012) Mapping of quick clay using sounding methods and resistivity in the Göta River valley. In: Coutinho RQ, Mayne PW (eds) Geotechnical and geophysical site characterization, vol 4. CRC Press, Leiden, pp 1001–1008

Loke MH (2010) Res2DInv ver. 3.59.102 Geoelectrical imaging 2D and 3D. Instruction manual. Geotomo Software. www.geoelectrical.com

Long M, Donohue S, L'Heureux JS, Solberg IL, Rønning JS, Limacher R, O'Connor P, Sauvin G, Rømoen M, Lecomte I (2012) Relationship between electrical resistivity and basic geotechnical parameters for marine clays. Can Geotech J 49:1158–1168

Lundström K, Larsson R, Dahlin T (2009) Mapping of quick clay formations using geotechnical and geophysical methods. Landslides 6(1):1–15

Reiser F, Dahlin T, Rønning JS, Solberg IL (2010) Resistivity modeling for clay layer characterisation, possibilities and limitations. NGU Report 2010.047. Available at www.ngu.no

Reite A, Sørensen E (1980) Støren. Kvartærgeologisk kart 1621 II, M 1:50 000. Geological Survey of Norway (NGU) (in Norwegian)

Reynolds JM (2011) An introduction to applied and environmental geophysics, 2nd edn. Wiley-Blackwell, Chichester

Sandven R, Vik A (2011) Ny E6 Haga-Skjerdingstad. Utredning for kommuneplan. Datarap-port grunnundersøkelser. Beskrivelse av grunnforhold. Multiconsult Report r414622-1 (in Norwegian)

Sandven R, Vik A, Rønning S, Christensen SO, Tørum E (2012) Detektering av kvikkleire fra ulike sonderingsmetoder. Multiconsult Report 415559 – Detection of quick clay, NIFS-project task 6.4.2 (in Norwegian)

Solberg IL, Dalsegg E (2012) Resistivitetsmålinger for løsmassekartlegging i Kaldvelladalen og ved Fallan i Melhus kommune, Sør-Trøndelag. Data og tolkninger. NGU Report 2012.013 (in Norwegian). Available at www.ngu.no

Solberg IL, Rønning JS, Dalsegg E, Hansen L, Rokoengen K, Sandven R (2008) Resistivity measurements as a tool for outlining quick-clay extent and valley-fill stratigraphy: a feasibility study from Buvik, Norway. Can Geotech J 45:210–225

Solberg IL, Hansen L, Rønning JS, Dalsegg E (2011) Veileder for bruk av resistivitetsmålinger i potensielle kvikkleireområder. Versjon 1.0. Geological Survey of Norway Report 2010.048 (in Norwegian). Available at www.ngu.no

Solberg IL, Hansen L, Rønning JS, Haugen ED, Dalsegg E, Tønnesen JF (2012) Combined geophysical and geotechnical approach for ground investigations and hazard zonation of a quick-clay area, Mid Norway. Bull Eng Geol Environ 71(1):119–133. doi:10.1007/s10064-011-0363-x

Solberg IL, Hansen L, Rønning JS, Dalsegg E (this volume) Applications of 2D resistivity measurements for quick-clay mapping in Mid Norway. In: Proceedings of 1st international workshop on landslides in sensitive clays (IWLSC), Quebec28–30 October 2013

Chapter 17
Mapping of Quick Clay by ERT and CPT-R in the Göta Älv River Valley

Torleif Dahlin, David Schälin, and Johannes Tornborg

Abstract Within clay prone areas it is important to know if and to what extent quick clay is present. In Sweden undisturbed sampling and laboratory investigations are used to determine clay sensitivity and identify quick clay. Recent research for mapping of quick clay areas with the resistivity method is based on the fact that low pore-water salinity constitutes a criterion for quick clay formation. Electrical resistivity tomography (ERT) is an established geophysical method that provides an overview of the resistivity in large volumes, but resolution decreases rapidly with depth. CPT-R probes (Cone Penetration Test with Resistivity module) on the other hand can register the resistivity with very high resolution continuously along the probed depth. In a known quick clay area along the Göta river in Gothenburg, Sweden, both ERT and CPT-R were used to evaluate possible synergy effects when the results were combined. Field investigations comprised 11 CPT-R soundings and five parallel ERT profiles, each 400 m long. The results gave good agreement between the methods down to depths of 10–15 m. Below that the CPT-R probe generally registered lower values than the ERT inverted resistivity models. The results from ERT comply with the value of 6 Ωm as a lower threshold value for possible quick clay formation in Sweden. From CPT-R measurements this value was found to be approximately 3 Ωm. The discrepancy appears to be related to sulphide content and may have measurement technical explanations or be due to the differences in chemical composition. The results show that a combination of ERT and CPT-R provides an efficient approach for mapping of possible quick clay volumes. It can be an integral part of an optimized detailed geotechnical investigation program,

T. Dahlin (✉)
Engineering Geology, Lund University, Lund, Sweden
e-mail: torleif.dahlin@tg.lth.se

D. Schälin
Swedish Geotechnical Institute (SGI), Linköping, Sweden

J. Tornborg
Skanska, Stockholm, Sweden

J.-S. L'Heureux et al. (eds.), *Landslides in Sensitive Clays: From Geosciences to Risk Management*, Advances in Natural and Technological Hazards Research 36,
DOI 10.1007/978-94-007-7079-9_17, © Springer Science+Business Media Dordrecht 2014

and form a decision basis for where to do undisturbed sampling for laboratory analyses. It also opens possibilities for correlation between resistivity and mechanical parameters via the CPT-R results.

Keywords Quick clay • ERT • CPT-R • Sweden • Mapping

17.1 Introduction

During recent years research has been carried out on the use of electrical resistivity tomography (ERT) measurement as a tool for mapping of potential quick clay volumes (see for example Rankka et al. 2004). ERT measurement is an established geophysical method that provides an overview of the resistivity in large volumes, but resolution decreases rapidly with depth (Loke 2003). ERT has supported the development of CPT probes with a resistivity module that can register the resistivity with very high resolution continuously (in this article referred as CPT-R probes).

The key objective of the survey presented in this study has been to evaluate resistivity mapping primarily as a tool for mapping of quick clay and possible synergy effects when the results of ERT and CPT-R are combined (Schälin and Tornborg 2009). The possibility to correlate resistivity values obtained from the CPT-R soundings to geotechnical parameters such as for example friction ratio, liquid limit and sensitivity is investigated. An additional objective of the study is to propose how to implement resistivity measurements in geotechnical investigation programs where the existence and extent of quick clay is important.

To gather measurement data for the study, CPT-R and ERT measurements were carried out in an area known to contain volumes of quick clay and where an extensive geotechnical field investigation campaign had been carried out in connection with the planning process for improving the highway and railroad system.

17.2 Site Investigation Methods

17.2.1 ERT

The main principle of ERT measurements is that a constant current is transmitted into the ground across two electrodes. The electrical resistivity of the subsurface can then be inferred by measuring the voltage across another electrode pair. The main mechanism for current flow through soils and rocks is via electrolytic conduction (Loke 2003).

Inversion using the program Res2dinv was employed to create models of the resistivity distribution of the ground: Two approaches were tried; robust (L1-norm) inversion and smoothness-constrained least-squares (L2-norm) inversion. The differences between them are in how they handle the misfit between the measured

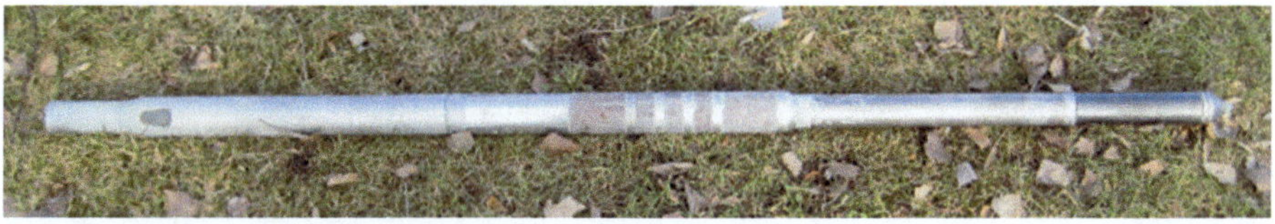

Fig. 17.1 The CPT-R probe used in this study. CPT, resistivity and battery modules comprise a probe with a total length of 1,125 mm

values and the corresponding values in the generated model. Robust inversion uses the absolute value to minimize misfit in the model and the other inversion method uses least-square optimization. L2-norm gives models with smooth transitions between layers of different resistivity whereas L1-norm can produce layers with piecewise constant resistivity and sharper transitions. A disadvantage of L2-norm inversion is that it is more sensitive to low quality data (Loke 2003).

Some pitfalls of interpretation of geophysical data are pointed out by Wisén et al. (2008). For example the suppression of hidden layers, non-uniqueness, equivalence and poor resolution are model pitfalls that can cause problems. Suppression of hidden layers means that thin layers or layers in between layers of similar resistivity will not show. Non-uniqueness means that there are in fact an infinite number of models that can fit the measured data. Equivalence is that the thickness and resistivity of a layer can be altered and still produce the same data.

17.2.2 CPT-R

The CPT-R combines the standard CPT, commonly used in geotechnical investigations in Sweden, with the geophysical method of electrical resistivity. By the addition of a resistivity module to the standard CPT probe, the electrical bulk resistivity of the soil and pore fluid can be assessed (Campanella et al. 1994).

The CPT-R consists of a CPT probe and an additional module, with current and voltage measurement electrodes, placed above the cone, filter media and friction sleeve. Measurement data are, using the probe in this study, transferred via acoustic signals running in the sounding rods to the surface. The CPT-R probe used within this study is shown in Fig. 17.1.

The resistivity module is equipped with an outer electrode pair for current excitation and an inner pair for measurements of voltage in a Wenner-a electrode configuration. Electrodes are hardened nitrided steel rings with a width of 10 mm that are separated by 30 mm insulating plastic. Measurements are made using a 200 Hz square wave transmitted current. The module used in this study is manufactured by Geotech AB.

Due to the large diameter of the probe in relation to the distance between the electrode rings the geometry factor for point electrodes is not valid. The resistivity module was therefore calibrated in media of known resistivity. Calibration factors were derived by submerging the module into water of known resistivity.

Fig. 17.2 Aerial photograph overlooking the investigation site from the south. The approximate locations of the ERT profiles (400 m long and 5 m apart) have been indicated in *yellow* (Photograph provided by the BViV project organization)

Contrary to an ERT measurement, an inversion procedure is not needed and the measured resistivity with CPT-R is the true value for sufficiently thick layers.

17.3 Site Description

The site chosen for the field investigations is located approximately 10 km north of the Göteborg city centre. The area is enclosed by highway 45 as well as the Norge-Vänerbanan railroad to the east and the Göta River to the west. To the east of the highway a rock outcrop rises to approximately 80 m above the flat surface of the investigation site. See Fig. 17.2 for an overview of the area.

Demands for improved highway and railroad standards have been the reason for performing a large number of geotechnical field investigations at the site during recent years. Documentation of the geotechnical engineering properties within the investigation site is thus of good quality. Previous investigations comprise soundings (CPT, static pressure sounding, vane testing), disturbed and undisturbed sampling, laboratory analyses and measurements of pore water pressure.

Results from geotechnical field investigation show that the soil profile generally consists of 0.3 m of topsoil which overlays ca 0.5 m of dry crust. Underneath the crust there is a thick layer of soft glacial and postglacial marine clay which overlays coarser material with varying thickness. The thickness of the clay deposit gradually decreases from 30 to 40 m in the south of the test site to bedrock outcropping under a thin layer of coarse grained fill in the north. In a direction perpendicular to the river the thickness of the clay deposit increases slightly towards the middle of the valley. The bulk density of the clay is around 15–16 t/m^3. The water content varies from approximately 80–90 % in the soil profile and the liquid limit between 50 and 70 % with the water content values generally being higher than the corresponding liquid limit at sampling level. Results from laboratory tests have shown that clay below a depth of 10 m generally is found to have a sulphide content.

In general the pore water pressure is classified as artesian based on previous measurements. At depths of 30 m below the ground surface the pore water pressure at some locations rises to ca 3 m above the level of the ground surface.

17.4 Quick Clay

In Sweden quick clay is defined as clay with a sensitivity above 50 and a remolded shear strength of less than 0.4 kPa. The sensitivity is the ratio of the undisturbed and remolded undrained shear strength, determined according to Swedish practice.

Due to deglaciation and deposition of glaciomarine and marine deposits followed by isostatic rebound clay sediments rose above present sea level. As a result the sediments become subject to leaching and changes in pore water ion concentration. The leaching of salt and the change in ion concentration, according to Rankka et al. (2004), can occur by percolation of rainwater through the deposit, water seepage due to artesian pore water pressures, ground water flow in permeable soil or rock or by diffusion of salts towards zones of lower ion concentrations. Torrance (1974) found that a salt content reduction below 2 g/l was a prerequisite for quick clay formation. The level of 2 g/l corresponds to a resistivity of 6–13 Ωm for clays in Sweden having a bulk density in the range of interest for quick clay formation (Rankka et al. 2004). This is based on laboratory studies on typical Swedish clay. The important resistivity limit for leached and unleached clays is thus 6–13 Ωm. In areas where the measured resistivity is above 6–13 Ωm the clay could have been leached and formed quick clay.

17.5 Site Investigations

ERT measurements were performed as five parallel profiles in the same direction as the railroad and the Göta River. The resistivity profiles were 400 m long from start to finish and the distance between each of the lines was 5 m. Profile 1 was located closest and approximately 8 m away from the existing railroad track. The short distance between the railroad and the river excluded the possibility to conduct

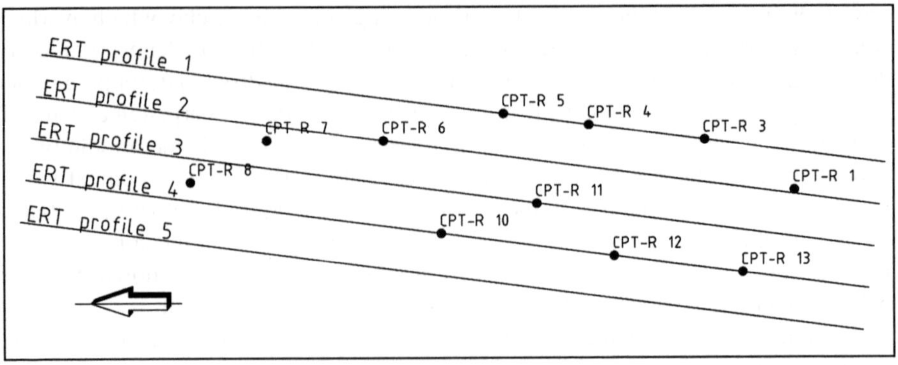

Fig. 17.3 Schematic outline of the investigation (perpendicular distance between ERT profiles are shown with a spacing which has been exaggerated four times)

measurements perpendicular to the parallel lines. To optimize the measuring speed and resolution the multiple-gradient array protocol was used as the electrode configuration (Dahlin and Zhou 2006). Electrodes were placed with a spacing of 5 m (in Profile 2 measurements were also conducted with 2 m electrode spacing). Two versions of the ABEM Lund Imaging System were used for the ERT depending on availability, one based on a Terraohm RIP924 as receiver-control unit, and another using a Terrameter SAS4000.

A total of 11 CPT-R soundings were carried out to depths varying between 12 and 32 m. CPT-R soundings were placed on the resistivity profiles, close to undisturbed sampling points from previous investigations. See Fig. 17.3 for a schematic outline of the investigation site.

17.6 Results and Interpretation

17.6.1 ERT

The results from smoothness constrained L2-norm inversion for ERT profiles 2 and 4 are presented in Fig. 17.4. As previously described, a large number of soundings have been conducted in earlier investigations at the site. Sensitivity values derived from previously made undisturbed samplings are included in Fig. 17.4, and displayed marker lines illustrate interpreted levels for top of friction material. No drillings have been made to confirm the location of the bedrock surface. Presumably the coarser material underneath the clay consists of till with varying thickness. Sensitivity values from sampling points have been divided into three categories, $S_t>50$ quick clay, $50>S_t>30$ highly sensitive and $30>S_t>8$ moderately sensitive, according to Swedish definition.

With the multiple-gradient configuration and an electrode spacing of 5 m the survey depth is approximately 75 m. At this site the CPT-R only reaches a depth that

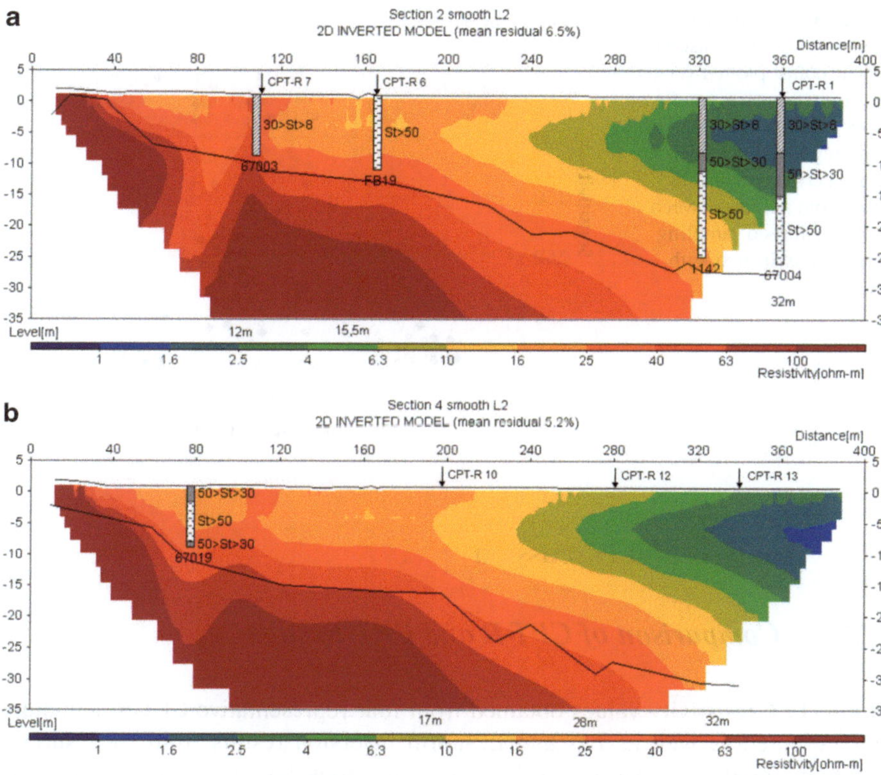

Fig. 17.4 ERT results using smoothness constrained inversion for (**a**) Profile 2 and (**b**) Profile 4 with north towards *left* of figures. Sensitivity values derived from undisturbed sampling are included in the figures as well as the location of RCPT tests at the profiles and marker lines illustrating interpreted level, based on previous soundings, for friction material. Note vertical exaggeration 3×

is half the depth reached with ERT. Therefore the resistivity plots were limited to show only the top 35 m. The results from ERT measurements indicate the bottom of the clay deposit sloping to the south. A volume of comparatively low resistivity is clearly observable towards the south.

17.6.2 CPT-R

In Fig. 17.5 the measured resistivity from all CPT-R tests are displayed together with their corresponding sensitivity values from undisturbed sampling. Sensitivity values are derived from undisturbed sampling located near the CPT-R soundings. No sampling on dry crust is included in Fig. 17.5 displayed below. Fig. 17.5 shows that for resistivity values below 2–4 Ωm the corresponding values of sensitivity lay solely below the limit of quick clay according to Swedish definition (S_t>50).

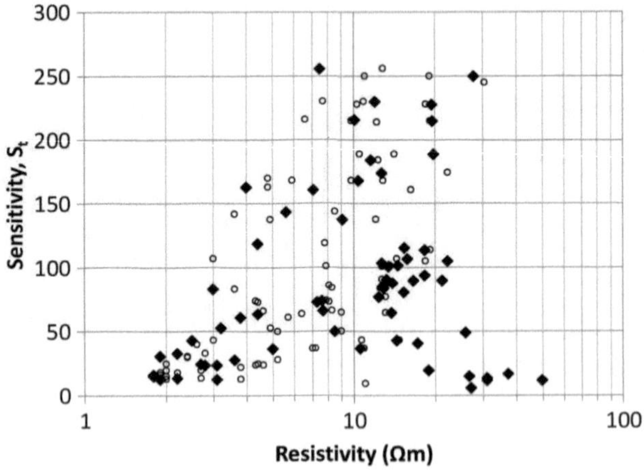

Fig. 17.5 Specific depth resistivity values measured in CPT-R points plotted versus sensitivity at corresponding depth in nearby undisturbed sampling locations. CPT-R points performed further away from and/or between two undisturbed sampling points are indicated with rings

17.7 Analysis and Discussion

17.7.1 Comparison of CPT-R and ERT Results

In Fig. 17.6 resistivity values obtained from four representative CPT-R soundings are presented as well as L1- and L2-norm inversion results from ERT surface resistivity models. Sensitivity values from previous investigation together with information regarding soil classification are also presented in the diagrams. The columns are used to display the generalized soil classification and the color descriptions from laboratory protocols.

Figure 17.6 indicate that CPT-R give the same trends on resistivity as ERT surface measurements, but with high resolution maintained at depth, yet with limited consistency with the soil classification. Inversion using the smoothness constrained L2-norm appears to produce the most consistent trends with respect to the specific depth CPT-R readings. Below depths of 15 m there is a difference in measured and interpreted values of up to 10 Ωm between CPT-R resistivity and ERT results from L2-inversion, which coincides with the sulphide content horizon.

With surface resistivity surveys, the need to invert data is essential. The inversion process minimizes the difference between the forward response of the generated model and the measured data. Resolution obtained from surface resistivity measurements is best near the ground surface and declines with depth. This could explain the better agreement with CPT-R measurements at depths shallower than 10–15 m below the ground surface. The survey results with electrode spacing of 2 m, see Fig. 17.6a, b show better resolution in the upper parts of the subsurface.

When comparing resistivity values from CPT-R and ERT measurements, they agree very well in the upper 10–12 m, as can be seen in the Fig. 17.6a–d, whereas CPT-R tends to give lower values at greater depths.

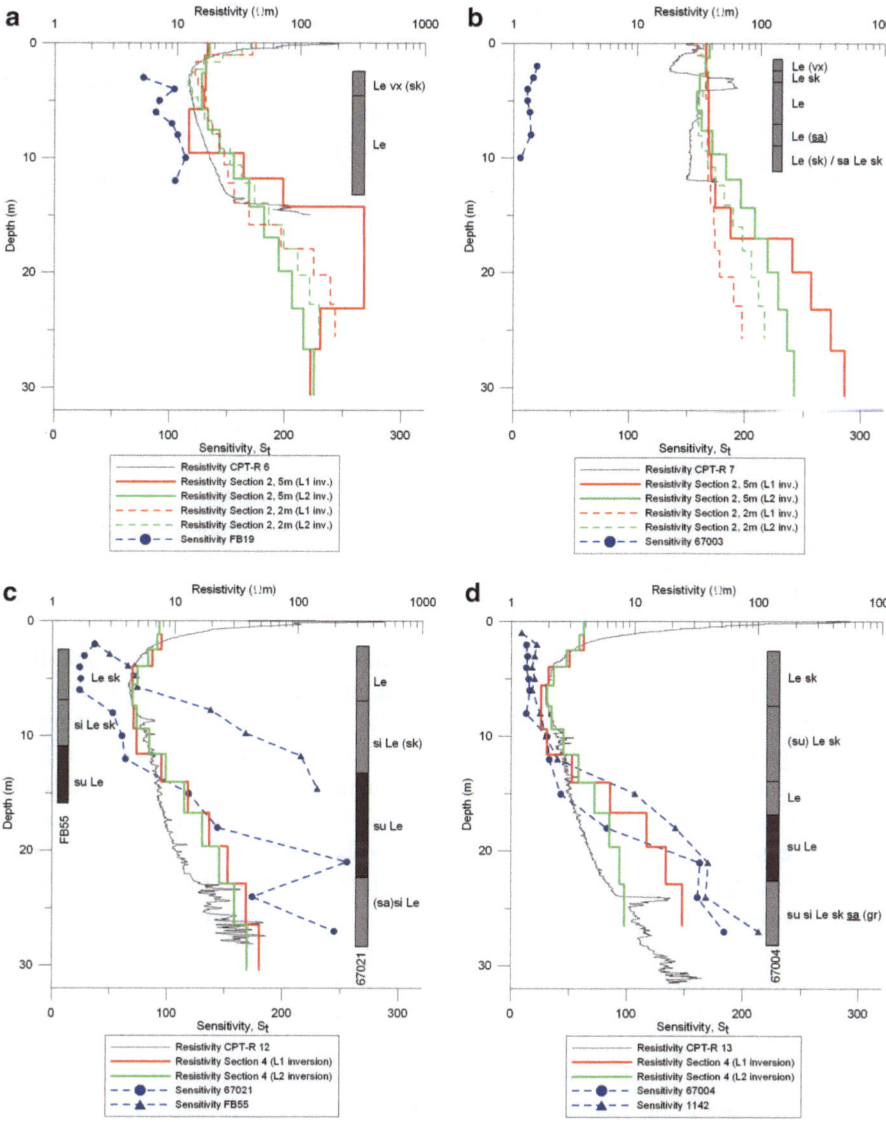

Fig. 17.6 Display of ERT resistivity, previous laboratory data (Swedish abbreviations *Le* clay, *sa* content of sand lenses, *si* silt content, *sk* shell content, *su* sulphide content, *vx* plant fragments) and CPT-R specific depth resistivity. Figures (**a**) and (**b**) CPT-R 6 and 7 located on ERT profile 2 and in the north of the investigation site. Figures (**c**) and (**d**) CPT-R 12 and 13 located on ERT profile 4 and in the south of the investigation site

17.7.2 ERT

With respect to data from previous investigations and the expected geological model, inversion using L2-norm produces the most consistent and reasonable resistivity models in this study. The reason as to why these inversion settings appear to give better results could be that the resistivity in the ground varies smoothly at this site, which works well with L2-norm inversion.

Studying the results from all five ERT profiles the nearby railroad does not seem to influence the resistivity model results since no significant changes were recognized between the profiles closest to and furthest away from the track. ERT measurements are however sensitive to installations in the ground, which is a clear limitation for the use of ERT in urban areas.

17.7.3 CPT-R

The resistivity results obtained during CPT-R soundings were compared to sensitivity values from nearby sampling points. In general, when CPT-R resistivity is below 3 Ωm corresponding values of sensitivities are solely below 50 (see Fig. 17.5). This is not consistent with the resistivity level of 6 Ωm, which Rankka et al. (2004) have given as the boundary level when quick clay formation should not occur. One reason for differences in measured resistivity compared to ERT might be equivalence in the inverted ERT models. 3D effects do not seem to be a major issue here due to the consistency between the parallel profiles.

It may also be that the sulphide content in the clay found at depths from around 10 m below the ground surface is the reason. If the sulphide creates significant IP effects it would affect the resistivity measured by CPT-R. The reason is that the CPT-R probe uses a much high frequency of the measurement signal than the ERT instrument, 200 Hz vs. 1 Hz, and this could lead to severe underestimation of the DC resistivity if there is significant IP effect (Fiandaca et al. 2012). It has also been suggested that the resistivity may decrease in the immediate vicinity of the probe as it is pushed into the ground and the material is disturbed in quick clay horizons. Söderblom (1958) presented laboratory results showing a decrease in resistivity when the material was disturbed.

It might also be that the sulphide content lowers the clay resistivity without affecting the sensitivity, thus lowering the resistivity threshold for quick clay. The discrepancy between methods below 10 m depth would then have to be explained by equivalence in the inversion.

The uncertainty around the reason for the lower threshold and the discrepancy between methods imply that further research is needed to understand this. This concerns the chemical composition of quick clay and understanding the effects and the connection between resistivity in a soil media and sensitivity. It also concerns technical measurement issues, where measuring combined resistivity and time-domain IP with a CPT-R probe using the same timing sequences as ERT equipment would reveal IP effects in a subtle way and show conclusively if this is an issue. A resistivity module with different electrode configurations giving different depths of investigation into the formation around the probe could be a future development if the resistivity changes when the clay is disturbed. Thus, when probing in quick clay, detecting the relative difference in resistivity between the disturbed soil closest to the probe and the undisturbed soil in the outer surrounding ground would be possible.

17.8 Conclusions

The results confirm previous reports on the potential of using ERT to scan large areas for possible quick clay. If the resistivity of marine clay is low enough it is safe to assume that it is non-quick, but if it is above the threshold it may be quick but this may also be due to different types of soils or clay that is non-quick. The threshold for possible quick clay suggested by CPT-R in this study is around 3 Ωm which is much lower than previous Swedish results. It is also lower than that of the ERT results in this survey that comply with the value of 6 Ωm as a lower threshold value for possible quick clay formation in Sweden. The discrepancy may have different explanations, where e.g. IP effects due to sulphide content would lead to underestimation of the resistivity by the CPT-R probe due to its high measurement frequency.

With ERT measurements it is possible to cover relatively large areas depending on the ground conditions of the investigation area. An ERT survey must always be combined with in-situ investigations to allow for proper interpretation and correlation with site specific conditions. Here the ERT results can be used to optimize the location for the detailed investigations to ascertain that they are made in representative locations. The results from the detailed investigation can then in turn be used to refine the ERT interpretation. Careful planning of geotechnical investigations which originates from a correct geological model, including ERT and CPT-R, can improve the quality and reliability of the geotechnical models and thus reduce risk and potentially cut project costs.

CPT-R is an attractive tool with high resolution maintained throughout the depth range with simultaneous registration of resistivity and mechanical parameters. It opens possibilities for correlation between electrical and mechanical properties if there are any. If the CPT is used for estimation of the sensitivity as suggested by Löfroth et al. (2012) it opens possibilities to use the ERT models as a base for extrapolation of these and to create 3D models of the sensitivity with a limited number of in situ samples for laboratory testing as verification.

Acknowledgments The study was carried out as a master thesis project in 2009. We are grateful to Håkan Garin (Geoverkstan / Banverket) for initiating the project, and his supervision together with Claes Alén (Chalmers University of Technology) for supervision. Thanks also to Per Lerjefors and Anders Hallingberg at Banverket who financed the study. We thank Andreas A Pfaffhuber for valuable comments on the manuscript and Richard Owen for language revision.

References

Campanella RG, Boyd TJ, Davies MP (1994) Use of the resistivity piezocone for the geotechnical and geochemical evaluation of a tailings impoundment. In: Proceedings of the 47th Canadian geotechnical conference, Halifax, pp 601–610

Dahlin T, Zhou B (2006) Multiple-gradient array measurements for multi-channel 2D resistivity imaging. Near Surf Geophys 4:113–123

Fiandaca G, Auken E, Christiansen AV, Gazoty A (2012) Time-domain-induced polarization: full-decay forward modeling and 1D laterally constrained inversion of Cole-Cole parameters. Geophysics 77(3):1–13

Löfroth H, Suer P, Schälin D, Dahlin T, Leroux V (2012) Mapping of quick clay using sounding methods and resistivity in the Göta River valley. In: Proceedings of the 4th international conference on geotechnical and geophysical site characterization (ISC'4), Porto de Galinhas, 15–18 September 2012

Loke MH (2003) Tutorial notes: lecture notes on 2D & 3D electrical imaging surveys. Revision date: 25th March 2003

Rankka K, Andersson-Sköld Y, Hultén C, Larsson R, Leroux V, Dahlin T (2004) Quick clay in Sweden, Report 65. Statens Geotekniska Institut, Linköping, 145 pp

Schälin D, Tornborg J (2009) Evaluation of CPT-R and resistivity measurements in quick clay area. Master's thesis 2009:32, Chalmers University of technology, Göteborg, 77pp

Söderblom R (1958) Saltsonden och dess användning vid bestäming av skred-botten vid Göta, Svenska Föreningen för Lerforskning, Nr. 9, pp 87–96

Torrance JK (1974) A laboratory investigation of the effect of leaching in the compressibility and shear strength of Norwegian marine clays. Géotechnique 24(2):155–173

Wisén R, Christiansen AV, Dahlin T, Auken E (December 2008) Experience from two inversion techniques applied in three cases of geotechnical site investigation. J Geotech Geoenviron Eng 134(12):1730–1742

Chapter 18
Geophysical Data Integration for Quick-Clay Mapping: The Hvittingfoss Case Study, Norway

Guillaume Sauvin, Isabelle Lecomte, Sara Bazin, Jean-Sébastien L'Heureux, and Maarten Vanneste

Abstract Quick-clay landslides are a known hazard in formerly glaciated coastal areas. Some of Norway's most densely populated areas are located in potential quick-clay zones and, hence, large efforts are devoted to map the distribution of quick clays. Here, we focus on one particular Norwegian site (Hvittingfoss, 100 km south-west of Oslo), which was remediated against potential sliding in 2008. A set of geophysical methods including Electrical Resistivity Tomography, P-wave seismic refraction tomography, S-wave seismic reflection profiling, and Ground Penetrating Radar, were jointly analysed and complemented with laboratory data and *in-situ* geotechnical measurements (i.e., CPTU, SCPTU and RCPTU) in order to establish a suitable, integrated and multi-disciplinary approach to map the extent of the quick-clay zone. Through careful integration and interpretation of the different data, the main

G. Sauvin (✉) • I. Lecomte
International Centre for Geohazards (ICG), P.O. Box 3930, N-0806 Oslo, Norway

Department of Geosciences, University of Oslo (UiO), P.O. Box 1072, Blindern 0316 Oslo, Norway

NORSAR, P.O. Box 53, N-2027 Kjeller, Norway
e-mail: Guillaume.Sauvin@norsar.no; Isabelle.Lecomte@norsar.no

S. Bazin
Norwegian Geotechnical Institute (NGI), P. O. Box 3930, Ullevål Stadion 0806 Oslo, Norway
e-mail: Sara.Bazin@ngi.no

J.-S. L'Heureux
Norwegian Geotechnical Institute (NGI), P. O. Box 3930, Ullevål Stadion 0806 Trondheim, Norway
e-mail: jsl@ngi.no

M. Vanneste
International Centre for Geohazards (ICG), P.O. Box 3930, N-0806 Oslo, Norway

Norwegian Geotechnical Institute (NGI), P. O. Box 3930, Ullevål Stadion 0806 Oslo, Norway
e-mail: Maarten.Vanneste@ngi.no

J.-S. L'Heureux et al. (eds.), *Landslides in Sensitive Clays: From Geosciences to Risk Management*, Advances in Natural and Technological Hazards Research 36, DOI 10.1007/978-94-007-7079-9_18, © Springer Science+Business Media Dordrecht 2014

deposits were identified. Both the clay deposit and the overlying sand layer were precisely imaged and their lateral variations were determined. The underlying moraine deposit and the bedrock were also identified, thereby yielding an idea of the preferential leaching paths. Considering the inherent complexity of quick-clay mapping, the collected data illustrate the benefit of an integrated approach, and emphasise the need for high resolution, proper imaging, calibration and ultimately joint inversion of the different data.

Keywords Quick clay • Landslides • Georadar • ERT • S-Wave seismic • CPT • Data integration

18.1 Introduction

Formerly glaciated margins are occasionally subject to destructive quick-clay landslides. According to Norwegian standards, quick clays are defined by a sensitivity (i.e., ratio of undrained shear strength in its undisturbed condition s_u and undrained remoulded shear strength s_{ur}) greater than 30 and a remoulded undrained shear strength less than 0.5 kPa (NGF 1975). The sediment was originally deposited in a marine environment and emerged following isostatic rebound and fall of the relative sea level during the Holocene. Long-term leaching of salt due to groundwater flow and percolating surface water affects clay-particle bonding and makes the soil highly susceptible to failure when disturbed (Rosenqvist 1953; Mitchell 1976; Brand and Brenner 1981). These quick-clay landslides may occur on low-angle slopes and with only small perturbations in stress conditions caused by e.g., human activity or erosion. The final extent of a landslide in clay is governed by several factors, including topography, stratigraphy and the clay sensitivity (e.g., Mitchell and Markell 1974; Tavenas et al. 1983; L'Heureux 2012). Information about the extent and thickness of the highly sensitive clay deposit or "quick" clay is thus required for landslide hazards mapping in the framework of planning and protection in such areas (Gregersen 2008).

In our experience, no single geophysical method yields the optimal information to accurately map the distribution of the quick-clay deposits. We therefore try to combine different geophysical methods (e.g., Electrical Resistivity Tomography – ERT, Multi-channel Analysis of Surface Wave – MASW, seismic refraction tomography – SRT, Ground Penetrating Radar – GPR) as well as geotechnical data (*in situ* measurements using CPTU, Seismic-CPTU – SCPTU and Resistivity-CPTU – RCPTU; laboratory tests) for a proper quick-clay assessment of a given site, as will be demonstrated here. Furthermore, we do recommend a more systematic use of geophysical data to fill the gaps between isolated geotechnical boreholes, the latter being the only mandatory approach to provide ground truth, but at the potential risk of missing information about the complexity of a site. As such, quick-clay mapping gradually moves towards 2D or pseudo-3D site characterization. The present study illustrates the advantages of such multidisciplinary investigation, as also supported by previous studies (Donohue et al. 2012; Sauvin et al. 2013).

18.2 Site Description

The study site is situated in Hvittingfoss, about 80 km southwest of Oslo, Southern Norway (Fig. 18.1). Soils in the area are dominated by marine and fjord deposits (clay and silt), as well as alluvial deposits (sand and gravel) (http://geo.ngu.no/kart/losmasse).

The top sand and gravel layer is up to 23-m thick on the Fossness plateau (Moholdt 2008). This alluvial deposit is related to an old meander wing that thins both northwards and southwards of the investigated area, and it overlays a thick layer of clay.

The marine deposit is a silty clay with some sand and gravel. Laboratory measurements on samples from borehole 2 (BH2 in Fig. 18.1) indicate that the clay fraction ranges from 18 to 30 %, and the plasticity index is lower than 10–15 down to 12-m depth. Layers of silt, sand and gravel lie underneath (Moholdt 2008). The marine deposit was interpreted as 'quick' as the penetration resistance curves are vertical or near vertical (Rygg 1988) and the sensitivity of the clay measured from the fall cone test on samples exceeds 200 in borehole 2 and 500 in borehole 5 located further south (Fig. 18.1).

Some of the boreholes (6 and 7, Fig. 18.1) indicate that the bedrock is reached at depths ranging from 19 to 30 m (Moholdt 2008). Bedrock crops out close to the waterfall just north of the investigated site. The bedrock types in the area are mainly syenite, quartz-syenite, romb porphyry and monzonite, quartz-monzonite (http://geo.ngu.no/kart/berggrunn).

Other boreholes indicate that most of the buildings east of the investigated area rest on the same marine and fjord deposits. Due to the active erosion effect of the river to the west, the area has a low factor of safety. For that reason, the site was

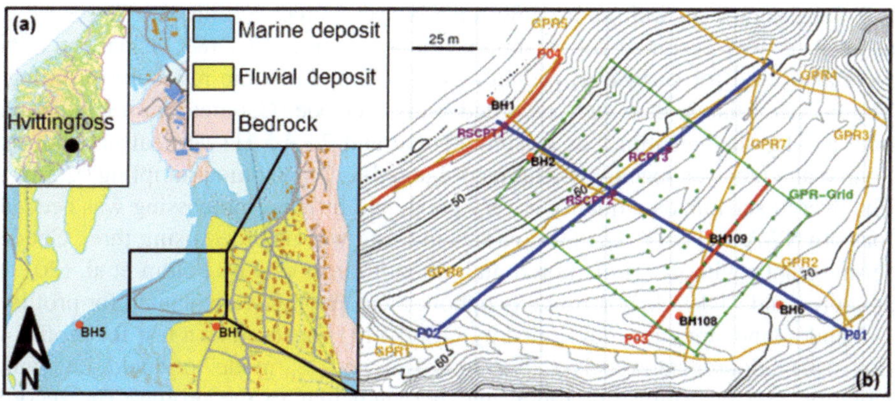

Fig. 18.1 (**a**) Quaternary geology map of the area. (**b**) Topographical map, with profiles localisation. ERT, GPR, Seismic P and S were recorded at profile 1 and 2 (*blue*), whereas only Seismic S and GPR were recorded at profile 3 and 4 (*red*). The *green box* and *dots* correspond to the GPR grid contour and 10 m grid spacing point. *Yellow lines* refer to 2D GPR profiles. RCPTU, SCPTU and boreholes location are also given

mitigated in 2008 by moving parts of the soil from the upper part downwards to the lower slope to stabilize (unload) the latter, and erosion from the river was prevented by adding boulders at the bottom of the slope.

18.3 Data and Methods

18.3.1 ERT Measurements

2D resistivity measurements were carried out using a Terrameter LS (ABEM) based on the Lund-system developed by Dahlin (1993). From previous geotechnical investigations, a simple initial model was defined and data acquisition was then planned according to the modelling results. Both gradient and Dipole-Dipole arrays were used with an electrode spacing of 2 m resulting in two 160-m long profiles (profile 1 and 2, Fig. 18.1b). Data quality is good with a very high signal to noise ratio, and nearly no data had to be removed or filtered prior to inversion. The apparent resistivity was inverted with the RES2DINV software (Loke 2010) using the L1-norm inversion optimization.

Generally, inversions converged to RMS errors of less than 5 % after 7 iterations. The 2D ERT profiles were then combined to generate pseudo-3D displays.

18.3.2 Seismic Measurements

Two series of seismic measurements were conducted at Hvittingfoss, i.e., one using a P-wave source as well as a dedicated S-wave source experiment. The P-wave acquisition was designed for seismic refraction tomography (SRT), but the data were also used for surface-wave analysis (SWA) and seismic reflection profiling. The P-wave seismic data were recorded using a Geode (Geometrics) seismograph with a 5-kg sledgehammer as seismic source and 24 4.5-Hz vertical geophones. Source and receiver spacing is 4 m, record length is 2 s and time sampling 0.25 ms. The overall seismic data quality is very good, and little pre-processing was needed for both SWA and SRT. The S-wave seismic data was recorded using three Geode seismographs with a seismic horizontal vibrator developed by Polom et al. (2011) as source and 71 12-Hz horizontal geophones. Geophones were planted for profiles 1–3 and a land streamer (developed by GEOSYM) was used as a test for profile 4 (Fig. 18.1). Source and receiver spacing is 1 m, and "S_H-mode" oriented, i.e., with vibration and recording horizontal and perpendicular to the profile. The record length is 11 s and time sampling 1 ms.

P-wave Seismic Refraction. Picking of first-arrival travel times for SRT was performed semi-automatically on raw data, and, following fine tuning of the initial velocity model, the inversion of the travel times was performed. The inversion

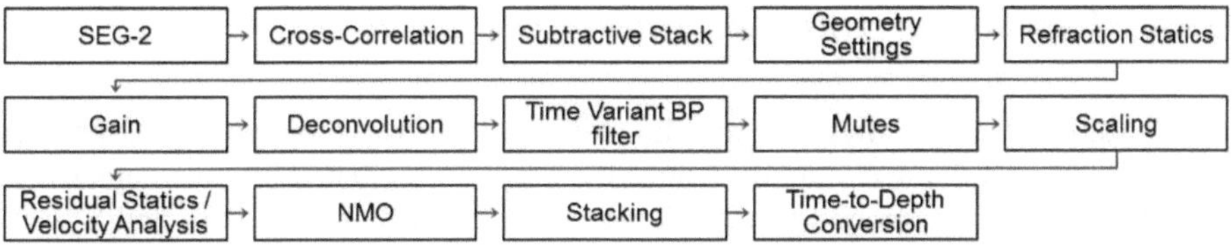

Fig. 18.2 Seismic processing flow used for the four S-wave seismic reflection profiles

algorithm in the latter is based on an iterative adaptation (Simultaneous Iterative Reconstruction Technique), and the final result is cross-checked by modelling with an eikonal solver.

S-wave Seismic Reflection. Seismic data processing is summarized in the diagram represented in Fig. 18.2. A similar processing was applied to S-wave data set acquired in Trondheim harbour and discussed in details in Sauvin (2009) and Polom et al. (2010); we hereby refer to these documents for more information on the different processing steps.

18.3.3 GPR Measurements

A Ramac (Malå) non-shielded 50-MHz rough-terrain antenna (RTA) was used for profiling, whereas standard Ramac 50-MHz non-shielded antennas were used for Common MidPoint (CMP) measurements. The GPR grid is 100 m by 70 m with 1 m spacing between in-lines and cross-lines, which gives 172 profiles. In order to obtain a representative velocity field, CMP acquisition was performed at every 10th grid point (10-m spacing) in both directions (Fig. 18.1). The GPR data pre-processing mainly consists of dewow filtering, subtracting DC shift, and applying gain. A velocity model is then built using diffraction hyperbola and/or velocity analyses on CMP gathers in order to perform time migration and depth conversion.

18.3.4 Geotechnical Investigations

Ground conditions in the study areas were previously investigated by different consulting companies (Moholdt 2008). The methods used include 54-mm piston samplers (with laboratory testing), Rotary Pressure Sounding, CPTU, Total Sounding, Rotary Sounding and Vane Shear Tests. A few 1D resistivity and S-wave velocity measurements were retrieved using RCPTU and SCPTU in order to link dynamic measurements with static measurements.

18.4 Results and Interpretation

The results of the ERT inversion are shown in Fig. 18.3a together with GPR profiles (Fig. 18.3b) and P-wave velocity from SRT (Fig. 18.3c). The horizons depicted in Fig. 18.3a, were interpreted from ERT and GPR profiles as well as P- and S-wave

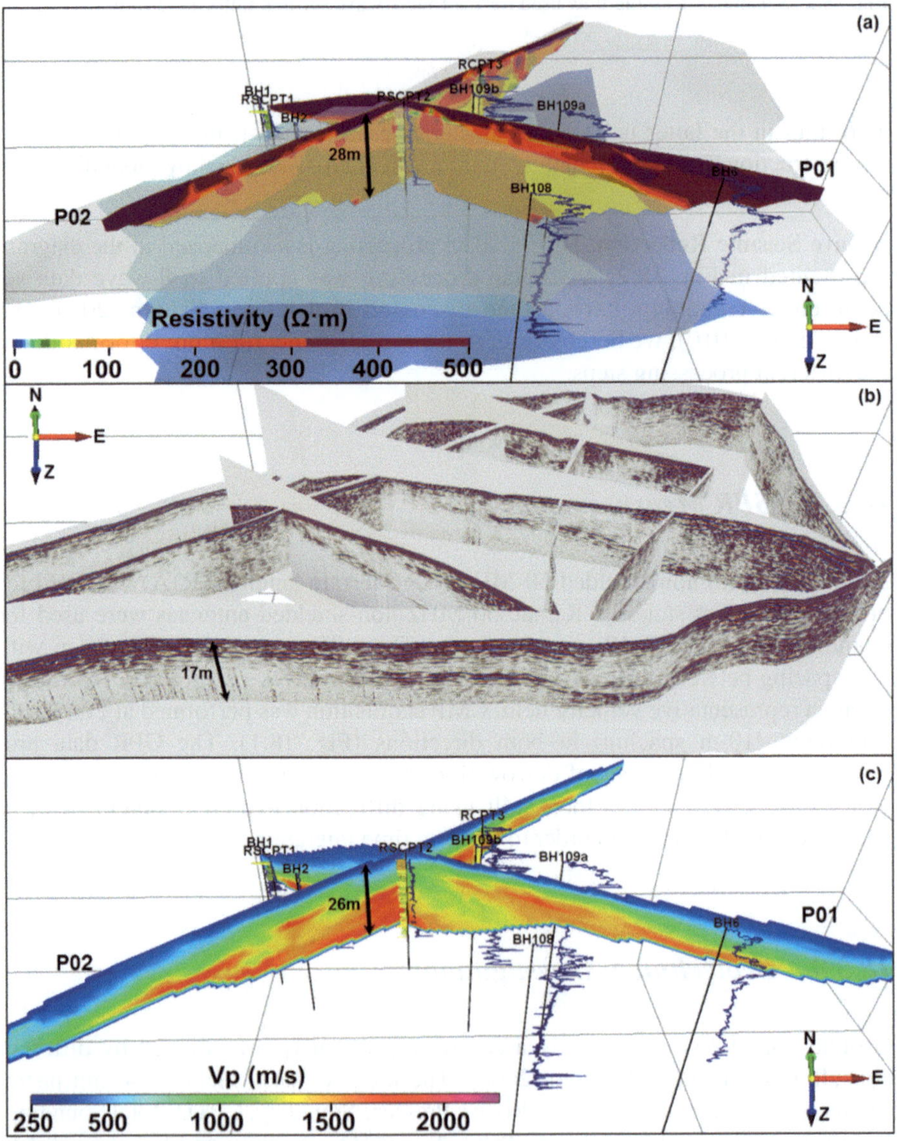

Fig. 18.3 (**a**) ERT profiles and interpretation of the sand/clay (*grey*), clay/moraine (*turquoise*) and bedrock (*blue*) interfaces. As the focus of this paper is on quick clay, the resistivity scale has a limited range; it should be noted, however, that the resistivity values detected in the upper coarse-grained deposit can be over 2,000 **Ω**m. (**b**) GPR profiles. (**c**) P-wave velocity fields derived from seismic tomography. Rotary pressure and RCPTU soundings are also displayed

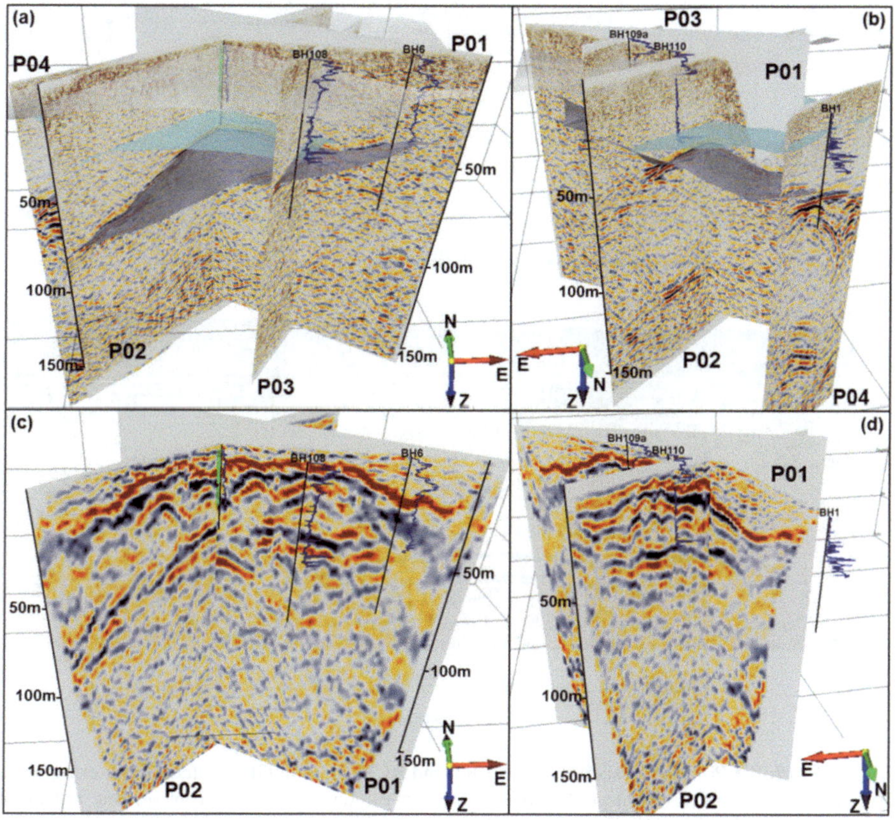

Fig. 18.4 (**a** and **b**) Seismic reflection S from 2 angles of view, together with interpretation. (**c** and **d**) Seismic reflection P from refraction data set, same angles of view. Rotary pressure soundings are also displayed

seismic reflection sections (Fig. 18.4). There is a reasonably consistent upper layer with high resistivity (ca. >300 Ωm), ranging in depth from 2 m in the middle of the profiles to 15 m towards the eastern end of profile 1, and which corresponds to the coarse-grained deposit. Below this layer, resistivity decreases rapidly to low values (ca. 15–100 Ωm). According to Solberg et al. (2012), resistivity values of clays within 10–80 Ωm may indicate leached clay and thus potentially quick clay. The ERT penetration depth is not sufficient to reach the underlying moraine deposit (as interpreted from geotechnical soundings). No resistivity values below 15 Ωm are observed, suggesting that most of the marine and fjord deposits have been leached in the vicinity of the two ERT profiles. Even though ERT inversion does not allow for the detection of sharp geological boundaries, comparison with rotary pressure sounding and GPR results indicate that the 100-Ωm resistivity contours could be associated to the sand/clay interface. Nonetheless, looking at ERT profile 2 and its corresponding GPR profile (Fig. 18.5a), it appears that the interpreted quick-clay layer is not homogeneous and includes non-quick clay zones with higher resistivity and better GPR depth penetration. These results suggest that quick-clay

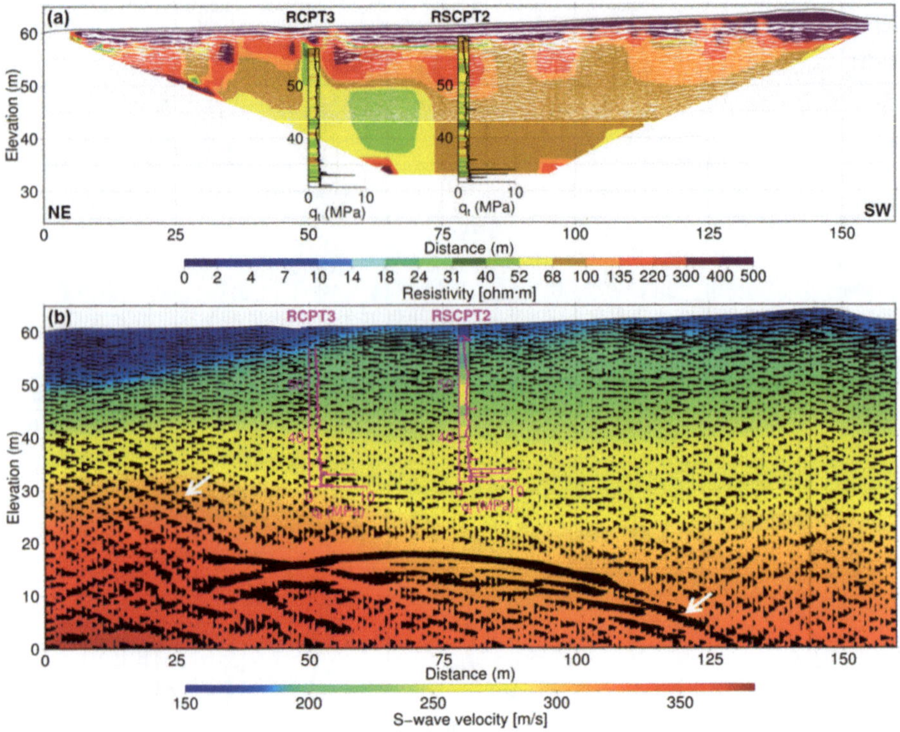

Fig. 18.5 Profile P02, (**a**) Resistivity from ERT and RCPTU, with GPR reflection superimposed. (**b**) Depth converted S seismic reflection, with S-wave interval velocity from seismic reflection velocity analysis in background and S-wave interval velocity from SCPT. The *white arrows* indicate the bedrock reflection. The corrected cone tip resistance (q_t) are also displayed

investigations using discrete rotary pressure soundings can be significantly complemented by using ERT profiles to interpolate in between, and extrapolate from these point-wise calibration points. This is further emphasized by the good consistency between resistivity values from RCPTU measurements with those from ERT (Figs. 18.5a and 18.6), thereby validating the ERT inversion results.

The 2D and 3D GPR data allow for a detailed mapping of the shallow sand layer, which is on average 15-m thick and thins towards the north-west (Fig. 18.3b). Distinct reflections with noticeable amplitude variations within the sand deposit suggest spatial heterogeneity (Figs. 18.3b and 18.5a). Highly-attenuated zones along the profiles correspond to sediments with higher clay/silt content. Amplitude variations could also be observed within the clay body and, as confirmed by geotechnical soundings (Moholdt 2008), are probably due to variations in grain size.

The P-wave seismic refraction sections (Fig. 18.3c) show a thin top layer of low velocity (ca. 250–750 m/s), i.e., the coarse-grained body, overlying a layer of higher velocity (ca. 1,250–1,750 m/s), i.e., the marine deposit, at a depth of 3–20 m. North-west of the profiles, in the deepest part, velocities up to 2,250 m/s may well be indicative of bedrock. In addition, the reflection-stacked section extracted from

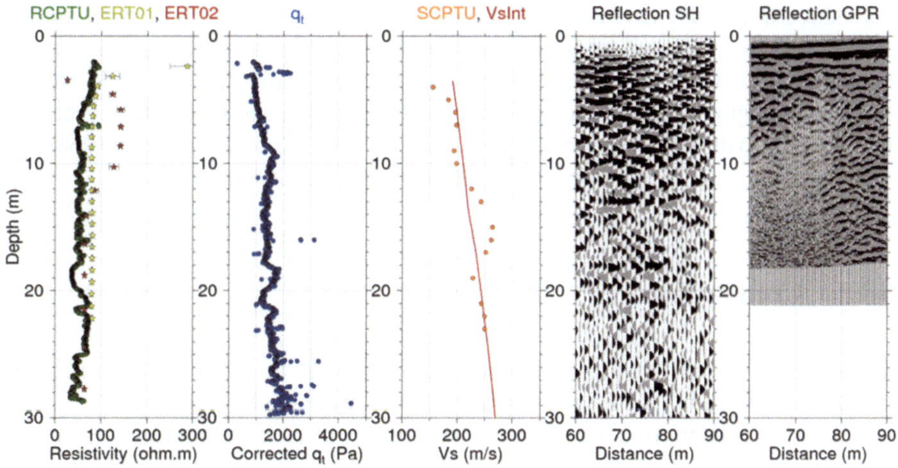

Fig. 18.6 RSCPTU02: resistivity from RCPT and ERT profile P01 and P02, q_t, interval velocity from SCPT and S-wave seismic reflection, S reflection and GPR reflection

P-wave seismic refraction data (Fig. 18.4c, d) presents both a reflection that could be interpreted as the bedrock and one as the sand/clay interface. This is further validated and detailed by the S-wave seismic reflection profiles (Fig. 18.4a, b), which have a higher resolution due to the inherent lower S-wave velocities, and give detailed structural information, even down to the bedrock. A clear distinction between the different deposits is therefore possible and imbedded layers within the clay deposits can also be tracked. The prominent event around 15 m elevation, in Fig. 18.5b is interpreted as the bedrock. The reflectivity is quite high due to the good signal coherency and certainly because of strong velocity contrast. More processing would be needed to better image this dipping reflector. The S-wave velocity fields extracted from the velocity analysis of the S-wave seismic reflection is consistent with the S-wave velocity derived from SCPTU (Figs. 18.5b and 18.6), and nearly no lateral variations are observed within the clay layer.

Correlation of the geophysical data with the existing rotary pressure soundings and CPTU is very good and allows for an accurate interpretation of the sand/clay (from GPR and ERT), and clay/moraine (from S-wave reflection seismic) interfaces as well as bedrock (from S- and P-wave reflection seismic), as shown in Fig. 18.4a, b. The geometry and thickness of the clay layer can therefore be evaluated – from 10 m towards the river to 30 m at borehole 109 – and might enhance the computation of the safety factor involved in slope stability assessment. Once more, this highlights the usefulness of geophysical measurements to interpolate/extrapolate between/from discrete 1D geotechnical soundings.

The clay layer overlays a moraine deposit and the inherent permeability of the latter increases the probability of leaching. The lower P-wave velocities within the clay layer correlate with higher resistivity and GPR amplitude variations (Figs. 18.4 and 18.5), suggesting grain-size variation. The GPR penetration in some part of the 'clay' layer indicates indeed the presence of 'sandy' material. Such permeable layers embedded in the clay may have enhanced the leaching of the surrounding clay. The S-wave seismic reflection makes up for the lack of resolution of the ERT data, which does not allow detailed correlation between resistivity variations and GPR images. Bedrock topography imaged from S-wave seismic gives a good idea of the preferential leaching path which influenced the nature of local drainage (Løken 1968) and therefore had a significant effect on the formation of quick clay at that site.

18.5 Conclusions

A number of geophysical methods were combined at the Hvittingfoss mitigated quick-clay site and promising preliminary results were obtained.

The structural information retrieved from all geophysical measurement may, for example, help locating preferential leaching paths, i.e., depending on bedrock topography, presence of underlying and/or embedded coarse-grained sediments, and thickness of the clay layers. Resistivity variations and GPR attenuation within clay layers could also reflect the variation of salt concentration in pore water and/or the variation in grain size. Combined with geotechnical soundings, it can also help directly tracking layers of high sensitivity. Regarding the interpretation, geophysical data can be used in the interpolation between the drill points, thus providing more realistic models for stability calculations. Since the profiles cover a relatively large area, it facilitates a more comprehensive understanding and its development, both in terms of geology and geotechnical engineering.

When geophysical measurements are used in an early phase of site investigations (which was not the case at Hvittingfoss) they may provide useful input to the placement of wells. This would also potentially limit the number of wells, and information from drilling operations could even be more valuable. It is recommended, however, that geophysical measurements should not be used as a substitute for geotechnical testing which will always be necessary.

Acknowledgments The authors are grateful to Ranajit Ghose for reviewing this paper. The authors also thank Gedco and Sandmeier for providing academic licences (Vista and ReflexW, respectively). The authors acknowledge Nadège Langlet, Sylvain Tissot, Christian Maskrey, Håkon Akerholt, Tor Overskeid, Karine Petrus, Mesay Geletu Gebre, Gunther Druivenga and Berit Paulsen for their help in the field, and the University of Oslo for some of the equipment. Guillaume Sauvin thanks the sponsors of his Ph.D. thesis: the Norwegian Public Roads Administration, the Norwegian National Railway Administration, the Norwegian Water Resources and Energy Directorate and International Centre for Geohazards.

References

Brand EW, Brenner RP (1981) Soft clay engineering. Elsevier, Amsterdam

Dahlin T (1993) On the automation of 2D resistivity surveying for engineering and environmental applications. Doctoral thesis, Lund University, Lund

Donohue S, Long M, O'Connor P, Eide-Helle T, Pffaffhuber AA, Rømoen M (2012) Geophysical mapping of quick clay: a case study from Smørgrav, Norway. J Near Surf Geophys, EAGE 10(3):207–219

NGF (Norsk Geoteknisk Forening) (1975) Retningslinjer for presentasjon av geotekniske undersøkelser, NGF (Norsk Geoteknisk Forening) (In Norwegian)

Gregersen O (2008) Program for økt sikkerhet mot leirskred – Metode for kartlegging og klassifisering avfaresoner, kvikkleire. Norwegian Geotechnical Institute Report 20001008-2 (3rd revision), 24p (In Norwegian)

L'Heureux JS (2012) A study of the retrogressive behaviour and mobility of Norwegian quick clay landslides. In: Proceedings of the 11th international and 2nd North American symposium on landslides, Banff

Loke MH (2010) Res2DInv ver 3.59.102. Geoelectrical imaging 2D and 3D. Instruction manual. Geotomo Software. http://www.geotomosoft.com/

Løken T (1968) Kvikkleiredannelse og kjemisk forvitring i norske leirer. Norwegian Geotechnical Institute, Oslo, Publication 75, pp 19–26 (In Norwegian)

Mitchell JK (1976) Fundamentals of soil behavior. Wiley, New York

Mitchell RJ, Markell AR (1974) Flowslides in sensitive soils. Can Geotech J 11:11–31

Moholdt R (2008) Fossnes på Hvittingfoss, Vurdering av skredfare og sikringstiltak. NGI report, 20071564-2 30

Polom U, Hansen L, Sauvin G, L'Heureux JS, Lecomte I, Krawczyk CM, Vanneste M, Longva O (2010) High-resolution SH-wave seismic reflection for characterization of onshore ground conditions in the Trondheim Harbor, Central Norway. In: Miller RD, Bradford JD, Holliger K (eds) Advances in near-surface seismology and ground-penetrating radar. SEG, Tulsa, pp 297–312

Polom U, Druivenga G, Grossmann E, Grüneberg S, Rode W (2011) Transportabler Scherwellenvibrator. Patent application DE 103 27 757 A1, Deutsches Patent- und Markenamt (in German)

Rosenqvist IT (1953) Considerations on the sensitivity of Norwegian quick-clays. Geotechnique 3:195–200

Rygg N (1988) Rotary pressure sounding: 20 years of experience. In: DeRuiter (ed) Proceedings of the pen testing, 1988, ISOPT-1, Balkema, Rotterdam, pp 453–457

Sauvin G (2009) S-wave seismic for geohazards: a case study from Trondheim Harbor. Diploma engineering thesis, University of Strasbourg, Strasbourg, 2009-6-1

Sauvin G, Lecomte I, Bazin S, L'Heureux JS, Vanneste M (2013) Towards geophysical and geotechnical integration for quick clay mapping in Norway. J Near Surf Geophys, EAGE 11. doi: 10.3997/1873-0604.2012064

Solberg IL, Hansen L, Rønning JS, Haugen E, Dalsegg E, Tønnesen JF (2012) Combined geophysical and geotechnical approach to ground investigations and hazard zonation of a quick clay area, Mid Norway. Bull Eng Geol Environ 71:119–133

Tavenas F, Flon P, Leroueil S, Lebuis J (1983) Remolding energy and risk of retrogression in sensitive clays. In: Symposium on slopes on soft clays, Linkoping, Swedish Geotechnical Institute Report, No. 17, pp 205–262

Chapter 19
Joint Acquisition and Processing of Seismic Reflections and Surface Waves in a Sensitive Clay Deposit in the Outaouais Region (Québec), Canada

Gabriel Fabien-Ouellet, Richard Fortier, and Bernard Giroux

Abstract The joint acquisition and processing of vertically polarized shear (*SV*) wave seismic reflections and surface waves during a seismic survey were carried out in Buckingham (Québec), near Ottawa, Canada, to characterize a thick (20–40 m) sensitive clay deposit. At the study site, the outcropping clay unit overlays a 20–50 m thick layer of sand and gravel and the bedrock depth reaches more than 90 m along the survey line. The seismic reflection survey using common-mid-point (CMP) inversion of *SV*-wave reflections allowed the localization of the clay-sand and sand-bedrock interfaces as well as the measurement of *SV*-wave velocities down to the bedrock contact. Velocity variations at depths less than 10 m could not be assessed due to the early reflections hidden by seismic arrivals such as surface waves. However, multi-channel analysis of surface waves (MASW) provided the variations in *S*-wave velocity from the surface down to a depth of 12 m at each CMP location. The joint acquisition and processing of *SV* reflections and Rayleigh waves provided a more complete and accurate 2D *SV* velocity model than both methods taken separately. To test the accuracy of the proposed approach, a multi-offset seismic piezocone penetration test (SCPTu) was performed along the survey line from the surface down to a depth of 25 m. The vertical variations in seismic velocities in sensitive clay as inferred from the *SV* seismic reflection survey and MASW are comparable to the SCPTu *S*-wave profile.

Keywords Seismic reflection survey • Vertically polarized shear waves • Multi-channel analysis of surface waves (MASW) • Vertical seismic profiling (VSP) • Multi-offset seismic piezocone penetration test (SCPTu) • Sensitive clay

---

G. Fabien-Ouellet (✉) • R. Fortier
Département de géologie et de génie géologique, Université Laval,
Québec City, QC, Canada
e-mail: gabriel.fabien-ouellet.1@ulaval.ca

B. Giroux
INRS-ETE, Québec City, QC, Canada

J.-S. L'Heureux et al. (eds.), *Landslides in Sensitive Clays: From Geosciences to Risk Management*, Advances in Natural and Technological Hazards Research 36,
DOI 10.1007/978-94-007-7079-9_19, © Springer Science+Business Media Dordrecht 2014

19.1 Introduction

Shear-wave velocity is a fundamental parameter for assessing the dynamic properties of soils in areas prone to earthquakes and landslides. This led the National Building Code of Canada to include shear-wave velocity at a 30-m depth (Vs_{30}) in the definition of the soil classification categories (Finn and Wightman 2003). Several methods are available for the measurement of Vs_{30}; the most common being vertical seismic profiling (VSP) and multi-channel analysis of surface waves (MASW). Both methods have serious drawbacks for very thick (>50 m) and soft soils, as deep drilling, needed for VSP, becomes cost prohibitive, and the depth of investigation of the active MASW method is somewhat limited. Recently, seismic reflection surveys using vertically polarized shear (*SV*) waves have been used in conjunction with the landstreamer technology to efficiently investigate thick clay deposits (Pugin et al. 2009). Heavy and expensive vibratory sources were used in these surveys to generate *SV*-waves. An alternative approach to evaluate Vs_{30} is proposed in this paper using the joint processing of *SV* reflections and Rayleigh waves acquired during a typical seismic reflection survey using a straightforward hammer impact on a steel plate as a seismic source. The MASW method is first used to process Rayleigh waves and obtain a velocity model for the first meters of soil. Those results are then integrated in the semblance analysis of the *SV* reflection data to produce a complete 2D velocity model of the overburden as well as a stacked seismic reflection profile. A case study on this alternative approach is presented for investigating a thick (>30 m) sensitive clay deposit in the Outaouais region (Québec), Canada. For this case study, the proposed approach was effectively used to assess a 2D shear-wave velocity model along a 1 km long survey line from the surface down to depths as great as 90 m. The velocity model is finally compared with the results of a multi-offset seismic piezocone penetration test (SCPTu) carried out at one location along the survey line to test its accuracy in a blind test.

19.2 Study Site

The study site is located in the Outaouais region (Québec), Canada, close to the city of Ottawa. This region is at high risk of landslides due to thick sensitive clay deposits and seismic activity (Aylsworth et al. 2000). Extensive studies were carried out in this region to assess these risks (Motazedian and Hunter 2008). Several seismic methods were used in these studies, in particular downhole seismic, MASW, seismic reflection and refraction, and spectral ratio methods.

The Quaternary geology of the Outaouais region consists of a thin veneer of till overlain with glaciofluvial sediments mainly composed of sands and gravels. This sequence is covered by a marine unit deposited by the Champlain Sea around 11,400 years. BP and composed of sensitive clay. The study area and the seismic line location are shown in Fig. 19.1. According to the logs of nearby wells, the thickness of the clay deposit ranges between 10 and 35 m while the thickness of the

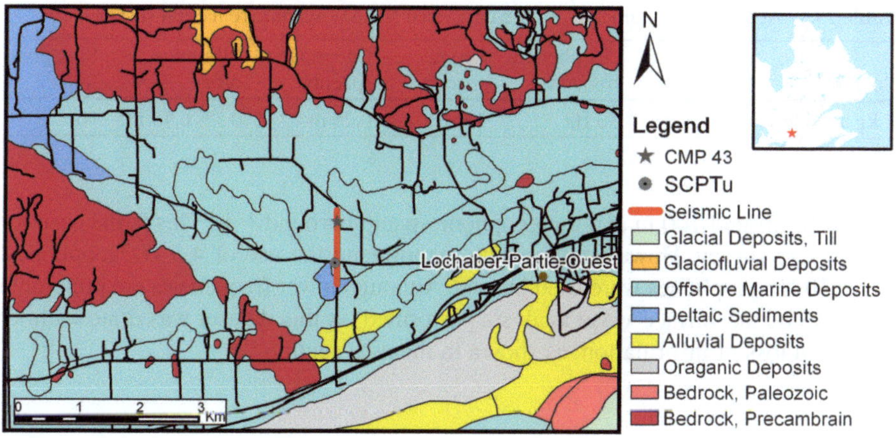

Fig. 19.1 Map of Quaternary deposits (Modified from Bélanger et al. 1997). The locations of the seismic reflection survey line, CMP gather no. 43 and SCPTu are identified by a thick *red line*, a *grey star*, and a *grey circle* respectively

glaciofluvial unit varies between 20 and 60 m. There is no till in contact with the bedrock. The depth to bedrock ranges from 40 to 90 m along the survey line. Due to the outcropping clay unit and the thick overburden, the study site is suitable for testing the proposed approach.

19.3 Combined MASW-SV Seismic Reflection Survey

19.3.1 Field Data Acquisition

The seismic reflection survey was designed primarily for the acquisition of *SV*-wave reflections. Parameters were chosen according to the walk-away method proposed by Steeples and Miller (1998). The survey parameters are given in Table 19.1. A straightforward hammer impact on a steel plate lying on the ground surface was used as a seismic source. The survey was carried out on the shoulder of a dirt road. Despite difficult source and geophone coupling on this kind of surface, a good signal was obtained for Rayleigh waves, *SV*- and *P*-wave reflections.

Although a single 24-channel Geode was used, 96 traces per shot position were acquired by revisiting each shot position four times as the spread cable was moved forward along the line. This configuration was required to meet the constraints of geophone spacing and spread length. Indeed, small geophone spacing is required for *SV* reflection surveys to avoid aliasing of reflection hyperbolas while long spread cable is needed for a better coverage of *P*-wave reflections. Moreover, this configuration meets the guidelines for the MASW method when the depth to bedrock is in excess of 50 m. In such case, according to Park et al. (2002), the maximum offset

Table 19.1 Acquisition parameters for the combined MASW-seismic reflection survey

Source	Stack	Geophone frequency	Source spacing	Geophone spacing	Minimum offset	Shot point fold
8 kg hammer	2	30 Hz	3 m	0.75 m	0.75–6.75 m	96

should be around 100 m and the geophone spacing around 1 m. This is the case for the present survey. However, they also recommend the use of 4.5 Hz geophones, while 30 Hz geophones were used instead. Although lower frequencies are preferable for MASW, the surface wave analysis performed on this dataset was done without too much loss of information, as shown in the next section.

19.3.2 MASW Processing

MASW processing was applied as described in Park et al. (1998). After specifying the geometry for each shot file, the traces were sorted by their common-middle-point (CMP) with a binning of 6 m, resulting in a 192 fold. With such a high fold, there is no loss using CMPs instead of shot gathers. These steps were performed with the help of CREWES MATLAB package (Margrave 2003) and in-house MATLAB codes. All the remaining processing steps were carried out using SeisImager/SW software package.

The fundamental mode of each CMP was handpicked and smoothed using a median filter. An initial model containing 15 layers ranging from the surface down to a depth of 12 m was assigned to each dispersion curve. They were inverted individually using the algorithm described in Xia et al. (1999), keeping only the layer thickness fixed. The resulting models were linearly interpolated to obtain a 2D velocity profile.

The dispersion curve for the CMP gather no. 43 is shown in Fig. 19.2b. The first mode appears clearly between 3 and 15 Hz and higher modes are also present. The dispersion curves are very smooth because of the high CMP fold. The estimated picking error is also shown in Fig. 19.2b and its effect on the inverted model is shown in Fig. 19.2c. The mean variation between the inverted models due to picking errors is 4 %, slightly higher than the picking error itself (3 %).

Although the response of 30 Hz geophones is not flat below 30 Hz, it did not hinder too much the MASW processing. First, each frequency is normalized on the dispersion curves, so amplitude attenuation is not a concern, as long as the signal can be detected. Moreover, spurious frequencies generated by calculation artefacts would not generate the continuous and sharp dispersion patterns seen in Fig. 19.2b. All those facts support the claim that frequencies as low as 3 Hz were effectively detected on the dispersion curves. The use of 4.5 Hz geophones could help the detection of lower frequencies. However, it is doubtful that a hammer can produce frequencies much lower than 2 Hz, and the gain in penetration depth would not be enough to reach the bedrock.

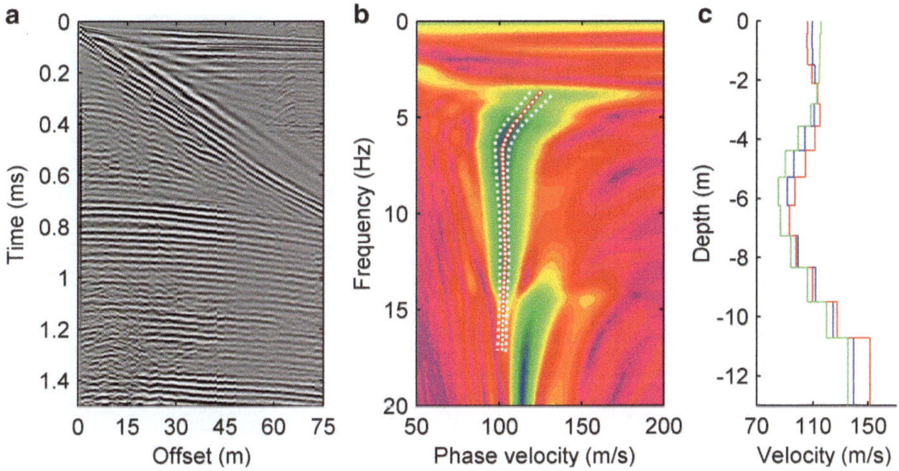

Fig. 19.2 CMP gather no. 43 (**a**), its dispersion curve (**b**) with the selected first mode (*orange circles*) and the picking error (*dotted white lines*). The inverted model is shown in *blue* in (**c**). The maximum and minimum models due to picking error are shown in *red* and *green* respectively

Table 19.2 CMP processing of *SV* seismic reflections

Processing steps	Comments
1. Geometry	Position and elevation of the shots and geophones
2. Bad trace removal	Automatic filter based on S/N ratio
3. Surface wave filter	Based on the dispersion of surface waves
4. Scaling	AGC with a 200 ms window
5. CMP binning	Bin length: 6 m
6. MASW model V_{RMS} conversion	Included in step 7
7. Semblance analysis	Each CMP was processed
8. Velocity conversion	V_{RMS} conversion to interval velocity

19.3.3 SV Seismic Reflection Combined Processing

The CMP data processing flow for *SV* seismic reflections is summarized in Table 19.2. Two programs were used: CREWES MATLAB package for steps 1, 2, 3, 6 and 8 and GEDCO Vista for the remaining steps. The surface wave filter in step 3 is based on the dispersion of surface waves and more details on its performance will be published later. The processing flow was designed to produce a 2D *SV* velocity section; so post velocity analysis steps leading to a stacked section are not given and detailed herein.

The semblance analysis of the CMP gather no. 43 is given in Fig. 19.3. The quality of the semblance panel is judged very good for this example since each reflection has a quite isolated semblance peak. *SV* reflections can be observed from 150 ms to the end of the record. The bedrock reflection is at about 750 ms. All reflections at later times are multiples and are not included in the velocity analysis. This is

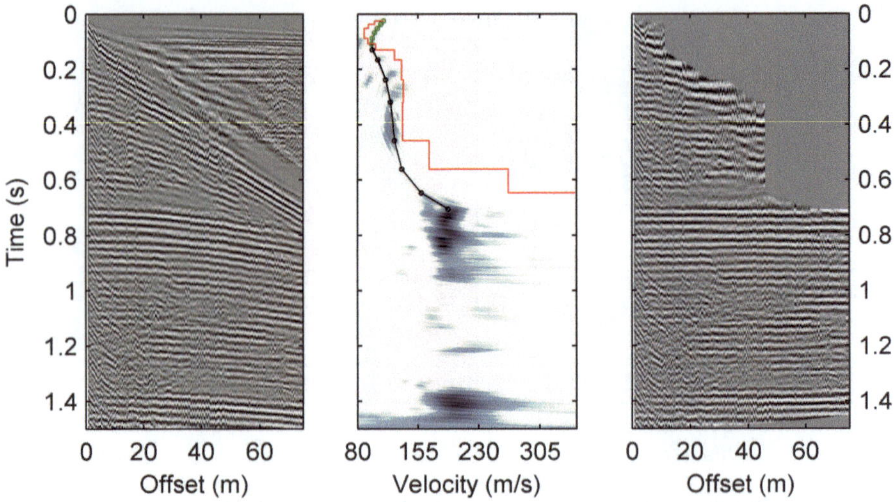

Fig. 19.3 CMP gather no. 43 (*left*), its semblance analysis (*middle*) and the NMO corrected gather (*right*) with a 100 % stretch mute. Selected RMS velocities are shown in *black* and the interval velocities are shown in *red*. The *green circles* are the velocities derived from the MASW method

supported by the velocity inversion after 750 ms on the semblance panel, which is unrealistic for reflections below the bedrock. *P*-wave reflections are located at the top-right of the gather and are clearly separated from *SV*-wave reflections. The processing of both *SV*- and *P*-waves reflections on the CMP gathers can therefore be performed.

Because there is no distinguishable reflection before 150 ms, earlier velocities in Fig. 19.3 are derived from the MASW results as mentioned in step 6 of the processing flow (Table 19.2). In the present case, MASW can give velocities from the surface down to a depth of about 12 m. This limited depth of investigation is caused by the very low seismic velocities in sensitive clay, which yield a small wavelength even at low frequencies. *SV* reflections are distinguishable from about 150 ms or at a depth of about 10 m for an average velocity of 130 m/s. Therefore, both methods are complementary and their joint acquisition can provide a velocity profile from the surface down to the bedrock contact.

To combine CMP reflection processing and the MASW method, the MASW velocity model was first converted to RMS velocities in time at each CMP location using the Dix formula (Yilmaz 2001). Semblance analysis was then carried out using these velocities before the first reflection. The combined RMS velocity profiles were converted back to interval velocities for the final result.

19.3.4 Interpretation

Results from the MASW survey are shown in Fig. 19.4a. As previously mentioned, due to the very low shear wave velocity in sensitive clay, the MASW method has a limited depth of investigation even if the frequency content of Rayleigh waves is

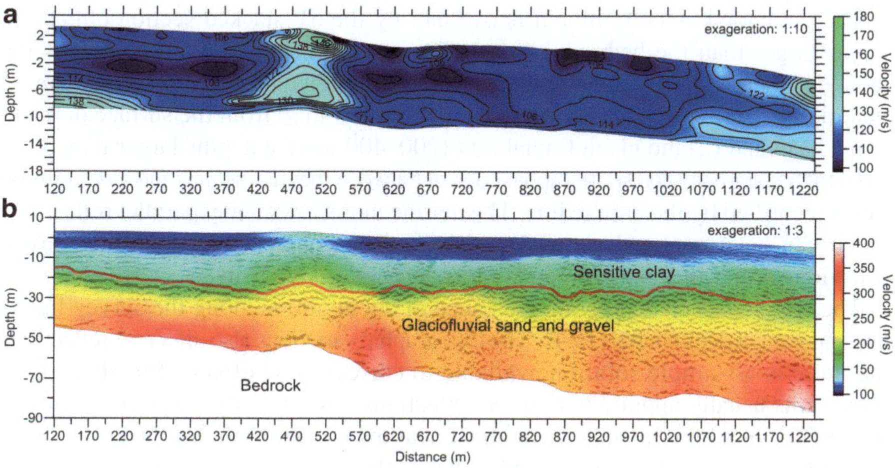

Fig. 19.4 *SV*-wave velocity models obtained from (**a**) MASW alone and (**b**) combined MASW and *SV* reflection processing of the seismic reflection survey. The stacked seismic section overlays the combined velocity profile. Both colormaps are the same but the vertical exaggerations are different due to the large difference in depth for both models

quite low. Because the stratigraphic contacts between the marine unit, the glaciofluvial unit and the bedrock are deep, they cannot be detected using MASW. Moreover, frequencies higher than 15–20 Hz could not be picked for the first mode due to the very high energy content of the higher modes. This limits the resolution of the inversion for depths smaller than 2 m and the accuracy of the velocity layers above 2 m can be questioned. The use of the higher modes in the inversion could solve both of these problems as well as improve the overall accuracy of the velocity model (Xia et al. 2003), especially since very energetic higher modes up to 80 Hz are visible in every dispersion panel for this survey.

Nevertheless, the MASW results are very useful to characterize the variations in shear wave velocity near the ground surface. There is a velocity inversion at a depth of about 4 m due to the contact between the over-consolidated clay layer caused by the freeze thaw cycle and the undisturbed soft clay at depth. The increase in shear wave velocity between depths of 6–12 m is caused by the clay consolidation due to the overburden pressure. Two high velocity anomalies are apparent in the model (Fig. 19.4a): one around 500 m and the other from 1,020 m to the end of the survey line. Both anomalies are likely artefacts associated to topographic variations. No corrections were applied for the topographic variations which are known to affect velocity estimations of MASW (Bodet et al. 2004).

The combined model in Fig. 19.4b extends down to the bedrock contact. As no reflections below the bedrock could be observed, the bedrock velocity cannot be estimated in this survey. This is usually the case in engineering seismology and can be considered a disadvantage compared to seismic refraction survey or MASW. However, the superior depth of investigation and resolution of reflection seismic are major advantages in comparison to the previous methods.

The combined velocity model is overlaid by the *SV* stacked section clipped at depths larger than the bedrock contact. The stacked section greatly improves the geologic interpretation of the velocity model. Two distinct velocity layers can be identified in Fig. 19.4b: (1) the marine unit (90–200 m/s) from the surface down to about 30 m and (2) the glaciofluvial unit (200–400 m/s) at depths larger than 30 m down to the bedrock. The stratigraphic contact between the marine unit and glaciofluvial unit is identified by the orange line. The marine unit shows many parallel reflections which can be interpreted as thin sand beds. The coherency of the glaciofluvial reflections drops after 600 m indicating coarser gravel deposits.

The low velocity anomaly at a distance of about 500 m along the survey line is also apparent at depth in Fig. 19.4b. The topographic variations also affect *SV* reflections and there is no simple solution available to correct these effects. Standard static corrections usually applied to *P*-wave reflections fail for *S*-wave reflections due to the absence of a high impedance contrast near the surface (Yilmaz 2001).

The combined model can be used to assess the seismic hazard parameters of the site. First, this model can give an estimate of VS_{30}. For the velocity model in Fig. 19.4b, the average VS_{30} is 147 m/s, which corresponds to soft clay (Finn and Wightman 2003). The fundamental site resonance period T related to the seismic amplification can also be calculated from the relation $T=4H/V_{av}$ (Bard and Bouchon 1980) where H is the overburden thickness and V_{av} is the average velocity of the soil. The fundamental site period ranges from 1.19 to 1.89 s for the study site.

19.4 Multi-offset Seismic Piezocone Penetration Test

19.4.1 Data Acquisition and Processing

A multi-offset SCPTu was carried out at a distance of 895 m from the beginning of the survey line (Fig. 19.1) to verify the accuracy of the combined velocity model. The offsets ranged from 1 to 10 m and the receiver depth interval was 1 m. The equipment and the settings used to acquire the data are similar to the ones described in LeBlanc et al. (2006). Two seismic sources were tested: (1) a VIBSIST-20 source based on the swept impact seismic technique (Park et al. 1996) and (2) a 3.6 kg hammer impact on a steel plate lying on the surface of the ground. Only the results with the hammer are presented because they are less affected by noise and their frequency content is similar to the VIBSIST-20 data.

Examples of raw and processed SCPTu shot gathers are given in Fig. 19.5. Two types of noise can be identified in these gathers: ground roll and very high-frequency arrivals. The later is assumed to be sound waves guided by the pushing rods as supported by the high velocity and high frequency of these arrivals. A low-pass filter of 600 Hz was applied to remove this noise. To suppress the ground roll, the signals from the two horizontal accelerometers H1 and H2 embedded in the penetrometer

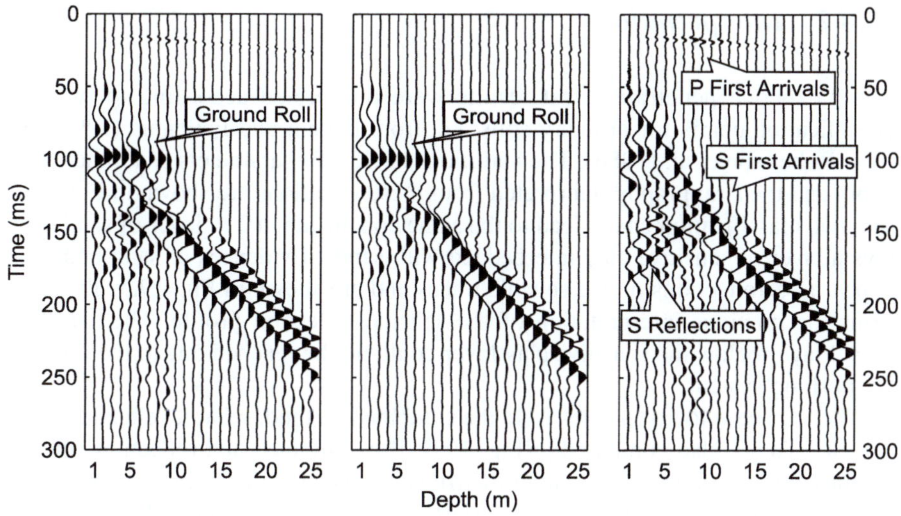

Fig. 19.5 SCPTu shot gathers for a lateral offset of 10 m. The first two gathers are for the two horizontal H1 and H2 accelerometers embedded in the penetrometer. The ground roll noise is completely suppressed in the third gather after the combination of the H1 and H2 gathers

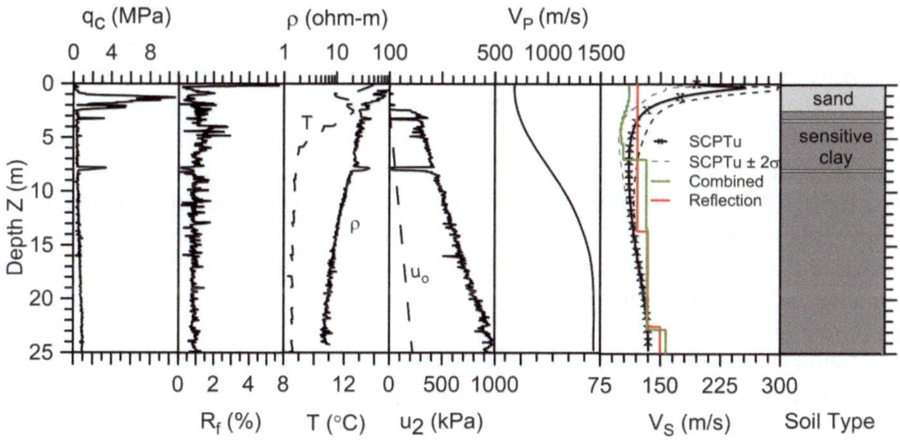

Fig. 19.6 SCPTu logs: cone resistance (q_c), friction ratio (R_f), electrical resistivity (ρ), temperature (T), pore pressure (u_2), *P*- and *S*-waves velocities. The stratigraphic column is interpreted from the SCPTu logs

were combined using the formula $H_{comb} = H_1 - 0.6\ H_2$. The first arrivals were then picked and wavepath eikonal traveltime inversion was carried out using Rayfract (Schuster and Quintus-Bosz 1993). The resulting velocity models were averaged over the length of the multi-offset profile to produce *P*- and *S*-waves vertical seismic profiles (Fig. 19.6).

19.4.2 Interpretation

The piezocone penetration test results were interpreted according to two soil classification charts for ordinary soils, i.e. from Eslami and Fellenius (1997) and from Robertson et al. (1986). The variations of SCPTu parameters as a function of depth and the stratigraphy interpreted from these variations are shown in Fig. 19.6. The two classification charts lead to the same conclusion: a 2 m layer of sand overlays a layer of sensitive clay interbedded with sands. Evidences of sand beds also appear as *S* reflections on the seismic traces and are shown in Fig. 19.5. This confirms the previous interpretation of the seismic reflection profile. The sand and gravel layer below the sensitive clay layer could not be reached due to the limits of the pushing system.

The *S*-wave velocity profiles from the combined method, seismic reflection alone and SCPTu are also compared in Fig. 19.6. The combined and reflection velocity profiles come from the processing of CMP gather no. 151. This CMP is the closest to the SCPTu location. The smooth appearance of the velocity curves for the SCPTu comes from the averaging of the tomographic 2D models not shown herein. This averaging of the multi-offset data improves the accuracy of the 1D velocity profile. To show the dispersion of the data in the 2D model, the velocity profile more or less two times the standard deviation is also plotted in Fig. 19.6.

The three velocity profiles are in good agreement. Using the SCPTu as the reference, the mean absolute errors (MAE) are 15 and 11 % and the root mean square errors (RMSE) are 22 and 19 % for the combined and the reflection profiles respectively. The higher value of the RMSE in both cases is caused by the first 2 m of the models: neither shows the high velocity superficial sand layer. This is to be expected for the reflection model, as it can only give an average velocity for the first few meters. As for MASW, the lack of frequencies higher than 20 Hz for the fundamental mode can explain the misfit above 2 m. Moreover, both measurements were not exactly taken at the same location: the surface survey was carried out on the road shoulder and the SCPTu was acquired in the adjacent orchard, so surface conditions may be different in both locations.

Below 2 m, the MASW and SCPTu models agree very well as the MASW model actually falls inside the standard error limits of the SCPTu. Both models also show the same trend. By removing data above 2 m, RMSE falls at 13 % and MAE at 12 %. This agrees very well with the expected error between MASW and VSP measurements obtained by Xia et al. (2002).

19.5 Discussion and Conclusions

In the case study presented herein, the MASW survey failed to provide the *S*-wave velocity profile from a depth of 12 m down to the bedrock contact. The depth of investigation is so limited that the *S*-wave velocity at 30-m depth Vs_{30} cannot be assessed. Far from being an exceptional case, this is typical for areas with thick sensitive clay deposits. To avoid this limitation, the combined processing of Rayleigh waves and *SV*-wave reflections is proposed as a solution. Surface waves and *SV*-waves are generated during a typical seismic reflection survey and they can be simultaneously acquired with the proper field parameters. Special care must be taken to avoid aliasing of *SV*-wave reflections by reducing the geophone spacing. To maintain proper spread length, the record of at least 48 live channels with an engineering seismograph is recommended.

Other solutions are available to measure shear wave velocity profiles when the overburden depth exceeds 50 m but they are not as effective as *SV*-wave reflection surveys. Downhole seismic survey is expensive to achieve as the depth becomes too great and seismic information is available at only one location. As for passive MASW surveys, which can go to a depth as great as 100 m, they do not provide as accurate stratigraphy of the soil as reflection seismics. In addition to the velocity assessment, the processing of the *SV*-wave reflections leads to a stacked seismic profile providing the location of stratigraphic contacts along the survey line. Moreover, this can be used to calculate the fundamental site resonance period.

Even in the case of shallow bedrock, *SV*-wave reflections should be considered in MASW records in sensitive clay and their processing would bring better constraints on stratigraphic contacts and add redundancy to velocity estimations. In fact, all seismic arrivals recorded in a typical MASW survey, such as Rayleigh waves, first arrivals, *P*- and *SV*-waves reflections can be processed. In all likelihood, combining the information of these separate techniques gives a far more robust earth model.

Acknowledgments We thank Y. Vincent, S. Bérubé and M. El Baroudi for their help during the acquisition of the geophysical data in the field. We are grateful to Dr. Maarten Vanneste from the NGI for his constructive reviews. The research fund to support this project was provided by the Department of Environment of Québec for the acquisition of data on groundwater in the Outaouais region for the sustainable management of this natural resource.

References

Aylsworth JM, Lawrence DE, Guertin J (2000) Did two massive earthquakes in the Holocene induce widespread landsliding and near-surface deformation in part of the Ottawa Valley, Canada? Geology 28(10):903–906. doi:10.1130/0091-7613(2000)28<903:DTMEIT>2.0.CO;2

Bard PY, Bouchon M (1980) The seismic response of sediment-filled valleys. Part I. The case of incident SH waves. Bull Seismol Soc Am 70:1263–1286

Bélanger JR, Moore A, Prégent A (1997) Surfacial geology, digital map, Thurso Quebec (31G/11). Geological Survey of Canada, Ottawa, Open File Issue 3477

Bodet L, Abraham O, Bitri A, Leparoux D, Côte P (2004) Effect of dipping layers on seismic surface waves profiling: a numerical study. SAGEEP 17:1601. doi:10.4133/1.2923306

Eslami A, Fellenius BH (1997) Pile capacity by direct CPT and CPTu methods applied to 102 case histories. Can Geotech J 34(6):886–904. doi:10.1139/t97-056

Finn WDL, Wightman A (2003) Ground motion amplification factors for the proposed 2005 edition of the National Building Code of Canada. Can J of Civil Eng 30(2):272–278. doi:10.1139/l02-081

LeBlanc AM, Fortier R, Cosma C, Allard M (2006) Tomographic imaging of permafrost using three-component seismic cone penetration test. Geophysics 71(5):H55–H65. doi:10.1190/1.2235876
Margrave GF (2003) Numerical methods of exploration seismology with algorithms in MATLAB. Department of Geology and Geophysics, University of Calgary, Calgary, 225 p
Motazedian DMD, Hunter JHJ (2008) Development of an NEHRP map for the Orleans suburb of Ottawa, Ontario. Can Geotech J 45(8):1180–1188. doi:10.1139/T08-051
Park CB, Miller RM, Steeples DW, Black RA (1996) Swept impact seismic technique (SIST). Geophysics 61(6):1789–1803. doi:10.1190/1.1444095
Park CB, Miller RD, Xia JB (1998) Imaging dispersion curves of surface waves on multi-channel record. SEG Expanded Abs 17(1):1377–1380
Park CB, Miller RD, Miura H (2002) Optimum field parameters of an MASW survey. Exp. Abs SEG-J Tokyo, 17–18 May, pp 22–23
Pugin AJM, Pullan SE, Hunter JA (2009) Multicomponent high-resolution seismic reflection profiling. The Leading Edge 28(10):1248–1261. doi:10.1190/1.3249782
Robertson PK, Campanella RG, Gillespie D, Greig J (1986) Use of piezometer cone data. Use of in situ tests in geotechnical engineering (GSP 6). ASCE, Reston, pp 1263–1280
Schuster GT, Quintus-Bosz A (1993) Wavepath eikonal traveltime inversion: theory. Geophysics 58(9):1314–1323. doi:10.1190/1.1443514
Steeples DW, Miller RD (1998) Avoiding pitfalls in shallow seismic reflection surveys. Geophysics 63(4):1213–1224. doi:10.1190/1.1444422
Xia J, Miller RD, Park CB (1999) Estimation of near-surface shear-wave velocity by inversion of Rayleigh waves. Geophysics 64(3):691–700. doi:10.1190/1.1444578
Xia J, Miller RD, Park CB, Hunter JA, Harris JB and Ivanov J (2002) Comparing shear-wave velocity profiles inverted from multichannel surface wave with borehole measurements. Soil Dyn Earthq Eng 22(3):181–190. http://dx.doi.org/10.1016/S0267-7261(02)00008-8
Xia J, Miller RD, Park CB and Tian G (2003) Inversion of high frequency surface waves with fundamental and higher modes. J Appl Geophys 52(1):45–57. http://dx.doi.org/10.1016/S0926-9851(02)00239-2
Yilmaz Ö (2001) Seismic data analysis: processing, inversion, and interpretation of seismic data, Investigations in geophysics no. 10. SEG Books, Tulsa

Chapter 20
Empirical Geophysical/Geotechnical Relationships in the Champlain Sea Sediments of Eastern Ontario

H.L. Crow, J.A. Hunter, A.J.M. Pugin, S.E. Pullan, S. Alpay, and M. Hinton

Abstract Geophysical and geotechnical data are presented from different sites in eastern Ontario where variable geotechnical properties of Champlain Sea sediments ('Leda Clays') are found. Sites range from thick "undisturbed" silts and clays, to "disturbed" geologically similar soils (earthquake triggered landslides and other deformed materials). High-resolution seismic profiles provide stratigraphic context for some of the boreholes drilled in the study area. Downhole geophysical logs from 14 boreholes are compared to core sample measurements of porosity, sensitivity, and porewater conductivity to develop useful empirical relationships. According to these relationships, silt and clay sediments can be sensitive or quick when formation conductivity drops below 100 mS/m. Conversely, silts and clays with elevated conductivities (>250 mS/m) are rarely sensitive. Salinity values calculated from porewater conductivity indicate sensitive or quick behaviour may be expected in leached soils when salinity drops below 2 g/l.

Keywords Sensitive clays • Geophysical surveys • Eastern Ontario • Conductivity • Porewater salinity • Empirical equations • Landslide hazard

20.1 Introduction

The clayey silts of the Champlain Sea, locally known as Leda Clays, have been intensively studied in Canada since the 1950s because of their unusual geotechnical behaviours (e.g. Eden and Crawford 1957; Gadd 1986; Torrance 1988; Law and Bozozuk 1988). For many years, the Geological Survey of Canada (GSC) has

H.L. Crow (✉) • J.A. Hunter • A.J.M. Pugin • S.E. Pullan • S. Alpay • M. Hinton
Natural Resources Canada, Geological Survey of Canada,
601 Booth Street, Ottawa, ON, Canada K1A 0E8
e-mail: hcrow@nrcan.gc.ca

J.-S. L'Heureux et al. (eds.), *Landslides in Sensitive Clays: From Geosciences to Risk Management*, Advances in Natural and Technological Hazards Research 36,
DOI 10.1007/978-94-007-7079-9_20, © Springer Science+Business Media Dordrecht 2014

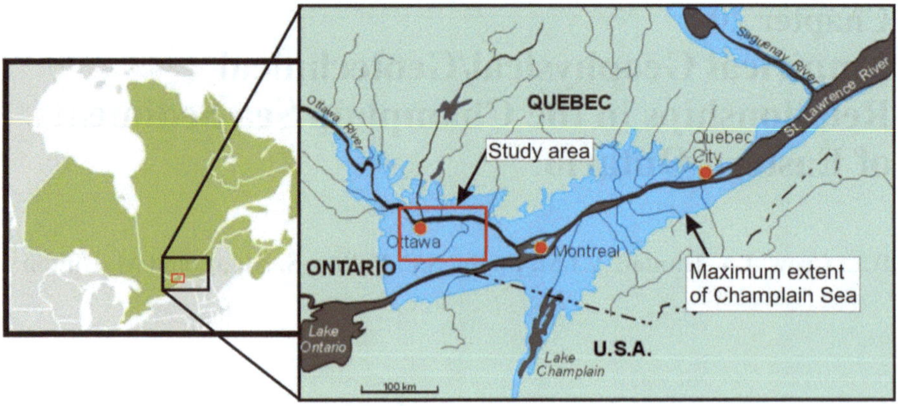

Fig. 20.1 Eastern Ontario study area shown within the extent of the Champlain Sea inundation dating from 11,500 years BP (^{14}C dating)

collected and processed shallow surface geophysical (seismic, electromagnetic, resistivity) and downhole geophysical data in the Champlain Sea basin within Eastern Ontario (Fig. 20.1). The focus has been on understanding the evolution of basin sedimentation (e.g. Cummings et al. 2011; Medioli et al. 2012), detecting the presence of ground disturbance (e.g. Pullan et al. 2011), and measuring overburden properties for geohazard and groundwater studies (Benjumea et al. 2003; Pugin et al. 2007, 2009; Hunter et al. 2010; Crow et al. 2011). Hyde and Hunter (1998) specifically looked at the relationship between conductivity and salinity in the Champlain Sea sediments east of Ottawa, and Calvert and Hyde (2002) looked at assessing landslide hazard in the Ottawa area using EM and resistivity methods.

Combining the existing borehole dataset with new borehole and surface seismic data acquired over the past 10 years, this paper presents an updated series of empirical equations relating geophysical measurements (shear wave velocity, and formation conductivity) to basic geotechnical parameters (porosity, bulk density, sensitivity, and salinity) within this region. Three case studies from differing soil sites (sensitive versus non-sensitive, disturbed versus non-disturbed), are also presented to highlight the importance of establishing the geological context surrounding the borehole site using high resolution seismic reflection profiling techniques.

20.2 Methods

The seismic sections and boreholes presented in this paper are a subset of a more extensive dataset collected by the GSC in this region. Sites were chosen to represent a range of end member geotechnical properties in the area, from undisturbed and non-sensitive (e.g. GSC-BH-JSR-01), to undisturbed and sensitive (e.g. NRC-MO) to disturbed and non-sensitive (e.g. JA04-02) (Fig. 20.2). Soils which are interpreted as "undisturbed" are those which have not been affected by mass movement triggered

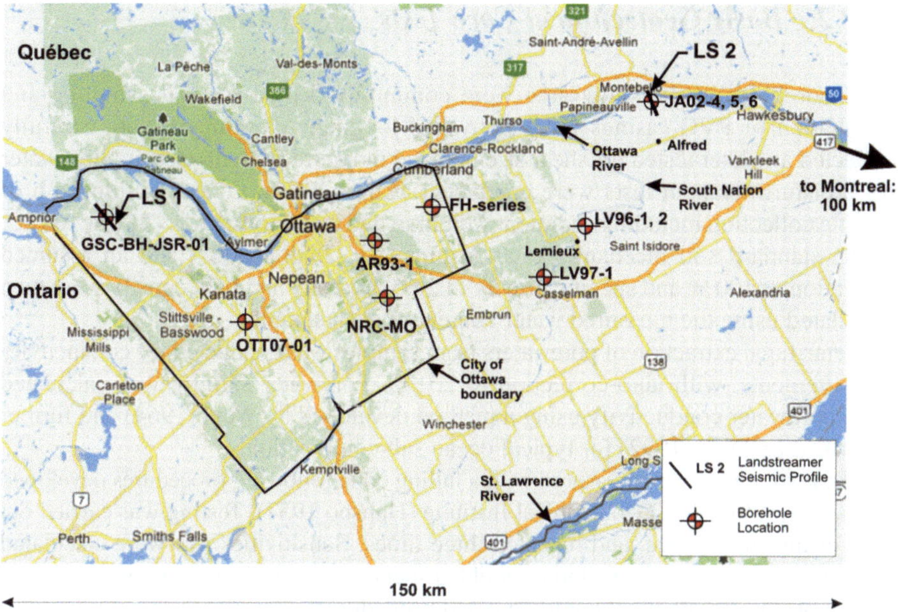

Fig. 20.2 Surface seismic profiles (LS 1 & 2) and borehole locations within the study area selected to represent a range of Leda Clay conditions (Map background © Google Maps 2013)

by earthquakes or landsliding, although these soils may have been reworked since their original deposition.

20.2.1 Surface Seismic Methods

Currently, the GSC collects seismic profile data using a vibrating seismic source (a Minivib) towing a landstreamer array (an in-house development), which provides a very high resolution seismic reflection data acquisition system. This array is designed for use along gravel or paved roads, and is composed of 24–72 sleds, each mounted with three orthogonal geophones. The Minivib is a low-impact vibrating seismic source capable of operating in vertical or horizontal modes to generate P (compressive) wave, and polarized S (shear) wave modes. Vibrations are programmable in time (seconds) and frequency (10–350 Hz) depending on the site conditions and depth of investigation (e.g. Pugin et al. 2009, 2013).

This seismic reflection system is particularly effective in imaging structure in the Champlain Sea sediments with sub-metre resolution in shear wave mode. Processed profiles indicate continuous reflectors in areas of undisturbed layering, while a lack of coherence can indicate large-scale ground disturbance caused by landsliding or paleoearthquakes. For geotechnical and geohazard investigations, this subsurface structural information provides important geological context at a site which would not otherwise be available through drilling alone.

20.2.2 Basic Geotechnical Core Tests

GSC boreholes in the study area were commonly drilled for soil sampling and installation of PVC casings for geophysical logging. All samples were carefully collected and later tested for shear strength and moisture content. Gravimetric water (or moisture) content tests were conducted at the GSC in Ottawa within 48 h of sample collection including sediment weighing and oven drying for >24 h at 105 °C. Using standard soil mechanics relationships (e.g. Budhu 2007) and an assumed saturation of 100 % and specific gravity of 2.65, porosity and bulk density were then calculated using the measured water content of each sample.

Centrifugal extraction of porewaters from samples was conducted for conductivity measurements, with auto-correction to 25 °C. Porewater salinity was calculated from porewater conductivity using a method developed by Weyl (1964) and further described by Dera (1992) for typical ocean salt composition.

Shear strengths were measured with a falling cone following procedures developed by the Royal Swedish Geotechnical Institute (Hansbo 1957). Testing was carried out on a clean sample face, and repeated three times. Sensitivities (S_t) were calculated from the ratios of undisturbed to remoulded shear strengths. Tests were performed on all samples using identical equipment and procedures.

20.2.3 Borehole Geophysical Logging

The GSC routinely uses downhole logs in Champlain Sea sediments to calibrate geophysical surface profiles (seismic or electromagnetic), measure presence or lack of soil conductivity (potential sensitivity indicator), identify coarser grained intervals in silty/clayey units (possible pathways for fresh water), and interpret variations in porosity, density, and shear modulus. Together these parameters help predict Leda Clay behaviour for hazard studies. The downhole data used to develop the empirical relationships presented here include induction logs (formation conductivity) and downhole S-wave velocities. Magnetic susceptibility is also important to assess the presence of sand. The systems and procedures developed for overburden logging methods used by the GSC can be found in Hunter et al. (1998, 2007), Douma et al. (1999), and Medioli et al. (2012).

The induction logs were collected using Geonics EM39 apparent conductivity and EM39S magnetic susceptibility logging probes. On-site calibrations were performed with the tools prior to each run using known calibration points which could be adjusted depending on the conductivity ranges encountered in the materials. S-wave logs were obtained using a clamped, 3-component downhole receiver with 15 Hz geophones, and a 16 lb sledgehammer striking a loaded I-beam surface plate or a Minivib coded source. Seismic data were recorded by a Geometrics 24-channel Geode seismograph.

20.3 Selected Site Studies

20.3.1 Case 1: Undisturbed and Non-sensitive Champlain Sea Sediments West of Ottawa

In 2008, the GSC drilled a 97 m borehole in Champlain Sea sediments (GSC-BH-JSR-01, see Fig. 20.2). High resolution seismic profiling had delineated an elongate, structurally controlled, NW-SE trending bedrock depression within the Champlain Sea basin, and the borehole was sited in a deep part of this bedrock depression (Fig. 20.3). The seismic section also indicated that the silts and clays had not been disturbed by large-scale mass movement since deposition, as evidenced by laterally-continuous, flat-lying reflectors. Combined analyses of the seismic data, core samples, and downhole geophysical logs indicated that the sedimentary succession of the basin represents a typical deglaciation sequence. The sediment package consists of a basal, coarse-grained unit (esker), overlain by silt-rich, marine sediments (Units B & C), capped by a high porosity sequence of silts and clays (Unit D) (Medioli et al. 2012).

Of geotechnical importance at this site are the ranges of measured material properties from various periods of Leda Clay deposition. The seismic section shows that a laterally continuous, low velocity unit exists in the upper 19 m of the ground surface (Unit D). Core testing revealed elevated porosities (60–70 %), relatively low bulk densities (1.50–1.60 g/cm^3), and sensitivities below 25. Within the underlying marine clays (19–62 m, Unit C), seismic data indicate a very uniform material with S-wave velocities increasing with depth. The pore water conductivities are nearing

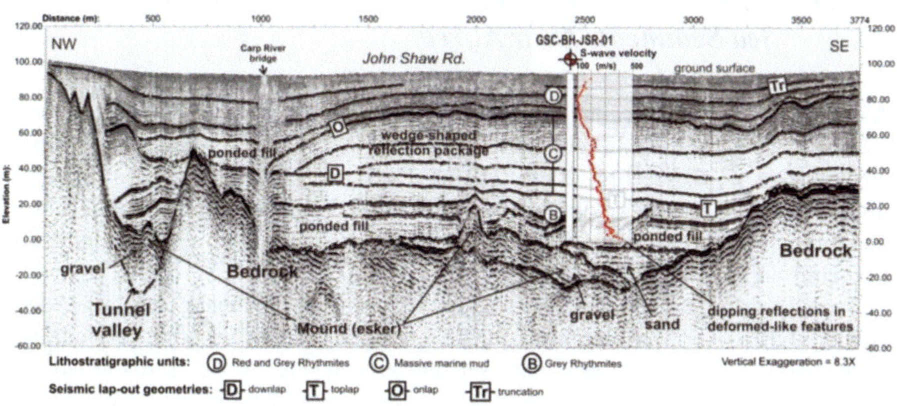

Fig. 20.3 3.8 km S-wave seismic section in west Ottawa (LS 1 in Fig. 20.2). Section shows laterally continuous reflectors indicating undisturbed conditions since the time of deposition. The downhole S-wave velocity log is displayed to the *right* of the borehole location (Modified from Medioli et al. 2012)

values of present-day sea waters (~45 mS/cm). In this interval, formation conductivities reach 1,000 mS/m, and sensitivities are normal to low ($S_t < 20$). The thick sequence of silt and clay near the base of the borehole (62–96 m, Unit B) is characterized by decreasing conductivity, increasing sensitivity, and diminishing salt content. Figures and detailed depositional interpretation can be accessed online (Medioli et al. 2012).

20.3.2 Case 2: Undisturbed and Sensitive Champlain Sea Sediments in Ottawa

As part of a large seismic microzonation project for the City of Ottawa undertaken by the GSC and Carleton University between 2006 and 2010, a 30 m-deep GSC borehole was drilled for core analyses and downhole logging (OTT07-01, see Fig. 20.2). Soil was found to be undisturbed (as evidenced by flat beds and rhythmites), but contained extremely soft intervals, with sensitivities reaching 100. In these intervals, downhole formation conductivities were found to be less than 100 mS/m (Hunter et al. 2010).

In the south end of Ottawa a 20 m-deep borehole (NRC-MO) forms part of the Canadian Geotechnical Research Site No. 1 (McRostie and Crawford 2001). Here, sensitivities exceed 100 in the upper 15 m of the borehole and moisture contents can exceed 80 %. Conductivity data have not been collected in this borehole, but GSC-measured S-wave velocities range between 50 and 150 m/s, making it among the lowest measured S-wave velocities in the Ottawa area.

20.3.3 Case 3: Disturbed and Non-sensitive Champlain Sea Sediments Near Alfred

As part of the Ottawa Valley Landslide Project, conducted by the GSC between 1994 and 2004, several boreholes were drilled in a 46 km^3 area north of Alfred, ON (JA02-4, 5, 6; see Fig. 20.2). The area exhibits near-surface deformation of marine sediments (silts, clays, and sands) where evidence suggests that a high-magnitude earthquake (>M6.5) 7060 years BP induced widespread ground disturbance throughout the region (Aylsworth et al. 2000).

Prior to drilling, GSC seismic surveys mapped the existence of a large, deep, bedrock basin (~180 m maximum depth) underlying the disturbed area (Benjumea et al. 2003). As shown in the P-wave seismic section (Fig. 20.4), the disturbed zone is characterized by a lack of coherence in the near-surface reflections. Borehole locations were chosen to intersect the deepest part (JA02-4), the margin (JA02-6), and outside (JA02-5), the bedrock basin.

Borehole JA02-4 was positioned where surface disturbance was most evident. Core examination indicated that clay deformation ranges from brittle shear to plastic

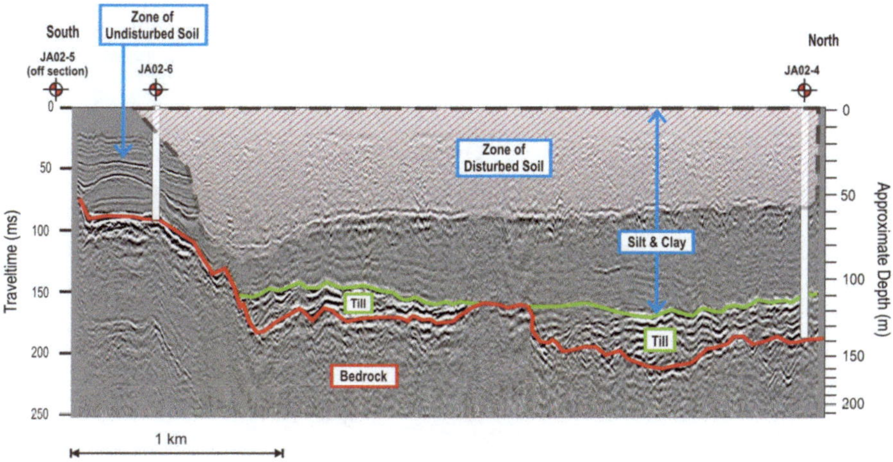

Fig. 20.4 P-wave seismic section collected near Alfred, ON, labelled as LS 2 in Fig. 20.2. Undisturbed beds in southernmost end of section become disturbed over the deep basin

deformation, and contains evidence of flowing. Sand dykes also intrude into the clay, and evidence of liquefaction was seen in the cores and in near-by excavated trenches. Formation conductivities measured in the disturbed zones generally ranged from 50 to 200 mS/m, and falling cone tests carried out on the cores indicated that the clay was not sensitive (Aylsworth et al. 2003).

20.4 Geotechnical/Geophysical Correlations

Downhole geophysical logs and geotechnical test results from the six boreholes described in these case studies (and eight additional boreholes in the region) were examined to develop some empirical relationships between the various measured parameters. Only data collected in silts and clays were considered.

20.4.1 Porosity and Bulk Density Versus Shear Wave Velocity

Semi-log plots of derived porosity and bulk density as functions of S-wave velocity (Vs) are shown in Fig. 20.5. GSC microzonation studies carried out in the City of Ottawa showed that the average Vs in the upper 40 m of the ground surface is quite variable (150 ± 30 m/s, one std dev) and is correlated to depth, porosity, and density (Hunter et al. 2010). The scatter in the plots below reflects the local variability in Vs and water contents throughout the wider region. No clear correlation between Vs and disturbed-undisturbed ground was found in this dataset.

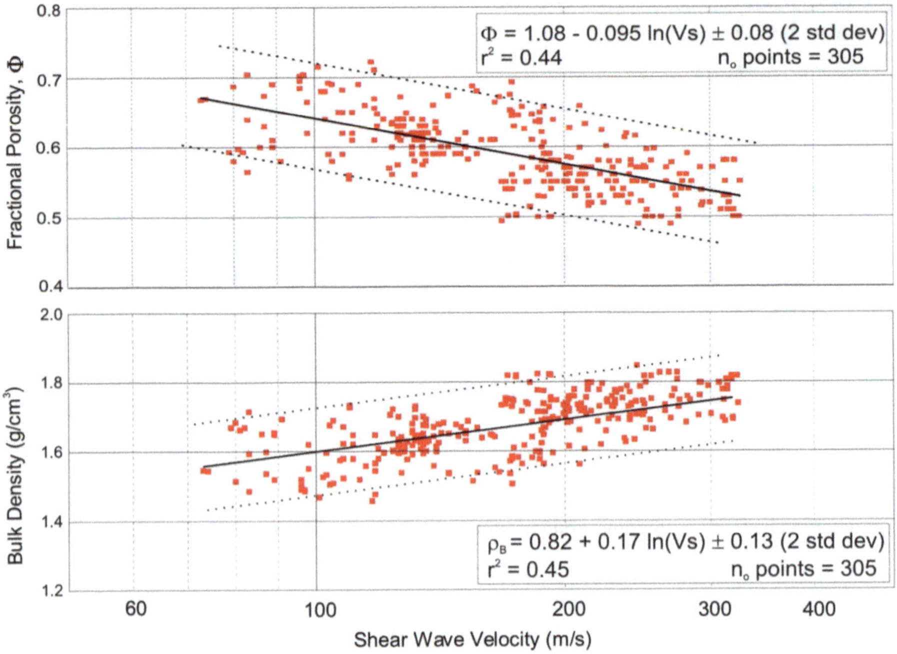

Fig. 20.5 Cross plots of porosity and bulk density versus shear wave velocity for Champlain Sea sediments (Data from JA02-4, JA02-5, JA02-6, NRC-MO, LV96-1, LV96-2, AR93-1, and GSC-BH-JSR-01)

20.4.2 Conductivity, Salinity, and Sensitivity

Changes in porewater salinity within the Champlain Sea sediments are the primary cause of variability in formation conductivities. It is well known that the leaching of saline porewaters from marine silts and clays has been correlated with sediment sensitivity (e.g. Torrance 1975, 1988). Although salinity alone is not an indicator of sensitivity, it can be used in areas where the sediments have been leached to predict sensitivity.

Empirical equations relating downhole formation conductivity to porewater conductivity and salinity are presented in Fig. 20.6. These updated plots are an extension of the initial work of Hyde and Hunter (1998) performed west of Ottawa using boreholes LV96-1, -2, and AR93-1.

Sensitivity versus formation conductivity for both undisturbed and disturbed sediments, as interpreted from seismic profiling and core examination, are shown in Fig. 20.7. Soils with elevated conductivities ($\sigma_f > 250$ mS/m) are generally not sensitive ($S_t < 20$) and probably have not undergone significant leaching of their saline porewaters since deposition. Conversely, soils which are undisturbed but of low conductivity ($\sigma_f < 100$ mS/m) are commonly highly sensitive or quick; these soils are generally interpreted to be leached of their original porewaters. However, it must be noted that deposition of silts and clays in fresher waters can also produce formation

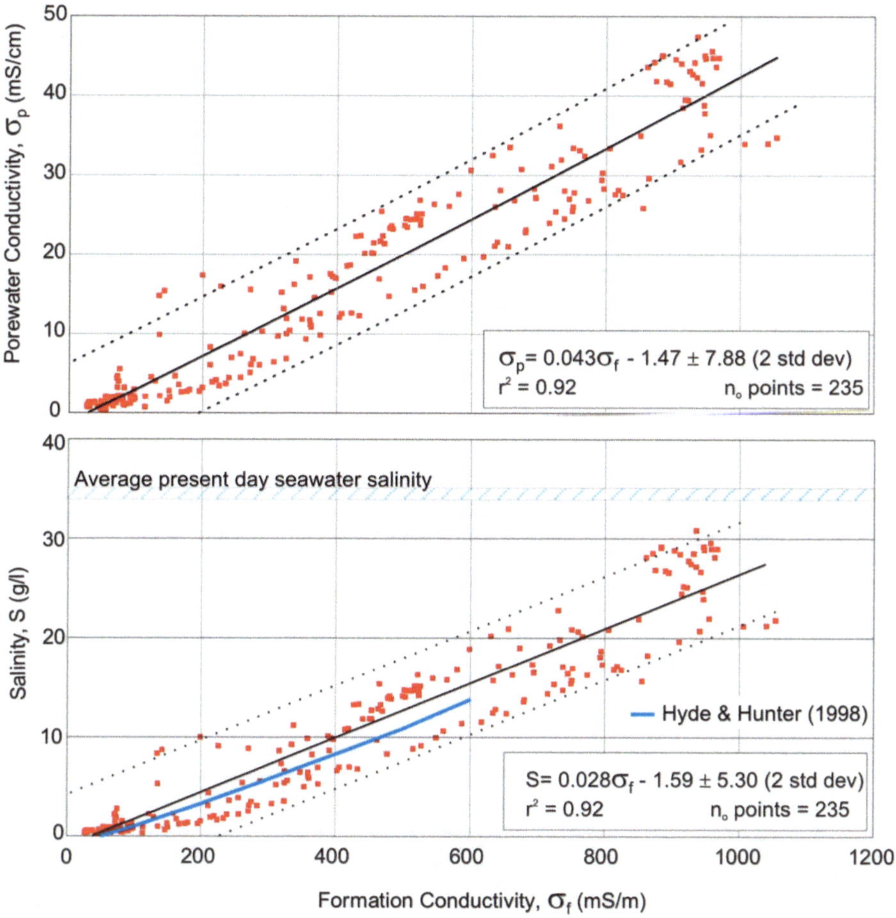

Fig. 20.6 Porewater conductivity and salinity as functions of formation conductivity using data from boreholes JA02-5, JA02-6, LV96-1, LV96-2, LV97-1, AR93-1, and GSC-BH-JSR-01. Note that the downhole data were collected at ground temperatures of 8±0.5 °C (as measured by downhole temperature logs), whereas the porewater conductivities are presented corrected to a standard temperature of 25 °C in the laboratory. Also note that porewater conductivities are shown in mS/cm, and formation conductivities are presented in mS/m to respect the scales at which the data are collected

conductivities <100 mS/m without an associated high sensitivity; therefore, sensitivity must be confirmed through shear strength testing.

Disturbed sediments reside in a low sensitivity-low conductivity cluster. These data come from the uppermost 40 m of the ground surface where flowing sands were encountered. Although the original conductivities of the porewaters (and thus sensitivities) pre-disturbance are unknown, it is likely that ground shaking and remoulding of the sediments has further reduced the sensitivity of the soils. If these sediments were highly sensitive and/or quick pre-disturbance, it suggests that the

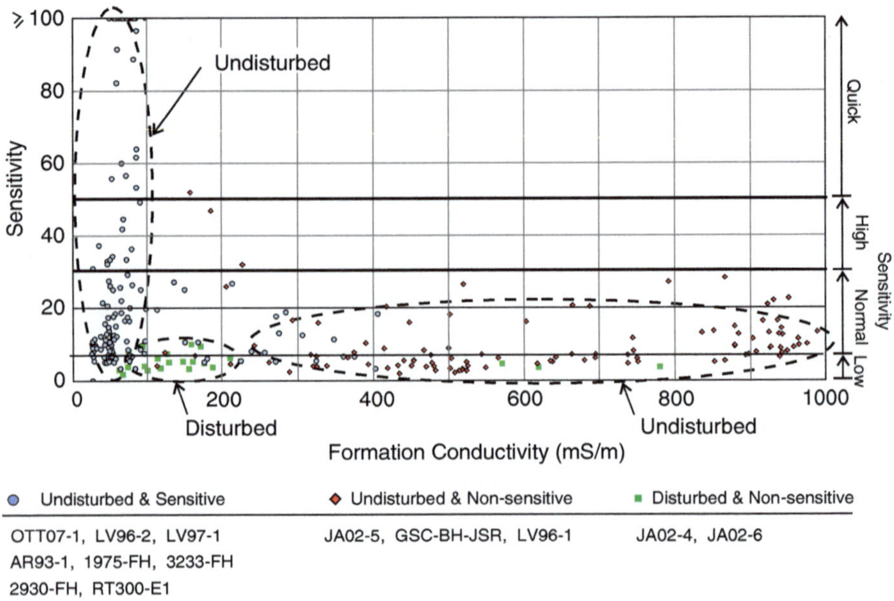

Fig. 20.7 Sensitivity as a function of formation conductivity in Champlain Sea sediments (Sensitivity scale from Broms and Stål 1980)

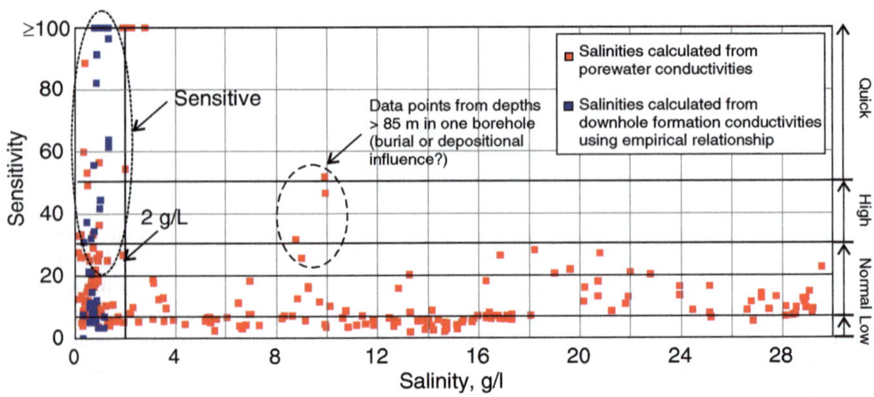

Fig. 20.8 Sensitivity versus salinity using data from JA02-4, JA02-5, JA02-6, AR93-1, LV96-1, LV96-2, LV97-1, OTT07-01, FH-series (3 holes), GSC-BH-JSR-01, and RT300-E1. 2 g/l marks the threshold between predicted sensitive and non-sensitive behaviour in marine soils which have been leached of their original saline porewaters (Sensitivity scale from Broms and Stål 1980)

threshold formation conductivity between sensitive and non-sensitive may in fact be closer to 200 mS/m for some deposits in this region.

Sensitivity is plotted as a function of porewater salinity in Fig. 20.8. In this dataset, highly sensitive or quick behaviour may be expected once the salinity drops below 2 g/l, which is in agreement with the findings of Torrance (1988) for this region and several quick clay sites in Norway.

20.5 Conclusions

A series of empirical relationships between geotechnical and geophysical parameters have been presented for the post-glacial Champlain Sea sediments of eastern Ontario. In addition to depth of burial, S-wave velocities have been shown to vary with porosity (and density). Formation conductivity can be related to salinity content as well as threshold sensitivity, although sensitivity can only be confirmed through sample shear strength testing on undisturbed samples. According to downhole and sample test data, silt and clay sediments with formation conductivities of less than 100 mS/m can be highly sensitive or quick, although exceptions exist for fine-grained soils deposited in fresher water environments. Silts and clays with elevated conductivities (>200 mS/m) are rarely sensitive ($S_t < 20$). Based on the salinity values calculated from porewater conductivities, highly sensitive or quick behaviour may be expected in leached soils where the salinity drops below 2 g/l.

These cohesive sediments show widespread variability in properties, which can be further influenced by post depositional disturbance. Site assessments using seismic reflection surveys are needed to identify disturbance on a regional scale (100s of m to km), and to identify variation of bedrock depth (i.e. occurrence of basins). This has permitted the GSC to carefully select drill targets to intersect features of interest, and interpret complex depositional environments.

Acknowledgements The authors wish to thank Drs. Richard Fortier of Université Laval and Greg Oldenborger of the GSC for their reviews of this paper. We also thank Jan Aylsworth for contributing lab sample results from her studies in the South Nation/Alfred areas, Krystal Aubry for her collation and further study of these results, and Claudia Moore for her porewater extractions and conductivity measurements of the JSR borehole samples. This work was made possible through funding from the GSC's Public Safety Geoscience, Landslide Hazard Reduction, and Groundwater Geoscience Programs. This work represents ESS contribution number 20120459.

References

Aylsworth JM, Lawrence DE, Guertin J (2000) Did two massive earthquakes in the Holocene induce widespread landsliding and near-surface deformation in part of the Ottawa Valley, Canada? Geology 28:903–906

Aylsworth JM, Hunter JA, Gorrell GA, Good R (2003) Borehole investigation of earthquake-induced soft-sediment deformation at Lefaivre: a contribution to the Ottawa Valley Landslide Project. In: Proceedings of the GeoHazards 2003, 3rd Canadian conference on Geotechnique and natural hazards, Edmonton, p 359

Benjumea B, Hunter JA, Aylsworth JM, Pullan SE (2003) Application of high resolution seismic techniques in the evaluation of earthquake site response. Tectonophysics 368:193–209

Broms BB, Stål T (1980) Landslides in sensitive clays. In: Proceedings of the international symposium on landslides, vol 2, New Delhi, pp 39–66, 7–11 Apr 1980

Budhu M (2007) Soil mechanics and foundations. Wiley, New York

Calvert HT, Hyde CSB (2002) Assessing landslide hazards in the Ottawa Valley using electrical and electromagnetic methods. In: Proceedings of the symposium on the application of geophysics to engineering and environmental problems, Environmental and Engineering Geophysical Society, Las Vegas

Crow HL, Hunter JA, Motazedian D (2011) Monofrequency in situ damping measurements in Ottawa area soft soils. J Soil Dyn Earthq Eng 31:1669–1677

Cummings DI, Gorrell G, Guilbault JP et al (2011) Sequence stratigraphy of a glaciated basin fill with a focus on esker sedimentation. Geol Soc Am Bull 123:1478–1496

Dera J (1992) Marine physics, vol 53, Elsevier oceanography series. Elsevier, Amsterdam/New York

Douma M, Hunter JA, Good RL (1999) A handbook of geophysical techniques for geomorphic and environmental research. In: Gilbert R (ed) Open file 3731. Geological Survey of Canada, Ottawa, pp 57–68

Eden WJ, Crawford CB (1957) Geotechnical properties of Leda Clay in the Ottawa area. In: Proceedings of the 4th international conference of the international society of soil mechanics and foundation engineering, London, pp 22–27

Gadd NR (1986) Lithofacies of Leda Clay in the Ottawa Basin of the Champlain Sea. Paper 85–21, Geological Survey of Canada, Ottawa

Hansbo S (1957) A new approach to the determination of the shear strength of clay by the fall cone test. In: Proceedings: Royal Swedish Geotechnical Institute, Stockholm, No. 14

Hunter JA, Burns RA, Good R et al (1998) Downhole seismic logging for high-resolution reflection surveying in unconsolidated overburden. Geophysics 63:1371–1384

Hunter JA, Burns RA, Good RL et al (2007) Borehole shear wave velocity measurements of Champlain Sea sediments in the Ottawa-Montreal region, GSC Open File 5345. Geological Survey of Canada, Ottawa

Hunter JA, Crow HL, Brooks GR et al (2010) Seismic site classification and site period mapping in the Ottawa area using geophysical methods. Open file 6273, Geological Survey of Canada, Ottawa

Hyde CSB, Hunter JA (1998) Formation electrical conductivity-porewater salinity relationships in quaternary sediments from two Canadian sites. In: Proceedings of the symposium on the application of geophysics to engineering and environmental problems, Environmental and Engineering Geophysical Society, Chicago, pp 499–510

Law KT, Bozozuk M (1988) Engineering problems in Leda Clay. In: Proceedings of the International Conference on Engineering Problems of Regional Soils (ICEPRS), Beijing, pp 775–792

McRostie GC, Crawford CB (2001) Canadian Geotechnical Research Site No. 1 at Gloucester. Can Geotech J 38:1134–1141

Medioli BE, Alpay S, Crow HL et al (2012) Integrated datasets from a buried valley borehole, Champlain Sea basin, Kinburn, ON. Current Research Report no. 2012–3. Geological Survey of Canada. ftp://ftp2.cits.rncan.gc.ca/pub/geott/ess_pubs/289/289597/cr_2012_03_gsc.pdf Accessed 1 Feb 2013

Pugin AJM, Hunter JA, Motazedian D et al (2007) An application of shear wave reflection landstreamer technology to soil response of earthquake shaking in an urban area, Ottawa, Ontario. In: Proceedings of the symposium on the application of geophysics to engineering and environmental problems, Environmental and Engineering Geophysical Society, Denver, pp 885–896

Pugin AJM, Pullan SE, Hunter JA (2009) Multicomponent high-resolution seismic reflection profiling. Leading Edge 28(10):1248–1261

Pugin AJ-M, Brewer K, Cartwright T, Pullan SE, Perret D, Crow H, Hunter JA (2013) Near surface S-wave seismic reflection profiling – new approaches and insights. First Break 31(2):49–60

Pullan SE, Pugin AJM, Hunter JA, Brooks GR (2011) Mapping disturbed ground using compressional and shear wave reflection sections. In: Proceedings of the symposium on the application of geophysics to engineering and environmental problems, Environmental and Engineering Geophysical Society, Charleston

Torrance JK (1975) On the role of chemistry in the development and behaviour of the sensitive marine clays of Canada and Scandinavia. Can Geotech J 12:326–335

Torrance JK (1988) Mineralogy, pore-water chemistry, and Geotechnical behaviour of Champlain Sea and related sediments. In: Gadd NR (ed) The late quaternary development of the Champlain Sea Basin, Special Paper. Geological Association of Canada, St. John's, pp 259–275

Weyl PK (1964) On the change in electrical conductance of seawater with temperature. Limnol Oceanogr 9(1):75–78. In: Limnology and oceanography. American Society of Limnology and Oceanography, Lawrence. ISSN 0024–3590

Part IV
Slope Stability, Modelling and Progressive Failure

Chapter 21
The Effect of Deformation Rate in Progressive Slope Failure

Anders Samstad Gylland

Abstract For understanding the processes acting during the triggering and evolution of a landslide in sensitive clay, it is of importance to account for the material response in the failed state. This includes the rate of strain softening and the level of residual resistance. Through a review of new experimental data it is illustrated here that these properties depend on the displacement rate in the case of shear band formation. Higher rate increases the brittleness both in terms of a steeper softening response and a lower residual level. These observations are attributed to processes of local drainage and dissipation of pore pressure gradients in the vicinity of the shear band. Viscous effects, which yield higher resistance for higher rates of deformation, are also apparent. These aspects of material failure in sensitive clay are utilized in the context of progressive slope failure in order to explain observations from case records and to quantify the impact of the rate dependent softening response on the resistance of a slope. The strong brittleness of sensitive clay at high displacement rates under shear band formation motivates a recommendation of using a design approach close to a first yield criterion if the construction situation might include such high rates.

Keywords Slope stability • Progressive failure • Sensitive clay • Strain softening • Strain localization

---

A.S. Gylland (✉)
Department of Civil and Transport Engineering, Norwegian University of Science and Technology (NTNU), Trondheim, Norway

SINTEF Building and Infrastructure, Trondheim, Norway
e-mail: anders.gylland@ntnu.no

J.-S. L'Heureux et al. (eds.), *Landslides in Sensitive Clays: From Geosciences to Risk Management*, Advances in Natural and Technological Hazards Research 36,
DOI 10.1007/978-94-007-7079-9_21, © Springer Science+Business Media Dordrecht 2014

21.1 Introduction

When dealing with construction work and slope stability considerations in deposits of marine clay in Norway, the main questions and challenges often relate to the presence of sensitive and quick clays. In principle, these clays are not more susceptible to landslides compared to non-sensitive clays. The concern relates to the large extent and severity of the potential events. Historic quick clay slides like the ones at Verdal (1893), Bekkelaget (1953) and Rissa (1978) as well as the more recent at Kattmarkveien (2009), Lyngen (2010) and Esp (2012) are examples of how a relatively small initial slide can trigger a landslide that covers large areas of nearly flat terrain. In Norway, the total land area covered by marine clay deposits is in the order of 5,000 km^2. The clays in about 20 % of this area are classified as highly sensitive or quick (Karlsrud et al. 1984). Valuable residential and industrial areas as well as farmland lie on these deposits.

For undrained conditions, sensitive clays show a reduction in shearing resistance after the peak resistance is reached. This is known as strain-softening, a phenomenon exemplified and discussed later in the paper. In this situation one can argue that a progressive failure mechanism is dominating during the triggering process and the evolution of a failure event. For further understanding of the failure processes in slopes of sensitive clays, it is then vital to describe and quantify the material behaviour both before and after material failure occurs. Relevant questions are: What is the rate of strength loss after reaching the peak undrained resistance? How does this parameter affect the resistance and failure evolution?

21.1.1 *Scandinavian Sensitive Clays*

The term sensitive clay is used here to cover a class of marine clays deposited in salt water during the last glaciation in Scandinavia that receded from the coastal regions of Norway about 12,000 years ago. The sensitivity relates to post depositional leaching of salt ions due to infiltration of meteoric water during the period of glasio-isostatic uplift (e.g. Rosenqvist 1953, 1966).

Sensitive clays are often classified according to their sensitivity $s_t = s_u/s_r$ where s_u is the peak undrained shear strength and s_r is the remoulded shear strength. Quick clay is a sub category of sensitive clays and is defined in Norway by s_r <0.5 kPa. In Norway, these clays are typically normally to moderately overconsolidated. Because of the open structure, the natural water content is above the liquid limit. A set of isotropically consolidated undrained compression (CIUc) triaxial tests from the Tiller quick clay test site of NTNU are included in Fig. 21.1 to exemplify the shear response of these clays. A more thorough classification of quick clay in both geological and geotechnical terms is available in Karlsrud et al. (1984).

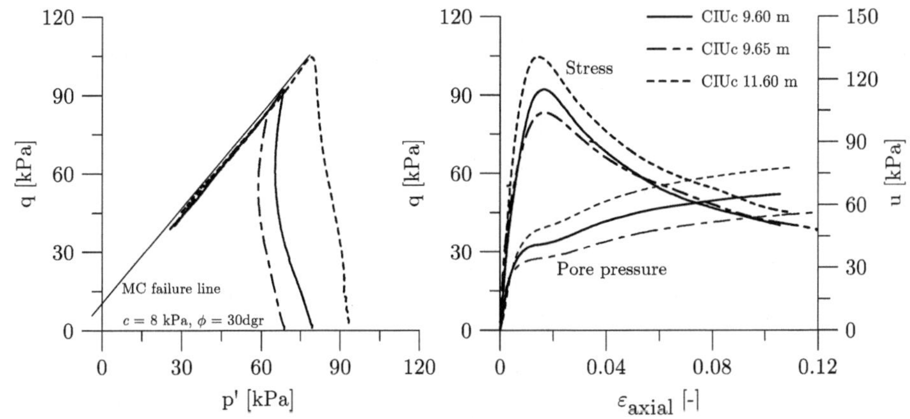

Fig. 21.1 CIUc Triaxial tests with 300 kPa backpressure at an axial strain rate of 3 %/h from the Tiller test site. Consolidated to en equivalent isotropic in-situ stress state (Adapted from Gylland 2012)

21.1.2 Strain Softening and Strain Localization

Sensitive clays display strain-softening behaviour under undrained shear as shown in Fig. 21.1. When sheared under undrained conditions there is no possibility of volume change and the contractant material behaviour induces excess pore pressures. In turn the effective stress, and thus the shear strength, is reduced as the stress state is forced to move down along the Mohr-Coulomb failure state.

As several other geomaterials, sensitive clay exhibits strain localization. This phenomenon is linked with the strain softening behaviour and is characterized by the emergence of one or more shear bands in an initially homogeneously deforming sample under monotonic loading (Fig. 21.2). Once developed, all further deformation tends to concentrate in these zones or bands (Desrues and Viggiani 2004). The underlying mechanisms are related to material instability (Hill 1958; Drucker 1951; Rudnicki and Rice 1975).

21.1.3 Progressive Slope Failure

The term progressive can be defined as a process that is advancing in space and time; proceeding step by step as a set of falling dominos. In the context of slope stability, progressive failure is linked to strain-softening material behaviour and implies non-uniform stress and strain conditions.

Progressive failure can occur in several settings, and although each slide event is special and often combines several mechanisms, some common features can be extracted in order to make classifications. Based on Karlsrud et al. (1984), four main modes are identified: downward progressive failure, retrogressive failure, upward progressive failure and lateral spread.

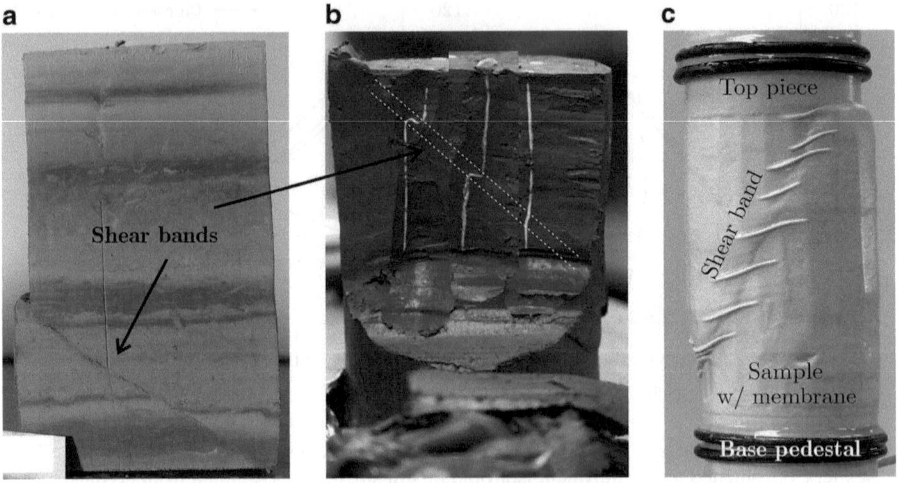

Fig. 21.2 Examples of strain localization in quick clay; (**a**) Cut surface from Ø54 mm sample, partly dried; (**b**) shear band as detected by latex-cores in a localized Ø54 mm sample; (**c**) shear band as obtained in the modified triaxial cell used in the experiments of Gylland et al. (2013)

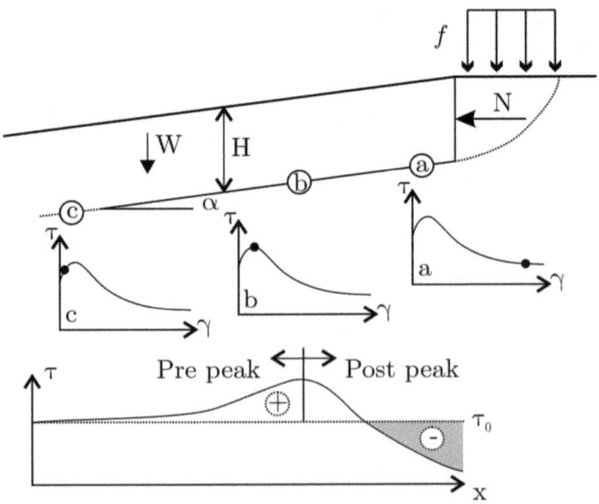

Fig. 21.3 Downward progressive failure mechanism

This paper focuses on downward progressive failure as illustrated in Fig. 21.3 based on Bernander (2011). The slope is loaded uphill and N is the horizontal force due to the distributed load, f. Three points are studied in terms of stress and strain along the potential failure surface. At the considered stage of loading, point **a** has reached the residual strength, point **b** is at the peak level and point **c** is in the stress-increasing hardening regime. This is also illustrated in the graph showing shear stress distribution along the failure surface. The plus-sign indicates increased resistance and the minus-sign indicates loss of resistance in relation to the initial shear stress level (τ_0).

In contrast to Limit Equilibrium simulations, analyses of progressive failure do not only consider the value of a peak strength, but also the effect of soil stiffness and the deformation softening response.

21.2 Material Response of Sensitive Clay in the Failed State

Gylland (2012) and Gylland et al. (2013) presented work on failure in sensitive clay which is reviewed here. The study focused on the global response in the case of shear band formation and local properties of the shear band such as structure and thickness. Two experimental setups were utilized; the shear vane (Fig. 21.4a) and a modified triaxial cell (Fig. 21.4b) where an instrumented low friction sled was installed at the sample base to allow for the formation of planar shear bands and monitoring of the localization process. The shear vane experiments were performed on site with over-coring and retrieval of the sheared soil element. In the triaxial experiments, high quality block samples and Ø75 mm steel tube piston samples were used. The applied deformation rate was varied in both setups in order to investigate potential effects related to local drainage on the peak and post-peak properties of sensitive clay. Quick clay from the Tiller test site of NTNU, located 3 km south of Trondheim, Norway, was used in all experiments.

Strain localization and the formation of distinct shear bands occurred for all rates in the modified triaxial device. An example of the shear stress-displacement response is shown in Fig. 21.5a. It is seen that increased rate of deformation gives higher peak resistance and a steeper softening response. The ultimate level appears

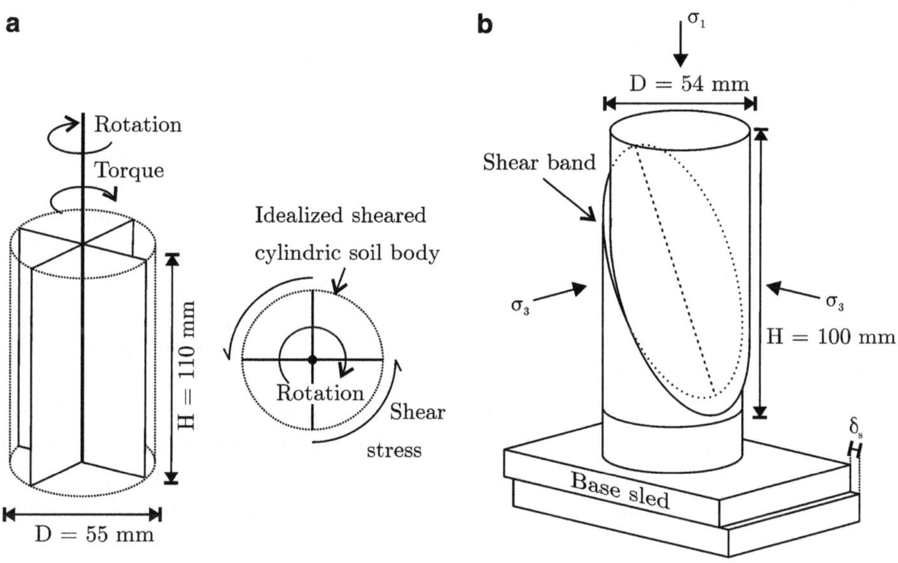

Fig. 21.4 (**a**) The shear vane; (**b**) idealized triaxial sample with a shear band (See also Fig. 21.2c)

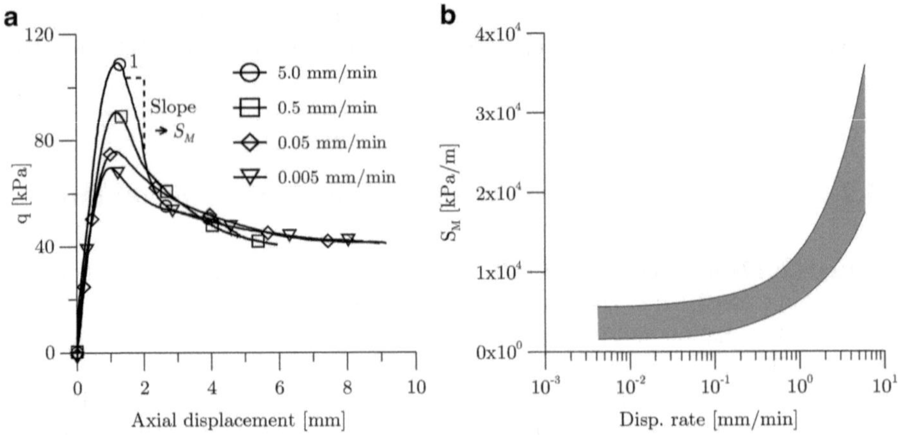

Fig. 21.5 (**a**) Experimental response from samples with a shear band in the triaxial experiments of Gylland et al. (2013). Block samples from same depth; (**b**) range of the softening parameter S_M as a function of the applied axial displacement rate in the triaxial experiments of Gylland et al. (2013)

to be independent of the applied rate. However, the shear vane experiments suggest that the residual level also is rate dependent with lower resistance for higher rate. The observations are consistent with the experimental work of Bernander and Svensk (1982) and Bernander et al. (1985) on a soft Swedish clay and the numerical work of Jostad et al. (2006) and Thakur (2011) in their analyses of strain localization in sensitive clays.

A softening parameter defined as $S_M = 0.5\,\mathrm{d}q / \mathrm{d}\delta_{sb} = \mathrm{d}\tau / \mathrm{d}\delta_{sb}$ (see Fig 21.5a) is utilized to study the rate of strain softening systematically. In the relationship, $q = 2(\sigma_1 - \sigma_3)$ and δ_{sb} is the relative displacement over the shear band. S_M then denotes the reduction in shear resistance in with increasing relative displacement over the shear band. S_M is plotted against the applied axial displacement rate in Fig. 21.5b. For the displacement rate of 5 mm/min the post-peak response is highly brittle; S_M is above 10,000 kPa/m and 50 % of the capacity is lost over a relative shear displacement less than 2 mm.

The increase in peak shear strength with increasing deformation rate can be related to intrinsic viscosity which is a well-known phenomenon in soft clays (Graham et al. 1983; Sheahan et al. 1996). The increased brittleness, both in terms of increased rate of softening and lower ultimate level, can be attributed to the effects of local drainage in vicinity of the shear band. In a situation with a shear band in sensitive clay, the band is the zone of structure collapse where excess pore pressure develops. The material outside the shear band unloads elastically with close to no pore pressure generation. Depending on the rate of pore pressure generation and the consolidation properties of the material, the situation leads to the formation of internal gradients of pore pressure for globally undrained boundary conditions.

On the local material level, high rate implies little time for dissipation to take place. The result is a rapidly increasing pore pressure level in the shear band and a correspondingly rapid loss of shear resistance. In the case of low rates, there is more time for pore water dissipation and the resulting pore pressure level in the shear band will be lower. This yields a less dramatic softening behaviour and a higher residual level.

This hypothesis regarding material failure in sensitive clay is strongly supported by the presented experimental findings and implies an indirect rate-dependency of the post-peak softening response through local consolidation of excess pore pressure in vicinity of the shear band.

The study of Gylland (2012) also included a comprehensive microscopic study of the shear bands involving the use of thin sections and polarized light microscopy as well as advanced scanning electron microscopy (Electron Micro Probe Analyzer) and micro-CT scans. These analyses show that the shear band has lower porosity compared to the outside material. This is a direct observation of soil contractancy and implies migration of water from the zone of shear to the outside. In turn, this indicates the existence of pore pressure gradients.

21.3 Slope Stability

The introductory sections illustrated how the rate of strain softening comes in as an important parameter when studying progressive slope failure. It has further been demonstrated how the softening behaviour of sensitive clay depends on the applied rate of deformation. In this section, it is studied how these two aspects influence the magnitude of external loading that can be applied to a slope and the eventual failure evolution.

21.3.1 Rate Effects on Slope Resistance

The impact of the rate of strain softening on the slope resistance is investigated in Fig. 21.6 using the finite difference method of Bernander (2011). An infinite slope based on Appendix A of Bernander (2000) has been analysed and S_M varied for otherwise constant parameters. The slope is illustrated in Fig. 21.6b. The results are interpreted in terms of the total slope resistance and the relative contribution of material in the failed state to this value. Figure. 21.6b, c show examples of shear stress distributions along the potential failure surface for two values of S_M. Material in the local post-peak range, at failure, is represented by ΔN_F and the total resistance is defined by $N_{cr} = \Delta N_H + \Delta N_F$. For $S_M = 35$ kPa/m, material in the failed state contributes about 37 % to the total whereas for $S_M = 1{,}035$ kPa/m the same contribution is below 3 %. The full set of results is shown in Fig. 21.6a where the Tuve-material parameters are interpreted based on Appendix A of Bernander (2000) and can be

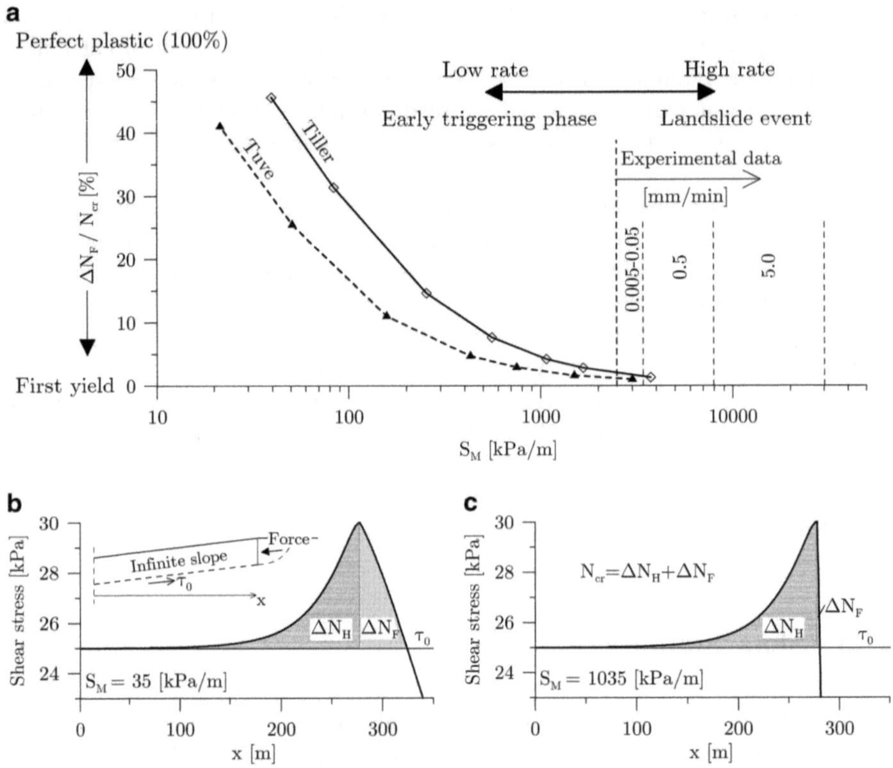

Fig. 21.6 (**a**) Contribution of material at failure relative to the total slope capacity as a function of the softening parameter S_M. S_M ranges as obtained in the triaxial experiments are included; (**b**) shear stress distribution for S_M =35 kPa/m (τ_0=25 kPa), Tuve material parameters; (**c**) shear stress distribution for S_M =1,035 kPa/m, Tuve material parameters

considered as soft in terms of material stiffness. The Tiller data is based on the experimental findings of Gylland (2012) and represents higher stiffness. A value of 0 % in Fig. 21.6a represents a first yield criterion in terms of slope resistance. 100 % reflects perfect plastic material behaviour (no strain softening). From Fig. 21.6a it is seen that the total slope resistance has a significant contribution from material at failure for materials of low brittleness in terms of S_M. However, for increasing values of S_M this contribution is strongly reduced.

The range of S_M for the involved rates of the triaxial experiments performed by Gylland et al. (2013), as shown in Fig. 21.5b, are included in Fig. 21.6a. No adjustments for intrinsic strain rate effects have been applied to the peak strength in the simulations. The work performed herein and similar findings by Lunne and Andersen (2007) suggest that viscous rate effects are negligible at rates below 0.05 %/h (the 0.005 and 0.05 mm/min tests). For the 0.5 and 5.0 mm/min rates the increase in peak strength with respect to the 0.005 mm/min rate is in the order of 20 and 30 % respectively. This is outside what was possible to model in terms of S_M and corresponding rates in the simulations of Fig. 21.6.

The experimental data is in the upper range of the simulations and indicates a resistance contribution of material in the post-peak range in the order of 3 % for the low rate tests and possibly less for the higher rates. However, the closed consolidation boundaries of the triaxial samples might over-predict S_M if compared to in-situ conditions with a larger consolidation volume (Gylland et al. 2013). Also, the direct (simple) shear mode, which can be assumed to dominate the failure process in a downward progressive landslide, does in general show less severe strain-softening compared to the active mode (Lacasse et al. 1985). The effective softening behaviour will also be less dramatic if strain localization does not occur. In total, this implies that there is a possibility for sensitive clay to show less critical softening behaviour in terms of S_M than obtained in the modified triaxial setup. Also, as discussed by Bernander and Gustaas (1984), the deformation rate can be expected to vary throughout the failure process. In the early triggering stages the rates are low and the corresponding S_M should be low. Transferring such considerations to Fig. 21.6 one can interpret a contribution in the range of 10–15 % of material in the failed state to the total resistance.

The findings of the above study can be summarized by the following statement: if high rates are expected to be present in the construction situation, utmost care should be taken in utilizing any high extent of resistance from quick clay at failure when considering the load bearing capacity of a slope.

21.3.2 *Failure Evolution*

Case records of landslides often report of a time lag between the identified triggering action and the failure event. There are also reports of animals being restless in the time preceding the actual event (Broms and Stal 1980). These observations can be attributed to a process of stepwise stress redistribution in the early triggering phases. If the load is applied rapidly, the response of the soil is an increase in resistance due to intrinsic rate-effects. But as time passes, the resistance reduces and local failure can occur. This involves local strain-softening and an increase in rate. The resistance increases and restrains the initiating failure. But again, the resistance goes down with time and the process repeats. For each local failure, the global resistance of the slope is reduced. Unless this process stabilizes, the "point of no return" is passed where the actual resistance of the slope starts to decrease. This leads to rapid acceleration and high rates.

21.4 Conclusions and Perspectives

- Recent experiments support the hypothesis that the brittleness of sensitive clay, in the case of shear band formation, depends on the acting displacement rate. Increased rate yield (1) more rapid strength loss once material failure takes place and (2) a lower ultimate level. These effects are attributed to local drainage of excess pore pressure in the vicinity of the shear band.

- In the case of downward progressive slope failure, the brittleness of the material is one of the governing aspects. As this behaviour is a function of the displacement rate, it follows that the rates involved in the triggering and evolution of a downward progressive landslide is of importance for assessing the resistance of the slope and potential failure development.
- The strong brittleness of sensitive clay at high displacement rates under shear band formation motivates a recommendation to use a design approach close to a first yield criterion if the construction situation might include such high rates.
- The simulations presented in this paper have several limitations in both the adapted methodology and slope geometry, but still the results gives an order of magnitude of the involved parameters. The accuracy of the considerations can be improved by adding data from advanced back calculations of case records.
- To further quantify the impact of progressive failure in natural slopes, it is recommended to expand the experimental database on failure in sensitive clays. It is of particular interest to study the response under direct (simple) shear conditions.

Acknowledgments Dr. Stig Bernander (Luleå University of Technology) is acknowledged for valuable comments and suggestions in reviewing this paper. The presented work is based on the PhD study of the author performed at NTNU with Adj. Prof. Dr. Hans Petter Jostad (Norwegian Geotechnical Institute) and Prof. Dr. Steinar Nordal (NTNU) as supervisors. The paper was made possible by support from the National research program 'Natural hazards: Infrastructure, Floods and Slides (NIFS)' initiated by the Norwegian Public Roads Administration, Norwegian Water Resources and Energy Directorate and Norwegian National Railways Administration.

References

Bernander S (2000) Progressive landslides in long natural slopes. MSc thesis, Luleå University of Technology

Bernander S (2011) Progressive landslides in long natural slopes: formation, potential extension and configuration of finished slides in strain-softening soils. PhD thesis, Luleå University of Technology

Bernander S, Gustaas H (1984) Considerations of in situ stresses in clay slopes with special reference to progressive failure analysis. In: Proceedings of the 9th international symposium landslides, Toronto, pp 417–430

Bernander S, Svensk I (1982) On the brittleness of clays. In: Symposium on soft clays, Linköping, pp 99–112

Bernander S, Svensk I, Holmberg G, Bernander J, Isacsson K (1985) Shear strength deformation properties of clays in direct shear tests at high strain rates. In: Proceedings of the 11th international conference soil mechanics foundation engineering, San Francisco, pp 987–990

Broms B, Stal T (1980) Landslides in sensitive clays. In: Proceedings of the international symposium landslides, vol 2, New Delhi, pp 39–66

Desrues J, Viggiani G (2004) Strain localization in sand: an overview of the experimental results obtained in Grenoble using stereophotogrammetry. Int J Num Anal Meth Geomech 28:279–321

Drucker DC (1951) A more fundamental approach to plastic stress-strain relations. In: Proceedings of the 1st US National Congress of applied mechanics, Chicago, pp 487–491

Graham J, Crooks JHA, Bell AL (1983) Time effects on the stress-strain behaviour of natural soft clays. Géotechnique 33(3):327–340
Gylland AS (2012) Material and slope failure in sensitive clays. PhD Thesis, The Norwegian University of Science and Technology
Gylland AS, Jostad HP, Nordal S (2013) Experimental study of strain-localization in sensitive clays. Accepted for publication in Acta Geotechnica. DOI: 10.1007/s11440-013-0217-8
Hill R (1958) A general theory of uniqueness and stability in elastic-plastic solids. J Mech Phys Solid 6:236–249
Jostad HP, Andresen L, Thakur V (2006) Calculation of shear band thickness in sensitive clays. In: Proceedings of the 6th European conference numerical methods in geotechnical engineering, Graz, pp 27–32
Karlsrud K, Aas G, Gregersen O (1984) Can we predict landslide hazards in soft sensitive clays? Summary of Norwegian practice and experiences. In: Proceedings of the 4th international symposium landslides, vol 1, Toronto, pp 107–130
Lacasse S, Berre T, Lefebvre G (1985) Block sampling of sensitive clays. In: Proceedings of the 11th international conference soil mechanics foundation engineering, vol 2, San Francisco, pp 887–892
Lunne T, Andersen KH (2007) Soft clay shear strength parameters for deepwater geotechnical design. In: Proceedings of the 6th international conference on offshore site investigation and geotechnics, London, pp 151–176
Rosenqvist IT (1953) Considerations on the sensitivity of Norwegian quick-clays. Géotechnique 3:195–200
Rosenqvist IT (1966) Norwegian research into the properties of quick clay: a review. Eng Geol 1(6):445–450
Rudnicki JW, Rice JR (1975) Conditions for the localization of deformation in pressure-sensitive dilatant materials. J Mech Phys Solid 23:371–394
Sheahan TC, Ladd CC, Germaine JT (1996) Rate-dependent undrained shear behavior of saturated clay. J Geotech Eng ASCE 122(2):99–108
Thakur V (2011) Numerically observed shear bands in soft sensitive clay. Geomech Geoeng 6(2):131–146

Chapter 22
Failure Mechanism of Spreads in Sensitive Clays

Ariane Locat, Serge Leroueil, and Hans Petter Jostad

Abstract Through detailed case studies from the literature it is suggested that a sensitive clay spread is formed by propagation of a failure surface in an intact slope and dislocation of the soil mass in horsts and grabens. It is proposed that the initiation and propagation of the failure surface can be explained by progressive failure mechanism. According to this failure mechanism, failure is initiated near the toe of the slope and the strain-softening stress-strain behaviour of sensitive clays is used to redistribute shear stress along the quasi-horizontal shear zone. The failure propagates inside the deposit reducing the horizontal stress. Active strength of the soil may be mobilised, explaining the dislocation of the soil mass above the shear zone in horsts and grabens. A numerical procedure is used to back calculate the 1994 spread at Sainte-Monique, Québec, Canada, involving slightly over-consolidated sensitive clay. The initiation and extent of the failure surface observed on site are explained by a soil having large brittleness during shear and large-deformation shear strength close to the remoulded shear strength of the soil.

Keywords Spread • Progressive failure • Strain softening behaviour • Large landslide • Sensitive clay • Brittleness

A. Locat (✉) • S. Leroueil
Department of Civil Engineering and Water Engineering,
Laval University, Québec City, QC, Canada
e-mail: ariane.locat@gci.ulaval.ca

H.P. Jostad
Norwegian Geotechnical Institute, Oslo, Norway
e-mail: hpj@ngi.no

J.-S. L'Heureux et al. (eds.), *Landslides in Sensitive Clays: From Geosciences to Risk Management*, Advances in Natural and Technological Hazards Research 36,
DOI 10.1007/978-94-007-7079-9_22, © Springer Science+Business Media Dordrecht 2014

22.1 Introduction

Spreads are one type of large landslides occurring in sensitive clays. According to Cruden and Varnes (1996) landslide classification, they are characterised by debris formed of blocks having horst and graben shapes (Fig. 22.1a). Those landslides cover large areas and occur very rapidly, apparently without any warning. They constitute 38 % of large landslides occurring in the province of Québec, the rest being earthflows (57 %) and unidentified large landslides (5 %; Chap. 7 by Demers et al., this volume). They represent a threat to population and infrastructures located on sensitive clays. Traditional stability analyses using limit equilibrium method give too large factors of safety for this type of landslide and justify the need for a failure mechanism explaining this particular type of landslide. The objectives of this paper is to (*i*) briefly describe characteristics of spreads from detailed case studies in literature, (*ii*) suggest a failure mechanism for spreads in sensitive clays, and (*iii*) present the application of this failure mechanism to the Sainte-Monique spread. The paper ends on conclusion from this first application of progressive failure to a spread.

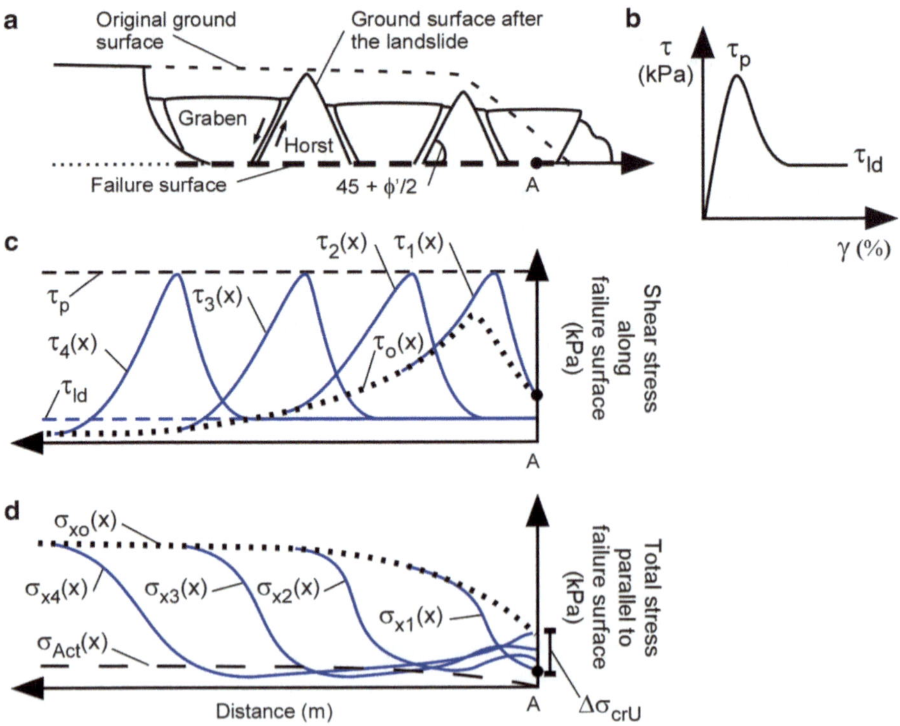

Fig. 22.1 Schematic representation of progressive failure through time as a function of distance from the slope toe (time is not explicitly represented on this Figure). Figure shows (**a**) geometry and failure surface; (**b**) soil behaviour; (**c**) shear stress along failure surface and (**d**) total stress parallel to the failure surface (Modified from Locat et al. 2011)

22.2 Spreads in Sensitive Clays

Spreads in sensitive clays are characterised by a crater filled with horsts and grabens (Fig. 22.1a). Grabens have typically flat horizontal surface and horsts are blocks having sharp tips pointing upward, creating high ridges in the crater. One of the first detailed study of a spread is the analysis by Odenstad (1951) of the spread that occurred at Sköttorp, Sweden, in 1946. Seven ridges, formed by horsts separated by grabens, were identified in the crater of this landslide. The detailed observations of these morphological structures led Odenstad (1951) to conclude that this landslide occurred by the dislocation of the soil mass in horsts and grabens in a mostly translational movement involving no rotational failure.

Similar observations were made by Carson (1979) with the detailed morphological study of the 1978 spread at Sainte-Madeleine de-Rigaud, Québec, Canada. Piles were being driven near the crest of the initial slope, for the construction of an electric pylon, when the ground movement occurred. The movement stopped about 50 m behind the position of the crest of the initial slope creating four ridges formed by horsts in the crater of the landslide. Detailed examination of the morphology showed that (*i*) grabens had flat surfaces inclined from 0 to 14° to the horizontal, (*ii*) horst had stratification close to the horizontal, and (*iii*) sides of horsts were inclined between 55 and 64° to the horizontal, which is very close to the active failure surface inclination during active failure of clay ($45+\phi'/2$, as shown in Fig. 22.1a). This indicated to Carson (1979) that horsts and grabens resulted in dislocation of the soil mass and translated and subsided in the remoulded shear zone below, which is in accordance with Odenstad (1951).

Grondin and Demers (1996) investigated the spread that occurred in 1989, in the municipality of Saint-Liguori, Québec, Canada. The movement was estimated by witnesses to last from 30 s to 5 min. Grondin and Demers (1996) state that even if many horsts were present in the scarp of the landslide, the debris were almost intact, only few cracks were present, and it was possible to walk across the crater from the front to the back scarp. Given the eye witness account of the event and observed morphology, they conclude that this landslide occurred in only one large movement and not by a retrogressive mechanism in which individual horst and graben would slide.

Demers et al. (2000) described the spread that occurred in 1996 at Saint-Boniface, Québec, Canada. This spread had a width of 950 and 150 m of retrogression. Up to ten ridges formed by horsts were observed in the crater. Some of these ridges were continuous over a length of 700 m. The failure surface was detected with CPTUs performed inside the crater and was found to be close to horizontal. The authors conclude that horsts and grabens had moved along an almost horizontal surface and that the idea normally proposed regarding retrogression by successive circular slips can no longer be considered as the main failure mode leading to large movements in the clayey soil of Eastern Canada.

The failure surface was also detected with CPTU at the spread that occurred at Saint-Barnabé-Nord, Québec, Canada, in 2005 (Locat et al. 2008). The investigation shows that the failure surface was located at the toe of the slope and continuous over 140 m in the deposit with about 4° inclination. Observation of these quasi-horizontal failure surfaces in spreads indicates that the process of dislocation is the result of a propagating horizontal failure surface and not the other way around.

More recently, the analysis of the spread that occurred in 2010, in the municipality of Saint-Jude, Québec, Canada, shows that, in this case, the failure surface developed at a depth of 2.5 m under the river bed and propagated 100 m horizontally and then went up about 10 m to propagate an additional 50 m (Locat et al. 2012b). This indicates the occurrence of two failure events during this landslide and reveals that spreads can be complex and need detailed investigations in order to understand the kinematic involved during failure.

According to these observations, and as proposed by Locat et al. (2011) and Quinn et al. (2011), it seems that spreads result from two distinct processes: (*i*) initiation and propagation of a quasi-horizontal failure surface by progressive failure mechanism and (*ii*) dislocation in horsts and grabens of the soil mass above it. As stated by Leroueil et al. (2012), it is generally considered that dislocation of the soil mass above the failure surface follows at some distance the development of the shear zone. The following section describes how progressive failure mechanism can be used to explain the formation of spreads in sensitive clays.

22.3 Failure Mechanism of Spreads: Progressive Failure

Figure 22.1 presents a schematic representation of progressive failure through time, starting with initial conditions (time 0), initiation of progressive failure (time 1), and up to the end of failure propagation (time 4). Figure 22.1a presents the initial geometry of the slope formed by river erosion, typical of most slopes in eastern Canada where spreads do occur. In addition, the final geometry of a spread, as well as the potential failure surface (assume horizontal) is shown.

In a progressive failure analysis (Skempton 1964; Bjerrum 1967; Kovacevic et al. 2007; Bernander 2011; Leroueil et al. 2012), the soil is assumed to exhibit a strain-softening stress-strain behaviour including peak shear strength (τ_p) and lower large-deformation shear strength (τ_{ld}) mobilised with increasing shear strain (γ; Fig. 22.1b).

The shear stress before failure ($\tau_o(x)$) along the potential failure surface is represented by the dotted line in Fig. 22.1c. It is near zero further up slope where the ground surface is close to the horizontal and increases where the slope is steeper. The strength properties (τ_p and τ_{ld}) are assumed constant along the potential failure surface (dashed lines in Fig. 22.1c).

Figure 22.1d presents the average total horizontal stress in the soil mass above the shear zone before failure (dotted line, $\sigma_{ox}(x)$). The magnitude of which is essentially a function of the depth of the failure surface, the coefficient of earth pressure at rest (K_o) in the deposit, and the geometry of the slope. An active strength ($\sigma_{Acr}(x)$) is assumed for this soil mass above the failure surface (dashed line in Fig. 22.1d).

Erosion or a small slide near the toe of the slope, at point *A* for example (Fig. 22.1), increases shear along the potential failure surface and can be assumed to initiate failure, i.e. to locally increase the shear stress to the shear strength. The shear stresses along the failure surface at different time *i* during failure progression ($\tau_i(x)$) are shown by full lines in Fig. 22.1c. Total shear stresses during failure progression ($\sigma_{ix}(x)$) are shown by full lines in Fig. 22.1d. Under this given disturbance at point *A*, the peak shear strength of the soil can be mobilised. Under further shear strain, the strength of the soil decreases toward its large-deformation shear strength, and failure propagates along the developing failure surface. At some point, the soil can lose enough shear strength that progressive failure is initiated. This condition is illustrated by time 1 in Fig. 22.1c, d ($\tau_1(x)$ and $\sigma_{x1}(x)$, respectively). According to Bernander (2011) and Locat et al. (2011), a critical unloading ($\Delta\sigma_{crU}$) can be defined as necessary unloading that has to be applied at point *A* for initiation of instability along the developing failure surface. This defines a critical condition beyond which failure propagates under no additional disturbance than $\Delta\sigma_{crU}$.

During failure propagation, the redistribution of shear stress through time ($\tau_2(x)$ and $\tau_3(x)$ in Fig. 22.1c) generates changes in horizontal stress in the slope ($\sigma_{x2}(x)$ and $\sigma_{x3}(x)$ Fig. 22.1d). The magnitude of the decrease in horizontal stress grows as failure propagates, thus reducing total horizontal stress further upslope. The failure propagates inside the deposit, as $\Delta\sigma_{crU}$ is distributed along the shear zone, and reaches horizontal ground where the initial shear stress ($\tau_o(x)$) is lower compared to the large-deformation shear strength of the soil (τ_{ld}; see time 4 in Fig. 22.1c, d). At some time, the total horizontal stress may decrease sufficiently to mobilise the active resistance of the soil above the failure surface ($\sigma_{Act}(x)$ in Fig. 22.1d) resulting in an active failure. In the case illustrated in Fig. 22.1, the failure propagates and results in an active failure of the slope occurring between times 3 and 4. Under this global failure process, the soil mass above the failure surface extends and dislocates into horsts and grabens that are translated downslope as they partly subside into the remoulded clay of the shear zone. Grabens and horsts dislocate along failure surfaces having angles close to $45+\phi'/2$ with the horizontal, corresponding to an active Mohr-Coulomb's failure (Fig. 22.1a). The back scarp of the slide is formed by the last graben created during the landslide. Dislocation in horsts and grabens is also certainly favored by the progression of the failure, the remoulding in the shear zone, and consequently the lack of support for the soil mass above. An important consideration here is that, in this conceptual model, the failure propagates essentially independently of the dislocation of horsts and grabens.

In order to test the conceptual model of failure mechanism for spreads, a numerical approach was developed by Locat et al. (2012a). The numerical approach was applied to the Sainte-Monique spread (Locat et al. 2013). The next section briefly describes this landslide and the soil involved, and presents the results of the application of the progressive failure analysis to this spreads.

22.4 Example of Case Study: The 1994 Sainte-Monique Spread

22.4.1 Description of the 1994 Landslide, Soil Properties and Slope Conditions Before Failure

On April 21st 1994 a landslide occurred in the municipality of Sainte-Monique, 130 km north-east of Montreal in the province of Québec, Canada. Figure 22.2 presents a cross-section of this landslide. The height of the slope was about 16.6 m, its inclination was about 24°, and the retrogression distance of this landslide was about 100 m,

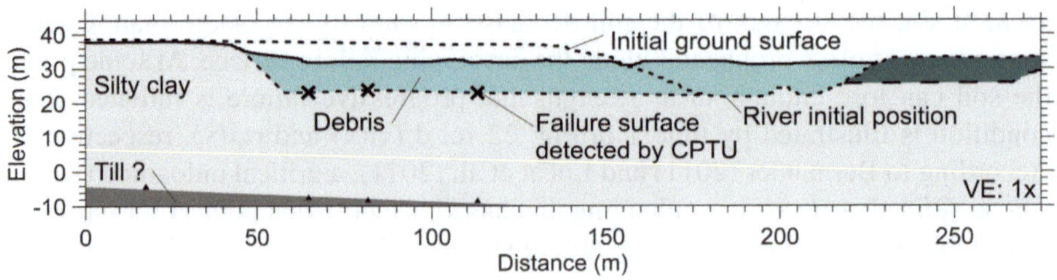

Fig. 22.2 Cross-section of the Sainte-Monique spread. Note: Horsts and grabens are not visible in the 1994 topography due to the low resolution of the topographic data (Modified from Locat et al. 2013)

taken from the initial crest to the back scarp of the landslide. The landslide debris were blocked by the opposite side of the brook and stayed mainly inside the landslide crater with very little lateral flow. The soil mass dislocated in blocks of more or less intact material having horst and graben shapes. This enabled to qualify it as a spread according to Cruden and Varnes (1996) landslide classification. A detailed field investigation of this landslide was carried out by the Ministry of Transportation of Québec (MTQ) in 2003 and 2004 and analysis of this landslide, study of the shear behaviour of the clay involved in it, as well as stability analyses were performed at Université Laval. Details of this work are presented by Locat (2012a) and Locat et al. (2013).

The soil involved in the landslide is a sensitive grey silty clay with traces of sand (% grain size < 2 μm varies between 72 and 85 %). The liquidity index varies from 1.1 to 1.4. The undrained shear strength varies from 25 to 40 kPa between the ground surface down to the toe of the slope. The remoulded shear strength is 0.7 kPa on average. The soil is slightly overconsolidated (overconsolidated ratio is around 1).

The failure surface was located comparing CPTUs performed in the debris of the landslide with CPTUs performed in intact soils. As shown in Fig. 22.2, the failure surface was located at an elevation 1.3 m higher than the toe of the slope. The failure surface was therefore almost horizontal and was located at an elevation close to the toe of the initial slope.

Direct simple shear tests (DSS) test is believed to simulate the conditions of shear occurring along the failure surface during progressive failure and gives information about the stress-strain behaviour of the soil during failure propagation. Three DSS tests were performed on samples from around 25.5 m of elevation and consolidated around 93.5 kPa (in situ vertical effective stress). Peak shear strengths of 38.3 kPa on average were measured, at about 2 % strain, and large-deformation strength, obtained at a strain of about 30 %, were about 19.6 kPa ($0.5\tau_{peak}$) on average.

Locat et al. (2013) evaluated the stability of the slope before failure in drained conditions and calculated a safety factor of 1.25 for a critical failure surface. The long-term stability of the 1992 slope, before the 1994 landslide, was therefore low.

Considering the entire failure that occurred in 1994, Locat et al. (2013) calculated a safety factor of 2.99 in an undrained analysis using the failure surface shown in Fig. 22.2. Therefore, the limit equilibrium cannot explain the entire failure that occurred at Sainte-Monique in 1994 and another failure mechanism is required.

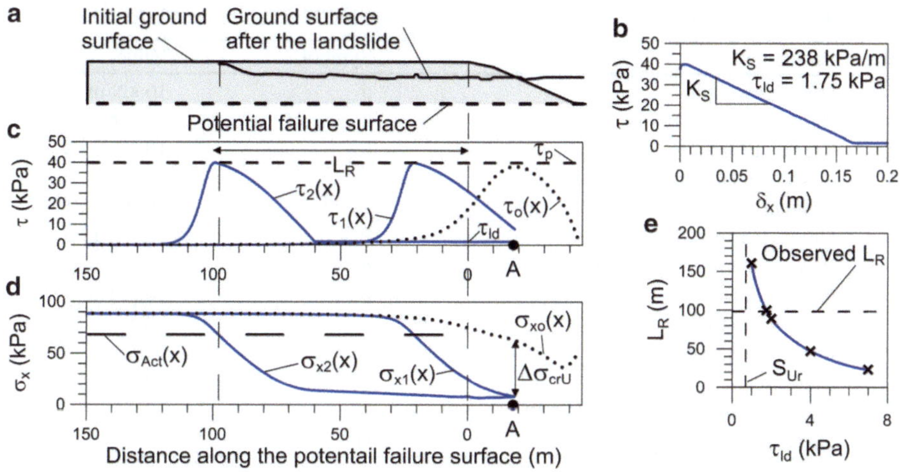

Fig. 22.3 Results of the progressive failure analysis of the Sainte-Monique spread for K_S equal to 238 kPa/m. Figure shows (**a**) geometry and failure surface; (**b**) soil behaviour used; (**c**) shear stress and (**d**) average horizontal total stress along failure surface before failure at time 0 (τ_o and σ_{xo}), at failure initiation at time 1 (τ_1 and σ_{x1}), and at the end of propagation at time 2 (τ_2 and σ_{x2}); and (**e**) effect of τ_{ld} on L_R (Modified from Locat et al. 2013)

22.4.2 Application of Progressive Failure Mechanism

22.4.2.1 Numerical Procedure

The modeling of spread is done in two steps: (*i*) calculation of initial stresses in the slope with the finite element software PLAXIS 2D 2010 (PLAXIS Manuals 2011) by unloading of river valley; and (*ii*) modeling of the initiation and propagation of progressive failure with the finite element code BIFURC (Andresen and Jostad 2004; Jostad and Andresen 2002).

The geometry used for this analysis is presented in Fig. 22.3a. Similarly to what is shown in Fig. 22.2, the failure surface is horizontal and at the elevation of the toe of the initial slope.

In step one, the initial conditions in a horizontal deposit were calculated using the K_o procedure in PLAXIS. To model the formation of the river valley, the cluster forming the river valley were deactivated in one step in drained conditions using the drained calculation option. For this analysis, the Hardening soil model was used along with the soil properties given in Table 22.1. Readers are referred to PLAXIS Manuals (2011) to learn more about this soil model and the numerical procedure. The shear stress ($\tau_o(x)$) along the potential failure surface and corresponding average total horizontal stress ($\sigma_{xo}(x)$) in the soil above the potential failure surface are output from this first calculation step.

In the second step, the initial shear stresses from the first step ($\tau_o(x)$ and $\sigma_{xo}(x)$) were input in BIFURC (Andresen and Jostad 2004; Jostad and Andresen 2002), a finite element program developed at the Norwegian Geotechnical Institute (NGI).

Table 22.1 Soil properties used for the Hardening Soil Model (drained) in PLAXIS

Parameter	Symbol	Value
Soil unit weight	ρ	16 kN/m^3
Hydraulic conductivity	$k_x = k_y$	10^{-9} m/s
Secant stiffness for primary deviatoric loading	E_{50}^{ref}	10,000 kPa
Tangent oedometer stiffness	E_{oed}^{ref}	10,000 kPa
Unloading/reloading stiffness	E_{UR}^{ref}	30,000 kPa
Power for stress dependent stiffness	m	1
Reference stress	p_{ref}	100 kPa
Poisson's ratio	ν_{ur}'	0.25
Lateral stress coefficient	K_o^{OC}	0.5
Cohesion	c'	7.5 kPa
Friction angle	ϕ'	30°
Dilatancy angle	ψ	0°

Here are the main characteristics of this program used to model the initiation and propagation of the failure surface in space:

- 3-noded truss elements were used to model the soil above the shear zone with an elastic behaviour. These elements are subjected to a change in horizontal stress during progressive failure.
- 6-noded isoparametric (no thickness) interface elements were used to represent the horizontal displacement at the top of a shear zone (δ_x). The soil model for these elements is a stress-displacement behaviour defined by peak shear strength (τ_p), large-deformation strength (τ_{ld}) and corresponding horizontal displacements for a given shear zone thickness t (δ_p and δ_{ld} where $\delta_x = \gamma$ x t).
- Advanced solution algorithm that reduces the horizontal stress at a given point, enables to determine the critical unloading triggering the failure, and distributes the change in shear stress, horizontal stress and the displacements along the mesh (Andresen and Jostad 2004; Jostad and Andresen 2002).

More details about this modeling procedure are given in Locat (2012) and Locat et al. 2012a. For the present case study, a stiffness of 13,000 kPa was given to the soil above the potential failure surface. The behaviour of the shear zone is presented in Fig. 22.3b. Peak shear strength (τ_p) of 40 kPa was used, which comes from the undrained shear strength profile measured on site at the failure surface depth. The strain at which the peak shear strength is mobilised (γ_p) was put at 2 %, which is similar to the strains at which the peak shear strength is mobilised in DSS tests. For the present analysis, a shear zone thickness (t) of 0.3 m was assumed, giving displacement at the peak shear strength of 0.006 m ($\delta_p = \gamma_p \times t = 0.02 \times 0.3$ m). The shear band thickness (t) is not easy to assume and more study on shear behaviour and shear strain localisation is necessary in order to validate this parameter.

The post-peak behaviour is defined by the large-deformation shear strength (τ_{ld}), defining the sensitivity of the soil, and its corresponding horizontal displacement (δ_{ld}), defining the brittleness of the soil. It is assumed that large-deformation shear strength should vary between the DSS shear strength mobilised at 30 % strain ($0.5\tau_p = 0.5 \times 40$ kPa = 20 kPa) and the remoulded shear strength (0.7 kPa).

The brittleness is the rate at which the strength decreases from the peak to its large-deformation shear strength under increasing shear strain. Locat et al. (2012a) introduced a shear zone softening parameter (K_S) to quantify the brittleness of the post-peak behaviour of the shear zone with stress-displacement behaviour (Fig. 22.3b). The higher K_S, the steeper is the shear behaviour after the peak shear strength and the more brittle is the shear zone. This modulus is influenced by the stress-strain behaviour of the soil and the thickness (*t*) of the shear zone. These last parameters (τ_{ld} and K_S) have been varied in the analysis in order to examine their influence on failure initiation and propagation and to see which range of value explains the observed landslide.

22.4.2.2 Results

An example of results of the analysis is presented in Fig. 22.3c, d. These figures presents the shear stress ($\tau_i(x)$) and average total horizontal stress ($\sigma_{xi}(x)$) along the potential failure surface before failure (time 0), at failure initiation (time 1) and at the end of propagation (time 2). It can be seen that before failure, the initial shear stress ($\tau_o(x)$ dotted line Fig. 22.3c) along the potential horizontal failure surface is unevenly distributed with a maximum close to the intact strength of the soil (τ_p) at point *A*. The failure was therefore initiated at this point in the numerical procedure. It can also be seen that the initial total horizontal stress ($\sigma_{xo}(x)$) is maximal in horizontal ground behind the crest and decreases toward the toe of the slope.

For the example shown in Fig. 22.3, it was supposed that τ_{ld} of 1.75 kPa was mobilised at a 55 % strain (γ_{ld}), which gives, for a shear zone thickness of 0.3 m, a shear displacement at large-deformation (δ_{ld}) of 0.165 m ($\delta_{ld} = \gamma_{ld}$ x $t = 0.55 \times 0.3$ m = 0.165 m). This gives a shear zone softening parameter (K_S) of 238 kPa/m (Fig. 22.3b). It can be seen that for a K_S of 238 kPa/m, a critical unloading ($\Delta\sigma_{crU}$) of 55 kPa is needed to initiate the progressive failure ($\sigma_{x1}(x)$ in Fig. 22.3d). With these parameters, the failure retrogresses up to a distance (L_R) of 100 m, measured from the crest of the slope to the maximum distance where the peak shear strength is mobilised ($\tau_2(x)$ and $\sigma_{x2}(x)$ in Fig. 22.3 c, d). This is similar to retrogression distance observed on site (Fig. 22.3a).

Figure 22.3e presents the effect of the large remoulded shear strength on the retrogression distance. It can be seen that the lower is the τ_{ld} the larger is L_R and that for the soil behaviour used in this analysis (Fig. 22.3b), a large-deformation shear strength of 1.75 kPa was needed for the failure to propagate of 100 m in the deposit. High brittleness is therefore needed to explain the initiation of the failure by progressive failure and low large-deformation shear strength (close to the remoulded shear strength) is necessary to explain the extent of the observed spread.

The active strength of the soil (σ_{Act}) has been plotted in Fig. 22.3d from the left boundary to the crest of the slope (where the height of the slope is constant). It can be seen that, at the end of failure propagation, and well before, the average total horizontal stress ($\sigma_{x2}(x)$) is below the active strength ($\sigma_{Act}(x)$) from the toe of the slope up to the back scarp of the 1994 landslide (Fig. 22.3c). The soil mass above the shear zone is therefore able to stretch and to break into horsts and grabens. Although the numerical procedure used in this study is focusing on the initiation and propagation of the progressive failure, explaining the observed landslide, it is possible to use the results of the analysis to discuss the formation of horsts and grabens.

22.5 Conclusion

Detailed studies of spreads show that the failure surface of this type of large landslide is initiated near the toe of the slope and propagates quasi-horizontally in the intact deposit. Dislocation of the soil mass follows the formation of the failure surface and forms horsts and grabens, typical of spreads in sensitive clays.

It is proposed that progressive failure mechanism explains the initiation and propagation of the failure surface and the dislocation of the soil mass in horsts and grabens. Progressive failure implies that the soil exhibiting a strain-softening behaviour, instability is initiated by a critical unloading near the toe of the slope and active failure of the soil mass forms horsts and grabens.

The progressive failure mechanism has been applied to the spread that occurred at Sainte-Monique, in 1994. The main conclusions from this application are:

- Before failure, the initial shear stress at the level of the failure surface was unevenly distributed with a maximum value close to the intact strength of the soil. The stability of the slope was therefore precarious.
- Stress-strain behaviour of the soil needed to back calculate the failure indicates that the soil needs a brittle and sensitive behaviour for the failure to be initiated.
- Results of the modeling indicate that large-deformation shear strength explaining the extent of the observed landslide retrogression is closer to the remoulded shear strength of the soil than the shear strength reached at 30 % strain (end of test) in DSS tests.
- The reduction of the horizontal stress during failure propagation seems large enough to explain the formation of horsts and grabens by active failure of the soil mass.

To continue the study of progressive failure mechanism and its application to spreads in sensitive clays, the failure mechanism should be tested on other spreads. Also, tests involving sufficient remoulding energy should be performed on sensitive clay to validate post peak stress-strain behaviours used in progressive failure analyses. As the brittleness of the shear zone is largely influenced by its thickness, which is generally undefined, more knowledge on shear zone thickness and shear strain localisation from laboratory and field experiments are needed to validate the shear zone thickness used in the numerical model. Finally, modeling in two dimensions would enable to properly model the dislocation of the soil mass in horsts and grabens.

Acknowledgements The authors are grateful to the Ministère des Transports du Québec for the field and laboratory data they provided and for their interest in this project. The Norwegian Geotechnical Institute is recognised for the technical and financial support they offered. The authors would also like to acknowledge the financial contributions of the The Natural Sciences and Engineering Research Council of Canada and the Fonds québécois de recherche sur la nature et les technologies. In addition, Dr. D. Perret is kindly acknowledged for the valuable review of this paper he provided.

References

Andresen L, Jostad HP (2004) Analyses of progressive failure in long natural slopes. In: Proceeding of the 9th symposium on numerical models in geomechanics – NUMOG IX, Ottawa. A.A. Balkema, Leiden, pp 603–608, 25–27 August 2004

Bernander S (2011) Progressive landslides in long natural slopes, formation, potential extension and configuration of finished slides in strain-softening soils. PhD thesis, Department of Civil and Mining Engineering, Luleå University of Technology

Bjerrum L (1967) Progressive failure in slopes in overconsolidated plastic clay and clay shales. Terzaghi Lecture. J Soil Mech Found Div ASCE 93(5):3–49

Carson MA (1979) Le glissement de Rigaud (Québec) du 3 Mai 1978: Une interprétation du mode de rupture d'après la morphologie de la cicatrice. Géographie physique et Quaternaire 33(1):63–92

Cruden DM, Varnes DJ (1996) Landslides types and processes. In: Turner AK, Schuster RL (ed) Landslides investigation and mitigation, Special report 247, Transportation, Research Board, National Research Council, National Academy Press, Washington, DC, pp 37–75

Demers D, Robitaille D, Perret D (2000) The St. Boniface landslide of April 1996: a large retrogressive landslide in sensitive clays with little flow component. In: Proceedings of the 8th international symposium on landslides, Cardiff, Thomas Telford, London, pp 447–452, 26–30 June 2000

Grondin G, Demers D (1996) The Saint-Liguori flakeslide: characterisation and remedial works. In: Balkema KS (ed) Proceedings of the 7th international symposium on landslides, Trondheim, vol 2, Rotterdam, pp 743–748, 17–21 June 1996

Jostad HP, Andresen L (2002) Capacity analysis of anisotropic and strain-softening clays. In: Proceedings of the NUMOG VIII. Rome, pp 469–474

Kovacevic N, Hight DW, Potts DM (2007) Predicting the stand-up time of temporary London Clay slopes at Terminal 5, Heathrow Airport. Géotechnique 57(1):63–74

Leroueil S, Locat A, Eberhardt E, Kovacevic N (2012) Keynote lecture: progressive failure in natural and engineering slopes. In: Proceedings of the 11th international and 2nd North American symposium on landslides, Banff, pp 31–46, 3–8 June 2012

Locat A (2012) Rupture progressive et étalements dans les argiles sensibles. PhD thesis, Département de génie civil et de génie des eaux, Université Laval

Locat A, Leroueil S, Bernander S, Demers D, Locat J, Ouehb L (2008) Study of a lateral spread failure in an eastern Canada clay deposit in relation with progressive failure: the Saint-Barnabé-Nord Slide. In: Proceedings of the 4th Canadian conference on Geohazards, Presses de l'Université Laval, pp 89–96, 20–24 May 2008

Locat A, Leroueil S, Bernander S, Demers D, Jostad HP, Ouehb L (2011) Progressive failures in Eastern Canadian and Scandinavian sensitive clays. Revue Canadienne de géotechnique 48(11):1696–1712

Locat A, Jostad HP, Leroueil S (2012a) Numerical modeling of progressive failure and its implication to spreads in sensitive clays. Accepted for publication in the Can Geotech J

Locat P, Demers D, Robitaille D, Fournier T, Noël F, Leroueil S, Locat A, Lefebvre G (2012b) The Saint-Jude landslide of May 10, 2012, Québec, Canada. In: Proceedings of the 11th international and 2nd North American symposium on landslides, Banff, pp 635–640, 3–8 June 2012

Locat A, Leroueil S, Fortin A, Demers D, Jostad HP (2013) The 1994 landslide at Sainte-Monique, Quebec: Geotechnical investigation and application of the progressive failure analysis. Submitted to the Can Geotech J

PLAXIS Manuals (2011) PLAXIS 2D 2010. PLAXIS bv. Delft

Odenstad S (1951) The landslide at Sköttorp on the Lidan River, February 2, 1946. In: Royal Swedish Institute proceedings, vol 4, pp 1–40

Quinn PE, Diederichs MS, Rowe RK, Hutchinson DJ (2011) A new model for large landslides in sensitive clay using a fracture mechanism approach. Can Geotech J 48(8):1151–1162

Skempton AW (1964) 4th Rankine lecture: long-term stability of clay slopes. Géotechnique 14(2):77–102

Chapter 23
How Well Do We Understand the Undrained Strain Softening Response in Soft Sensitive Clays?

V. Thakur, H.P. Jostad, H.A. Kornbrekke, and S.A. Degago

Abstract Many geomaterials exhibit a reduction in strength after attaining their peak strength states. This behavior, also known as strain softening, has commonly been attributed to reductions in the cohesion and/or friction angle of the materials. This holds true especially when these strength parameters are evaluated at a very large strain levels. However, for the strain levels achieved by standard undrained triaxial tests on soft sensitive clays show that the friction angle and cohesion of these materials remain almost unchanged, even when they display a significant reduction in post-peak shear strength under undrained condition. In such conditions, the increase in shear-induced pore pressure is observed to be responsible for the observed undrained softening in soft sensitive clays. This paper elaborates on this aspect based on the behavior of soft sensitive Norwegian clays.

Keywords Strain softening • Cohesion softening • Friction softening • Block samples • Triaxial tests

---

V. Thakur (✉)
Geotechnical and Landslide Division, Norwegian Public Roads Administration (NPRA), Oslo, Norway
e-mail: vikas.thakur@vegvesen.no

H.P. Jostad
Norwegian Geotechnical Institute, Oslo, Norway

H.A. Kornbrekke
Norwegian University of Science and Technology, Trondheim, Norway

S.A. Degago
Norwegian Public Roads Administration (NPRA), Oslo, Trondheim, Norway

J.-S. L'Heureux et al. (eds.), *Landslides in Sensitive Clays: From Geosciences to Risk Management*, Advances in Natural and Technological Hazards Research 36,
DOI 10.1007/978-94-007-7079-9_23, © Springer Science+Business Media Dordrecht 2014

23.1 Background

In the soft sensitive clay deposits, landslides are particularly destructive, due to the possibility of small landslides initiating a fast and extensive retrogression process, which may involve massive soil movements on the order of millions of cubic meters. All the available evidence from landslides in soft sensitive clays shows that slides are preceded by the subsequent development of a continuous sliding surface over a large area. Such type of slide mechanism in soft sensitive clays is related to their strain softening behavior under undrained condition. Strain softening is characterized as a monotonous decrease in shear strength of the materials with increasing deformation once the peak shear strength is attained. After peak strength is attained, two states can be identified for convenience in design purpose. The first one is the fully softened post-peak or post-rupture state and this state is reached with a strain level of around 10–20 % (e.g. Lacasse et al. 1985; Berre 1986; Burland 1990; Burland et al. 1997; Sandven and Sjursen 1998; Lunne et al. 1999; Sandven et al. 2004; Lunne et al. 2006; Berre et al. 2007). This state can easily be determined from standard laboratory tests, triaxial test being the common one, and it is very crucial in assessment of first time slides and progressive failures (e.g. Bjerrum 1955; Bjeruum 1961; Skempton 1964; Bishop 1967; Karlsrud et al. 1996; Bernander 2000; Leroueil 2001; Jostad and Andresen 2002; Andresen and Jostad 2007; Locat et al. 2008; Quinn et al. 2011; Mesri and Huvaj-Sarihan 2012; Jostad et al. 2013). The second state is referred to as a residual state and it is achieved at a very large strains. Ring shear tests and reversal shear box test are used to study achieve a residual state. The strength parameters at residual state are especially important in assessment of slide reactivation (e.g. Skempton 1964; Bromhead and Dixon 1986; Stark and Eid 1994; Stark and Contreas 1996).

In this work the strain softening response of Norwegian soft sensitive clays, as observed in standard triaxial tests are studied. There have been several suggestions as to what governs post-peak strength reduction (e.g. Taylor 1937; Bjeruum 1961; Skempton 1964, 1970, 1977; Eigenbrod 1972; Vaughan and Walbancke 1973; Janbu 1985; Chandler 1966; Kenny 1967; La Gatta 1970; Bishop 1971; Lo and Lee 1973; Lefebvre 1981; Lupini et al. 1981; Tavenas and Leroueil 1981; Tavenas et al. 1983; Locat and Leroueil 1988; Burland 1990; Leroueil et al. 1996; Burland et al. 1997; Bernander 2000; Fell et al. 2000; Leroueil 2001; Jostad et al. 2006; Thakur 2007, 2011; Mesri and Huvaj-Sarihan 2012; Thakur and Degago 2012; Jostad et al. 2013). In this paper, the governing mechanisms that lead to the observed softening are explored in light of existing literature and high quality laboratory tests conducted during this study. In doing so, this paper focuses on laboratory strain level up to 10–15 % where a softening stage can be fully established.

23.2 Inception of the Concept of Strain Softening

In 1937, Taylor reported that once a soil attains its peak strength, its resistance to further shear deformation in the post-peak regime is often reduced. In 1964, Skempton drew attention to this phenomenon by studying overconsolidated clays. Skempton emphasized that the values of the shear strength parameters determined from conventional tests do not necessarily bear any relation to their values at the

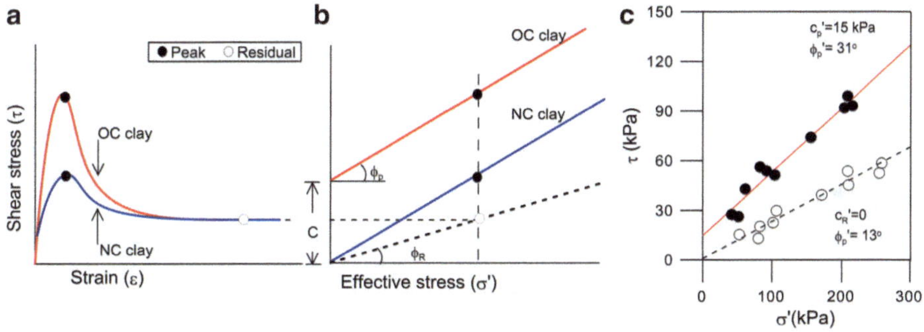

Fig. 23.1 Strain softening in clays under drained conditions: (**a**) shear stress–stain relationship, (**b**) corresponding strength envelope at the peak and residual states, and (**c**) frictional and cohesion softening observed along the shear plane (After Skempton 1964)

time of failure. He suggested that the post-peak reduction in the shear strength of such clays is associated with reductions in the friction angle (φ) and cohesion (*c*). These reductions are referred to as friction softening and cohesion softening. Skempton proposed two strength envelopes with corresponding φ and *c* values that represent the peak and residual states, as shown in Fig. 23.1a, b. Skempton assessed this model experimentally using test results on clay samples from a landslide at Walton's Wood in Staffordshire (see Fig. 23.1c).

Ever since the pioneering work by Skempton (1964) the concept of strain softening has been considered synonymous with softening of the friction angle (φ) and cohesion (c) for all types of geomaterials. Several researchers, e.g., Bjerrum (1961), Chandler (1966), Kenny (1967), La Gatta (1970), Bishop (1971), Lupini et al. (1981), Burland (1990), Viggiani et al. (1994), and Burland et al. (1997), have measured φ and c softening in a variety of geomaterials. The bonds between particles that confer the effective cohesion in these materials are usually destroyed by some minor deformations, leading to a total absence of effective cohesion. According to Burland (1990), Burland et al. (1997) and Vermeer et al. (2004), φ decreases to residual values much more slowly than c does, and thus, cohesion softening is more dramatic than friction softening. Notably, laboratory studies by Viggiani et al. (1994), Burland et al. (1997) and Hight et al. (2002) have confirmed the concept of friction and cohesion softening in stiff clays and soft rocks. Because the long-term stability (drained behavior) of overconsolidated clays is considered more critical than short-term stability, shear-induced excess pore pressure is not considered directly in strain softening.

Bjerrum (1961) suggests similarity in the failure mechanism for soft sensitive clays and for loose sands. He relates an increase in pore pressure with increasing strain, which may cause a decrease in the post-peak shear strength as a result of diminished effective stress in soft sensitive clays. Janbu (1985) suggests that the effective shear strength parameters, φ and c, of soft sensitive clays are unique and independent of whether the clays are loaded under drained or undrained conditions and whether the soil is at its peak or post-peak state. Recently Bernander (2000), Thakur et al. (2005), Jostad et al. (2006), Thakur (2007, 2011), and Gylland et al. (2012) have addressed that the post-peak shear strength reduction in soft sensitive clays is governed by the shear induced pore pressure rather than to a reduction in the

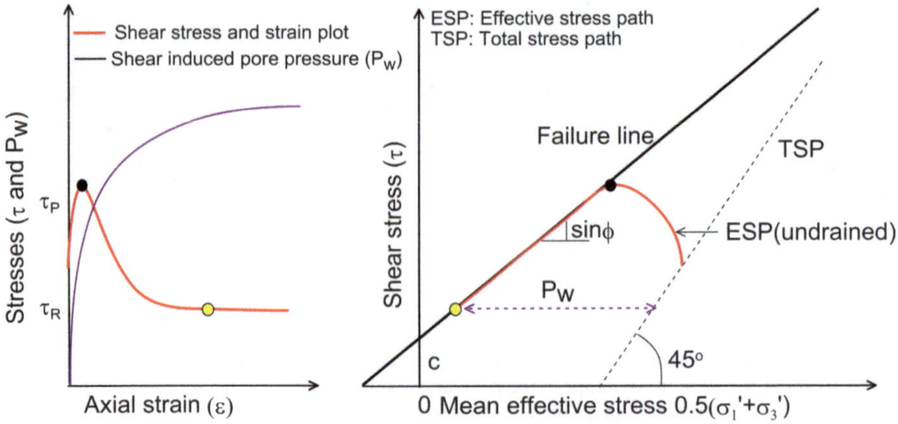

Fig. 23.2 Idealization of undrained strain softening in soft sensitive clays seen at the laboratory strain level up to 20 %

values of the strength parameters (*c* and φ). Based on these studies an idealization of undrained strain softening is presented in Fig. 23.2 where the undrained effective stress path (ESP) is following a unique failure line when subjected to undrained shearing between shear stress at the peak (τ_p) and the post-peak (τ_R) between 10 and 20 % strain. The resulted undrained strain softening is related with the increasing shear induced pore pressure (P_w) and thereby diminishing effective stress.

It must be emphasized that reductions in *c* and φ are possible even when soft sensitive clays are subjected to very large strains. This has been shown to be the case when Stark and Eid (1994) and Stark and Contreas (1996) performed a constant volume ring shear testing on low sensitive Drammen plastic clay. Stark and Contreas (1996) further suggests that a complete residual (remolded) state may occur when specimen are sheared to several hundred millimeters corresponding to several hundred percent strain inside the specimen.

Design parameters for assessment of stability analysis, especially for first time failures, are normally interpreted from laboratory strain levels up to 10–20 %. It is, therefore, important to understand the mechanisms responsible for softening observed in undrained triaxial tests and the corresponding nature of strength states. Considering this, the objective of this paper is to present and elaborate upon this hypothesis using triaxial laboratory test results.

23.3 Laboratory Tests on Rissa Clay

In connection with this study, a series of triaxial tests under undrained conditions were carried-out in a soft sensitive Norwegian clay also known as Rissa clay. Undrained strain softening in soft sensitive clays is discussed in light of the laboratory observations for the Rissa clay along with six other soft sensitive Norwegian clays reported in the literature.

Table 23.1 Engineering properties of the Rissa clay obtained from block samples

Properties	Values
Sampling depth (*H*) [m]	3–5
Clay fraction (<2 μm) [%]	40–43
Water content (*w*) [%]	28–38
Plasticity index (I_P) [%]	8–10
Liquidity index (I_L) [–]	1.8–2.0
Undisturbed undrained shear strength (c_{ui}) [kPa]	13–18
Remolded undrained shear strength (c_{ur}) [kPa]	0.69–1.45
Sensitivity (S_t) [–]	13–20
Overconsolidation ratio (*OCR*) [–]	2–3
Salt content (g/l)	2–6

23.3.1 Test Details

Rissa clay has been extensively studied with respect to assessing the stability of a natural slope in Rissa, Norway. Samples were obtained from a location near the famous Rissa landslide which occurred in 1978. The test specimens were extracted from high-quality block samples (230 mm in diameter and 450 mm high). The results of routine characterization and index tests carried out on the samples are summarized in Table 23.1. Anisotropically consolidated and undrained triaxial compression (CAUc) tests were carried out at strain rates varying from 0.1 to 4.5 %/h.

23.3.2 Test Results

The results of the CAUc tests on the Rissa clay samples are presented in Fig. 23.3. The effective stress paths are plotted in terms of the mean effective stress $\sigma'_m = 0.5\,(\sigma'_1 + \sigma'_3)$. The maximum shear stress $\tau = 0.5\,(\sigma'_1 + \sigma'_3)$ is normalized with respect to the in-situ effective overburden pressure σ'_{vo}. Shear stress and pore pressure development with increasing axial strain are plotted in Fig. 23.3. Table 23.2 presents a summary of the CAUc test results; along with the corresponding overconsolidation ratio (*OCR*) values obtained from continuous rate of strain (CRS) oedometer tests.

23.3.3 Evaluation of the Test Results

The test results shown in Fig. 23.3 are evaluated by considering different aspects of strain softening. The effective stress paths presented in Fig. 23.3a show no reduction in the effective shear strength parameters (c and φ) between the peak shear stress (τ_p) and the post-peak shear stress (τ_R), which is similar to what is shown in Fig. 23.2. This observation holds true for all of the tests, despite a significant range in the deformation rates (0.1–4.5 %/h). Moreover, the strength envelopes for the Rissa clay shown in Fig. 23.3a are seen to depend on the strain rates. This dependency has been noted for other sensitive clays as well (Leroueil 2001).

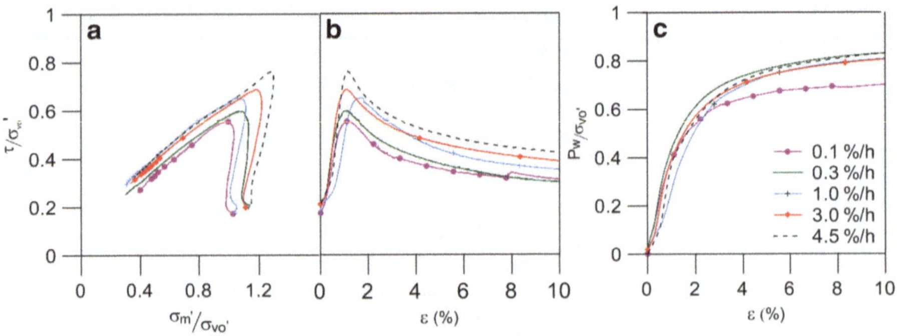

Fig. 23.3 CAUc test results for the block samples of Rissa clay

Table 23.2 Results of CAUc tests on Rissa clay[a]

Rate [%/h]	Depth [m]	σ_{vo}' [kPa]	ε_v [%]	ε_p [%]	OCR[–]	c_p [kPa]	$tan\phi_p$ [–]
0.1	4.07	46.60	1.86	1.21	2.22	5.1	0.51
0.3	3.64	42.80	1.73	1.22	2.29	5.8	0.48
1.0	4.07	46.60	1.40	1.51	2.22	9.6	0.46
3.0	3.75	43.80	1.41	1.07	2.26	9.7	0.43
4.5	3.64	42.80	1.57	0.96	2.29	10	0.41

[a]ε_v = Volume change during the consolidation phase, ε_p = axial strain at the failure

The shear stress–shear strain characteristics of the Rissa clay samples shown in Fig. 23.3b display a significant strain softening response. The peak and post-peak undrained shear strengths, normalized with respect to the in-situ effective overburden pressure, vary between 0.55–0.75 and 0.3–0.42, respectively. The post-peak undrained shear strengths were measured at a 10 % axial strain where the undrained shear strength is reduced by half.

The pore pressure response, normalized with respect to σ'_{vo} in Fig. 23.3c, shows a continuous increase in pore pressure with increasing axial strain (ε), which may be attributed to contraction of the clay. The pore pressure responses shown in Fig. 23.3c are also nearly identical for all of the samples tested, despite the range of deformation rates (0.1–4.5 %/h). These test results for the Rissa clay support the hypothesis that undrained softening in soft sensitive clays is not related to a reduction in the values of the strength parameters (c and φ) but rather to the generation of shear-induced excess pore pressure.

23.4 Discussion

23.4.1 Sample Disturbance

The behavior during shear tests of soft sensitive clays is affected by sample disturbance. The effect of sample disturbance on strain softening is shown in Figs. 23.4a–c, using laboratory test results obtained for Onsøy clay for test specimens obtained

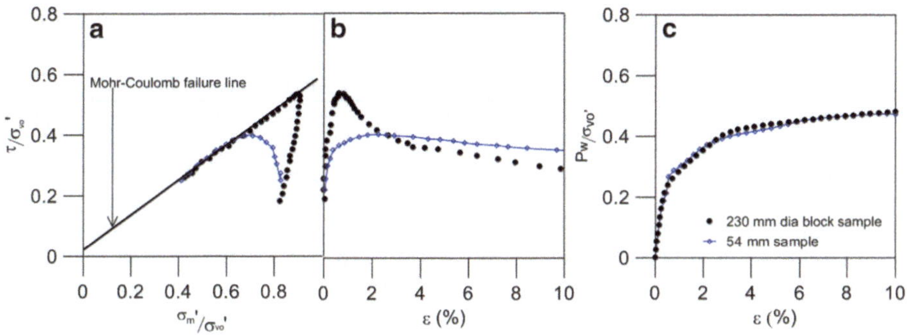

Fig. 23.4 Effect of sample disturbance on CAUc test results for Onsøy clay (Adapted from Lunne et al. 2008)

from high quality 230 mm (block) samples and conventional 54 mm (tube) samples. The results indicate that the test specimens obtained from the block samples exhibited significantly higher peak shear stress (τ_p) values at lower failure strains (ε_p) than those obtained using a conventional tube sampler. Therefore, it is possible that sample disturbance influences τ_p and ε_p. Nonetheless, the strength parameters *c* and φ are unique for each test, regardless of sample quality.

The pore pressure plots for the Onsøy clay (Lunne et al. 2008) are nearly identical for the high-quality 230 mm (block) sample and the conventional 54 mm sample. However, the normalized P_w corresponding to τ_p for the block sample is 0.25, which is much lower than the value of 0.4 observed for the 54-mm sample. The normalized pore pressure generated between τ_p and the end of the test is approximately 0.25 for the block sample and approximately 0.1 for the 54-mm sample. These results suggest that the degree of strain softening that occurs is positively related to the amount of shear-induced pore pressure generated between the peak and post-peak states.

23.4.2 Strain Softening in Soft Sensitive Norwegian Clays

A series of CAUc tests was conducted on high-quality block samples of a variety of soft sensitive Norwegian clays collected from more than seven different locations. A summary of the test results is presented in Table 23.3. The test results for six different sensitive clays are shown in Figs. 23.5a–f. Each of the clays exhibits a unique Mohr–Coulomb failure line between the peak and the post-peak shear strength, regardless of the sample diameter (54, 75 or 230 mm). No softening in the cohesion or the friction angle was observed at laboratory strain levels for any of these clays. These results confirm that the hypothesis of pore-pressure-induced strain softening proposed by Bjerrum (1961) and Janbu (1985) explains the post-peak strength reduction of soft sensitive clays in the range of 10–20 % strain better than the concept of *c* and φ softening.

The test observations presented in Fig 23.5 lead to a question if the concept of cohesion and friction softening is applicable to soft sensitive clays at triaxial strain

Table 23.3 Characterization of selected soft sensitive Norwegian clays as reported in the literature

Block samples	Depth [m]	OCR [–]	w [%]	w_L [%]	I_p [%]	S_t [–]	c_p [kPa]	ϕ_P [°]	References
Drammen clay	5–10	1.5	51–54	51–54	27–29	7	5	29	Lunne and Lacasse (1999)
Emmerstad clay	4–6	3–4.5	37–44	28–33	6–11	>100	6	35	Lacasse et al. (1985)
Kløfta clay	6–12	1.9	28–32	37–34	12	100	5	28	Long et al. (2009)
Lierstranda clay	12–16	1.4	35	38	17	10	3	30	Lunne et al. (1997)
Onsøy clay	7–15	1.4–2	61–65	61–74	34–41	3–7	2	33.5	Lunne et al. (2008)
Rissa clay	3–5	2.2	28–38	17–19	9.1	16.1	9.6	27.3	Kornbrekke (2012)
Tiller clay	8–13	1.4	35–45	25–30	3–8	>200	8	30	Gylland et al. (2012)

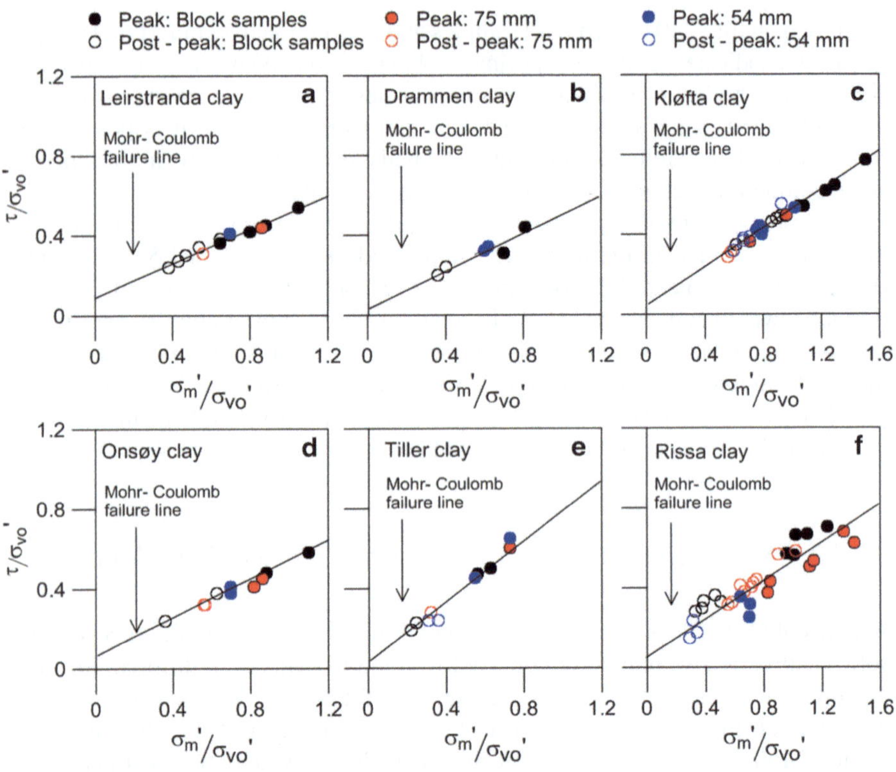

Fig. 23.5 Peak strengths and post-peak shear strengths corresponding to10 % axial strain obtained from CAUc tests on six different soft sensitive clays from Norway

levels. To answer this question, one must examine the origin of the concept of strain softening by Skempton (1964). This concept was originally proposed in the context of drained conditions for overconsolidated clays, also referred to as long-term conditions. Skempton (1964) proposed the concept of cohesion and friction softening with respect to characterization of the long-term stability of overconsolidated clays under drained conditions. Failure in overconsolidated clays may occur considerably later than the initiation of strain softening because cohesion and friction softening are progressive phenomena. However, in the case of soft sensitive clays, a small load applied under undrained conditions can build up a significant amount of shear-induced pore pressure that may initiate an instantaneous progressive failure in the material. The amount of strain softening therefore depends on the negative dilatancy of the material, because the contraction of the material will dictate how much excess pore pressure can develop during undrained shearing (Jostad et al. 2006; Thakur 2007). Coupling between the plastic normal strain ($\varepsilon_n{}^p$) and the plastic shear strain (γ^p) should be taken into account in quantifying pore-pressure-induced strain softening. This coupling is related to the dilatancy angle ψ:

$$tan\psi = -\frac{\Delta\varepsilon_n^p}{\Delta\gamma^p} \tag{23.1}$$

The laboratory test results show that the relationship between pore pressure increase and axial strain is hyperbolic in nature. For simplicity, this relationship can be described using a hyperbolic function:

$$P_w = -\frac{2A_p P_a}{\left(A_p + \frac{P_a}{\gamma^p}\right)} \tag{23.2}$$

where A_p and P_a are parameters used to fit a hyperbolic curve to the measured shear-induced pore pressure. The dilatancy angle is subsequently calculated to obtain the shear-induced pore pressure under perfectly undrained conditions. Jostad et al. (2006) suggests that undrained strain softening in soft sensitive clays can be modeled based on Eq. (23.2), along with suitable failure criteria and a hardening function. Cohesion and friction softening is responsible for strength loss in soft sensitive clays at large strain levels as shown by Stark and Eid (1994) and Stark and Contreas (1996) for Drammen plastic clay.

23.4.3 Strain Softening in Highly Overconsolidated Sensitive Norwegian Clays

Sensitive clays could exhibit a high overconsolidation states. To examine strain softening in overconsolidated sensitive clays, Lacasse et al. (1985) conducted CAUc testing of Emmerstad clay. The test results are presented in Fig. 23.6. The test specimen had an OCR $\approx$ 4.5 and S_t > 100. Under undrained triaxial shear test, the overconsolidated Emmerstad clay exhibited cohesion softening. This indicates that highly overconsolidated Norwegian sensitive clays may exhibit friction and cohesion softening under undrained conditions even at laboratory strain levels. However,

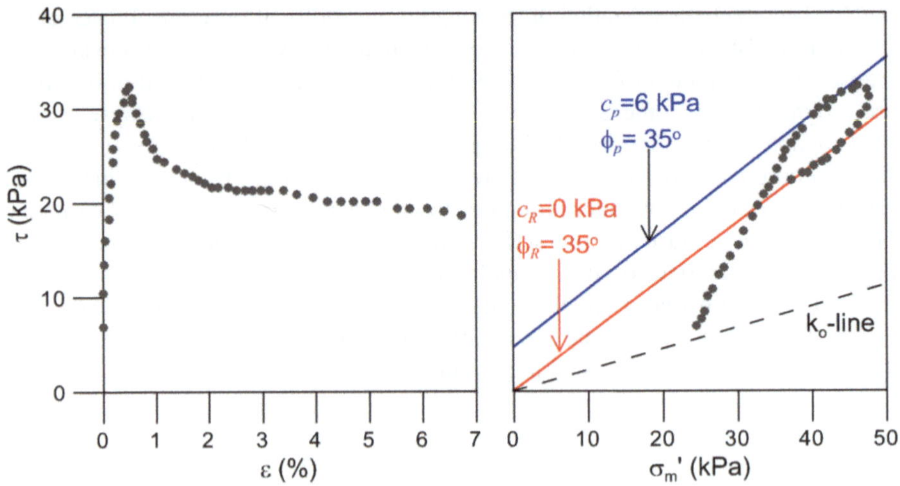

Fig. 23.6 CAU_C test results for Emmerstad clay (Adapted from Lacasse et al. 1985)

according to Terzaghi and Peck (1967), testing under drained conditions is more suitable for materials with OCR > 4–8. From this perspective, drained tests would be more reliable in describing strain softening in soils such as the Emmerstad clay.

23.5 Concluding Remarks

Assessment of stability of slopes for first time failure is a commonly encountered task in standard geotechnical routine. Design parameters for such analysis are normally interpreted from laboratory strain levels up to 10–20 %. For this strain range, undrained triaxial tests on soft sensitive clays show a marked reduction in strength. In this work, undrained strain softening response at a laboratory strain level, corresponding to standard triaxial test, is studied for soft sensitive Norwegian clays. Interpretation of CAUc test results of six different soft sensitive clays clearly indicated that undrained soft sensitive clays exhibit strain softening due to shear-induced pore pressure and not due to cohesion and friction softening at laboratory strain level up to 10–20 %. It is also observed that the concept of cohesion and friction softening is more applicable to overconsolidated clays than to soft sensitive clays at laboratory strain levels under undrained conditions. However, for a very large strain levels cohesion and friction softening may also be observed for soft sensitive clays and this must be addressed in the future. Further research is also needed to determine whether the hypothesis of strain softening due to increased shear-induced pore pressure can be generalized to all types of soft sensitive clays.

Acknowledgments National research program "Natural hazards: Infrastructure for floods and slides (NIFS)", by the Norwegian Public Roads Authority, Norwegian Water Resources and

Energy Directorate and Norwegian National Railways Administration. The authors wish to acknowledge Prof. Rolf Sandven, Multiconsult AS, for the review and the constructive feedback on this paper.

References

Andresen L, Jostad HP (2007) Numerical modelling of failure mechanism in sensitive soft clays – application to offshore geohazards. Offshore Technology Conference, Huston, pp 1–7

Bernander S (2000) Progressive landslides in long natural slopes. Licentiate thesis, Luleå

Berre T (1986) Effect of sampling disturbance on undrained static triaxial tests on plastic Drammen clay, NGI report 56001-3. Norwegian Geotechnical Institute, Oslo

Berre T, Lunne T, Andersen KA et al (2007) Potential improvements of design parameters by taking block samples of soft marine Norwegian clays. Can Geotech J 44(6):698–716

Bishop AW (1967) Progressive failure – with special reference to the mechanism causing it. In: Proceedings of the geotechnical conference on shear strength properties of natural soils and rocks, Oslo, pp 142–150

Bishop AW (1971) The influence of progressive failure on the choice of the method of stability analysis. Géotechnique 21:168–172

Bjerrum L (1955) Stability of natural slopes in quick clay. Géotechnique 5(1):101–119

Bjerrum L (1961) The effective shear strength parameters of sensitive clays. In: Proceedings of the 5th international conference soil mechanics foundation engineering, Paris, pp 23–28

Bromhead EN, Dixon N (1986) The field residual strength of London clay and its correlation with laboratory measurements, especially ring shear tests. Géotechnique 36(3):449–452

Burland JB (1990) 30th Rankin lecture: on the compressibility and shear strength of natural clays. Géotechnique 40(3):329–378

Burland JB, Rampello S, Georgiannou VN et al (1997) A laboratory study of the strength of four stiff clays. Géotechnique 47(3):491–514

Chandler RJ (1966) The measurement of residual strength in triaxial compression. Géotechnique 16(3):181–186

Eigenbrod KD (1972) Progressive failure in overconsolidated clays and mudstones. PhD thesis, University of Alberta

Fell R, Hungr O, Leroueil S et al (2000) Keynote lecture – geotechnical engineering of the stability of natural slopes, and cuts and fills in soil. In: GeoEng 2000, vol 1, Melbourne, pp 21–120

Gylland A, Jostad HP, Nordal S (2012) Failure geometry around a shear vane in sensitive clays. In: Proceedings of the 16th Nordic geotechnical meeting, vol 1, Copenhagen, pp 103–110

Hight DW, McMillan F, Powell JM et al (2002) Some characteristics of London Clay. In: Proceedings of the international symposium on characterisation and engineering properties of natural soils, vol 2, pp 851–908

Janbu N (1985) Soil models in offshore engineering. Géotechnique 35(3):241–281

Jostad HP Andresen L (2002) Bearing capacity analysis of anisotropic and strain softening clays. In: Proceedings of the 8th international symposium NUMOG VIII, pp 469–474

Jostad HP, Andresen L, Thakur V (2006) Calculation of shear band thickness in sensitive clays. In: 6th numerical methods in geotechnical engineering, vol 1, pp 27–32

Jostad HP, Fornes P, Thakur V (2013) Effect of strain softening in design of fills in gently inclined areas with soft sensitive clays. In: Advances in natural and technological hazards research. Springer, Quebec city, ISSN: 1878–9897 (Print) 2213–6959 (Online)

Karlsrud K, Lunne T, Brattlien K (1996) Improved CPTU interpretation based on block samples. In: Proceedings of the 12th Nordic Geotechnical conference, vol 1, Iceland, pp 195–201

Kenny TC (1967) Slide behavior and shear resistance of a quick clay determined from a study of the landslide at Selnes, Norway. In: Proceedings of the Geotechnical conference, vol 1, Oslo, pp 57–64

Kornbrekke HA (2012) Stability evaluation of Rissa clay slopes based on block samples. Master's thesis, Norwegian University of Science and Technology

La Gatta DP (1970) Residual strength of clay and clay-shales by rotation shear tests, vol 86, Harvard soil mechanics series. Harvard University Press, Cambridge, MA

Lacasse S, Berre T, Lefebvre T (1985) Block sampling of sensitive clays. In: International conference of soil mechanics and foundation engineering, vol 2, pp 887–892

Lefebvre G (1981) Strength and slope stability in Canadian soft clay deposits. Can Geotech J 18:420–442

Leroueil S (2001) Natural slopes and cuts: movement and failure mechanisms. Géotechnique 51(3):197–243

Leroueil S, Locat J, Vaunat J et al (1996) Geotechnical characterisation of slope movements. In: Proceedings of the 7th international symposium on landslides, Trondheim, pp 53–74

Lo KY, Lee CF (1973) Analysis of progressive failure in clay slopes. In: Proceedings of the 8th international conference soil mechanics foundation engineering, vol 1, Moscow, pp 251–258

Locat J, Leroueil S (1988) Physicochemical and mechanical characteristics of recent Saguenay Fjord sediments. Can Geotech J 25:382–388

Locat A, Leroueil S, Bernander S et al (2008) Study of a lateral spread failure in an eastern Canada clay deposit in relation with progressive failure: the Saint-Barnabé-Nord slide. In: Proceedings of the 4th Canadian conference on geohazards: from causes to management, Quebec, pp 89–96

Long M, El Hadj N, Hagberg K (2009) Quality of conventional fixed piston samples of Norwegian soft clay. ASCE J Geotech Geoenviron Eng 135(2):185–198

Lunne T, Lacasse S (1999) Geotechnical characteristics of low plasticity Drammen clay. In: Tsuchida T, Nakase A (eds) International symposium on characterisation of soft marine clays. Balkema, Rotterdam, pp 33–56

Lunne T, Berre T, Strandvik S (1997) Sample disturbance effects in soft low plastic Norwegian clay. In: Proceedings of the conference on recent developments in soil and pavement mechanics, pp 81–102

Lunne T, Berre T, Strandvik S (1999) Sample disturbance effects in soft low plastic Norwegian clay, Publication 204. Norwegian Geotechnical Institute, Oslo

Lunne T, Berre T, Andersen KH (2006) Effects of sample disturbance and consolidation procedures on measured shear strength of soft marine Norwegian clays. Can Geotech J 43:726–750

Lunne T, Berre T, Andersen KH et al (2008) Effects of sample disturbance on consolidation behaviour of soft marine Norwegian clays. In: International conference on site characterization, vol 3, Taipei, pp 1471–1479

Lupini JF, Sinner AE, Vaughan PR (1981) The drained residual strength of cohesive soils. Géotechnique 31(2):181–213

Mesri G, Huvaj-Sarihan N (2012) Residual shear strength measured by laboratory tests and mobilized in landslides. ASCE J Geotech Geoenviron Eng 138(5):585–593

Quinn PE, Diederichs MS, Rowe K et al (2011) A new model for large landslides in sensitive clay using a fracture mechanics approach. Can Geotech J 48(8):1151–1162

Sandven R, Sjursen M (1998) Sample disturbance results from investigations in an overconsolidated clay, ISC'98, vol 1, Atlanta, pp 409–418

Sandven R, Ørbech T, Lunne T (2004) Sample disturbance in highly sensitive clay. In: Proceedings of the 2nd international conference on geotechnical and geophysical site characterization, pp 1861–1868

Skempton AW (1964) Long-term stability of clay slopes. Géotechnique 14(2):75–102

Skempton AW (1970) First-time slides in overconsolidated clays. Géotechnique 20(3):320–324

Skempton AW (1977) Slope stability of cuttings in Brown London clay. In: Proceedings of the ninth international conference on soil mechanics foundation engineering, vol 3, Tokyo, pp 261–270

Stark TD, Contreas IA (1996) Constant volume ring shear apparatus. GTJ 19(1):3–11

Stark TD, Eid HT (1994) Drained residual strength of cohesive soils. J Geol Eng 120(5):856–871

Tavenas F, Leroueil S (1981) Creep and failure in slopes in clays. Can Geotech J 18(1):106–120

Tavenas F, Flon P, Lerouil S et al (1983) Remolding energy and risk of slide retrogression in sensitive clays. In: Proceedings of the symposium slopes on soft clays, Linköping, pp 423–454

Taylor DW (1937) Stability of earth slopes. J Boston Soc Civil Eng 24(3):197–246

Terzaghi K, Peck R (1967) Soil mechanics in engineering practice. Willy, New York

Thakur V, Nordal S, Jostad HP, Andresen L (2005) Study on pore water pressure dissipation during shear banding in sensitive clays. In: 11th international conference on computer methods and advances in geomechanics, IACMAG, Turino, Italy, 4, pp 289–296

Thakur V (2007) Strain localization in sensitive soft clays. PhD thesis, Norwegian University of Science and Technology

Thakur V (2011) Numerically observed shear bands in soft sensitive clays. Int J Geomech Geoeng 6(2):131–146

Thakur V, Degago SA (2012) Quickness of sensitive clays. Géotechnique Lett 2(3):87–95

Vaughan PR, Walbancke HJ (1973) Pore pressure changes and the delayed failure of cutting slopes in overconsolidated clay. Géotechnique 23(4):531–539

Vermeer PA, Vogler U, Septanika EG et al (2004) Modelling strong discontinuities in geotechnical problems. In: Proceedings of the 2nd international symposium on continuous and discontinuous modelling of Cohesive Frictional Materials CDM 2004, Stuttgart, pp 381–394

Viggiani G, Finno R, Harris W (1994) Experimental observations of strain localisation in plane strain compression of a stiff clay. In: Proceedings of the 3rd international workshop on Localis and Bifurc theory for soils and rocks, pp 189–198

Chapter 24
Effect of Strain-Softening in Design of Fills on Gently Inclined Areas with Soft Sensitive Clays

Hans Petter Jostad, P. Fornes, and V. Thakur

Abstract The effect of strain softening in geotechnical design of fills in areas with soft sensitive clays is studied by a large number of finite element analyses. The reduction in undrained shear strength with increasing shear strain after the peak value will reduce the maximum fill height before failure compared with a perfectly plastic material. The finite element program Plaxis together with the material model NGI-ADPSoft are used in this study. A non-local strain formulation is used in NGI-ADPSoft to overcome the crucial problem of mesh dependent results typical for this type of problems. The effect of brittleness is then fully controlled by input parameters. The material properties are taken from NGI's database of undrained shear test results on high quality block samples. The effect of strain-softening is quantified by establishing a scaling factor $F_{softening}$ that gives the ratio between the calculated capacity without and with the effect of softening. The purpose is then that the peak undrained shear strength of sensitive clays simply can be divided by this factor before used in conventional limit equilibrium analyses with a strain independent (perfectly plastic) assumption to indirectly account for the effect of brittleness.

Keywords Quick clay • Strain softening • FEM • Numerical simulation • Progressive failure • Slope stability

H.P. Jostad (✉) • P. Fornes
Norwegian Geotechnical Institute (NGI), Oslo, Norway
e-mail: hpj@ngi.no

V. Thakur
Geotechnical and Landslide Division, Norwegian Public Roads Administration (NPRA), Oslo, Norway

J.-S. L'Heureux et al. (eds.), *Landslides in Sensitive Clays: From Geosciences to Risk Management*, Advances in Natural and Technological Hazards Research 36,
DOI 10.1007/978-94-007-7079-9_24, © Springer Science+Business Media Dordrecht 2014

24.1 Introduction

In geotechnical design of construction works in areas with sensitive and quick clays it is important in some way to take into account that the capacity or safety factor will be overestimated if it is calculated using undrained peak shear strengths. Based on NGI's experience from back-calculation of failures in soft sensitive clays the effect of softening is found to typically reduce the capacity by 10–15 %. NGI's current practice is therefore to reduce the peak undrained shear strength by a factor, varying between 0 and 15 %, dependent on the sample quality, shear direction (active, passive or simple shear), sensitivity, clay content, water content and remolded shear strength. This takes for instance into account that the results from high quality samples show significantly more brittle response than results from disturbed samples (e.g. Lunne et al. 2006).

The Norwegian national amendment to Eurocode 7 (2008) states in table NA.A.4 that the material coefficient should be increased when progressive failure is possible. However, it does not give guidelines of how much it should be increased.

According to the Norwegian Public Road Administration's Handbook 016 the effect of brittle post peak behavior is accounted for by increasing the required material coefficient from 1.4 to 1.5 or from 1.5 to 1.6 (about 7 % increase in γ_M) depending on the consequence of failure. This increase in γ_M is independent on whether the peak shear strength is based on results from high quality samples or not.

In the guidelines for evaluating overall stability of areas with quick clays by Norwegian Water Resources and Energy Directorate (NVE) more detailed descriptions of how to account for brittle soil behavior are given. The required material coefficient $\gamma_M > 1.4$ for construction works that reduces the safety factor of the area directly involved by the construction work. This means that the calculated average mobilized shear stress τ along the most critical slip surface in the area with reduced stability due to the construction work must be less than the characteristic shear strength divided by the material coefficient $\gamma_M = 1.4$. In addition a simplified strain compatibility approach should be utilized, where the same shear strain is assumed along the critical failure surface. This means that the shear strain that gives the maximum average undrained shear strength from anisotropically consolidated compression (CAUC), extension (CAUE) and direct simple shear (DSS) tests for each soil layer should be used to establish the characteristic anisotropic undrained shear strengths. For very brittle post peak behavior it is also recommended that the peak undrained compression shear strength is reduced by 15 %. Brittle materials are here defined as materials with a sensitivity of more than 15 and remolded undrained shear strength of less than 2 kPa.

The purpose of this paper is to evaluate the effect of strain softening based on finite element analyses. These results may then later be used to come up with new guidelines for how to perform safe design in areas with sensitive and quick clay.

24.2 Characteristic Behaviour of Soft Sensitive Clay

24.2.1 *Strain Softening*

The characteristic behavior of Norwegian sensitive and quick clays is that they due to the unstable structure of the skeleton exhibit contractant behaviour under increasing shear deformation. Under drained condition the soil will basically reduce the pore volume during shearing. However, since the clay generally is fully saturated and has a very low permeability, this tendency of pore volume reduction will gradually transfer effective stresses from the soil skeleton to excess pore pressure. At a critical shear strain, which is clay type, strain path and rate dependent, this reduction in effective stresses will result in significantly reduced shear strength. This means that the undrained non-linear shear stress-shear strain relationship (τ-γ) first increases to its maximum value before it gradually decreases towards its residual (fully remolded) value at very large shear strains. This is illustrated in Fig. 24.1 for an idealized undrained DSS test.

This non-linear shear stress-strain relationship depends on the stress or strain path. The behavior is generally totally different during undrained triaxial compression tests (increasing axial strain) and undrained triaxial extension tests (decreasing axial strain). The stress-strain relationship is also strain rate dependent.

In an effective stress based model this softening behavior is for instance accounted for by including destructuration into the elasto plastic formulation as in S-CLAY1 (Wheeler et al. 2003).

24.2.2 *Shear Band Thickness*

One consequence of the strain softening behavior is that the material becomes locally unstable and the continuing deformation within this area is not unique. When the peak undrained shear strength is mobilized, further shear deformation reduces the shear strength (as shown in Fig. 24.2). When considering an arbitrary

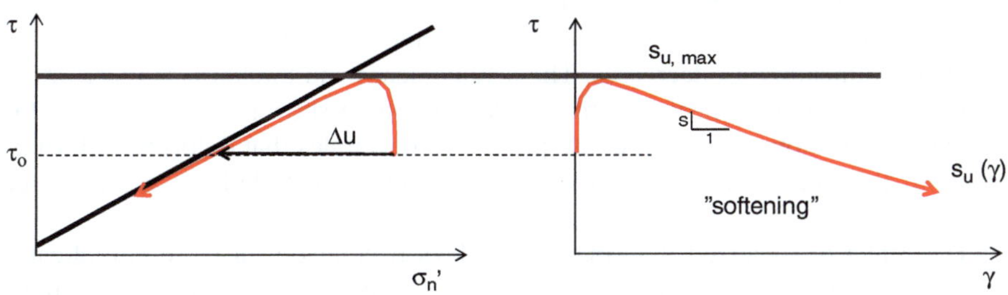

Fig. 24.1 Illustration of softening behavior during undrained DSS shear deformation of a sensitive soft clay. Effective stress path ($\tau - \sigma_n'$) to the left and shear strain γ versus shear stress τ to the right

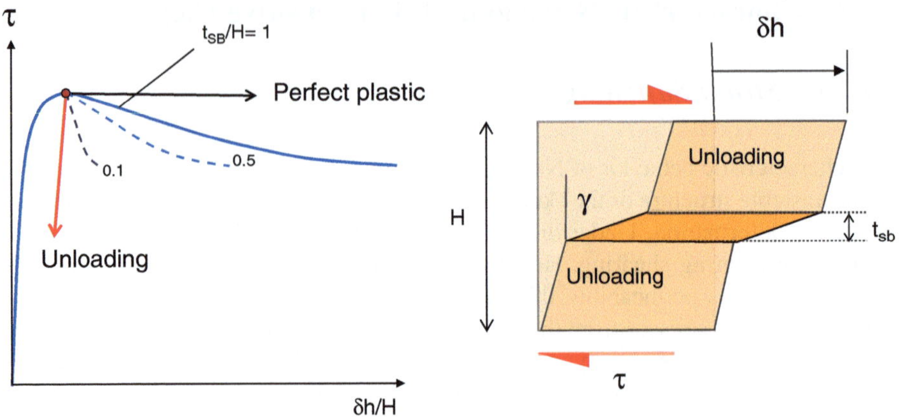

Fig. 24.2 Illustration of the effect of shear band thickness t_{sb} on the resulting softening behavior of an element with height H

soil element with height H under uniform horizontal shear deformation, the resulting strain softening curve may be as shown by the solid blue curve. However, if the deformation is controlled by a prescribed horizontal displacement at the top of the element, the resulting strain may be distributed non-uniformly over the height H. This means combination of further plastic deformation within a shear band with thickness t_{sb} and elastic unloading within the material outside. The total horizontal displacement δ_h at the top of the element is then given by the failure shear strain γ_f at the peak undrained shear strength s_u (where the deformation is still uniform), the elastic shear strain γ_e given by the secant unloading shear modulus G_o, the slope of the softening curve S (here taken as a secant value from s_u to the actual shear stress level $s_u(\gamma)$):

$$\delta h = \delta h_f - \delta h_e + \delta h_s = \gamma_f \bullet H - \frac{s_u - s_u(\gamma)}{G_o} \bullet (H - t_{sb}) + \frac{s_u - s_u(\gamma)}{S} \bullet t_{sb}$$

Based on the equation above and Fig. 24.2, the actual shear stress versus average shear strain response of the soil element becomes more and more brittle as the shear band thickness t_{sb} decreases. The actual shear band thickness is therefore as important as the slope S of the strain-softening curve shown in Fig. 24.1. At the same time, the slope S has to be interpreted from samples that deforms uniformly, to avoid that the effect of localized deformation is included twice (Andresen et al. 2002).

The effect of localized deformation in clays becomes even more complicated when considering effects as local drainage (e.g. Jostad et al. 2006 and Gylland et al. in press).

24.3 Numerical Method

24.3.1 Finite Element Method

The Finite Element Method is well suited to be used in capacity analyses of clays, because the method automatically satisfies:

- Compatibility between displacements and strains
- Advanced non-linear stress-strain relationships described by a material model
- Equilibrium between driving forces and resulting internal stresses
- Boundary conditions

However, the finite element program needs a robust solution algorithm in order to follow the non-linear response of the soil. In addition, for strain softening materials some kind of regularization method (as described in Sect. 24.3.2) is required in order to avoid numerical problems and mesh dependency.

24.3.2 A Brief Description of NGI-ADPSoft

NGI-ADPSoft (Grimstad and Jostad 2010; Grimstad et al. 2012) is a special user-defined version of NGIADP (Grimstad et al. 2010), which is implemented as a standard material model into Plaxis (www.plaxis.nl). The model is a total stress based model that describes the undrained anisotropic behavior of clays during shear deformation. The model uses input data from anisotropically consolidated undrained triaxial compression tests, constant volume direct simple shear tests and undrained triaxial extension tests. The input data is peak undrained shear strength s_{up} and residual shear strength s_{ur} and corresponding shear strains γ_p and γ_r along the shear stress – shear strain curves from these tests, as shown in Fig. 24.3.

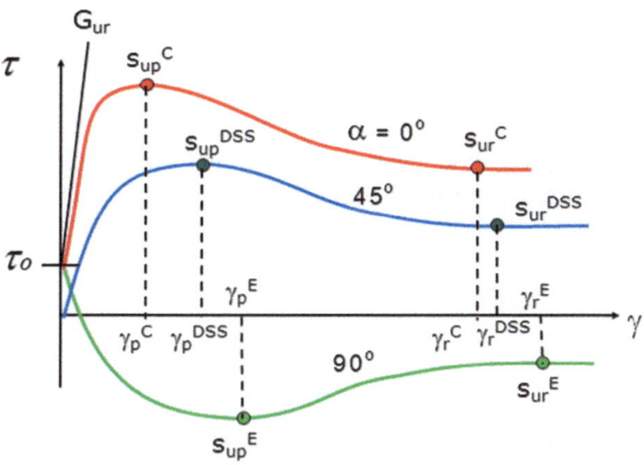

Fig. 24.3 Input parameters used in the material model NGI-ADPSoft

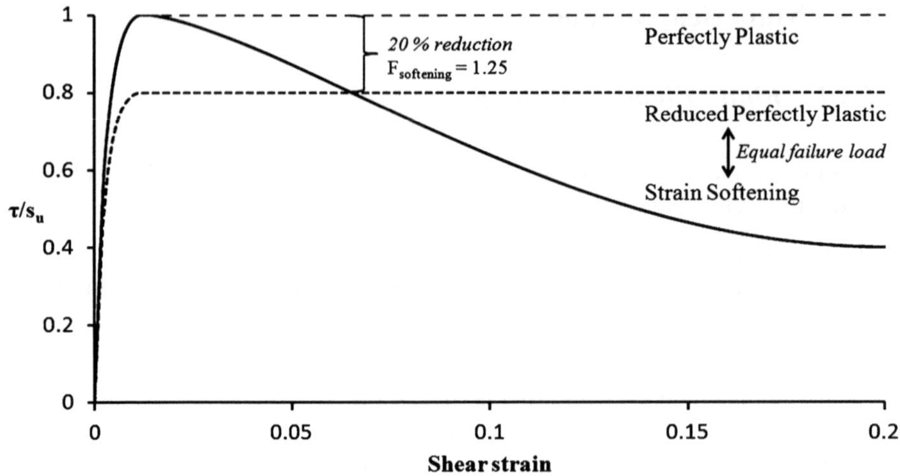

Fig. 24.4 Strain-softening behavior and perfectly plastic behavior that gives the same bearing capacity

In addition, these curves starts from the initial shear stress τ_0 with the slope given by the initial shear modulus G_0. In the following the plane strain "A=Active" behavior is assumed to be equal to the triaxial compression behavior, and correspondingly, the plane strain "P=Passive" behavior is assumed to be equal to the triaxial extension behavior.

As shown in Fig. 24.4, the model describes the non-linear behavior up to the peak undrained shear strength followed by a gradual reduction in the shear strength towards the residual shear strength. By interpolation between the three input curves the model describes the general 3D anisotropic behavior of the clay that depends on the actual orientation of the maximum shear deformation.

The so-called "non-local strain" (Eringen 1981) is used to regularize the mesh dependency in the softening regime. The plastic strain rate in a stress point (Gaussian integration point) is then replaced by an integrated weighted average strain within a specified zone around the point. This means that the plastic strain and resulting reduction in the shear strength during softening becomes mesh independent. Instead, it is controlled by input parameters. In this way the shear band thickness and resulting brittleness (as discussed in Sect. 24.2.2) are controlled by the input data. More information about the non-local strain formulation used in NGI-ADPSoft is given in (Grimstad and Jostad 2010; Grimstad et al. 2012).

24.4 Analyses

The following steps are used to establish the correction factor $F_{softening}$:

1. Running an analysis with softening behavior up to the maximum load (unstable condition where progressive failure is initiated)

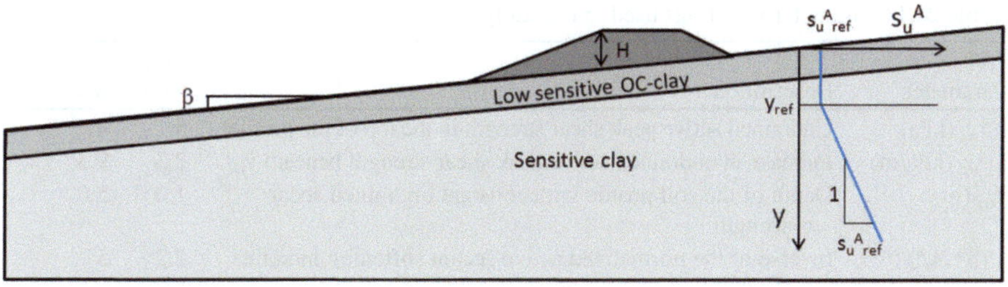

Fig. 24.5 Road embankment

2. Increasing the loads up to the same load level now with the same material without softening behavior
3. Reducing the peak undrained shear strength in a c/phi-reduction phase until global failure (starting from end of Step 2)

The calculated scaling factor in Step 3 gives the correction factor $F_{softening}$ as shown in Fig. 24.4. A large number of different combinations of input parameters are analyzed.

24.4.1 Road Embankment in Gently Inclined Sensitive Clay

The correction factor $F_{softening}$ is calculated in a large parametric study of a typical road embankment in a gently inclined slope of very sensitive clay as shown Fig. 24.5. The width of the embankment at the top is 8 m. The slope inclination β is varied between 4 and 8°. The soil condition consists of normally consolidated highly sensitive (quick) clay beneath layer of a low sensitive over-consolidated clay.

24.4.2 Shear Strength Profile

The shear strength profile is defined by an idealized variation of the undrained active strength with depth. The variation is defined by a constant shear strength $s_u{}^A{}_{ref}$ down to the depth y_{ref} below the terrain, followed by a linearly increasing shear strength $s_u{}^A{}_{inc}$ with depth (as shown in Fig. 24.5). These three parameters are then varied, as shown in Table 24.1, in order to cover typical soil conditions in areas with sensitive marine clay in Norway.

The ranges of the parameter values are based on the ratio of undrained shear strength to the effective vertical stress, which is adjusted for typical variation in the ground water level.

Table 24.1 Parameter variations used in the study

Parameter	Description	Min value	Max value
$s_u{}^A{}_{ref}$ (kPa)	Undrained active peak shear strength in the top of the profile	5.1	42
$s_u{}^A{}_{inc}$ (kPa/m)	Increase in undrained active peak shear strength beneath y_{ref}	2.3	9.3
y_{ref} (m)	Depth of the soil profile with constant undrained shear strength	1.0	5.0
$1/(S^A/s_u{}^A)$ (%)	Inverse of the normalized active secant softening modulus, $\gamma^A{}_r - \gamma^A{}_f$	2.2	57
t_{sb} (m)	Effective shear band thickness	0.1	1.2
β (°)	Terrain inclination	4	8

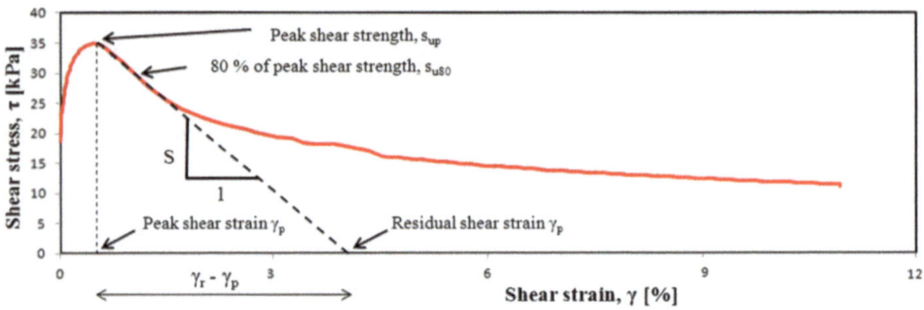

Fig. 24.6 Idealization of the slope of the softening curve

24.4.3 Softening Behavior

As discussed in Sect. 24.2.2, the brittleness is governed by the slope S of the shear stress-shear strain curve in the softening regime, the shear band thickness t_{sb}, the effects of local drainage from the shear band to the surrounding materials and the strain rate. As for instance shown by Bernander and Svensk (1981) and Gylland et al. (in press), the softening behavior becomes more brittle upon increased deformation rate.

Based on similar analyses with softening by e.g. Jostad and Grimstad (2011), it is found that the critical strain at which the progressive failure starts to develop is generally far from being large enough to remold the clay. It is only the initial part of the softening curve found from standard undrained shear tests that is of interest. The remaining part of the curve towards the fully remolded strength governs the post-failure movements of the slide after this critical condition. In order to simulate the entire failure process of retrogression and run-out the entire softening curve toward the remolded shear strength must be modeled properly.

In order to make it easier to evaluate the effect of the slope of the softening curve, an idealized linear softening curve as shown by the dashed line in Fig. 24.6 is used in the analyses. Based on results from a large number of undrained shear tests on block samples of sensitive clays (NGI 2011), the slope $S = s_{up}/(\gamma_r - \gamma_p)$ is varied as given in Table 24.1.

24.4.4 Dependent Input Parameters

In order to avoid unrealistic combinations of input parameters, the interdependency of some of the input parameters are taken into account. For instance, the peak undrained DSS strength is given as a factor times the average of the peak active and passive undrained shear strength, $s_u^{DSS} = \zeta \cdot (s_u^P + s_u^A)/2$, where the factor ζ, is varied between 0.8 and 1.2 based on results in NGI's block sample data base. Similar relationships are used to establish the entire anisotropic stress-strain relationship.

24.4.5 Parametric Study

The most important parameter variations are given in Table 24.1. The value of shear band thickness and maximum brittleness is limited by the element mesh used and the variation of t_{sb} is therefore chosen accordingly. In addition, all the other soil parameters are varied within realistic ranges based on data in NGI's block sample data base.

24.4.6 Monte Carlo Method

For rather complex problems with many input parameters, as the one considered in this study, it is unrealistic to simulate all possible combinations of these parameters. The effect of different combinations is therefore studied using the Monte Carlo method (Metropolis and Ulam 1949). Furthermore, in order to increase the number of the most realistic combinations some of the parameters are assigned a probability distribution. These distributions are established based on data in NGI's block sample data base.

24.4.7 Plaxis Model

The finite element model used in the analyses is 200 m wide, see Fig. 24.7. In order to reduce the computer time the mesh is only refined locally beneath the embankment where the softening zone initially develops. The 15 node triangular element is used in the mesh. The nodes at the bottom boundary are totally fixed, while distributed stresses in equilibrium with the soil weight are applied at the lateral boundaries.

24.5 Results

Figure 24.8 shows the results from all (ca. 500) successful analyses. About 100 cases ended up no results due to unrealistic combinations of input parameters. The reduction factor $F_{softening}$ is plotted against the failure load (in kPa) obtained from the

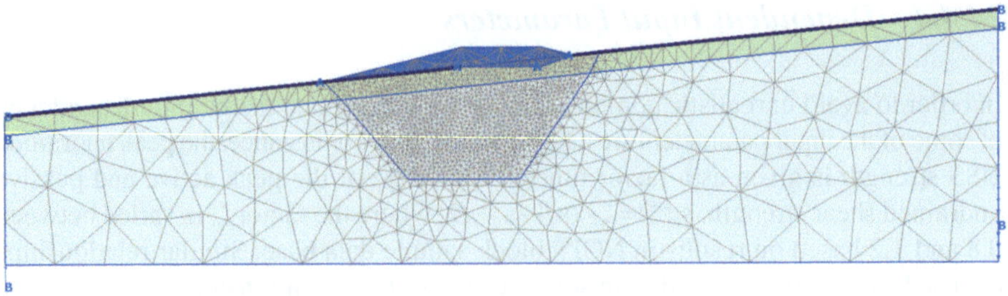

Fig. 24.7 Plaxis-model with slope inclination β=6°

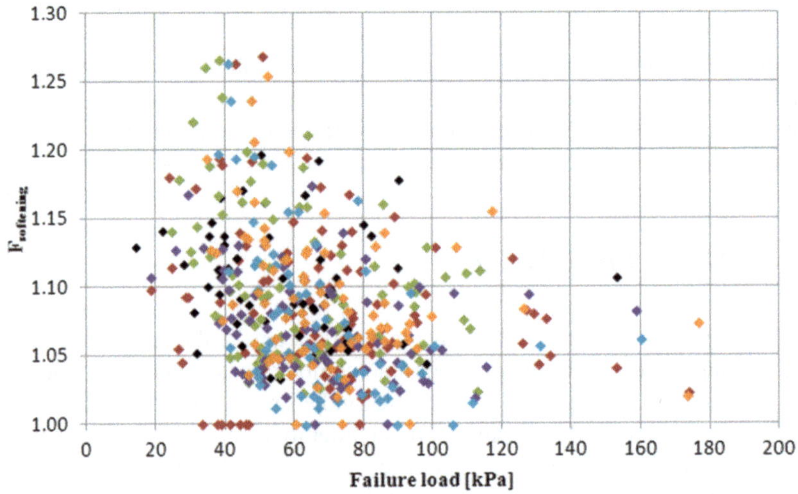

Fig. 24.8 Calculated effect of softening (correction factor $F_{softening}$) versus failure load (in kPa) from all successful analyses

analyses with softening. The failure load gives the maximum embankment height (multiplied by the unit weight of the fill material) that can be constructed. The calculated average effect of softening is 9 % ($F_{softing}=1.09$). For only 2.5 % of the cases is $F_{softing}>1.20$, and for 12 % of the cases is $F_{softing}>1.15$.

From Fig. 24.8 it is seen that the effect of softening is increasing with decreasing failure load. This means that the effect of softening is increasing with increasing initial shear mobilization (increasing slope inclination, reducing K_o and reducing peak undrained shear strength) together with increasing brittleness (steeper initial post peak stress-strain curve and reduced shear band thickness).

It is also found that calculated "first yield" (fill height where the peak shear strength is first mobilized at any place in the sensitive clay) underestimates the calculated average capacity with softening by 6 %. This means that if "first yield" is used as a safety criterion, the material factor is further increased compared to the effect of softening by 6 % in average.

Furthermore, the results suggest that strain compatibility is not valid for this type of failure. At the peak load, the undrained shear strength is not mobilized in the passive zone.

24.6 Conclusions

The effect of strain softening in geotechnical design of construction works in areas with soft sensitive clays is studied by a large number of finite element analyses. The effect of strain-softening is quantified by establishing a scaling factor $F_{softening}$ that gives the ratio between the calculated capacity without and with the effect of softening. The calculated average ratio is 9 %, while only 2.5 % of the cases have a ratio that is larger than 1.2. The peak undrained shear strength of sensitive clays can then be divided by $F_{softening}$ before used in conventional limit equilibrium analyses with a strain independent (perfectly plastic) assumption to indirectly account for the effect of post peak reduction in the undrained shear strength.

The authors will continue this research by including more complex geometries and long sliding surfaces to see the effect of softening on the overall stability of natural slopes. In addition the results will be linked to calculated probability of failure.

Acknowledgments The authors would like to thank the Norwegian Public Road Administration and the Norwegian Water Resources and Energy Directorate that partly have funded this research work. A special thank is also given to Dr. Grimstad at Oslo and Akershus University College of Applied Sciences who has developed the automated calculation procedure in Plaxis which was essential for performing this large parametric study. Finally, the authors wish to acknowledge Prof. Steinar Nordal at NTNU and Prof. Minna Karstunen at Chalmers University of Technology for the review and constructive feedback on this paper.

References

Andresen A, Jostad, HP, Høeg K (2002) Numerical procedure for assessing the capacity of anisotropic and strain-softening clay. In: Proceedings of the 5th world congress on computational mechanics – WCCMV, Wien

Bernander S, Svensk I (1981) On the brittleness of clays. In: Symposium on soft clays, Linkøping, pp 99–112

Eringen AC (1981) On non-local plasticity. Int J Eng Sci 19:1461–1474

Eurocode 7: Geotechnical design (2008) NS-EN 1997-1:2004+NA:2008. ICS 91.010.30; 93.020

Grimstad G, Jostad HP (2010) Undrained capacity analyses of sensitive clays using the nonlocal strain approach In: 9th HSTAM international congress on mechanics vardoulakis mini-symposia, Limassol, 12–14 Juli 2010

Grimstad G, Andresen L, Jostad HP (2012) NGI-ADP: anisotropic shear strength model for clay. Int J Numer Anal Method Geomech 36:483–497. doi:10.1002/nag.1016

Gylland AS, Jostad HP, Nordal S (in press) Experimental study of strain localization in sensitive clay. Accepted by Acta Geotechnica

Jostad HP, Grimstad G (2011) Comparison of distribution functions for the nonlocal strain approach. In: Proceedings of the 2nd international symposium on computational geomechanics, Cavtat-Dubrovnik, Kroatia

Jostad HP, Andresen L, Thakur V (2006) Calculation of shear band thickness in sensitive clays. In: Proceedings of the sixth European conference on numerical methods in geotechnical engineering, Graz, pp 27–32

Lunne T, Berre T, Andersen KH, Strandvik S, Sjursen M (2006) Effects of sample disturbance and consolidation procedures on measured shear strength of soft marine Norwegian clays. Can Geotech J 43:726–750, Erratum, 44: 111

Metropolis N, Ulam S (1949) The Monte Carlo method. J Am Stat Assoc 44(247):335–341

Norwegian Geotechnical Institute (NGI) (2011) Database for tests on high quality block samples on clay. Summary of compressibility, strength and deformation parameters in relation to index properties. Report 20051014–1, 2011, Oslo

Wheeler SJ, Näätänen A, Karstunen M, Lojander M (2003) An anisotropic elasto-plastic model for soft clays. Can Geotech J 40(2):403–418

Chapter 25
Effective Stress Based Stability Analysis of Normally Consolidated Clays

Tim Länsivaara, Juho Mansikkamäki, and Ville Lehtonen

Abstract This article discusses undrained effective stress stability analyses on normally consolidated soft clay. A key issue in such analyses is that the yield induced pore pressure is properly accounted for. Yielding of natural clays is affected amongst other factors on the shape and size of an anisotropic yield surface and the deformation properties of the clay, including creep. These then have a large effect also on the undrained capacity of the soil. The implementation of yield induced pore pressure into stability analyses conducted by both limit equilibrium method (LEM) and finite element method (FEM) is presented. Calculation examples are given for a full scale embankment failure experiment using various methods to include the yield induced pore pressure. It is shown that it is possible to account for such effects even in simple LEM analysis to obtain a realistic failure load, while in advanced FEM analysis it is possible to also account for time effects.

Keywords Slope stability • Clays • Undrained analysis • Effective stress • FEM • LEM

25.1 Introduction

Undrained stability analyses are commonly conducted as total stress analyses using undrained shear strength. For example in Finnish engineering practice embankment stability is typically calculated using undrained shear strength determined by vane

T. Länsivaara (✉) • J. Mansikkamäki • V. Lehtonen
Tampere University of Technology, Tampere, Finland
e-mail: tim.lansivaara@tut.fi

J.-S. L'Heureux et al. (eds.), *Landslides in Sensitive Clays: From Geosciences to Risk Management*, Advances in Natural and Technological Hazards Research 36,
DOI 10.1007/978-94-007-7079-9_25, © Springer Science+Business Media Dordrecht 2014

shear testing. However, this often results in unrealistically low calculated factors of safety. The issue is largely due to unreliable measurements (Ukonjärvi 2013, yet to be published) and will be left out of the scope of this paper.

Undrained effective stress analysis offers an alternative for total stress stability analyses. The uncertainty regarding strength parameters in effective stress analyses is arguably smaller than in total stress analyses based on undrained strength form the vane test. However, the key issue in any undrained effective stress analysis is how to reliably estimate the excess pore pressure. The in-situ pore pressure can surely be measured and the excess pore pressure from a load can be estimated quite reliably, but this alone is not enough. In soft clays the yielding of the clay induces pore pressure that also needs to be accounted for.

In this paper yield induced excess pore pressure will firstly be discussed. Methods to account for it in undrained effective stress stability analysis will be presented both for limit equilibrium method (LEM) and finite element method (FEM).

25.2 Yield Induced Pore Pressure

In undrained loading soft clays may exhibit higher pore pressures than the actual increase of stress. Such pore pressure build up might also occur in the shear and passive zones in stability analysis, without any direct external loading on these zones. Herein this phenomenon is addressed as yield induced pore pressure. It can be explained in the following manner.

When clay yields on the wet side of the yield surface, the breakdown of the clay structure causes a tendency for large volumetric compression. As the permeability is low the water cannot dissipate, and undrained conditions occur with buildup of excess pore pressure. In terms of soil modeling, the tendency to large positive plastic volumetric straining needs to be compensated by negative elastic straining, which is possible only by a reduction in effective stresses. The amount of excess pore pressure that is built up depends on both the properties of the clay and the rate of loading. It is well known that creep, or viscosity, influences much on soft clay behavior. In oedometer tests higher loading rates result in higher stresses for the same amount of compression, resulting in rate dependency of the preconsolidation pressure. Similar behavior can also be found in any test done to determine shear strength of the clay. Higher rates give higher undrained shear strength values. While the undrained shear strength increases with rate of loading, the effective strength parameters are not influenced as shown e.g. by Janbu and Senneset (1995). The reason for this is that higher loading rates result also in a smaller buildup of pore pressure. To summarize, lower loading rates are associated with smaller yield boundaries and development of more creep straining. In undrained conditions this results in higher yield induced pore pressure than for a higher loading rate, affecting thus also on the undrained capacity. A principal illustration of the phenomena is given in Fig. 25.1.

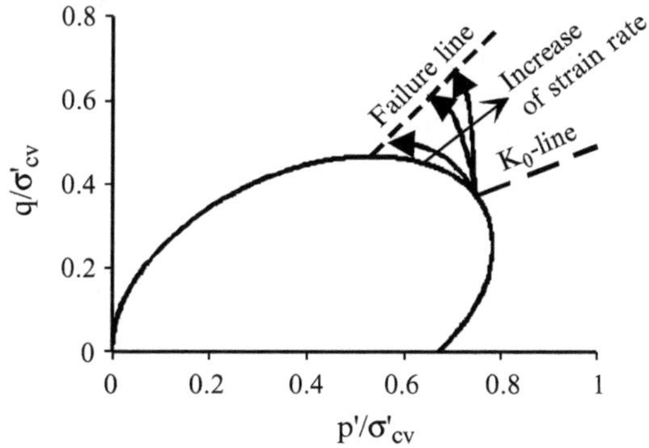

Fig. 25.1 Illustration of yield-induced pore pressure for a normally consolidated clay

25.3 Effective Stress LEM Analyses

25.3.1 Excess Pore Pressure and Strength at Failure in LEM

An advantage of a total stress analysis is that the mobilized shear stress is compared to the actual, yet simplified, undrained shear strength at failure (provided that it is correctly determined and modeled). In effective stress analyses this is not as straightforward, as one should account for the pore pressure state at failure.

Traditionally, for any mobilized equilibrium shear stress τ_e the factor of safety F is calculated against a failure stress state τ_f at the same effective normal stress:

$$F = \frac{\tau_f}{\tau_e} \tag{25.1}$$

This formulation disregards further pore pressure developments from the equilibrium state to failure. As a result, the factor of safety is calculated according to a shear strength that can never be attained due to further pore pressure buildup. This causes the implicit overestimation of strength and factor of safety as discussed e.g. by Leroueil et al. (1990) and Tavenas et al. (1980) and shown in Fig. 25.2. In principle such formulation can give a correct factor of safety only for F = 1.0.

Ideally the equilibrium shear stress τ_e would be compared to the shear strength at failure τ_{fu}. In LEM this is not quite straightforward as the equilibrium equations implicitly tie the shear stress and shear strength to a certain effective normal stress.

A suggestion to achieve a more realistic definition for F is to universally apply *failure state pore pressure* regardless of the actual in-situ pore pressure. This allows the equilibrium shear stress τ_e to be compared to the "actual" shear strength τ_{fu} that can be mobilized at failure. Here two calculation methods are presented where the pore pressure conditions at failure state are considered. Both methods model the stress path from the initial in-situ state to failure to approximate the yield induced

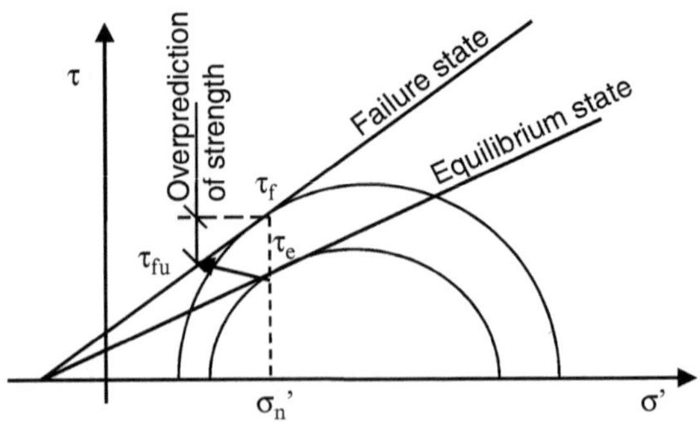

Fig. 25.2 Implicit overestimation of shear strength in undrained effective stress analyses (Länsivaara et al. 2011)

pore pressure. The first method r_u' is quite simplified, while the second method MUESA accounts for stress history and anisotropy and also takes into account the pore pressure caused by external loading.

25.3.2 Method r_u'

The method r_u' is in essence a simple pore pressure parameter for LEM. It is used to calculate yield-induced excess pore pressure (caused by tendency for contracting shearing behavior) for embankments on normally consolidated clays.

Typical Finnish clays are in their natural state only slightly overconsolidated. In the case of old embankments the soil below is then usually normally consolidated, while the clay on either side has retained the slight overconsolidation. Typical topographies on the Finnish railway network are quite level, especially in the case of clay deposits. The method r_u' thus assumes normal K_0-consolidation.

It is assumed that the initial stress state is located on the initial yield surface at the K_0NC-line. The effective stress path from the initial stress state to failure corresponds to the maximum yield-induced pore pressure that could occur in undrained loading with a low loading rate.

To estimate its value one needs an approximation of the initial yield surface. As shown by Länsivaara (1999) it is possible to estimate an anisotropic yield surface by assuming that the associated flow rule is valid for K_0-conditions by only knowing the friction angle and preconsolidation pressure of the clay. This is then further utilized to describe the difference in mean effective stress between the K_0 and failure states. This is finally utilized by introducing an effective stress pore pressure parameter r_u' as:

$$r_u' = \frac{u_{ey}}{\sigma_{v0}'} = \frac{1}{\sigma_{v0}'} f(\sigma_{v0}', \varphi) \tag{25.2}$$

where u_{ey} is the yield induced pore pressure. The graphical solution for r_u' is given in Fig. 25.3.

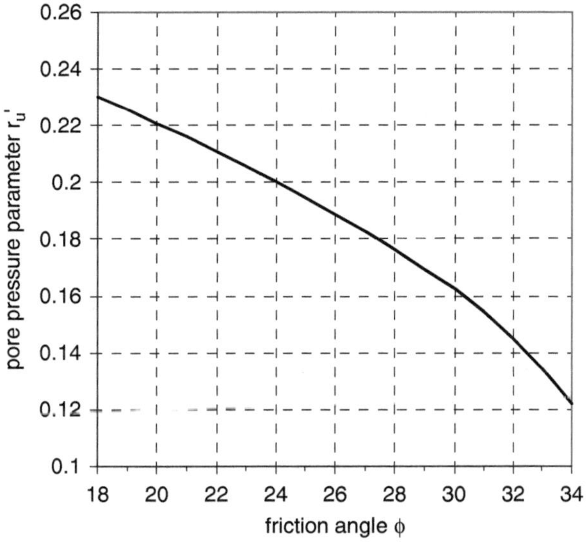

Fig. 25.3 Effective stress pore pressure parameter r_u' as function of friction angle (Länsivaara 2010)

While the parameter r_u' is strictly valid only for active loading of NC clay, in typical situations it is a good compromise. Based on the anisotropy of the yield surface it would be natural to assume larger excess pore pressure on the passive part of the slip surface. However, the soil on the side of the embankment is usually slightly overconsolidated, which counteracts this issue. For the active part the method provides in principal the maximum yield induced pore pressure.

As r_u' represents only the yield-induced excess pore pressure, the pore pressure component from external loading needs to be separately accounted for.

25.3.3 Method MUESA

The method MUESA ("Modified Undrained Effective Stress Analysis") is a more complex method used to estimate the excess pore pressure response of soft clays in LEM. In its current formulation it takes into account effects of anisotropy, overconsolidation and general (non-triaxial) stress states at failure. The method is partly based on the method UESA proposed by Svanø (1981).

In MUESA the excess pore pressure response is divided into two stress change components:

$$\Delta u = \Delta p - \Delta p' \tag{25.3}$$

This equation is simply derived from the principle of effective stresses. Δp is the component of excess pore pressure caused by changes in total confining stress. The effective stress component $\Delta p'$ represents a deviatoric response caused by the shearing and yielding of the clay structure. A close analogy would be the deviatoric component in Skempton's pore pressure equation (Skempton 1954).

MUESA assumes that the initial state is a cross-anisotropic K_0-stress state. Soft clay deposits in Finland are usually quite level, so K_0-consolidation is generally a good assumption. For an old railway or road embankment the initial state is calculated without traffic load, and the mobilized stress state contains the traffic load.

Iteration is used for solving: An initial assumption for excess pore pressure is made, and the desired limit equilibrium method is used to calculate stresses on a given slip surface. The output from the calculation is then used to calculate a new result for Δu. The iteration continues until Δu converges (typically after 3…5 cycles).

MUESA accounts universally for failure pore pressure as described in Sect. 25.3.1. The traditional approach of applying mobilized pore pressure has been under development, but at the time of writing this approach has issues with convergence and general robustness.

The total stress component Δp is calculated iteratively from limit equilibrium results using basic continuum mechanics. A three-dimensional stress formulation is used so that non-triaxial stresses can be considered. The direction of the principal stresses is defined so that the angle between the critical slip surface and σ'_3 is $45° + \varphi'/2$. The intermediate principal stress can be calculated assuming total stress changes that result from plane strain conditions during loading.

The effective stress component $\Delta p'$ is calculated by approximating the effective stress path from the initial stress state to failure. This is done using the anisotropic initial yield surface of the constitutive soil model S-CLAY1 (Wheeler et al. 2003).

As in the calculation of parameter r_u', the effective stress path is assumed to follow the initial yield surface to failure. The failure envelope used here is the Drucker-Prager criterion (corresponding to triaxial compression). The effective mean stress at failure (p'_f) is found at the intersection of the yield surface and the Drucker-Prager failure envelope in principal stress space. The assumption of following the initial yield surface (no hardening is modeled) results in an upper bound value of excess pore pressure.

It should be noted that at the time of writing the method MUESA is still under constant development. Hardening will be implemented in the model for better approximation of the effective stress path and pore pressure.

25.4 Effective Stress FEM Analyses

25.4.1 Material Models for Yield Induced Pore Pressure

As yield induced pore pressure is associated with yielding of the clay, a minimum requirement for the model would be that it includes hardening plasticity. An elastic, perfectly plastic model cannot describe prefailure pore pressure buildup properly, as the stress path in the elastic region follows a constant p' value.

Simple hardening plasticity models are often based on the Cam Clay model and utilize an isotropic yield surface. One such model quite much used is the Soft Soil model in the Plaxis finite element program. A detailed description of the model is not presented here, but can be found in the Plaxis manuals. One key feature of the model separating it from the Cam Clay model is that the yield surface is not directly fixed to the failure surface. The shape of the yield surface is adjusted by a parameter M, which is not used to describe the failure line as in the Cam Clay model. This way the model can be adjusted to give realistic K_0 values. However, the disadvantage of this is that the shape of the yield surface in the upper part near failure line is not representative. If this is then used in undrained analysis, the result will be a too low increase of pore pressure resulting in an over prediction of capacity. To overcome this, one might choose parameter M to match with the friction angle so the model would resemble more the Cam Clay model, to have a better description of yield induced pore pressure.

The yield surface in the Soft Soil model is isotropic. As natural clays have an anisotropic yield surface, a further improvement would be to include the anisotropy in the model. One such model is the S-CLAY1S model. It is not yet in commercial use but the different versions of the model are well documented and verified (Wheeler et al. 2003; Karstunen et al. 2005). Compared to the Soft Soil model, the S-CLAY1S model is rather complex as it accounts anisotropy, rotational hardening and destructuration. It is quite laborious to accurately define all the needed parameters in the model. However, the influence of some model parameters is rather minor in stability analyses. Recent studies by the authors indicate that the parameters can be determined from standard laboratory tests with sufficient accuracy.

As discussed in Sect. 25.2, yield induced pore pressure is also influenced by rate, or creep properties. A natural addition would thus be to include this also in the soil model. As a third alternative, EVP-SCLAY1S model (Karstunen and Yin 2010) is used to analyze the time-dependency of undrained creep.

25.5 Calculation Examples

A calculation example is presented of an embankment failure experiment conducted in Perniö, Finland in 2009. A shallow railway embankment on soft clay was brought to failure by quickly loading shipping containers with sand. The goal was to simulate a heavy train coming to a standstill on the track. The embankment failed at a load level of 87 kPa (2.5 m wide train load) after a total loading time of 30 h (Lehtonen 2011).

The basic soil parameters of Perniö clay are shown in Table 25.1. The calculation parameters that are used for modeling depend on the specifics of each material model. The modified creep index μ^* shown in Table 25.1 refers to the Soft Soil Creep model used in Plaxis.

Table 25.1 The basic soil parameters of the Perniö clay (mean values)

γ_{sat} (kN/m^3)	e_0	λ	κ	μ^*	K_0	φ' (°)	c′ (kPa)	S_u (kPa)	POP (kPa)
15.0	2.3	0.55	0.05	0.01	0.55	25.0	0.0	10.0–15.0	10–20

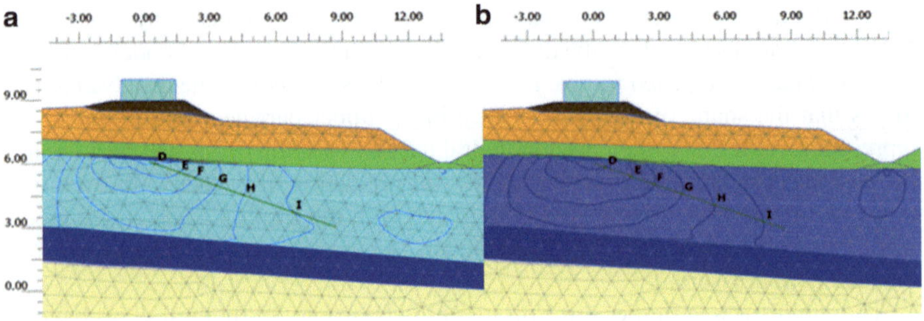

Fig. 25.4 Soft soil (**a**) an d S-CLAY1S (**b**). Excess pore pressure with 60 kPa loading

Table 25.2 Excess pore pressure (kPa) on contour lines

A: 45	C: 35	E: 25	G: 15	I: 5
B: 40	D: 30	F: 20	H: 10	

25.5.1 Finite Element Analysis

In Fig. 25.4 the excess pore pressure development in SS and S-CLAY1S FE analysis is shown for a train load of 60 kPa. The legend for the excess pore pressure contour lines is shown in Table 25.2. The same contours are used throughout the study.

As shown in Fig. 25.4, with the 60 kPa load, excess pore pressure is developing as a bubble from below the load. According to the Soft Soil model, lateral distance to the 15 kPa contour line is 3.3 and 7.4 m to the 5 kPa contour line from the center line of the track. In S-CLAY1S analysis the distances are 4.2, 7.8 m respectively.

As shown in Fig. 25.5a, according to the field measurements corresponding measured values in this loading stage were 0.5 and 6.0 m respectively. Hence the FEA models tend to overestimate excess pore pressure in this case. Deviation is emphasized just under the external load, where calculated excess pore pressures are 30–34 kPa and measured values only 15–17 kPa. The measured increase of the excess pore pressure is quite small compared to the external load, as the overconsolidation of the saturated clay layer is only 10–20 kPa.

The reason why the pore pressures were smaller than expected was the short time period and fast rate of loading. As the strength and pore pressure response of clay is highly time-dependent, the pore pressure did not have enough time to build up and therefore the measured values were small. This time-dependent behavior could be possible to take into account if a soil model including creep is used. In this case, the model EVP-SCLAY1S (Karstunen and Yin 2010) was used for time-dependent

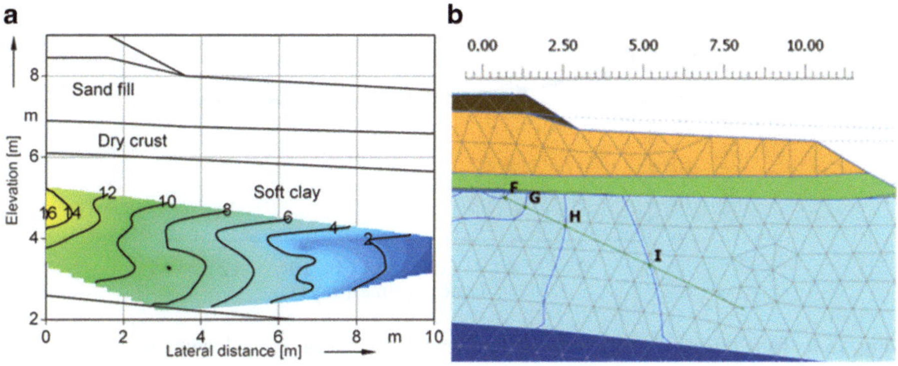

Fig. 25.5 Measured excess pore pressures (**a**) and modeled pressures with EVP SCLAY1S model with 60 kPa load (**b**)

Table 25.3 Failure loads (kPa) with different material FEA models applied to the soft clay layer

	q_f (kPa)
Soft Soil (φ)	76.1
Soft Soil (K_0)	92.1
S-CLAY1S	79.2
EVP-SCLAY1S	90.0/87.0

analysis. Figure 25.5b shows the calculated excess pore pressure after 60 kPa loading, when creep properties of the Perniö clay are accounted for. Otherwise the parameters are identical with the S-CLAY1S model.

When the Fig. 25.5a and b are compared, the conclusion can be drawn that the modeled values corresponds very well with the measured values. The difference to the analysis without time effect is significant. The failure load in the test was 87 kPa. As the Soft Soil and S-CLAY1S -models do not account the time effects they underestimate the failure load by 5–15 kPa. For the long-term loading situation the models gave a good approximation.

The model EVP-SCLAY1S in turn gave a good correlation for the excess pore pressure throughout the whole failure test even though it slightly overestimated the failure load.

The results of the FE analysis are shown in Table 25.3. In the analysis the train load is gradually increased to the stage where failure occurs. The ultimate loads of each material model or method are shown in the table. Soft Soil (φ) indicates the method where the parameter M is set to match to the friction angle while the Soft Soil (K_0) represents the default input values. It is shown that the models SS (φ) and the S-CLAY1S slightly underestimate and the model SS (K_0) slightly overestimates the failure load. One should notice that the relatively fast loading time clearly affected to the ultimate load in the field test. If loaded slowly enough the failure load would be less than 80 kPa. Therefore it can be said that the SS (φ) and the S-CLAY1S – models gave a very good approximation when the time effects are not accounted for.

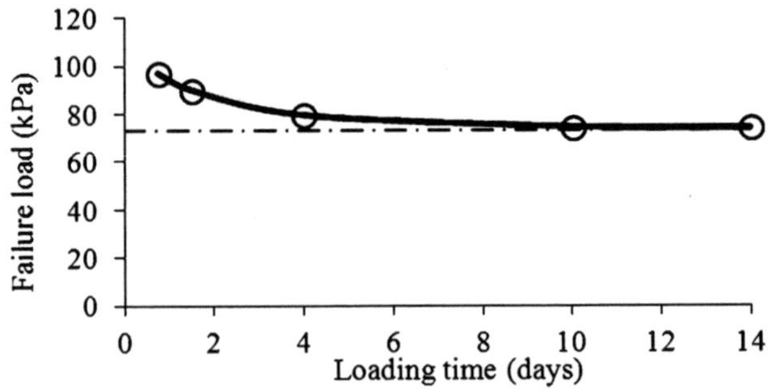

Fig. 25.6 Time effect. The calculated failure load is dependent on the loading time in the EVP-SCLAY1S analysis

For the time dependent EVP-SCLAY1S analysis two different approaches were applied. First the load was gradually increased in the same rate as in the field. In that analysis failure load was found to be 90.0 kPa. For the latter analysis the load was gradually increased to 87 kPa as it was during the field test and then put on hold. During this time excess pore pressure builds up and failure occurs. In the field test the waiting period needed for failure after reaching 87 kPa load was 2 h. In the EVP-SCLAY1S analysis the waiting time was 10 days. This indicates that even if the model predicted excess pore pressure increase during the loading very well, it was not able to accurately model time dependent yield induced pore pressure without considerable stress increase. This kind of modeling situation is not the main purpose of the model as it is developed for the long term creep settlements, but still the results were promising.

In Fig. 25.6 the effect of loading rate for the calculated failure load is shown. The analyses are conducted using EVP-SCLAY1S model. The loading time indicates the overall duration of the loading process. In the field test the duration was 30 h, which corresponds to the second calculation result (circular marker at 90 kPa). Although this analysis is an approximation, it is established that the time effects had a remarkable role in the field test.

A corresponding set of LEM calculations was conducted using a given slip surface shape that is based on the actual failure surface. The calculation did not include an optimization for the slip surface as this is not yet possible with MUESA, but the given slip surface is nevertheless very close to the most critical one.

The limit equilibrium method used for all calculations is Morgenstern-Price with $f(x)=1$ (i.e. Spencer's Method) The calculations were a typical total stress analysis using the measured profile of undrained shear strength, as well as undrained effective stress calculations with the two discussed methods.

As the method r_u' only models the yield-induced component of excess pore pressure, the excess pore pressure resulting from external loading (u_q) was modeled separately with the parameter $r_{uq}=0.5$ (i.e. $u_q=0.5q_{app}$). This parameter value resulted in a realistic shear stress distribution along the slip surface.

An overconsolidation of POP = 20 kPa in the soft clay layer was modeled in the method MUESA. As the method also takes into account the excess pore pressure from the increase in total mean stress (Δp) under the load, there was no need to model this separately. The calculated failure loads for LEM are given in Table 25.4.

Table 25.4 Failure loads (kPa) from LEM calculations

	q_f with given slip surface (kPa)	q_f with optimization (kPa)
LEM $\varphi=0$ s_u	71.8	62.5
LEM c'-φ' $r_{uq}+r_u'$	88.0	87.1
LEM c'-φ' MUESA	83.0	–

The effective stress LEM failure loads are very close both to FEA results and the actual failure load of 87 kPa (time effects notwithstanding). The effect of optimizing the slip surface must be noted, especially for the total stress calculation. It can also be argued that total stress calculation is subject to uncertainties regarding the determination of s_u, which was done with vane shear testing.

25.6 Conclusion

FEA can be used to accurately evaluate excess pore pressure response and stability conditions of soft clay by means of effective stress analysis. This however requires good knowledge of the yielding behavior of the clay.

Based on the Perniö failure test case, S-CLAY1S-model gave the most accurate approximation regarding the pore pressure and the displacements response, when the time effects are not accounted for. On the other hand, if the yield surface of the clay is known, the Soft Soil model is also usable for conducting accurate stability analyses. One should notice that neither simple soil models nor Soft Soil model are able to model failure induced pore pressure of very soft clays with default parameters.

To be able to conduct efficient undrained effective stress analyses in LEM, yield-induced excess pore pressure needs to be accounted for. In addition, the implicit overestimation of shear strength should be accounted for. The calculation methods presented in the paper model the effective stress path to failure to account for the undrained pore pressure response. They also implicitly apply failure pore pressure to reach proper shear strengths at failure. When excess pore pressure is modeled accurately enough, even LEM calculations can potentially be as accurate as FE analyses, as is shown by the calculation example.

Acknowledgments The authors would like to thank the Finnish Transport Agency for the funding of the research project, as well as the reviewer Dr. professor Hans Petter Jostad for his valuable comments.

References

Janbu N, Senneset K (1995) Soil parameters determined by triaxial testing. In: Proceedings of the 11th European conference on soil mechanics and foundation engineering, vol 3, Copenhagen, pp 101–106

Karstunen M, Yin ZY (2010) Modelling time-dependent behaviour of Murro test embankment. Géotechnique 60(10):735–749

Karstunen M, Krenn H, Wheeler SJ, Koskinen M, Zentar R (2005) Effect of anisotropy and destructuration on the behavior of Murro test embankment. Int J Geomech 5(2):87–97

Länsivaara T (1999) A study of the mechanical behavior of soft clay. Doctoral thesis, Department of Geotechnical Engineering, NTNU, Trondheim
Länsivaara T (2010) Failure induced pore pressure by simple procedure in LEM. In: Benz T et al (eds) Numerical methods in geotechnical engineering. Proceedings of the seventh European conference on numerical methods in geotechnical engineering Numge 2010, Trondheim, 2–4 June 2010, pp 509–514
Länsivaara T, Lehtonen V, Mansikkamäki J. (2011) Failure induced pore pressure, experimental results and analysis. In: 2011 Pan-Am CGS geotechnical conference, Toronto
Lehtonen V. (2011) Instrumentation and analysis of a railway embankment failure experiment. Research reports of the Finnish Transport agency 29/2011, Finnish Transport Agency, Helsinki
Leroueil S, Magnan J-P, Tavenas F (1990) Embankments on soft clays. Ellis Horwood, Chichester
Skempton AW (1954) The pore-pressure coefficients A and B. Geotechnique 4(4):143–147
Svanø G. (1981). Undrained effective stress analysis. Dissertation, NTH Trondheim
Tavenas F, Trak B, Leroueil S (1980) Remarks on the validity of stability analyses. Can Geotech J 17:61–73
Ukonjärvi P (2013) Rautatieympäristössä määritettävän suljetun leikkauslujuuden luotettavuus (In Finnish, title in English: reliability of measuring undrained shear strength in railway environment). Master's thesis, Tampere University of Technology (to be published)
Wheeler S, Näätänen A, Karstunen M, Lojander M (2003) An anisotropic elastoplastic model for soft clays. Can Geotech J 40:403–418

Part V
Hazard Assessment, Risk Management, Regulations and Policies

Chapter 26
An Overview of the Mapping of Landslide-Prone Areas and Risk Management Strategies in the Province of Québec, Canada

Janelle Potvin, Catherine Thibault, Denis Demers, and Chantal Bilodeau

Abstract The majority of the population of Québec (Canada) is settled in the St. Lawrence River Lowlands, where soils consist mainly of clays. In some areas, these clays are very sensitive, and are prone to the development of large retrogressive landslides. To identify areas that are potentially exposed to this hazard, the Québec government produces maps of areas prone to landslides in clayey soils. These maps allow the government authorities to take this risk into account in their operations. Furthermore, municipalities must include this mapping in their regulations for land use planning. The risk analysis for landslides in clayey soils is based on these maps. Different approaches are currently being developed in order to manage risks according to the types of landslides and the elements at risk (population, roads, etc.).

Keywords Mapping • Landslides • Risk management • Risk analysis • Quick clay • Risk assessment

26.1 Introduction

Approximately 90 % of the population of Québec is settled along the shores of the St. Lawrence and Ottawa rivers and in the Saguenay–Lac-Saint-Jean region, on lands that were once covered by ancient postglacial seas (Fig. 26.1). It is within these boundaries that one finds sensitive clays that are highly prone to the development of landslides. According to an inventory that was prepared by the Ministère des Transports du Québec (Fortin et al. 2008), 80 % of the recorded landslides are

J. Potvin (✉) • C. Thibault • D. Demers
Service de la Géotechnique et de la géologie (Geotechnique and Geology Branch), Ministère des Transports du Québec (MTQ), Québec City, QC, Canada
e-mail: Janelle.Potvin@mtq.gouv.qc.ca

C. Bilodeau
Service de la Géotechnique et de la géologie (Geotechnique and Geology Branch), Ministère de la Sécurité publique du Québec, Québec City, QC, Canada

J.-S. L'Heureux et al. (eds.), *Landslides in Sensitive Clays: From Geosciences to Risk Management*, Advances in Natural and Technological Hazards Research 36, DOI 10.1007/978-94-007-7079-9_26, © Springer Science+Business Media Dordrecht 2014

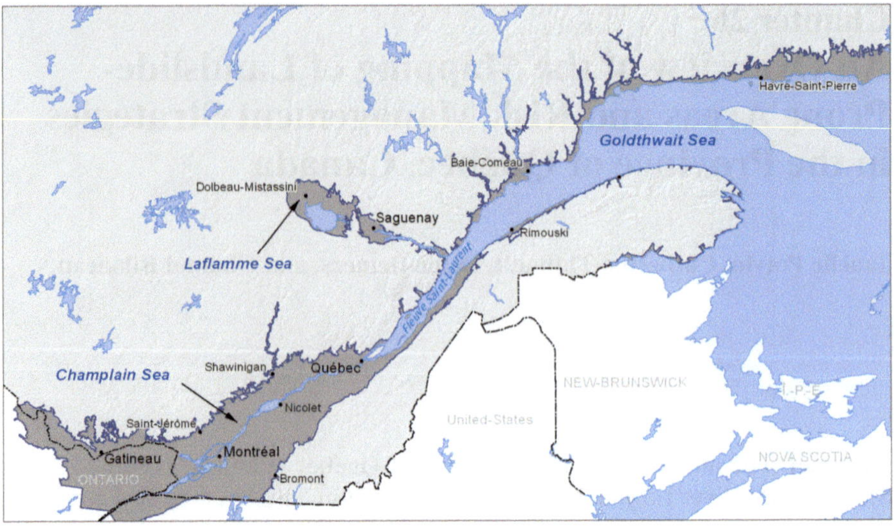

Fig. 26.1 Location of marine limits in southern Québec

Fig. 26.2 Typical example of the most common types of landslides in Québec and the context in which they occur, namely along the meanders of watercourses where erosion is active (Source MTQ)

concentrated within the marine transgression limits. For the most part, landslides in clayey soils occur on the banks of watercourses in meanders that are subject to erosion (Fig. 26.2). The vast majority of these are rotational landslides, with a slip surface of varying depth (Demers et al. 1999a, b). Although most of them have a retrogression distance of no more than approximately 10 m, they can still cause significant damage to property and infrastructures.

Fig. 26.3 Large retrogressive landslides of the flowslide (**a**) and spread (**b**) type (Source MTQ)

Deep rotational landslides, which occur in sensitive clays, can sometimes trigger very large movements. When their retrogression distance exceeds twice the height of the original slope, they are called "large retrogressive landslides". Their dimensions can reach several hundred metres, and can cause significant loss of life and property damage (Evans et al. 1997).

These large retrogressive landslides are of two main types, namely flowslides (Fig. 26.3a) and spreads (Fig. 26.3b). The two types differ in their failure mechanisms and characteristics (Chap. 7 by Demers et al., this volume) which may necessitate different approaches to risk management. On average, a large retrogressive landslide with an area greater than one hectare occurs every 2.3 years in Québec (Chap. 7 by Demers et al., this volume).

After the 1971 landslide of Saint-Jean-Vianney, which took the lives of 31 people, the government of Québec undertook a program to map landslide-prone areas, which extended over a period of approximately 15 years.

In the early 2000s, the Government of Québec reviewed its risk management approaches (Ministère de la Sécurité publique de Québec 2008) after some major natural disasters in the late 1990s (The 1996 floods in Saguenay and the 1997 ice storm in southern Québec). The Government of Québec recognized the urgent need to study landslides, especially those occurring in clayey soils. With several hundred per year, they are by far the common type of landslide in Québec. They can occur in the most densely populated regions of the province and cause great devastation.

Figure 26.4 illustrates the landslide risk management process in the province of Québec. It shows the three main steps in the process: setting the context, as described above, assessing the risks and treating the risks. It also shows the importance of communication and consultation, along with the monitoring and review activities that are required at the various stages of the process, depending on the measures implemented and the occurrence of extraordinary events. In addition, a number of measures are implemented in order to assist municipal or government authorities in managing risk on their territory more effectively, thus reducing the level of risk to the public and infrastructures (Demers et al. 2008; Thibault et al. 2008).

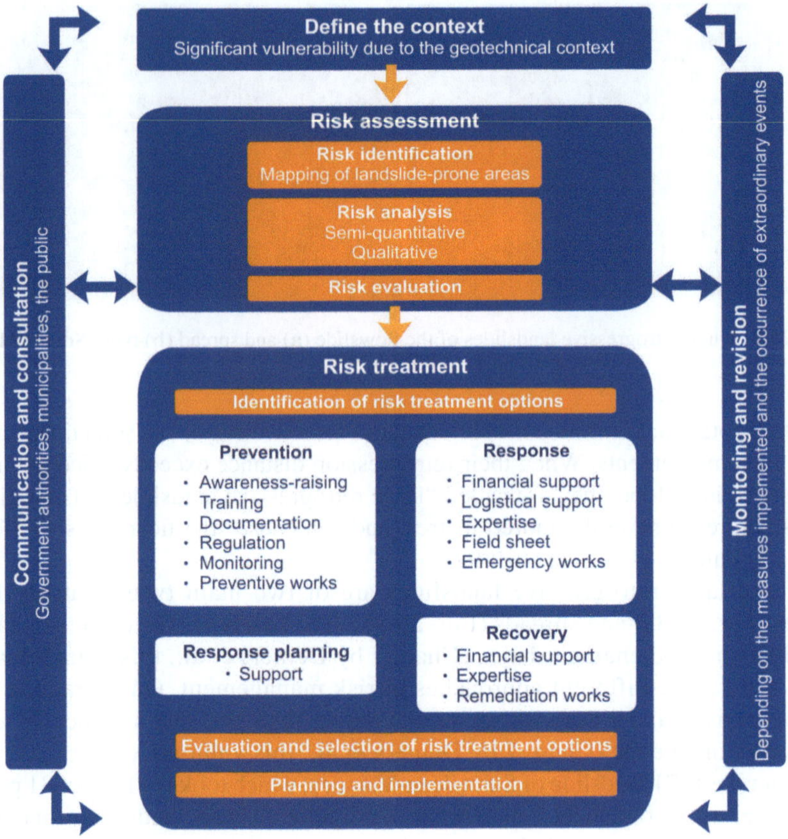

Fig. 26.4 The risk management process

26.2 Risk Assessment

Risk assessment forms the basis for decision making in subsequent stages of the greater process of risk management. It consists of three distinct sub-stages: identification, analysis and estimation of risks.

26.2.1 Risk Identification

Mapping of the landslide-prone areas is the basic tool used in the risk identification stage. Indeed, it is fundamentally important to identify the locations where landslides may occur in order to avoid building in these areas and to avoid activities that could impair slope stability.

Since 2003, the Government du Québec has charged the Ministère des Transports du Québec (MTQ) with mapping landslide-prone areas in clayey soils. Two actions were then undertaken on a priority basis. First, a comprehensive inventory of landslides

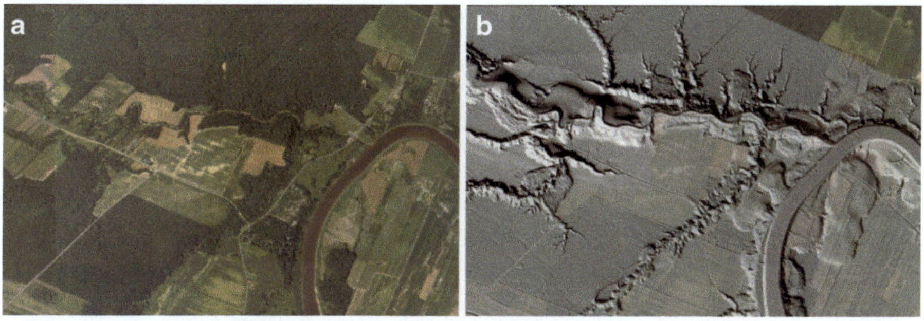

Fig. 26.5 Aerial photo alone (**a**), with overlaying of a digital terrain model, showing up landslide scars clearly (**b**)

was carried out (Fortin et al. 2008), by compiling and computerizing internal government records. It was completed by identifying landslide scars on aerial photos and lidar data. Many large retrogressive landslides were also investigated in order to complete the previously acquired information (Chap. 7 by Demers et al., this volume).

At the same time, digital terrain models were developed, first by means of conventional photo reconstruction, and more recently using airborne lidar covering a large area of the territory. Since 2003, an area of approximately 39,000 km^2 has been covered using this method. This represents 51 % of the entire area of southern Québec that was once covered by postglacial seas (Fig. 26.1). Airborne lidar not only provides very precise digital terrain model, it has proved to be of great help in filling out the landslides inventory (Fig. 26.5) and in carrying out a preliminary identification of the nature of the soils using geomorphological analysis.

The mapping of a region is based on the collection of large quantities of data from piezocone soundings, field vane tests, soil sampling drilling, laboratory testing and field observations. In addition to the evidence provided by the landscape, such as the presence of large retrogressive landslides scars within a sector, the identification of sensitive clays is primarily achieved by means of laboratory tests (fall cone and Atterberg limits). All of the available information, whether acquired specifically within the context of the mapping program or from the MTQ's geotechnical database, is transferred to a documentation map (Robitaille et al. 2002). This map also shows superficial deposits, slope boundaries and gradients established using lidar data, data from the landslide inventory, erosion zones, existing stabilization works, etc.

Analysis of all the data makes it possible to produce a landslide "susceptibility" map. The zones identified on these maps show a susceptibility level based on the type of soil that controls the slope failure, the type of potential landslide and the intensity of the erosion affecting the slopes.

The main purpose of the mapping exercise is to provide the regional and local authorities who are responsible for ensuring life and property safety within their territory with the tools they need to carry out the responsibilities that the government has assigned to them in this area (Act respecting Land Use Planning and Development (RSQ, c A-19.1)). In order to facilitate the use of the maps and the application of the regulatory framework, the various susceptibility classes are then grouped together into zones that are subject to the same land-use planning provisions. This map is

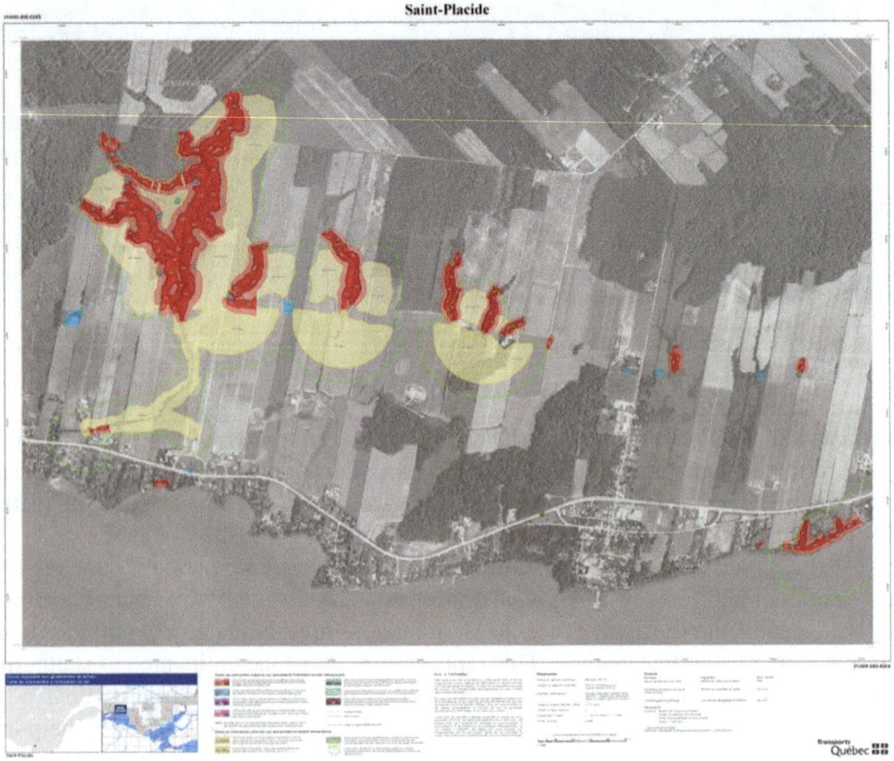

Fig. 26.6 Example of a landslide constraint map

called a landslide constraint map and is produced at a scale of 1/5,000 (Fig. 26.6). Pursuant to the requirements of the Act respecting Land Use Planning and Development, these maps and the regulatory framework pertaining to the control of land use in these zones must be incorporated into the land use and development plans developed by the regional authorities (Regional County Municipalities), and then into the local land use plans and regulations. The maps are distributed by the Géoboutique division of the Ministère des Ressources naturelles (http://geoboutique.mrn.gouv.qc.ca), and are made available to stakeholders and to the public.

26.2.2 Risk Analysis and Evaluation

Risk analysis is carried out on a regional level, based on the susceptibility map. A specific analysis is carried out for each type of possible landslide hazard for locations where vulnerable elements, such as residential homes and roads are exposed to these hazards. In these cases, hazard assessment is carried out by attempting to quantify the probability that a landslide will occur, based on the ancient landslide history and the number of events that are likely to occur in the region (Thibault et al. 2008).

The methodology for determining the probability of occurrence can vary depending on the type of potential landslide (superficial, rotational, or large retrogressive). For example, in the case of superficial landslides, the large number of cases allows for the calculation of a probability for each class of slope on a regional basis (Levasseur 2003).

The gravity of the consequences is then evaluated based on the spatial and temporal probability of the hazard, among other things. The results of the risk calculation are presented on a "Probability of occurrence – Gravity of consequences" graph using five risk categories (very low, low, moderate, high, very high).

Risk analyses must sometimes be carried out for areas for which no susceptibility map is available, including for the purpose of prioritizing the actions required to protect road stretches. In such cases, a qualitative approach is used, based on the presence of factors that can indicate, aggravate or cause a predisposition to the development of a landslide. A subjective weighting is applied to each of these factors in order to establish a susceptibility scale. The same approach is then followed in order to estimate the gravity of the consequences for the roads based on criteria such as the location of the road in relation to the potential danger, the daily traffic volume, and the relative importance of the segment. Risk is then evaluated by plotting the data on a five-level matrix.

The next step in the risk evaluation stage is to prioritize the zones that require to be treated in order to reduce the risk to a level that is deemed acceptable. The interventions can be aimed at reducing the hazard or the gravity of the consequences for a given anticipated event, or both at once. This stage is carried out in cooperation with all of the pertinent players: government authorities, municipal authorities and the public.

26.3 Risk Treatment

Risk treatment serves to identify all of the potential measures that can impact the four main aspects of civil protection, namely prevention, response planning, response and recovery (Ministère de la Sécurité publique du Québec 2008). At this stage, the various players must evaluate, select and implement measures aimed at reducing the risk to an acceptable level. To this end, the Government of Québec has adopted various measures in order to ensure that municipalities and government departments are better equipped for landslide risk management.

26.3.1 Preventive Measures

Prevention is considered to be an indispensable component of risk management, because it generally is the most effective and least costly. Taking the risks into account in land-use planning through mapping and application of relevant regulations helps

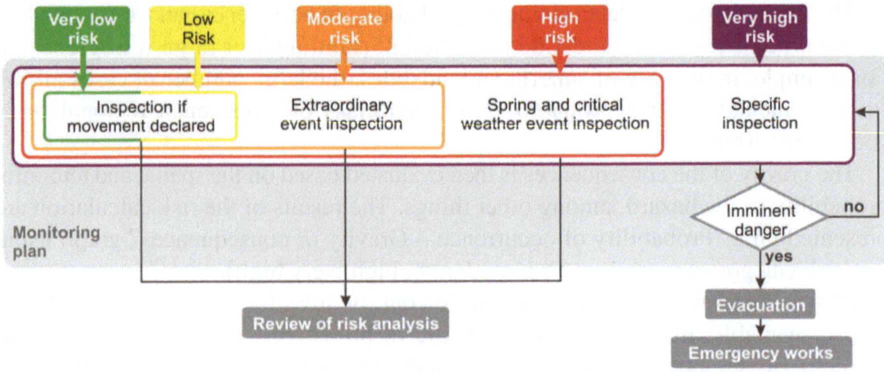

Fig. 26.7 Monitoring plan for superficial landslide-prone areas

to avoid an increase of the consequences associated with a potential landslide, mainly by prohibiting new construction in landslide-prone areas. However, prohibitions can be lifted, provided that a geotechnical expertise that meets the regulatory requirements is produced by a geotechnical engineer. (Lefebvre et al. 2008). By prohibiting inappropriate interventions, the implementation of a regulatory framework for built-up areas contributes to stop the increase of hazard in these areas. In fact, the MTQ's landslide inventory reveals that 40 % of cases where government assistance is required are caused by inappropriate interventions (Fortin et al. 2008).

Training sessions are offered to municipal and government workers who are required to take the map information into account and apply the regulations, so that they can comprehend the technical aspects underlying the mapping and regulatory framework. Raising awareness among the public is another indispensable step in the process. Although this is a municipal responsibility, given the complexity of the problem, the government provides technical support at public information meetings. A pamphlet was published to inform citizens who live in landslide-prone areas with respect to the types of potential hazards and common inappropriate interventions that can trigger a landslide (Government of Québec 2005).

When the risk analysis reveals that a number of residences in a given municipality are exposed to landslide hazard, the municipality is offered the option of implementing a slope-monitoring plan with technical support from the government. Figure 26.7 presents a monitoring plan adapted for zones prone to superficial landslides. It has been tested since 2007 in the city of Saguenay, where a landslide of this type resulted in the death of two young children in 1996 (Bouchard et al. 2008). Visual inspections of the slopes are conducted each year in order to identify any early signs of instability (cracks at the crest of a slope, bulging of the soil at the foot of the slope, degradation of retaining walls, etc.) that might have developed or worsened since the previous inspection. This innovative approach in Québec has led to the identification of approximately ten sites where the danger is imminent. As a result, the completion of stabilization works helped to prevent potential loss of life and significant property damage.

If a risk analysis indicates that an area that is prone to large retrogressive landslides presents a very high level of risk, the municipalities can carry out preventive works with technical and financial assistance from the Québec Government. The government assistance can cover between 50 and 90 % of the costs, depending on the situation. Since 2007, seven stabilization projects have been completed in six municipalities. It is estimated that every dollar invested in prevention has helped to protect the equivalent of five dollars worth of physical assets and most importantly, many lives.

The approach of the MTQ to prevention with respect to its road network includes, among other things, the development of tools such as checklists and documents combined with training courses. The courses are intended to ensure that road supervisors and any worker who is likely to be involved in the event of a landslide or the detection of early signs will be able to take the appropriate prompt action to ensure the safety of road users. In addition, these training courses raise awareness about inappropriate interventions that could cause slope instabilities among MTQ workers involved in construction and maintenance works on the provincial road network. In addition, the development of measures aimed at eliminating and reducing landslide risks at its source by means of preventive works is also included among the actions undertaken by the MTQ in its risk management plan. For example, emergency works have recently been carried out on some of approximately 40 problematic sites along 2 roads following the Richelieu River southeast of Montréal. Plans call for works on the other sites to be spread out over several years. In the meantime, these sites are subject to a monitoring program carried out by road supervisors. In light of this, these workers, who cruise the territory continuously, have received specific training in order to be able to recognize the early signs of landslides.

26.3.2 Response Planning Measures

The potential measures suggested for the response planning stage consist in implementing emergency measures plans so that the roles and responsibilities of all workers are clearly defined prior to any natural disaster. The staff of the regional offices of the Ministère de la Sécurité publique du Québec provides support to the municipalities in this exercise. Each regional office of the MTQ, on the other hand, must set up its own emergency measures plan, which must include a section dedicated to landslides. In addition, training courses are offered to first responders so that they would know how to respond in the event of a landslide (establishment of minimum safety perimeters, signs to watch for in order to avoid putting oneself at risk, field report sheets to be filled out, etc.).

26.3.3 Response Measures

In the event of a landslide threatening residences or municipal roads, the government of Québec offers expert assessment services in order to assess the situation and respond to the emergency and provides logistical support to municipalities. If the danger is imminent, emergency works such as temporary protection walls can be undertaken by the municipality in cooperation with the government.

In the case of MTQ infrastructures, procedures have been developed to allow local workers to quickly respond to an emergency situation while awaiting the intervention of geotechnical experts.

26.3.4 Recovery Measures

In terms of the recovery phase, the Québec government has a general financial assistance program intended for municipalities or citizens in the event of an imminent natural disaster (Ministère de la Sécurité publique du Québec 2012) based on the type of assets affected. Three options are available to an affected citizen: relocation of the principal residence, stabilization of the land or a moving allowance. Relocation of the residence is generally the least expensive solution. When the option involving improvement works is chosen, the works must be designed and recommended by a geotechnical engineer. In cases where the site is exposed to large retrogressive landslides, the study falls under the responsibility of government specialists. In such situations, which are almost always found along watercourses, the most commonly used technique remains the laying riprap at the foot of the slopes, which then plays a double role by serving as a counterweight and protecting against erosion of the banks.

In the event that a road or structure belonging to the MTQ is affected by a landslide, the geotechnical experts recommend the appropriate works to restore traffic flow on the road network as soon as possible.

26.4 Conclusion

The approach described in this article allows for a significant reduction of landslide risk. Its success depends on the measures put in place for all of the dimensions of civil protection (prevention, response planning, response and recovery), but especially prevention. It is essential to take action prior to an eventual natural disaster by mapping potential landslide zones and adopting appropriate regulations. This stage makes it possible to take action within the land development process, with the main goals of not increasing the number of vulnerable elements in the zones at risk and reducing the number of landslides triggered by inappropriate human intervention.

In order to be effective, risk management also requires ongoing training and awareness-raising activities for all workers and for the public, especially in light of the complexity of the problem.

Certain aspects still need to be explored and developed, including the impacts of the mapping program and the associated regulations on the real estate sector. In this regard, awareness rising among real estate professionals is a challenge that is still to be met. They must have all the information required to make the proper decisions in their work, without adversely affecting the public.

Finally, the approaches that the Government of Québec has implemented in recent years have led to great improvements in risk management and the attention paid to this subject by workers and the public.

Acknowledgements This publication has been authorized by the departmental authorities to whom the authors report. The authors would like to thank their many colleagues who took part in the discussions and in the development of this risk management process, as well as Didier Perret for revising the article.

References

Bouchard R, Michaud V, Demers D (2008) Le glissement de la rue McNicoll, 20 juillet 1996, Saguenay, Québec: causes et conséquences. In: Proceedings of the 4th Canadian conference on geohazards: from causes to management. Presse de l'Université Laval, Québec, pp 503–510

Demers D, Leroueil S, d'Astous J (1999a) Investigation of a landslide in Maskinongé, Québec. Can Geotech J 36(6):1001–1014

Demers D, Potvin J, Robitaille D (1999b) Gestion des risques de glissement de terrain liés aux pluies des 19 et 20 juillet 1996 au Saguenay – Lac-Saint-Jean. Report submitted to the Bureau de reconstruction et de relance du Saguenay – Lac-Saint-Jean, Ministère des Transports du Québec, Québec

Demers D, Robitaille D, Potvin J, Bilodeau C, Dupuis C (2008) La gestion des glissements de terrain dans les sols argileux au Québec. In: Proceedings of the 4th Canadian conference on geohazards: from causes to management. Presse de l'Université Laval, Québec, pp 519–526

Evans SG, Couture R, Chagnon J-Y (1997) Notes on major Leda clay landslides in the St. Lawrence Lowlands of Eastern Canada, 1615–1996. In: Proceeding of the 50th Canadian geotechnical conference, Ottawa, pp 839–846

Fortin A, Ouellet D, Demers D (2008) Développement au Ministère des Transports du Québec d'un portail informatique pour l'accès à des bases de données géotechniques. In: Proceedings of the 4th Canadian conference on geohazards: from causes to management. Presse de l'Université Laval, Québec, pp 169–174

Government of Québec (2005) Habiter dans les zones exposées aux glissements de terrain: Précautions à prendre en matière d'utilisation du sol. http://www.securitepublique.gouv.qc.ca/securite-civile/publications-statistiques-civile/zones-glissements-terrain.html

Lefebvre G, Demers D, Leroueil S, Robitaille D, Thibault C (2008) Slope stability evaluation: more observation and less calculation. In: Proceedings of the 4th Canadian conference on geohazards: from causes to management. Presse de l'Université Laval, Québec, pp 413–420

Levasseur M (2003) Contribution des systèmes d'information géographique à l'analyse quantitative de l'aléa « glissement de terrain » – Exemple d'application au secteur de la ville de La Baie, Québec. Master's thesis, INRS-E.T.E, Université du Québec, Québec, 213pp

Ministère de la Sécurité publique du Québec (2008) Gestion des risques en sécurité civile. Government of Québec, Québec. Available via http://www.securitepublique.gouv.qc.ca/securite-civile/publications-statistiques-civile.html

Ministère de la Sécurité publique du Québec (2012) Programme général d'aide financière lors de sinistres réels ou imminents. Available via http//www.securitepublique.gouv.qc.ca/securite-civile/aidefinanciere-sinistres/programmes-aide-sinistres/general/objet-general.html

Robitaille D, Demers D, Potvin J, Pellerin F (2002) Mapping of landslide-prone areas in the Saguenay region, Québec, Canada. In: Proceedings of the international conference on instability – planning and management, Ventnor, Isle of Wight, pp 161–168
Thibault C, Potvin J, Demers D (2008) Development of a quantitative approach for evaluating and managing the risk associated with highly retrogressive slides. In: GeoEdmonton'08 Organizing Committee (ed) Proceedings of the 61st Canadian geotechnical conference, Edmonton, pp 1055–1063

Chapter 27
Regulatory Framework for Road and Railway Construction on the Sensitive Clays of Norway

F. Oset, V. Thakur, B.K. Dolva, K. Aunaas, M.B. Sæter, A. Robsrud, M. Viklund, T. Nyheim, E. Lyche, and O.A. Jensen

Abstract Parts of the Norwegian roads and railway network are located in areas dominated by sensitive clays. Several new projects are facing challenges in areas with deposits of sensitive clay. The design codes and guidelines of the Norwegian Public Roads Administration (NPRA) and the Norwegian National Railways Administration (NNRA) have been developed over many years with respect to ground investigations and safety requirements. These regulations are specific for road and railway construction. In addition there are general design codes like the Norwegian Standards, and the regulations established by the Norwegian Water Resources and Energy Directorate (NVE) concerning the safety of larger areas adjacent to e.g. roads and railways. This paper points out the manner in which the regulations impose an increased attention to safety when sensitive clay is encountered. The need for further development of the regulations, based on recent studies and research, is also described.

Keywords Landslides • Sensitive clay • Stability regulations • Partial safety factor • Percentual improvement of safety factor • Risk prevention • Zoning plans

---

F. Oset (✉) • V. Thakur • B.K. Dolva • K. Aunaas
Geotechnical and Landslide Division, Norwegian Public Roads Administration (NPRA), Oslo, Norway
e-mail: frode.oset@vegvesen.no

M.B. Sæter • A. Robsrud • M. Viklund
Technology Division, Norwegian National Railways Administration (NNRA), Oslo, Norway

T. Nyheim • E. Lyche • O.A. Jensen
Landslides, Flood and River Management Division, Norwegian Water Resources and Energy Directorate (NVE), Trondheim, Norway

J.-S. L'Heureux et al. (eds.), *Landslides in Sensitive Clays: From Geosciences to Risk Management*, Advances in Natural and Technological Hazards Research 36,
DOI 10.1007/978-94-007-7079-9_27, © Springer Science+Business Media Dordrecht 2014

27.1 Introduction

Sensitive clays of Norway, when provoked by manmade or natural causes, have led to several landslide disasters throughout history. The most well-known are the landslides in Verdal (1893, 55 million m^3, 116 casualties) (Janbu et al. 1993) and Rissa (1979, 5 million m^3, 1 casualty). These landslides have occurred in highly sensitive clays also known as quick clays. In the last 40 years there has been approximately 1 or 2 slides per decade with a volume exceeding 500,000 m^3. Since 1996 no lives have been lost in quick clay landslides, but residential areas have been affected in varying degrees of destruction. Thakur et al. (2013) presents an overview over 33 large Norwegian landslides in sensitive clay slopes. On one hand, some of these landslides were caused by nature, e.g. erosion along rivers or canals, while in some cases it was difficult to single-out an obvious triggering factor. However, it appears that a majority of these landslides can be related to human activity. Within the framework of landslides in sensitive clays, the Norwegian governmental agencies e.g. NPRA, NNRA and NVE present regulatory frameworks to enable a decision making process for adopting appropriate design approaches regarding construction on sensitive clay deposits. These regulatory frameworks are focused on site investigations, design procedures and safety precautions for construction works. In addition, design principles for securing natural slopes are also addressed in the regulatory framework (Fig. 27.1).

Fig. 27.1 Quick clay landslide in Lyngen in Norway, September 2010 (Photo: F. Oset, NPRA)

A national R&D project called Natural hazards: Infrastructure for Flood and Slides (NIFS) has been initiated by NPRA, NNRA and NVE (NIFS 2012). More information can be found on www.naturfare.no. Challenges associated with sensitive clays are one of the projects within this R&D project. This article describes some of the issues the NIFS project expects to study during the project period. The paper first presents principles regarding requirements with partial safety factor for local stability and partial safety factor or percentual improvement regarding larger progressive slides. Thereafter, an attempt is made to describe the procedures concerning assessment of landslides in Norwegian landslide hazard mapping plans. This paper also highlights the on-going efforts to achieve harmonization of the guidelines. Finally, this paper also briefly looks at the challenges related to landslide assessments that can only be solved by research.

27.2 Historical Development of Regulations

In the 1970s, limit state design was introduced in the Norwegian construction standards. There was a need for a corresponding set of design regulations for geotechnical aspects. A committee established by the Norwegian Geotechnical Society provided a guideline that was published by the Norwegian Standardisation authorities in 1979 (NS 1979). In addition to earth pressure, bearing capacity etc., this guideline also gave the first national set of regulations for the evaluation of stability of slopes and fills. The guideline established two principles regarding the partial safety related to soil strength. First, the partial safety related to effective stress analyses was differentiated with respect to damage consequence and failure mechanism, as shown in Table 27.1. The effective stress analyses are performed using the effective stress parameters; cohesion (c') and the frictional angle (ϕ'). Such analyses are performed to study the long-term (or drained) stability of slopes. In 1973 Janbu proposed to use the effective stress analyses along with an appropriate pore pressure profile to compute the short-term stability of natural slopes. However, during the years, the total stress based undrained analyses to calculate short-term stability has been the most applied method. The undrained shear strength (c_u) is usually derived from confined triaxial tests. The principles in Table 27.1 have been adapted also for design based on undrained analyses.

Table 27.1 Partial safety factors for effective stress and total stress analyses according to Norwegian construction standards

	Failure mechanism		
Damage consequences	Dilatant	Perfectly plastic	Brittle and contractant[a]
Less serious	1.2	1.3	1.4
Serious	1.3	1.4	1.5
Very serious	1.4	1.5	1.6

[a]Sensitive clays exhibits brittle behaviour i.e. strain softening behaviour

Table 27.2 Partial safety factors for effective stress and total stress analyses according to NPRA regulations

	Failure mechanism		
Damage consequences	Dilatant	Perfectly plastic	Brittle and contractant
Less serious	1.2/1.4[a]	1.3/1.4[a]	1.4
Serious	1.3/1.4[a]	1.4	1.5
Very serious	1.4	1.5	1.6

[a]The Norwegian national application of Eurocode 7 from 2008 requires $\gamma_M > 1.4$ for total stress analyses

Secondly, the guideline stated that established practice in some cases had been to take precautions that reduced the calculated mobilised shear stress along potential slip surfaces with 10–20 %. This procedure for undrained total stress analyses has mostly been handled as a contingency procedure for roads, railways and construction works, until the introduction of the Norwegian Water Resources and Energy Directorate (NVE) regulations in 2008.

27.2.1 Regulatory Framework by the Norwegian Public Roads Administration (NPRA)

The NPRA regulation for geotechnical design in road construction was published in 1990, and has been developed to the present sixth edition (NPRA 2010). The principles in Table 27.2 have been maintained and adapted both for effective and total stress analyses. The main focus of the NPRA regulations from the beginning was the safety requirements regarding the local stability of construction works and the slopes directly affecting the road. The stability of adjacent terrain was addressed when the stability was obviously poor, but there was no systematic requirement for regarding the slopes not directly affected by the construction works for the road.

The possibility for percentual improvement was implemented in the NPRA regulations in 2006 as a contingency procedure. The illustration in Fig. 27.2 was introduced in 2010 to clarify the use regarding the overall stability of larger areas. However, the local stability related to construction elements still needs to satisfy a partial safety factor as shown in Fig. 27.3.

27.2.2 Regulatory Framework by the Norwegian National Roads Administration (NNRA)

The NNRA regulations (2013) for the design of railway structures are in a wiki-based framework which is revised on a yearly basis. Safety requirements for railway construction activities are described with an absolute safety factor. For the railway

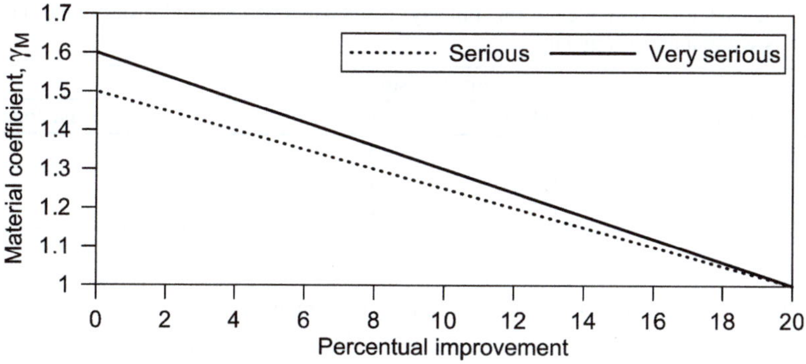

Fig. 27.2 NPRAs recommendations regarding the percentual improvement of material coefficients regarding the overall stability of larger areas based on the consequence category "serious" and "very serious"

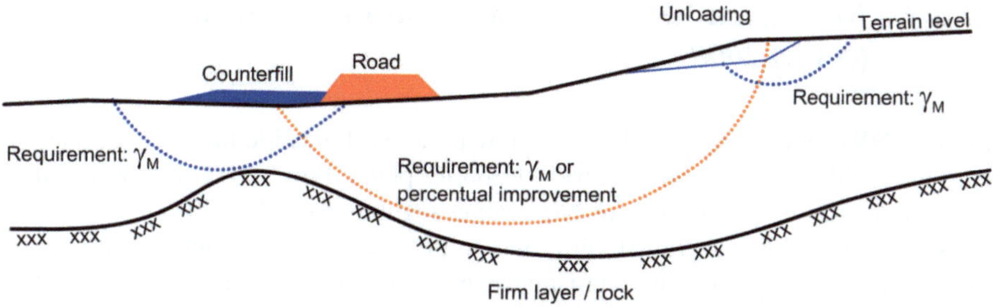

Fig. 27.3 Principle regarding requirements with partial safety factor for local stability and partial safety factor or percentual improvement regarding larger progressive slides (Adapted from NPRA 2010)

Table 27.3 Safety requirements for railway embankments analysed using drained parameters (c' and ϕ') and undrained (c_u) parameters including anisotropic behaviour of soil

	Failure mechanism		
Damage consequences	Dilatant	Perfectly plastic	Brittle and contractant
Less serious	1.2	1.3	1.4
Serious	1.3	1.4	1.5
Very serious	1.4	1.5	1.6

embankment the requirement in safety factor depends on which method of analysis used (drained or undrained), anisotropy, the consequence of failure and the failure mechanism. The required safety factor for railway embankments varies between 1.2 and 2.0 according to Tables 27.3, 27.4, and 27.5.

Presence of brittle clay will lead to a stricter requirement for the safety factor according to Tables 27.3 and 27.4. If the possibility of a progressive slide is present, safety for corresponding areas has to be investigated according to the NVE guidelines described below.

Table 27.4 Safety requirements for railway embankments analysed using undrained (c_u) parameters without anisotropy factors

	Failure mechanism		
Damage consequences	Dilatant	Perfectly plastic	Brittle and contractant
Less serious	1.40	1.55	1.70
Serious	1.55	1.70	1.85
Very serious	1.70	1.85	2.00

Table 27.5 Maximum slope inclination for railway construction

			Fine sand/silt		
Ground conditions	Stone, rock	Gravel, coarse sand	Dry	Layered, water dense	Clay
Maximum inclination	1:1.25	1:1.5	1:2	Special investigation	1:2

27.2.3 Norwegian Water Resources and Energy Directorate (NVE) 2008

In 2008 NVE published guidelines on how to assess landslide hazards in plans for land use, with emphasis on the level of zoning plans. From 2011 they were taken into use as official guidelines for Norwegian regulations (NVE 2011). The guidelines enforce that the overall stability has to be taken into consideration in zoning plans. Before the guidelines were approved, geotechnical consultants and local authorities only had to deal with this issue on the construction stage following the stage of zoning plans. This most often meant that overall stability was ignored in both planning and construction stages.

The method for assessing overall stability for zoning plans can be described as a step-by-step procedure. This procedure has to be carried out by a professional geotechnical consultant, specialized in evaluating these kinds of overall stability problems.

The natural prerequisite is that the zoning plan is beneath the marine limit. If this prerequisite is fulfilled then these steps are to be followed:

Step 1: Assess the possibility of marine clay

This step includes a geotechnical evaluation of the terrain by using existing maps showing quaternary deposits. If these maps indicate that marine clay might be present within or adjacent to the area of the zoning plan then the next step must be performed.

Step 2: Evaluating the terrain

The geotechnical consultant has to evaluate if the terrain shows slopes or ravines with inclination or height differences sufficient to lead to a progressive failures. If these criteria are considered to be present, then the next step must be performed.

Step 3: Finding and evaluating existing ground investigations

This step could also be done together with step 1, but it is here described as a separate exercise. The step implies that the geotechnical consultant gathers further knowledge regarding appearance of brittle clay within and adjacent to the area of the actual zoning plan, gathering information about existing ground investigations, executed in accordance to other projects. Field and laboratory testing shall be performed if ground conditions are not known a-priori. Placement of boreholes must be adjusted to the actual topography. Execution of boring and laboratory testing must fulfil requirements given in the guidelines and adjoining Norwegian standards for boring procedures. After analysing the existing ground conditions the geotechnical consultant should evaluate if the distribution of marine clay found in step 2 should be limited or extended, and in which levels beneath the ground surface marine clay could be found. The investigations might indicate if the marine clay found is to be characterized as brittle clay (defined as remoulded shear strength ≤ 2.0 kPa and Sensitivity ≥ 15 in the NVE presently valid guidelines).

Step 4: Limit and classify the area with potential danger for large landslide

The natural boundaries for where initial slides can occur and the area that may be affected by the development of the slide have to be mapped. At present time the run out area has to be assessed by examining existing literature and/or using simple empirical values from Karlsrud et al. (1985).

Step 5: Analysing stability and suggesting necessary preventive measures

This step includes calculations of overall and local stability, together with suggestions for preventive measures where this is found necessary to obtain a satisfactory level of safety. The NVE guideline suggests a partial safety factor of minimum 1.4 or by using percentual improvement of the initially calculated partial safety factor. The latter method implies that instead of reaching a partial safety factor of minimum 1.4, one can improve the initial safety factor for every potentially critical slip surface through the zone up to a certain level. The percentual improvement required depends on the initially calculated safety factor, and the consequence category.

It is important to state that percentual improvement could only be used by making geometrical changes (loading or unloading) in the surrounding area or by using lightweight materials (Fig. 27.4). For the most severe categories (category 3, and partly category 2) a third-party control of the project is demanded. The stability calculations have to be performed as both drained and undrained analyses. The use of conventional calculation programs using limit equilibrium methods or finite element methods are allowed. The use of percentual method when calculating progressive landslides lead to realization of more plans and projects, as this might reduce the cost of necessary preventive measures.

27.2.4 Eurocodes

The European Committee for Standardization (CEN) consists of members of European countries. This committee has developed a set of design rules which are to be used for the construction phase. These rules are to be unison for all countries

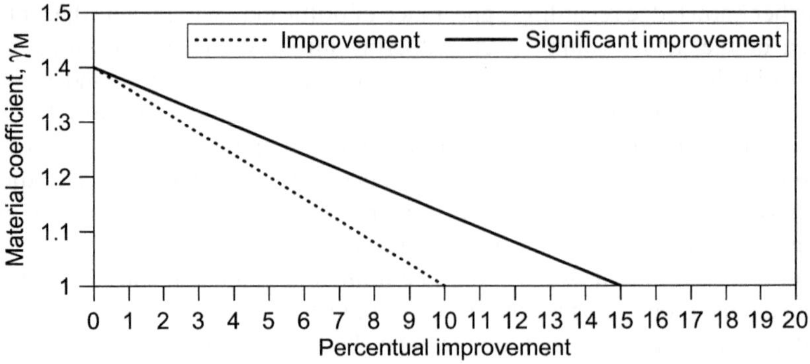

Fig. 27.4 NVEs recommendations regarding the percentual improvement of material coefficients. "Improvement" for consequences category 2, shown as a *dashed line* beneath, or "significant improvement" for consequences category 3. Consequence category 2 is applied for private or local roads, ditches, and smaller fillings etc., which have a negative impact on the overall stability. Consequence category 3 involves movement of people and important institutions for the community

Table 27.6 Partial factor for soil parameters

Soil parameters	Symbol	Values
Friction angle	$\gamma\varphi'$	1.25
Effective cohesion	γ_c'	1.25
Undrained shear strength	γ_{cu}	1.4
Unconfined undrained strength	γ_{qu}	1.4
Unit weight	$\gamma\gamma$	1.0

participating in the committee. In 2008 they were officially taken into use by the Norwegian authorities. Before the Eurocode came into use, Norway had its own set of national design rules for the construction phase. The Eurocode gives only a brief description of the fact that progressive failures have to be taken into account in design. Table NA.A.4. in the Eurocode, also presented in this paper as Table 27.6, indicates the use of an "absolute material factor" γ_M of minimum 1.25 on the friction angle and 1.4 for undrained shear strength. At the same time the Norwegian National Annex gives an opening for the use of percentual improvement, without any distinct requirements. Table NA.A.4. in Eurocode 7 suggests "*material factor is increased beyond the values shown in the table at risk for progressive fracture development in brittle materials considered to be present and when it is required to bring it in line with recognized practice for the stability calculation and the present issue*". The table further mention "*by analysis of overall stability as conditions appear without measures one may find a lower initial material factor then mentioned in table NA.A.4 shall be assessed in relation to the landslide hazard and stability. It is usually assumed that the constructive measures are implemented in a manner that provides unchanged or increased material factor, and so that the factors which may trigger a failure or a landslide are avoided.*" This corresponds with the principles of percentual improvement in the NVE regulations, but Eurocode 7 does not indicate how much increase in the material factor must be considered to account for the progressive failure mechanism.

27.3 Further Challenges in Harmonization of Regulations

Research institutions and consultants, like SINTEF, Multiconsult AS, Norwegian Geotechnical Institute (NGI) and Norwegian University of Science and Technology (NTNU) have been engaged to carry-out the academic and industrial research as a part of the NIFS project. An important challenge for the NIFS project is to provide the basis for a harmonization of the safety regulations for the three governmental authorities. Some of the addressed topics are mentioned below.

27.3.1 Equality Between Partial Safety Factors and Percentual Improvement

This is a central issue in the harmonization of regulations, and has been addressed in a report from SINTEF/Multiconsult (Tørum et al. 2012). Based on this report a concept will be further investigated, where local stability requirements are based on a partial safety factor and the larger progressive slides may be prevented by requirements either by percentual improvement or partial safety factor. The use of percentual improvement will probably be provided with a set of precautions, including requirements to the on-site control.

27.3.2 Anisotropy

Most commonly the use of anisotropy is based on empirical values. The NIFS project is aiming for a standard where the anisotropy conditions are evaluated from local conditions and the use of empirical values are an exception. The database for anisotropy related to plasticity index and other index parameters should be extended to get a better base for the empirical relations. The anisotropy related to principal stresses will also be investigated in the NIFS project. Some preliminary results from an on-going research related to use of anisotropy factors in soft sensitive Norwegian clays is presented by Fauskerud et al. (2012). The report is freely available at www.naturfare.no.

27.3.3 Effect of Strain Softening

Capacity or stability analyses performed using the classical limit equilibrium theory based calculation methods like the "method of slices" Bishop and Morgenstern (1960) and Janbu (1973) are based on the assumption that soil behaviour is "perfectly-plastic". However, for soft sensitive clays these classical theories have limited validity because capacity or safety factors are overestimated if the effect of strain-softening is not accounted in the analyses. The Norwegian code-of-practice suggests different material coefficients for construction activities on soft sensitive clay deposits. The guideline by the NPRA suggest an increase in the required material coefficient from 1.4 to 1.5 or to 1.6 (about 7–14 % increase) depending on the consequence category to account for the effect of strain softening. The guidelines for evaluating overall stability in areas with quick and sensitive clays by NVE suggest a material coefficient ≥1.4. The Norwegian National Annex to Eurocode 7 suggests an unspecific increase in the material coefficient from 1.4 to account for the effect of strain softening. Neither the NVE, nor the Eurocode 7 guidelines are specific about an increase in the material coefficient to account for the strain softening. Using some simplified finite element calculations, Jostad et al. (2013) suggests that capacity of Norwegian soft sensitive clays may be overestimated by an average of 6 % if the effect of strain softening is neglected. From this study it can be noted that the recommendation in the NPRA guideline is suitable. However, further studies are necessary to find a robust answer.

27.3.4 Post-failure Movement of Landslide and Landslide Potential in Sensitive Clays

According to the current practice in Norway, brittle clay is defined by a $S_t \geq 15$ and a $c_{ur} \leq 2.0$ kPa. Thakur et al. (2012) initiated a discussion on the definition of brittle clay. This study shows that we need to understand the mechanical properties of sensitive clay better, not only numerically but also physically. Thakur et al. (2013) presents a comprehensive overview over several parameters that may influence the extent of landslides, e.g. topography, stability number (Nc), remoulded shear strength (c_{ur}), liquidity index (I_L) and quickness (Q). The Norwegian landslide data as presented by Thakur and Degago (2012) and Thakur et al. (2013) supports the fact that large landslides with retrogression greater than 100 m are only possible when $c_{ur} < 1.0$ kPa or $I_L > 1{,}2$ or $Q > 15$ % and $N_c > 4$. These criteria are useful and can be used as indicators to assess the potential for occurrence of large landslides. However, the extent of a landslide with only an individual geotechnical parameter may not be sufficient. A complete stress–strain behaviour of soft sensitive clays must be accounted for in the calculation of the post-failure movement of landslides. The finding is supported by the Canadian landslide data by Mitchell and Markell (1974). A concept of defining remoulding energy for Norwegian sensitive clays is suggested by Thakur and Degago (2013) and Thakur et al. (2013). This concept shows that c_{ur}, S_t, and γ_r should be studied together to improve our understanding of the mechanical behaviour of brittle clays. The required remoulding energy provides a better basis for understanding the post-failure movements of landslides in brittle clays.

Acknowledgements The cooperative research program «Natural hazards: Infrastructure for Floods and Slides (NIFS)» by the Norwegian Public Roads Administration, the Norwegian National Rail Administration and the Norwegian Water Resources and Energy Directorate is acknowledged for the support. The authors also wish to express their gratitude to the reviewer Arnstein Watn for his valuable comments.

References

Bishop AW, Morgenstern NR (1960) Stability coefficients for earth slopes. Geotechnique 10(4):129–150

Fauskerud OA, Athanasiu C, Havnegjerde CR et al (2012) Use of anisotropy factors in stability calculations for brittle clay slopes. NIFS report nr 75. Norwegian Water Resources and Energy Directorate. ISBN:978-82-410-0863-4

Janbu N (1973) Slope stability computations. In: Hirschfeld RC, Poulos SJ (eds) The embankment dam engineering, casagrande volume. Wiley, New York, pp 47–86

Janbu N, Nestvold J, Røe Ø et al (1993) Leirras – årsaksforhold og rasutvikling. Særtrykk fra Verdalsboka Ras i Verdal, bind B (Clay slides, causes and development. Offprint from the book "Verdalsboka Ras i Verdal, bind B"). Verdal kommune. ISBN:8-9909500-8-5 (in Norwegian)

Jostad HP, Fornes P, Thakur V (2013) Effect of strain-softening in design of fills in gently inclined areas with soft sensitive clays. In: First international workshop on landslide in sensitive clays, Quebec

Karlsrud K, Aas G, Gregersen O (1985) Can we predict landslide hazards in soft sensitive clays? Summary of Norwegian practice and experiences. NGI Publication No. 158. Norwegian Geotechnical Institute. ISBN:82-546-0119-4

Mitchell RJ, Markell AR (1974) Flow slides in sensitive soils. Can Geotech J 11(1):11–31

NIFS (2012) Program plan 2012–2015 for etatsprogrammet: Natural Hazards – Infrastructure for flood and (in Norwegian). Published by Norwegian Water Resources and Energy Directorate, Olso. ISSN:1501–2832. Available at www.naturfare.no

NNRA (2013) Regulations for design of railway construction (in Norwegian). Published by Norwegian National Rail Administration, Oslo. Available at https://trv.jbv.no

NPRA (2010) Guidelines for geotechnical engineering in road constructions- Handbook 016. (in Norwegian). Published by Norwegian Public Roads Administration, Oslo. Available at http://www.vegvesen.no/Fag/Publikasjoner/Handboker

NS (1979) Sikkerhetsprinsipper I geoteknikk. Veiledning for bruk av grensetilstandsmetoden (in Norwegian). Published by Norges Byggstandardiseringsråd, Oslo

NVE (2011) Flaum- og skredfare i arealplanar-Regulation No. 2/2011. (in Norwegian). Published by Norwegian Water Resources and Energy Directorate, Oslo. ISSN:1501–9810

Thakur V, Degago SA (2012) Quickness of sensitive clays. Géotech Lett 2(3):87–95

Thakur V, Degago SA (2013) Disintegration energy of sensitive clays. Géotechn Lett 3:20–25. doi 10.1680/geolett.12.00062

Thakur V, Oset F, Aabøe R et al (2012) A critical appraisal of the definition of Brittle clays (Sprøbruddmateriale). In: Proceedings of the 16th Nordic geotechnical meeting, Copenhagen, 9–12 May 2012, vol 1, pp 451–462

Thakur V, Degago SA, Oset F et al (2013) Characterisation of post-failure movements of landslides in soft sensitive clays. In: First international workshop on landslide in sensitive clay, Quebec

Tørum E, Christensen S, Narjord H et al (2012) Likestilling mellom bruk av absolutt materialfaktor og prosentvis forbedring? SINTEF/Multiconsult Report SBF2012A0309. Norwegian Water Resources and Energy Directorate. ISBN:978-82-410-0860-3

Chapter 28
Risk Assessment for Quick Clay Slides – The Norwegian Practice

Bjørn Kalsnes, Vidar Gjelsvik, Hans Petter Jostad, Suzanne Lacasse, and Farrokh Nadim

Abstract The approach for the assessment of the risk associated with quick clay slides in Norway is a qualitative/semi-quantitative procedure developed as part of work for The Norwegian Water Resources and Energy Directorate. Slide areas are classified according to "engineering scores" based on an evaluation of the topography, geology and local conditions (to qualify hazard) and an evaluation of the elements at risk, persons, properties and infrastructure exposed (to qualify consequence). The risk score to classify the mapped areas into risk zones is obtained from the relationship $R_S = H_{WS} \times C_{WS}$, where R_S is the risk score, H_{WS} is the weighted hazard score and C_{WS} is the weighted consequence score. The risk matrix is divided in five risk classes. Guidelines for the implementation of the risk matrix are administered by NVE. In practice, the approach is used to make decisions on required mitigation measures to reduce the risk. The approach is simple and makes room for engineering experience and judgment. For detailed regional planning, slope stability calculations need to be made. Methods for quick clay slide stability calculations taking into account the brittle behaviour of the material are under development. This chapter provides an illustration of this development work.

Keywords Quick clay • Hazard • Risk • Codes • Slope stability • Strain-softening

---

B. Kalsnes (✉) • S. Lacasse • F. Nadim • H.P. Jostad
Norwegian Geotechnical Institute (NGI), Oslo, Norway
e-mail: bjorn.kalsnes@ngi.no; hpj@ngi.no

V. Gjelsvik
Norwegian Geotechnical Institute (NGI), Trondheim, Norway

J.-S. L'Heureux et al. (eds.), *Landslides in Sensitive Clays: From Geosciences to Risk Management*, Advances in Natural and Technological Hazards Research 36,
DOI 10.1007/978-94-007-7079-9_28, © Springer Science+Business Media Dordrecht 2014

28.1 Introduction

Quick clay landslides represent one of the major environmental challenges in Norway. They have been observed since the early settlers immigrated in the country. In the nineteenth and twentieth century, several damaging quick clay slides occurred. The Verdal slide in 1893 is the largest historical quick clay slide in Norway. It involved 55 million m^3 of clay and caused 116 fatalities, making it the most destructive natural catastrophe in Norway. Nadim et al. (2008) summarized the impact of natural hazards in the Nordic countries. Compared to many areas of the world, the number of fatalities caused by natural hazards in Norway is small, mainly due to low population density in exposed areas. The economic losses and the societal risks, however, are significant. In Norway, the main natural hazards are landslides, snow avalanche, floods and, to a lesser extent, earthquakes. Statistically, 10 large slides can occur in Norway in the next 50–100 years, each with possibly 20–100 deaths. There is no clear definition of a "large" slide. However, slides covering an area of 80,000–100,000 m^2 or involving a volume of 0.5–1 million m^3 are qualified as large.

About 5,000 km^2 of Norway are covered by soft marine clay deposits, with 20 % of this area on highly sensitive or quick clay. Landslides in quick clay frequently start without warning and turn into a fluid in minutes. Large quick clay slides involving several million m^3 occur at intervals of about 4 years (Aas 1981). The largest quick clay slide in Norway in the twentieth century occurred in Rissa, near Trondheim, in 1978. It covered an area of 330,000 m^2, and 5–6 million m^3 of clay poured out of the slide area. The Rissa quick clay slide started with the failure of a small fill close to a lakeside. The initial slide involved 200 m^3 of sediments. It grew to about 6 million m^3 in a few hours through a retrogressive sliding process.

After the quick clay slide in Rissa, a national plan for mapping and managing quick clay hazard and risk was initiated. Gregersen (2001) developed a method to classify and map the risk posed by potential quick clays. In 2009, the Norwegian Water Resources and Energy Directorate (NVE) was assigned the responsibility for managing all types of landslide risk (soil, rock and snow), including quick clay slides. NVE produced a guideline with special emphasis on stability issues for areas with quick or sensitive clays (NVE 2011). This paper outlines the Norwegian practice for handling the risk associated with quick clay landslides in Norway. The guidelines described cover all types of sliding that may occur in sensitive clays, including retrogressive and rotational slides.

28.2 Identification of Hazard Zones

Quick clay is normally defined as clay with remoulded undrained shear strength less than 0.5 kPa (see also Thakur and Degago 2012). Over the past 20 years, the Norwegian authorities have mapped about 1,800 potential hazard zones, prioritizing the most affected areas. The areas were evaluated and categorized with respect to their relative hazard, consequence and risk. The classification functions as a management and prioritizing tool to ensure sustainable development with respect to landslide hazard for the communities. In 2008, new guidelines for evaluating the overall stability in areas with quick and sensitive clays were developed by NVE. The regulations include requirements on e.g. extension and quality of site investigations, evaluation of landslide mechanisms and critical calculation profiles, choice of strength parameters, type of stability analyses, level of safety and effect of stabilizing measures. An important element in the guidelines from NVE is that hazard mapping shall be included in the construction planning process from the very start. The guidelines give directions for the quality and necessary amount of engineering work required for a specific project (Gregersen 2006).

28.3 Hazard Classes

The hazards level depends on topography, geological and geotechnical conditions, and changes at the site. The hazard levels are estimated on the basis of simple theoretical evaluations of the potential area that can be involved in a quick clay slide, in combination with back-calculations of a number of historical quick clay slides (Aas 1979). Based on these studies, the following topographical criteria were used for the mapping of the hazard:

- Evenly sloping terrain of inclination >15°
- "Ravines" with height differences of more than 10 m
- Maximum slide length of 15 times slope height difference

However, areas not fulfilling these criteria must also be analysed in the presence of quick clay slides, when building activities or other changes in topographical or loading conditions are considered. With the above criteria, the hazard is estimated with the help of Table 28.1. The weight on each hazard qualifies the importance of the factor relative to slope instability (or potential for causing loss). The hazard classes depend on topography, geology, geotechnical conditions and changes due to human activities, climatic effects, erosion, etc. The hazard is described as low, medium or high, depending on the weighted hazard score. The hazard classes are:

Low: Favourable topography and soil conditions; extensive site-specific data available; no erosion; no earlier sliding; no planned changes, or changes will improve stability.

Medium: Less favourable topography and soil conditions; uncertainty in site-specific data; active erosion; important earlier sliding in area; planned changes give little or no improvement of stability.

High: Unfavourable topography and soil characteristics; active erosion; extensive earlier sliding in area; planned changes will reduce stability.

The zones with weighted score between 0 and 17 in Table 28.1 (up to 33 % of maximum score) are mapped as "low hazard" and have low probability of sliding failure. The zones with weighted score between 18 and 25 (33–50 % of maximum score) are mapped as "medium hazard" and have a higher, though not critical, probability of sliding. The zones with weighted score between 26 and 51 (greater

Table 28.1 Evaluation of hazard score for slides in quick clay in Norway

		Score for hazard			
Hazard	Weight	3	2	1	0
Topography					
Earlier sliding	1	Frequent	Some	Few	None
Height of slope, *H*[a]	2	>30 m	20–30 m	15–20 m	<15 m
Geotechnical characteristics					
Overconsolidation ratio (*OCR*)	2	1.0–1.2	1.2–1.5	1.5–2.0	>2.0
Pore pressures[b]					
In excess (kPa)	3	>+30	10–30	0–10	Hydrostatic
Under pressure (kPa)	−3	>−50	−(20–50)	−(20–0)	Hydrostatic
Thickness of quick clay layer[c]	2	>H/2	H/2–H/4	<H/4	Thin layer
Sensitivity, S_t	1	>100	30–100	20–30	<20[e]
New conditions					
Erosion[d]	3	Active/sliding	Some	Little	None
Human activity					
Worsening effect	3	Important	Some	Little	None
Improving effect	−3	Important	Some	Little	None
Total score					
Maximum weighted score		51	34	17	0
% of max. weighted score		100 %	67 %	33 %	0 %

[a]For the quick clays in the study, inclination was identical for all slopes (1:3), and slope inclination was not included as a variable. In a general study, slope inclination should be added as a factor
[b]Relative to hydrostatic pore pressure
[c]In general, the extent and location of the quick clay are also important; "thin" layer is about 1 m
[d]Erosion at the bottom of a slope reduces stability
[e]For $S_t<20$, other factors need to also intervene to bring the hazard higher than low

than 50 % of maximum score) are mapped as "high hazard" and have a relatively high probability of sliding.

The existing guidelines do not prescribe where the pore pressures are to be measured. Pore pressures are rarely measured as part of the hazard and risk mapping, although this is gradually improving. The topography can be used to estimate pore pressures. The pore pressures on the slip surface are of most interest. Measurements are recommended at the foot of the slope and/or at depth at some distance upslope. Measurements at the top of the slope are preferably made at two levels, where the deepest is at a depth at least equal to the height of the slope, i.e. at about the same level as the measurement at the toe. Accessibility is also a major factor for the measurement.

28.4 Consequence Classes

Consequences are evaluated in terms of fatalities and environmental, financial and social losses. The consequences are described as *not severe*, *severe* or *highly severe*, and are derived through a scoring and weighting system similar to the one used

Table 28.2 Evaluation of consequence for slides in quick clay in Norway

		Score for consequence			
Possible damage	Weight	3	2	1	0
Human life and health					
Number of dwellings[a]	4	>5 Closely spaced	>5 Widely spaced	≤5 Widely spaced	0
Persons, industry building	3	>50	10–50	<10	0
Infrastructure					
Roads (traffic density)	2	High	Medium	Low	None
Railways (importance)	2	Main	Required	Level	None
Power lines	1	Main	Regional	Distrib. network	Local
Property					
Buildings, value[b]	1	High	Significant	Limited	0
Consequence of flooding[c]	2	Critical	Medium	Small	None
Total score					
Maximum weighted score		45	30	15	None
% of max. weighted score		100 %	67 %	33 %	0 %

[a]Permanent residents, in both sliding area and within run-out distance
[b]Normally no one on premises, but building(s) have historical or cultural value
[c]Slides may cause water blockage or dam overflow, flooding may cause new slides; there should be time for evacuation; damage depends on a complex interaction of several factors

for hazard. The evaluation of the consequences is done with the help of Table 28.2. The weight on each consequence qualifies the relative impact of the factor considered. The consequence classes are:

Not severe: no or small danger for loss of human life and costly damage.
Severe: potential life or property loss or important economical/social loss.
Highly severe: high exposure, potential large number of fatalities (residents or persons temporarily in area), or large economical or social loss.

Zones with weighted score between 0 and 6 (up to 13 % of maximum score) are classified as "not severe". There should be very few or no permanent residents in these zones. Zones with weighted score between 7 and 22 (13–50 % of maximum score) are classified as "severe". Zones with weighted score between 23 and 45 (greater than 50 % of maximum score) are classified as "highly severe". There would be a large number of persons, either as residents or temporarily in the area.

28.5 Risk Classes

Table 28.3 gives the scores for the risk classes used to map the quick clay zones in Norway. The risk score is the product of the weighted hazard and weighted consequence, where the weighted quantities are taken as the ratio of the maximum possible

Table 28.3 Risk classes for slides in quick clay in Norway

Risk class	1 (lowest)	2	3	4	5 (highest)
Risk weighted score (R_{WS})	0–160	167–600	628–1,900	1,906–3,200	3,200–10,000
R_{WS} (in % of max R_{WS})	0–1.6 %	1.6–6 %	6.3–19 %	19–32 %	32–100 %

Table 28.4 Activity matrix as a function of risk class

	Risk class			
Activity	1–2	3	4	5
Soil investigations	None	Consider additional *in situ* tests and pore pressure monitoring	Require additional *in situ* tests and pore pressure monitoring	Require additional *in situ* tests, pore pressure monitoring and laboratory tests
Stability analyses	None	None	Consider doing	Require
Remediation[a]	None	None	Consider doing	Require

[a]e.g. erosion protection, stabilizing berm, unloading, soil stabilization, moving of residents

total score. Using integers of the percentage, the maximum score becomes 100 × 100 or 10,000 and the lowest 0. The risk score is obtained as follows:

$$\text{Risk} = \text{Hazard} \times \text{Consequence}$$
$$R_{WS}\left(\text{in }\%\right) = H_{WS}\left(\text{in }\%\right) \times C_{WS}\left(\text{in }\%\right)$$

where R_{WS} = Weighted score for risk mapping (integer between 1 and 10,000)
H_{WS} = Hazard weighted score (integer between 0 and 100)
C_{WS} = Consequence weighted score (integer between 0 and 100)

28.6 Decision-Making

To make a decision on the need for additional soil investigations, new stability analyses or other remedial actions, Table 28.4 is used for quick clay areas in Norway.

28.7 Emergency Preparedness

NVE organized an exercise in preparedness in 2010 and the Norwegian Directorate for Civil Protection and Emergency Planning (DSB) in 2013. The first exercise simulated a quick clay landslide of national dimension with fatalities. The second assembled a group of experts to establish the premises for the national risk that could be posed by quick clay slides. Worst case scenario, estimates of hazard and vulnerability and valuation of the consequences were discussed by the different stakeholders involved. The results will become available in June 2013.

The NVE simulation in 2010 was made as realistic as possible with the participants not knowing what to expect and having a party simulating fatalities.

The participants were briefed of the exercise ahead of time, but they did not know the details of what was to happen. A majority of stakeholders were invited, from authorities, to police, private actors and the media. The exercise aimed at improving the parties' ability to respond under pressure in a complex context, and making decisions under critical conditions. The emergency routines, information channels and response tools in each of the participating organizations were tested. The exercise also tested who took responsibility for the decisions made, and whether the parties had the same understanding of the respective responsibility and roles.

An evaluation report was prepared with, among others, the following lessons learned: the respective roles and responsibility should be more clearly defined and communicated to all parties; not everyone received the required information in time and internal information within each of the organizations was insufficient; control of the information given to the media was problematic (it was not possible to check the wording with the journalists); making notes and looking at maps in a high pressure context proved very difficult; how could the private actors doing emergency work/repair ensure that they had a contract (and would be paid for their work); the authorities should create an emergency group, and invite wide participation; a logistical and communication tool should be developed for crisis situations and made available to all stakeholders; the need for frequent preparedness exercise, as the people change in each of the organisations involved; and some of the routines in the governmental handbooks needed to be updated.

28.8 Hazard and Risk Mitigation in Drammen

The city of Drammen, on the Drammen River, rests on a deposit of soft clay. Stability analyses close to the centre of the city indicated that some areas did not have satisfactory safety against a slope failure. Based on the results of the stability analyses and the factors of safety (FS) obtained, the area under study was divided into three zones as illustrated in Fig. 28.1 (Gregersen 2005):

- Zone I – FS satisfactory
- Zone II – FS shall not be reduced
- Zone III – FS too low, area must be stabilised

Figure 28.2 illustrates the mitigation done in Zone III: a counter fill was immediately placed in the river to support the river bank, and the factor of safety checked again. The counter fill provided adequate stability. In Zone II, no immediate mitigation was taken, but a ban was placed on any new structural and foundation work without first ensuring increased stability. Figure 28.3 illustrates three examples:

- if an excavation is planned, the clay will have to be stabilised with e.g. anchored sheetpiling or soil stabilisation, for example chalk-cement piles;
- if new construction is to be carried out, the engineer needs to check the possible effects of the change on the stability downslope: (1) adding a floor to a dwelling may cause failure because of added driving forces, or (2) new piling up slope will cause a driving force on the soil downslope.

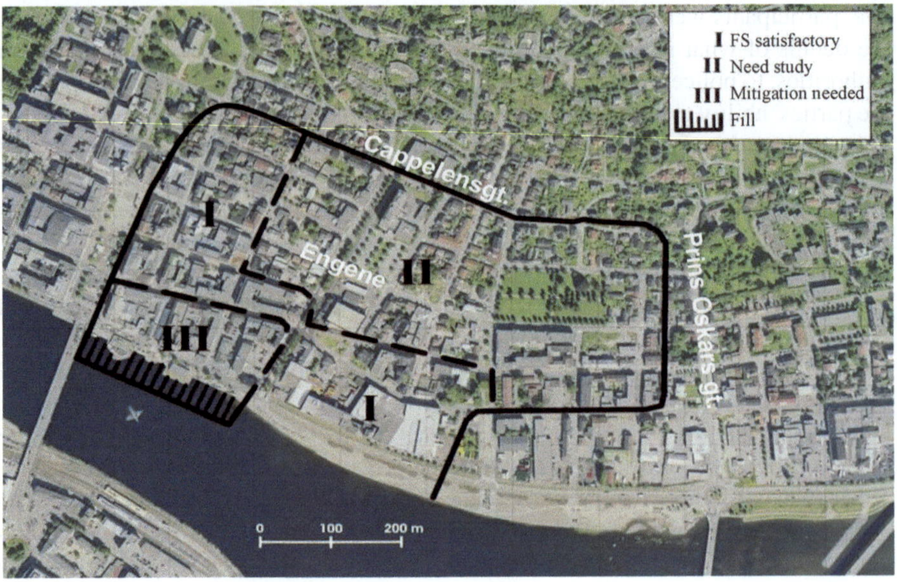

Fig. 28.1 Hazard zones (Gregersen 2008)

Fig. 28.2 Mitigation in Zone III in Drammen (Gregersen 2008)

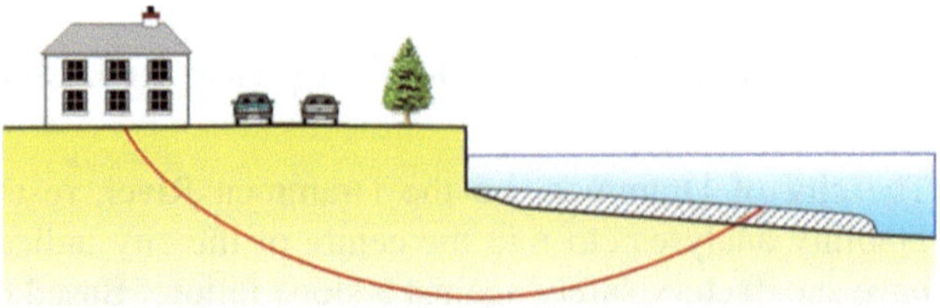

28.9 Stability in Areas of Sensitive and Quick Clays

28.9.1 Requirement

During the planning of fills and cuts in areas of sensitive clays, one needs to ensure an adequate safety against slope failure. This is done by calculating the average mobilized shear stress along different slip surfaces and comparing it with the corresponding characteristic average shear strength divided by a material coefficient, γ_M. NVE (2011) uses the term material coefficient rather than safety factor because of the partial safety coefficient practice in Norway and Europe.

The requirement in Norway for the material coefficient γ_M is 1.4 (NVE 2011). If the material coefficient in an area is less than 1.4, stabilizing measures should be undertaken to increase the material coefficient. However for a standing slope,

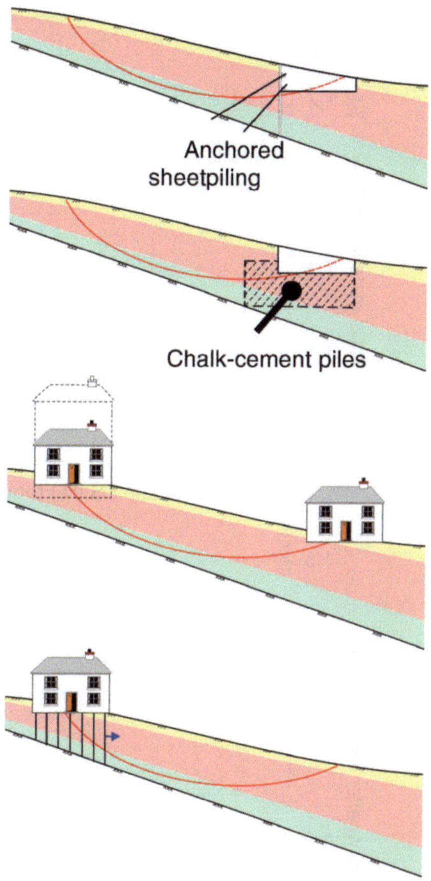

Fig. 28.3 Hazard and preventive measures in Zone II in Drammen

NVE allows an alternative, where the material coefficient should be increased, but not necessarily to as much as of 1.4.

Figure 28.4 illustrates the NVE requirement. The required γ_M is a function of the initial material coefficient and the improvement required. The improvement required depends on the hazard class (NVE 2011). The upper diagram in Fig. 28.4 gives the minimum required increase in material coefficient γ_M, (in %), for "substantial improvement" or "improvement" levels. The lower diagram provides the resulting required γ_M. The improvements are to be made through topographical modifications. As an example (lower diagram), for an initial γ_M of 1.2, an improvement from γ_M=1.2 to γ_M=1.26 and 1.29 is required by NVE (2011) for the two levels of improvement considered. Slip surfaces with a material factor γ_M of 1.0 under present conditions require an improvement to γ_M=1.10 and 1.15 for the two levels of "improvement".

The reason for allowing a material coefficient less than 1.4 is the fact that the slope is standing today which is a confirmation that the slope has a material coefficient of at least 1.0. Any improvement represents a real gain to the present

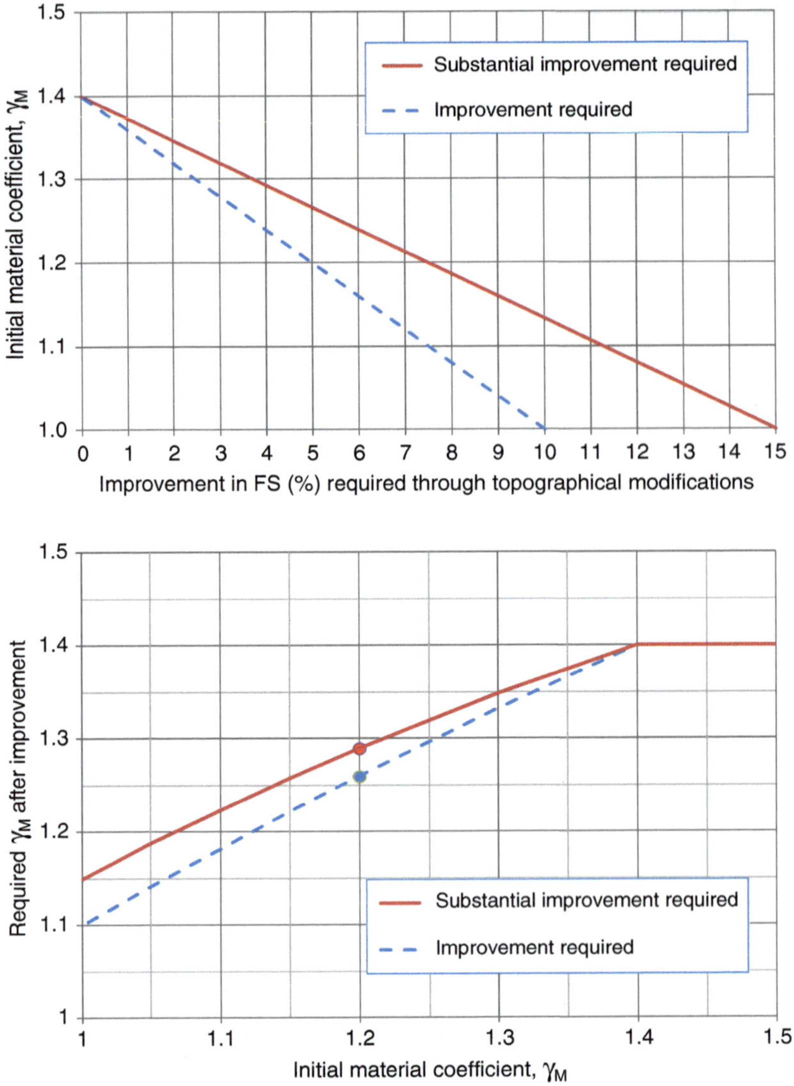

Fig. 28.4 Required increase in material coefficient (*upper diagram*, NVE 2011) and resulting required material coefficient (*lower diagram*, H. Heyerdahl, March 2013, personal communication)

safety of the slope (where the material coefficient has in itself some uncertainty). The NVE requirement needs to be satisfied for all potential slip surfaces. For sensitive clays, the peak undrained shear strength is reduced to account for strain-softening at large shear strains. The required reduction of the peak shear strength in triaxial compression is often taken as 15 % (NVE 2011).

The guidelines should include advice on how to evaluate slope stability near-shore and at the shoreline, as a number of slides have occurred, sometimes repetitively, with devastating consequences (L'Heureux et al. 2011; L'Heureux 2012).

28.9.2 Recent Developments

For brittle materials such as very sensitive and quick clays, the material coefficient can be overestimated if the peak undrained shear strength is used. The strain-softening behaviour needs to be taken into account (Jostad et al. 2013; Fornes and Jostad 2013).

The stability of long slip surfaces in brittle and sensitive soils cannot be calculated by classical limiting equilibrium methods. The calculated material coefficient will be overestimated for long slip surfaces to a greater degree than for local slip surfaces. Failure on long slip surfaces generally develops progressively in time and space. The shear strength along part of the slip surface reduces significantly, moving towards the remoulded shear strength, while other parts are still in the pre-peak, hardening regime. The peak shear strength is not representative for the shear resistance along the potential slip surfaces.

An example of such strain-softening effects can be found in the back-calculation of the Vestfossen slide.

28.9.3 Back-Calculation of the Vestfossen Slide

The slide involved about 50,000 m^3 of soil that propagated about 100 m in almost horizontal terrain until it stopped on the opposite side of the Vestfossen River. The geometry before and after failure illustrates a complex set of slip surfaces, from a long flat slip probably retrogressive surface to a rotational slip surface (Fig. 28.5). The central rotational slip surface is considered for the present paper. Figure 28.5 also shows the undrained shear strength (undisturbed and remoulded) from the field vane test (FV).

The failure was triggered by the new fill placed mid-slope when a new soccer stadium was to be built. During project planning, the slope was probably assumed to have sufficient safety margin because the new slope was not steeper than the original slope. However, when new limiting equilibrium analyses were done by Grimstad and Jostad (2011) with the results of triaxial compression, direct simple shear and field vane tests, the material coefficient was 1.06. The material coefficient calculated with the peak undrained shear strength along the failure surface gave a γ_M of 2.

The initiation and progressive failure were captured well by a large deformation finite element analysis with PLAXIS 2D (Grimstad and Jostad 2011), using the NGI-ADPSoft material model: γ_M was then 1.0.

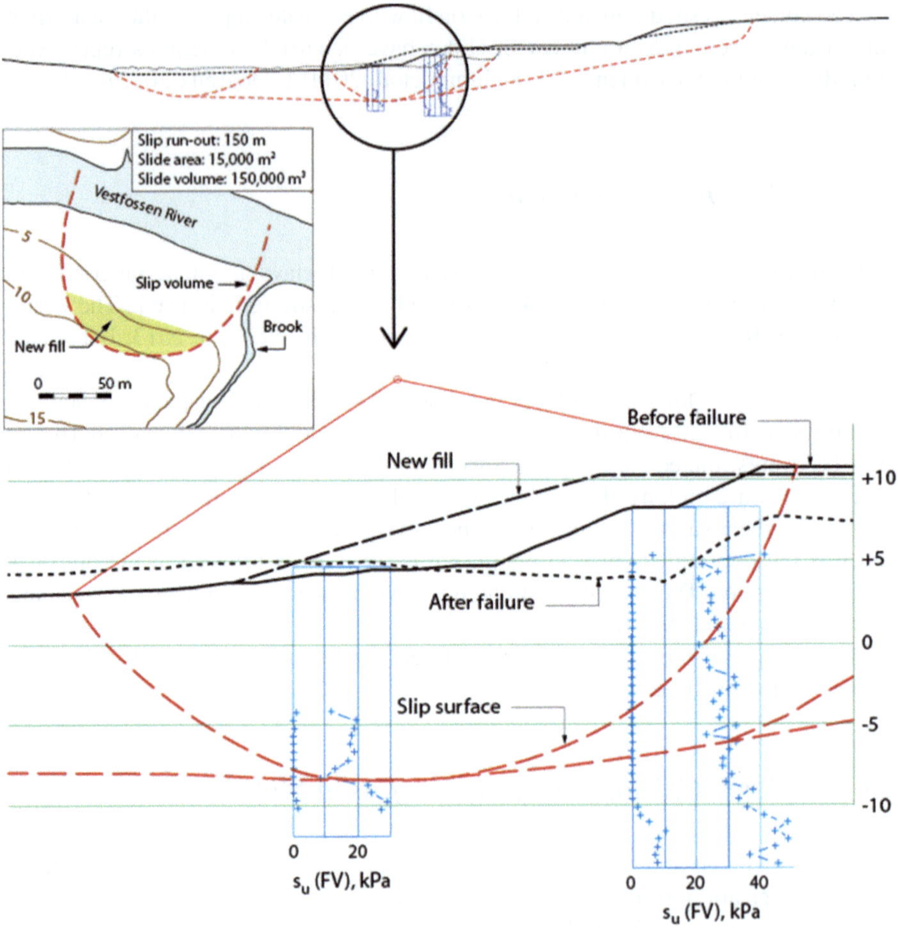

Fig. 28.5 Extent of the 1984 slide at Vestfossen (Grimstad and Jostad 2011; drwg from NGI files)

A paradox with the Vestfossen case is that with the present quick clay mapping criteria (Sect. 28.3), the Vestfossen slope would not have been mapped because its height is less than 10 m. And yet a multiple failure with long run-out occurred.

28.10 Summary and Conclusions

The paper presented the practice in Norway for ensuring stable slopes areas of sensitive and quick clays. The design requirements and risk qualification aim at helping the designer in finding practical and safe solutions. An easy-to-use approach was developed to qualify and partly quantify hazard, consequence and risk, where a good dose of engineering experience, expertise and judgment are necessary. An example of exercise in preparedness was also presented.

Stability calculations are usually done by limiting equilibrium approaches that account for horizontal, vertical and moment equilibrium. No commercial software that fully accounts for progressive failure is available today. Limit equilibrium methods will continue to be used in practice. In Norway, the consideration of progressive failure is done by reducing the triaxial compression peak shear strength by about 15 % while the peak direct simple shear and triaxial extension strengths are used in the analyses. This assumes that the undrained shear strength was obtained from samples of perfect quality, and is also meant to indirectly account for strain compatibility and time effects (NVE 2011).

Acknowledgments The authors wish to thank Dr. Denis Demers of Ministère des Transports, Québec, and Dr. Jean-Sébastien L'Heureux of NGI for their excellent review of the manuscript. The authors also thank Mr. Håkon Heyerdahl from NGI for his help with the manuscript.

References

Aas G (1979) Skredfare og arealplanlegging. Vurdering av faregrad og sikringstiltak. Ullensvang Hotell, Lofthus i Hardanger. Norske sivilingeniørers forening. NIF-kurs, Oslo

Aas G (1981) Stability of natural slopes in quick clays. In: Proceedings of the 10th international conference on soil mechanics and foundation engineering, vol 3, Stockholm, pp 333–338

Fornes P, Jostad HP (2013) A probabilistic study of an inclined slope in sensitive clay using FEA. In: 3rd international symposium on computational geomechanics (ComGeoIII), Krakow, 21–23 August 2013

Gregersen O (2001) Metode for kartlegging og klassifisering av faresoner, kvikkleire. Report 2000, 1008-2, Norwegian Geotechnical Institute, Oslo, 31 August

Gregersen O (2005) Program for økt sikkerhet mot leirskred. Risiko for kvikkleireskred Bragernes, Drammen. Report 2004 1343-1, Norwegian Geotechnical Institute, Oslo, 26 January

Gregersen O (2006) Construction work in clay hazard areas. New guidelines in geotechnical engineering. Geoteknikkdagen 2006, Oslo, pp 30.1–30.11

Gregersen O (2008) Kartlegging av skredfarlige kvikkleireområder. NGM 2008. In: Proceedings of the 15th Nordic geotechnical conference, Sandefjord, pp 178–186

Grimstad G, Jostad HP (2011) Stability analyses of quick clay using FEM and an anisotropic strength. In: NGM 2012, Proceedings of the 16th Nordic geotechnical conference, vol 2, Copenhagen, pp 675–680

Jostad HP, Fornes P, Thakur V (2013) Effect of strain-softening in design of fills in gently inclined areas with soft sensitive clays. In: 1st international workshop on landslides in sensitive clays, Québec

L'Heureux JS (2012) Vurdering av kartleggingsgrunnlaget for kvikkleire i strandsonen. NGI Report 20120754-01-R, Norwegian Geotechnical Institute, Oslo, 1 December

L'Heureux JS, Hansen L, Longva O, Eilertsen RS (2011) Landslides along Norwegian fjords, processes, cause and hazard assessment. In: Proceedings of the 2nd world landslide forum, Rome

Nadim F, Pedersen SAS, Schmidt-Thomé P, Sigmundsson F, Engdahl M (2008) Natural hazards in Nordic countries. Epis J Int Geoscie 31(1):176–184

NVE (2011) Retningslinjer 2/2011. Flaum- og skredfare i arealplanar. Rev.15, ISSN:1501-9810

Thakur V, Degago S (2012) Quickness of sensitive clays. Geotech Lett 2:81–88

good use of engineering experience, expertise and judgement are necessary. An example ... to ... procedures was also presented.

[illegible]

Chapter 29
Quaternary Geology as a Basis for Landslide Susceptibility Assessment in Fine-Grained, Marine Deposits, Onshore Norway

Louise Hansen, Fredrik Høgaas, Harald Sveian, Lars Olsen, and Bjørn Ivar Rindstad

Abstract For decades, mapping of Quaternary geology in Norway has been an important basis for hazard and risk assessment for landslides in highly sensitive clay (quick clay). One aspect of particular significance is information on the occurrence of fine-grained, marine deposits including clays. The marine limit (ML) defines a natural upper limit to these deposits, so issues concerning marine clays can be disregarded above ML. Below ML, a filtering of the Quaternary map information is needed to identify areas where clays are potentially present. Fine-grained, marine deposits are more frequently encountered below some deposit types than others, and the possibility of encountering marine clays within an area is defined as clay-deposit susceptibility. Stratigraphic information is needed for verification. Further landslide susceptibility assessment requires additional information on topography and ground conditions. Other geological information is also important such as on the distribution of landslide debris and landslide scars. Here, it is beneficial to use high-resolution, digital elevation models for detailed analysis of the terrain also below sea level in near-shore areas. Maps with filtered geological information can work as a supplement to maps produced during the ongoing quick-clay mapping program and as a help prioritize areas requiring further investigation.

Keywords Landslide susceptibility • Quaternary geology • Marine limit • Clay • Marine deposits • Quick clay • Sensitive clay • Clay-deposit susceptibility • Mapping

L. Hansen (✉) • F. Høgaas • H. Sveian • L. Olsen • B.I. Rindstad
Geological Survey of Norway (NGU), Trondheim, Norway
e-mail: louise.hansen@ngu.no

J.-S. L'Heureux et al. (eds.), *Landslides in Sensitive Clays: From Geosciences to Risk Management*, Advances in Natural and Technological Hazards Research 36,
DOI 10.1007/978-94-007-7079-9_29, © Springer Science+Business Media Dordrecht 2014

29.1 Introduction

Landslides in sensitive clays, or quick clay, are serious threats to Norwegian society and large landslides occur on a relatively regular basis (L'Heureux 2012). To reduce this threat, a national quick-clay mapping program has been running for decades to identify potential zones for large, quick-clay landslides. This mapping is important for municipal planning and protection of life and property. Around 1,750 quick-clay zones have been mapped in Norway till now. However, there are still significant challenges. Although many large areas have been mapped, several areas have not yet been addressed, and there is an increasing focus on smaller valleys and marginal areas along fjords and coasts. Another challenge is that quick clay can also be present outside the identified quick-clay zones that are designed to cover only areas where large landslides could potentially occur. However, smaller landslides can also be fatal. It is therefore important to communicate to decision-makers and the public about the limitations of the current quick-clay mapping program.

Quaternary maps, which are the basis for quick-clay mapping, contain information on the distribution of marine clays, landslide scars and other relevant geological parameters. The marine limit (ML) is a key parameter, and is a natural basis for general, landslide susceptibility assessment of fine-grained, fjord-marine deposits in Norway (Hansen et al. 2012a). The is also an important basis for landslide susceptibility in e.g. eastern Canada (e.g. Quinn et al. 2010). However, whereas the Champlain sea clays occur within one big basin the Norwegian fjord marine clays are generally distributed in numerous fjord valleys. This results in an uneven data coverage making it difficult to establish statistical trends for regional GIS analyses. In addition, many clay areas in Norway border fjords and sounds and subaqueous conditions are important.

Registrations of ML in Norway have been carried out for many decades leaving information scattered in publications, reports and maps. Registrations of ML have not formerly been systematically organized on a regional basis, and overviews of ML have not been detailed enough for practical purposes. For this reason the Geological Survey of Norway (NGU) initiated a project in 2011 to gather ML information on a national level to be stored in the national database of superficial deposits (Høgaas et al. 2012). About 1,750 ML points are registered pr. January 2013. A national overview of the new ML information is presented in Fig. 29.1a and a detail is shown in Fig. 29.1b. The new, high-resolution ML information will be accessible to the public via the internet (www.ngu.no).

The aim of this paper is to (1) give a brief overview of the present day Norwegian mapping procedures of areas that are potentially exposed to landslides in sensitive clays and quick clay, (2) explain the marine limit (ML) and show how it is mapped and modelled on a national level, (3) show how ML information can be used for landslide susceptibility mapping on a regional scale and define the concept of clay-deposit susceptibility, and (4) show how focused presentation of geological parameters can be used for landslide susceptibility assessment and the benefits of using high-resolution, digital elevation models (DEMs) for mapping.

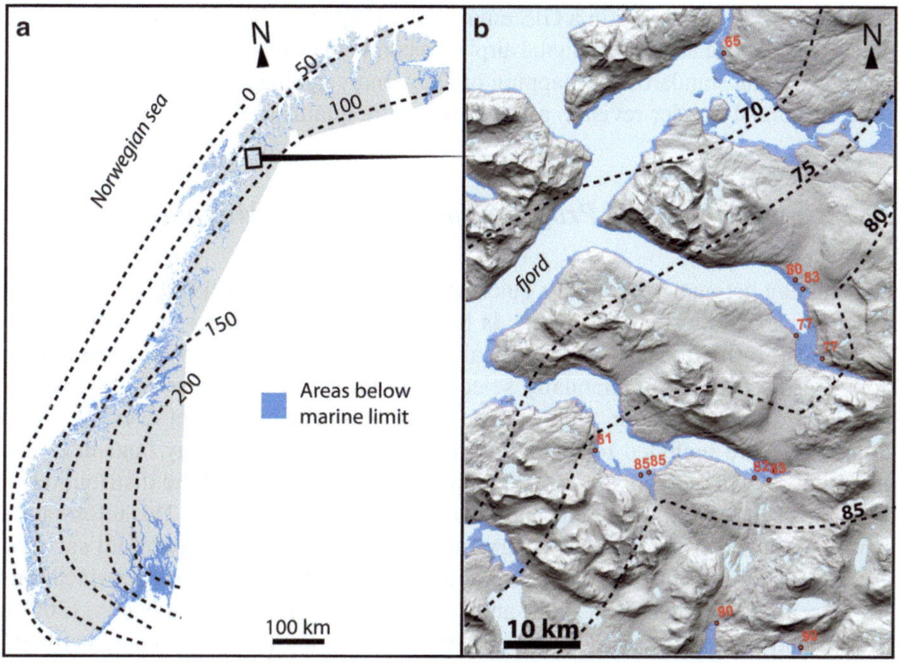

Fig. 29.1 (**a**) National overview of areas below the marine limit (ML) in Norway based on regional modeling of ML registrations. General isobases (*stippled*) of the marine limit in m a.s.l. are shown. (**b**) Example from northern Norway showing detailed ML isobases (*stippled*) modeled on the basis of ML registrations (*red numbers* in m a.s.l.). Areas below ML are shown in *blue*

29.2 Norwegian Mapping Procedures

29.2.1 Standards for Quaternary Mapping

According to Norwegian standards (Bergstrøm et al. 2001), Quaternary maps display genetic deposit types in colour representing e.g. fjord-marine deposits, beach deposits, fluvial deposits and glaciogenic deposits. Information also includes landforms, such as landslide scars, ravines and characteristic grain sizes, local stratigraphic information and several other features. These Quaternary maps, that are primarily produced by the Geological Survey of Norway, are based mainly upon air photo interpretation and extensive fieldwork. Mapping is carried out at various scales (e.g. 1:250,000, 1:50,000, 1:20,000 or 1:15,000). These maps are used for a broad range of purposes; such as, natural resource mapping, mapping of environmental conditions and as a basis for understanding the general landscape history. The map information is stored and managed at NGU in the national database of superficial deposits. Simplified versions of these maps are available on the web (www.ngu.no/kart/losmasse). Mapping activities and database updating is continuously carried out. New techniques allow mapping of Quaternary geology to be carried out through a digital workflow process in a GIS environment. Specifically, this process involves stereoscopic image analysis of digital airphotos combined with digital field registrations. The Norwegian standards for mapping of Quaternary geology, and sea-bed mapping, are continuously under revision at NGU to ensure quality and development.

29.2.2 The National Program for Quick-Clay Mapping

The ongoing national program for quick-clay mapping started in 1980. The mapping, which is carried out by consultants, is based primarily on Quaternary maps, topographical information and geotechnical drilling. The distribution of fine-grained, fjord-marine deposits is manually selected on the basis of Quaternary maps at a scale of 1:50,000 or better. Only areas over 10,000 m^2 with a relief over 10 m or a slope >1:15 are included to focus on areas where large quick-clay landslides can potentially occur (Gregersen 2008). Signs of active erosion and the distribution of landforms such as landslide scars and ravines are taken into account. This information can be extracted from the Quaternary maps, topographical data and field surveys. Procedures for identification of quick-clay zones and for hazard and risk classifications are described by Gregersen (2008) and summarized by Lacasse and Nadim (2009). Hazard and risk maps are available to the public via the Internett (www.skrednett.no). It should be noted that the onshore maps are not hazard maps *sensu stricto* since probability is not considered, but they can be regarded as special susceptibility maps focusing on large landslides.

29.3 The Marine Limit as a Basis for Regional Landslide Susceptibility Assessment

29.3.1 The Marine Limit (ML)

The marine limit (ML) marks the highest relative sea level in an area following the ice age and thereby sets a natural upper limit of the occurrence of unconsolidated, marine deposits such as fine-grained fjord deposits including clays. ML has always been an important parameter for Quaternary mapping as it sets an important framework for a general understanding of a landscape, its development, and of the overall distribution of deposit types. Assessment of ML is usually based on field observations. However, the regional context has to be taken into account when interpreting ML. The most important ways of identifying ML in the field is summarized below (Høgaas et al. 2012).

ML is commonly identified from high-lying glaciofluvial terraces along fjords and valleys. Such terraces represent early deltaic sedimentation often at bedrock constrictions in the newly deglaciated fjords while glaciers were still present (Fig. 29.2a, c). Prerequisites for accumulation of glaciofluvial deposits around ML

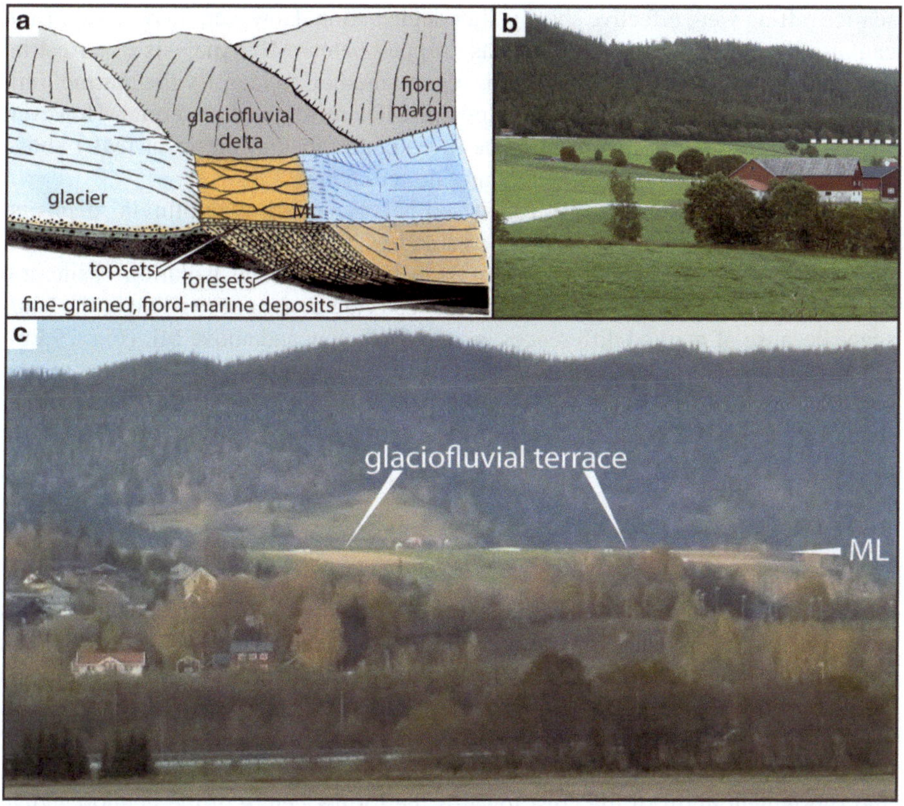

Fig. 29.2 (**a**) Sketch showing the construction of a glaciofluvial delta at the marine limit (ML) at the head of a fjord during overall glacier retreat (Modified from Andersen and Borns 1994). (**b**) Typical terrain dominated by fine-grained, fjord-marine deposits near the margin of a former fjord (*below stippled line*) near Åfjord, Mid Norway. Sediments directly *below the stippled line* are coarser-grained due to wave processes. Note the differences in land use above and below the *stippled line* (slightly below ML). (**c**) The glaciofluvial terrace in Gauldalen valley, Mid Norway, is a typical example of an emerged glaciofluvial delta deposited during deglaciation as illustrated in (**a**) (Photos: Louise Hansen)

are high rates of meltwater production and a local topography, which is not too steep for accumulation of sediments. The distal areas of smaller, glaciofluvial terraces often give the best estimate of ML since proximal parts are usually build up some metres above ML. Other features, from which the marine limit can be deduced, are the highest occurrence of raised beach deposits or wave-cut escarpments in unconsolidated deposits. The highest level of beach deposits is usually the wave level during storms, which may be several meters above the actual mean sea level. For this reason registrations of ML based on beach deposits can be slightly overestimated. A prerequisite is that topography is not too steep since this could prevent any beach deposits from accumulating around ML. Distinct marks of wave abrasion in bedrock are locally considered as a potential marker of ML, as frost action and sea-ice rafting were effective shortly after deglaciation. However, marks of abrasion can also develop after the sea level has dropped below ML. Analysis of terrestrial and marine fossils in lake sediments is a good method to identify the marine limit in areas where morphological signatures of ML are few and uncertain. However, several (paleo) lake basins need to be investigated to find the level (ML) above which lake basins do not contain any marine sediments.

By far, the most commonly registered ML types are glaciofluvial deposits, followed by abrasion marks in sediments and raised beach deposits (Høgaas et al. 2012). ML registrations based on abrasion in bedrock and isolation basins are relatively few. In general, emerged fjord deposits are commonly visible in the landscape because of marked differences in land use below and above ML (Fig. 29.2b). Detailed isobases for ML can be established in regions with good coverage of ML registrations (Fig. 29.1b). In this way ML can be stipulated in areas with no direct field evidence of ML.

29.3.2 Regional Trends of ML

General overviews of ML on a national scale have been available for many decades (Rekstad 1923; Holtedahl and Andersen 1960; Andersen 2000). The overviews show that glacioisostatic depression of the crust was greater than the eustatic sea level drop in many parts of Norway during the last glaciation (Jørgensen et al. 1995). The regional trend is a decline in ML towards the open sea in the north, west and southwest (Fig. 29.1a). This is explained by a greater glacioisostatic depression and subsequent uplift of the crust towards the former centre of the Fennoscandian ice sheet. The timing of deglaciation is another factor that controls the ML pattern. In Mid Norway and around Oslofjorden an early retreat of glaciers allowed the sea to inundate areas far inland before the glacioisostatic uplift could compensate. In other areas glaciers remained for a longer time whereby the area was already subjected to significant uplift prior to deglaciation and resulted in lower elevation of the ML (e.g. Aarseth and Mangerud 1974). It follows that the distribution of the ML does not necessarily show a linear pattern due to deglaciation patterns and the overall ML pattern is diachronous.

The marine limit most commonly corresponds to the relative sea level following deglaciation. The relative sea level was subsequently lowered to the present day level. However, the sea-level trends vary in different parts of Norway as exemplified by Svendsen and Mangerud (1987) and Romundset et al. (2011). In some outer regions the relative sea level shows a temporary rise during the Holocene before a continued fall to the present level. The various sea-level trends can be explained by varying rates of glacioisostatic uplift and eustatic sea level rise following deglaciation in different parts of Norway. Topography, deglaciation patterns and trends of regional relative sea level change have had an important control on the landscape development affecting the type of sediment successions produced in fjords and valleys (Corner 2006). A typical organization of deposit types in a Norwegian valley is illustrated in a schematic way in Fig. 29.3.

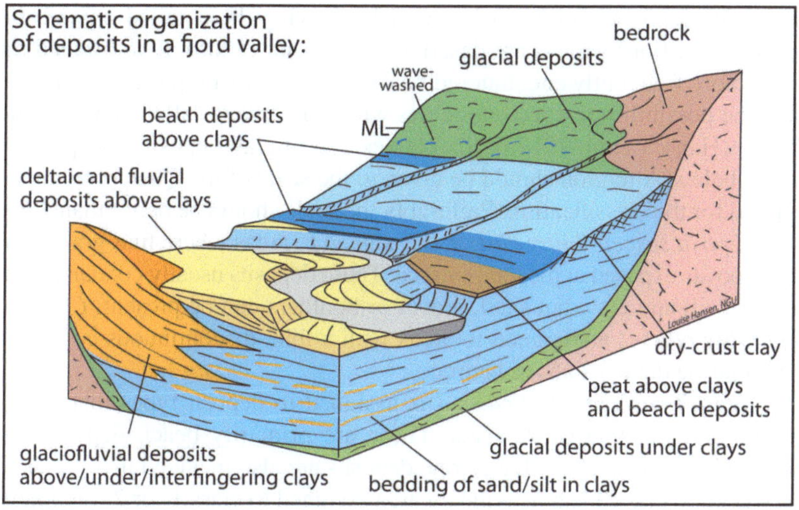

Fig. 29.3 Schematic 3D illustration showing a typical distribution of various deposit types in a Norwegian valley below the marine limit

29.3.3 Modelling of ML

Regional modelling of ML is carried out through interpolation of the registered ML points. A few 'artificial' control points were introduced in the marginal areas of the dataset to 'guide' the interpolation and avoid border effects. Interpolation is considered as an acceptable first-order approach for capturing areas with fine-grained marine deposits. This is not perfect, however, as ML is diachronous and ideal modelling should include the time aspect. Interpolation allows for ML isobases to be constructed for large regions of Norway at a relatively high resolution (see example in Fig. 29.1b). The actual trace of ML in the terrain is defined by the intersection between ML isobases and a DEM of 10 m resolution (Fig. 29.1b). Uncertainties in the estimation of ML are around 10 m. Further adjustments will improve the accuracy of the ML model.

29.4 Classification of Clay-Deposit Susceptibility

A major advantage of creating well-documented, regional overviews of ML is that it sets a framework for where fine-grained marine deposits can be encountered in the landscape. For example, it can be stated with confidence that primary, fine-grained marine deposits do not occur at elevations above ML. Below the level of ML the Quaternary maps show both areas with fine-grained, marine deposits exposed at the land surface and bedrock outcrops where such marine deposits are clearly excluded. Various deposit types commonly cover fine-grained marine deposits (Fig. 29.3). Comprehensive models showing the detailed organisation of deposit types in fjord valleys are described by Corner (2006). Fine-grained marine clays are more frequently encountered below some deposit types than others below ML and the possibility to encounter clays within an area is called clay-deposit susceptibility (Hansen et al. 2012a). Six classes 1–6 of clay-deposit susceptibility are described below. Attention should be given to the scale of the Quaternary maps that are applied for the classification. Preferably the scale should be better than 1:50,000.

Clay-deposit susceptibility class 1 is assigned to areas where fine-grained, fjord-marine deposits exposed at the land surface. These deposits usually contain significant amounts of silt and clay, but sand, gravel or boulders can be present. *Clay-deposit susceptibility class 2* is assigned to areas with a thin, discontinuous cover of fine-grained, marine deposits above bedrock.

Clay-deposit susceptibility class 3 is assigned to deposit types that frequently cover fine-grained, marine deposits. These are primarily beach deposits, fluvial deposits and peat (Fig. 29.3). Lacustrine deposits are also included in this class. If an area assigned to class 3 is bordering an area assigned to class 1 (or 2) the likelihood for encountering fine-grained marine deposits increases significantly. Beach deposits usually comprise a relatively uniform, continuous cover of sediments above fine-grained marine sediments (Fig. 29.3). Fluvial deposits, which include deltaic deposits, are often encountered in the central part of valleys. Their thickness is often greatest near the valley axis, and thins out towards the valley sides (Fig. 29.3). The fine-grained, marine deposits are often encountered below river level.

Clay-deposit susceptibility class 4 is assigned to deposit types that sometimes cover fine-grained, fjord-marine deposits. These are glaciofluvial deposits, wind deposits, anthropogenic deposits and certain types of glaciogenic deposits. Glaciofluvial deposits are usually relatively thick and are often present at the head of valleys or at bedrock constrictions. These deposits overlie, underlie or interfinger neighbouring fine-grained, fjord-marine deposits (Fig. 29.3). In a few cases glaciogenic deposits such as clayey till and proglacial ridges in the central part of valleys are found to cover fine-grained, marine deposits.

Clay-deposit susceptibility class 5 is assigned to deposit types that rarely cover fjord deposits. These include most types of glaciogenic deposits including till, glaciolacustrine deposits, and colluvium or scree. The glaciogenic deposits are often located at the margins of valleys and often cover the bedrock that may be exposed nearby. Neighbouring, fine-grained, marine deposits usually overlie the glaciogenic deposits. Colluvium or scree is usually located at the valley margins in association to steep slopes and bedrock. However, the lower part of these deposits can overlie, underlie or interfinger fine-grained fjord-marine deposits in the valley bottom. This is especially the case where the toe of larger, gently sloping cones is build into a lowland area.

Clay-deposit susceptibility class 6 is assigned to areas where marine deposits do not occur such as areas with exposed, humus-covered and/or weathered bedrock. Only small, insignificant patches of fine-grained, marine deposits below the scale of the applied Quaternary map may be present locally. Areas of thin, discontinuous glaciogenic deposits above bedrock are also included in this class.

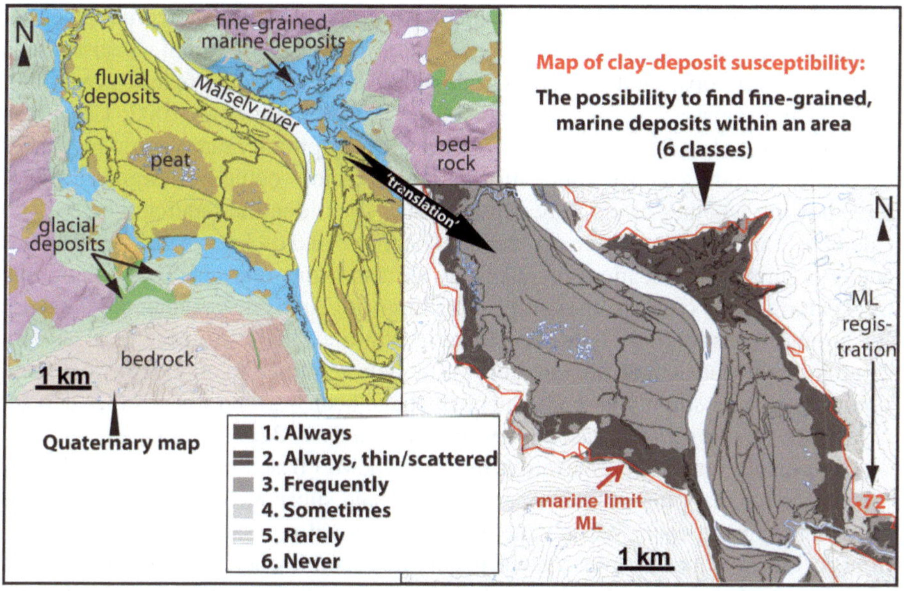

Fig. 29.4 Quaternary map displaying numerous deposit types in color (the most important types are indicated) that is 'translated' into a map showing six classes of clay-deposit susceptibility. The marine limit (ML) from regional modeling is shown as a *red line*

By assigning polygons to the different classes, the map may help to illustrate the areas where marine clay deposits are more likely to be encountered. An example from a valley in Northern Norway is presented in Fig. 29.4. The Quaternary map shows numerous deposit types whereas the clay-deposit susceptibility maps focus on possible clay occurrences. The map clearly indicates that fine-grained, marine deposits are most likely present across the entire valley and possibly below river level in the middle of the valley. The modelled marine limit shows the uppermost possible level of fjord-marine deposits. Polygons crossing the modelled ML boundary have been classified and are included in the map over clay deposit susceptibility (Fig. 29.4). This is in order to demonstrate uncertainties in ML estimations and the fact that some deposit types covering marine sediments extend above ML.

29.5 Additional Geological Parameters for Landslide Susceptibility Assessment

Regional overviews over the marine limit and maps that show clay-deposit susceptibility are useful for defining areas potentially susceptible to any size or type of failure in marine clays. Topographical information and information about ground conditions are needed for more detailed landslide susceptibility and hazard assessment. Other parameters in Quaternary geology are, however, also relevant for landslide susceptibility and hazard mapping because they are indicative of either stable or potentially unstable conditions. These include, for example, registrations of bedrock outcrops, landslide scars, landslide debris, erosional features including ravines, and registrations of active erosion (e.g. Robitaille et al. 2002; Gregersen 2008; Quinn et al. 2008, 2010). The clay-deposit susceptibility maps could be further refined by including stratigraphic information. Locations could be assigned a signature showing whether or not clay deposits have been identified and at which depth ('depth to clays'). Information on 'depth to bedrock' and on known clay thicknesses and/or occurrences/indications of quick clay are also highly relevant. The stratigraphic information could come from outcrops, drillings or geophysics. A national database for ground investigations is in the planning phase at NGU (Solberg et al. 2012). Coupling with subsurface information would help refining the clay-deposit susceptibility maps and show where clay deposits have been verified under other deposit types to be shown on web-based maps and ultimately in 3D.

High-resolution, digital elevation models (DEMs) based on LIDAR are highly favourable for mapping of Quaternary geology and subsequently also for landslide susceptibility assessment. Morphometric analyses are also possible (slope angle etc.). Subaqueous information must also be taken into account wherever possible (Hansen et al. in press). An example of a combined, high-resolution, onshore and subaqueous DEM at a resolution of 1m is shown in Fig. 29.5. The combined terrain model allows for detailed analysis of landforms e.g. to provide a detailed inventory of landslides and other processes. An additional advantage is that repeated mapping at this resolution will allow for changes in the terrain to be detected (Hansen et al. 2012b). Scars from different types of landslides can be distinguished and possible triggers can be interpreted (Fig. 29.5, see also Hansen et al. in press). Identification of major stable areas such as bedrock outcrops are also easily identified both on land and in the fjord (Fig. 29.5). The remaining areas are dominated by fine-grained, fjord deposits and by fillings along the shoreline. The large, onshore landslide scar shown in Fig. 29.5 appears on existing Quaternary maps (see Hansen et al. in press). However, the LIDAR data reveal more details and provides a possibility to better assess the age and the conditions of the landslide. It was possibly triggered around the time when an escarpment was eroded at the slide port (Fig. 29.5). Emergence curves could provide a time frame for wave reworking that affected the areas below the escarpment. Some of the subaqueous landforms are possibly linked to the landslide event and seismic data in the central fjord shows landslide debris near the area shown in Fig. 29.5 (Hansen et al. 2011). Slopes around the identified quick-clay zone are not affected directly by active erosion at present but processes under water could be unfavourable (Hansen et al. in press).

There are several possible improvements that could be of value for mapping of landslide susceptibility. An improved classification of fine-grained, marine deposits, that distinguishes between homogenous, stratified and diamictic glaciomarine sediments, or bioturbated marine sediments, could be beneficial as sensitive clays are usually restricted to the two first mentioned. Identification of intervals with permeable (sand) layers in clays is also relevant (Fig. 29.3). In addition, regional overviews of major glacier marginal positions during the overall deglaciation of Norway would help to indicate areas which were likely subjected to high, proglacial

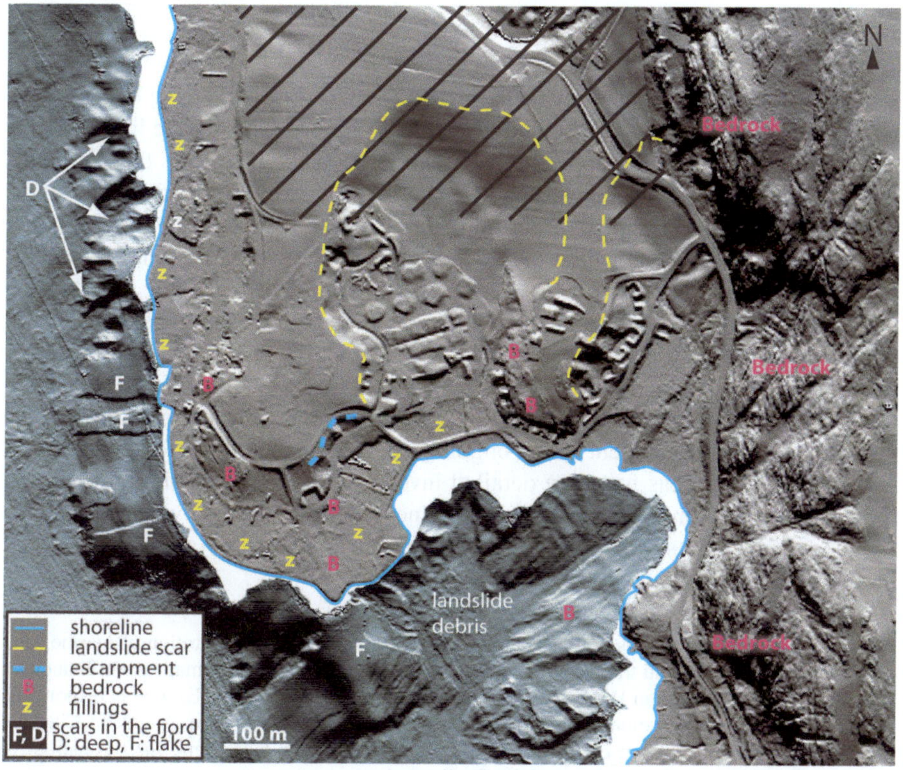

Fig. 29.5 Combined, shaded relief of high-resolution DEMs (1 m) from swath bathymetry in the fjord (*blue shading*) and LIDAR data on land along the shoreline of Drammens-fjorden, southern Norway. The fjord is down to 60 m deep in the shown area. The most shallow areas are shown in *white*. The detailed terrain information allows for identification of bedrock, unconsolidated deposits, and landforms such as landslide scars both on land and in the fjord. Examples of areas characterized by different processes are indicated. For more details on interpreted landforms in the fjord see Hansen et al. (in press). An identified quick-clay zone is shown (*cross-hatched*)

sedimentation rates and therefore forming thick, fine-grained deposits. Information on areas subjected to a temporary still stand or even a rise of relative sea level during an overall fall of relative sea level since deglaciation could also be valuable as an extra thick cover of beach or delta deposits over fine-grained marine deposits could be expected in these areas. Registration of areas where glacier ice is known to have overridden (glacio)marine deposits is also useful information as this indicates where overconsolidated or deformed deposits could be expected. Registrations of relict seabed surfaces on land also provide information on the consolidation history of an area. Other relevant parameters are bedrock types, grain size and mineralogical composition, groundwater chemistry and/or groundwater conditions because these parameters could potentially show correlation with sediment properties (e.g. see summary by Quinn et al. 2007). Analysis of landslide susceptibility could also include information on catchment scale such as drainage networks (Quinn et al. 2010).

29.6 Conclusions

Mapping of Quaternary geology is an important basis for discovering fine-grained, marine deposits that could be potentially prone to failure. The marine limit (ML) is a key feature as it sets the upper limit to marine deposits. It follows that problems associated with sensitive clays or quick clay can be excluded above ML. Areas below ML can be classified according to the possibility of encountering fine-grained, marine deposits called clay-deposit susceptibility. More detailed assessment of landslide susceptibility requires additional topographic, geological parameters indicative of stable and potentially unstable conditions and should include subsurface information such as geotechnical data. The proposed presentation of selected geological information can be used for screening of large areas with regard to marine clays and potentially unstable conditions. The mapping procedures can be used as a supplement to the ongoing quick-clay mapping program and as a tool for selecting areas for more detailed investigations. High-resolution elevation models, from onshore and subaqueous areas near the shoreline, provide a possibility for increasing the use of geological information for landslide susceptibility and hazard assessments.

Acknowledgements Projects on the establishment of a database for registration of marine limit information in Norway and the development of clay-deposit susceptibility maps were financially supported by the Norwegian Water Resources and Energy Directorate (NVE) who is responsible at government level to assist municipalities for the prevention of landslide disasters. Ane Bang-Kittelsen prepared LIDAR data for presentation. We are grateful to Marten Geertsema for a constructive review. Anne Liinamaa-Dehls corrected the English language.

References

Aarseth I, Mangerud J (1974) Younger Dryas end moraines between Hardangerfjorden and Sognefjorden, Western Norway. Boreas 3:3–22

Andersen BG (2000) Istider i Norge. Universitetsforlaget, Oslo, 216pp (in Norwegian)

Andersen BG, Borns HW (1994) The ice age world. Scandinavian University Press, Oslo, 208pp

Bergstrøm B, Reite A, Sveian H, Olsen L (2001) Feltrutiner, kartleggingsprinsipper og standarder for kvartærgeologisk kartlegging ved NGU. Internal report 2001.018, Norges geologiske undersøkelse (NGU), Trondheim, Norway (in Norwegian)

Corner GD (2006) A transgressive–regressive model of fjord–valley fill: stratigraphy, facies and depositional controls. In: Dalrymple RW, Leckie D, Tilman R (eds) Incised-valleys in time and space, vol 85, SEPM special publication. Society for Sedimentary Geology, Tulsa, pp 161–178

Gregersen O (2008) Program for økt sikkerhet mot leirskred – Metode for kartlegging og klassifisering av faresoner, kvikkleire. Norwegian Geotechnical Institute (NGI) report 20001008-2 (3rd revision), Oslo, Norway, 24p (in Norwegian)

Hansen L, L'Heureux JS, Longva O, Eilertsen RS (2011) Undersjøiske former og skredprosesser langs strandsonen i Drammensfjorden. NGU report 2011.003, Geological Survey of Norway, Trondheim, Norway (in Norwegian)

Hansen L, Sveian H, Olsen L, Høgaas F, Rindstad BI, Wiig T, Lyche E (2012a) The marine limit as a basis for mapping of landslide susceptibility in fine-grained, fjord deposits, onshore

Norway. In: Eberhardt E, Froese C, Turner AK, Leroueil S (eds) Landslides and engineered slopes: protecting society through improved understanding. 11th international symposium on landslides, Banff, Canada, June 2012. Taylor & Francis Group, London, pp 1833–1838

Hansen L, L'Heureux JS, Solberg IL, Longva O (2012b) Forebyggende kartlegging mot skred langs strandsonen i Norge. Oppsummering av erfaring og anbefalinger. NGU report 2012.046, Geological Survey of Norway, Trondheim, Norway (in Norwegian)

Hansen L, L'Heureux JS, Longva O, Eilertsen RS (in press) Mapping of subaqueous landforms for near-shore landslide susceptibility assessment in Norwegian fjords. In: Margottini C, Canuti P, Sassa K (eds) Landslide science and practise. Springer, Berlin

Høgaas F, Hansen L, Rindstad BI, Sveian H, Olsen L (2012) Database for registrering av marine grense (MG) i Norge. NGU rapport 2012.063 (in Norwegian)

Holtedahl O, Andersen BG (1960) Glacial map of Norway. 1: 2.000.000. Norges geologiske undersøkelse (NGU) Publication 208, Trondheim, Norway

Jørgensen P, Sørensen R, Haldorsen S (1995) Kvartærgeologi. Landbruksforlaget, 344pp (in Norwegian)

Lacasse S, Nadim F (2009) Landslide risk assessment and mitigation strategy. In: Sassa K, Canuti P (eds) Landslides – disaster risk reduction. Springer, Berlin/Heidelberg, pp 31–61

L'Heureux JS (2012) A study of the retrogressive behaviour and mobility of Norwegian quick clay landslides. In: Eberhardt E, Froese C, Turner AK, Leroueil S (eds) Landslide and engineered slopes: protecting society through improved understanding. Taylor & Francis Group, London, pp 981–988

Quinn PE, Hutchinson DJ, Rowe RK (2007) Towards a risk management framework: sensitive clay landslide hazards affecting linear infrastructure in eastern Canada. In: Turner AK, Schuster RL (eds) Landslides and society: proceedings of the first North American conference on landslides. Association of Environmental and Engineering Geologists, Denver, 398pp

Quinn PE, Hutchinson DJ, Diederichs MS, Rowe RK, Alvarez J (2008) Susceptibility mapping of landslides in champlain clay from a digital landslide inventory. In: Locat J, Perret D, Turmell D, Demers D, Lerouil S (eds) Proceedings of the 4th conference on geohazards: from causes to management, Presse de l'Université Laval, Quebec, 594pp

Quinn PE, Hutchinson DJ, Diederichs MS, Rowe RK (2010) Region-scale landslide susceptibility mapping using the weights of evidence method: an example applied to linear infrastructure. Can Geotech J 47:905–927

Rekstad J (1923) Norges heving etter istiden. Norges geologiske undersøkelse 96, pp 27, Trondheim, Norway (in Norwegian)

Robitaille D, Demers D, Potvin J, Pellerin F (2002) Mapping of landslide-prone areas in the Saguenay region, Quebec, Canada. In: McInnes RG, Jakeways J (eds) Instability – planning and management. Thomas Telford, London, pp 161–168

Romundset A, Bondevik S, Bennike O (2011) Postglacial uplift and relative sea level changes in Finnmark, Northern Norway. Quat Sci Rev 30:2398–2421

Solberg IL, Ryghaug P, Nordahl B, deBeer H, Hansen L, Høst J (2012) Nasjonal database for grunnundersøkelser (NADAG) – forundersøkelse. NGU rapport 2012.054, Trondheim, Norway (in Norwegian)

Svendsen JI, Mangerud J (1987) Late Weichselian and Holocene sea-level history for a cross-section of Western Norway. J Quat Sci 2:113–132

Chapter 30
Management of Quick Clay Areas in Slope Stability Investigations – The Göta River Valley

Helen Åhnberg, Hjördis Löfroth, and Karin Lundström

Abstract In order to limit the vulnerability of society to effects of the climate change, the Swedish Government in 2008 commissioned the Swedish Geotechnical Institute to map the risk of landslides along the Göta River. The Göta River valley is one of the areas most prone to landslides in Sweden. Since it is an area with highly sensitive clays, mapping of quick clay areas and estimation of possible landslide extent were important parts of the investigation. In the Göta River Commission, a correlation between the sensitivity and the total rod friction evaluated from CPTs and static pressure sounding tests was used as a complement to sampling for mapping of quick clay areas. If a landslide is initiated in an area with quick clay the probability is high that it will propagate and affect areas nearby. Within the Göta River Commission, estimates of how far a landslide would extend backwards was made taking into account the slope geometry and the sensitivity of the clay, partly based on analyses of earlier occurred landslides. A methodology was developed for rational management of quick clay areas in the stability investigations for the Göta River valley.

Keywords Göta River • Quick clay • Mapping method • Sensitive soil • Slope stability

30.1 Introduction

The Göta River valley is located in southwestern Sweden and runs from Lake Vänern in the north to Göteborg in the south, Fig. 30.1. The valley, which has a length of 90 km, is one of the areas in Sweden which has been most frequently

H. Åhnberg • H. Löfroth • K. Lundström (✉)
Swedish Geotechnical Institute (SGI), 581 93 Linköping, Sweden
e-mail: karin.lundstrom@swedgeo.se

J.-S. L'Heureux et al. (eds.), *Landslides in Sensitive Clays: From Geosciences to Risk Management*, Advances in Natural and Technological Hazards Research 36,
DOI 10.1007/978-94-007-7079-9_30, © Springer Science+Business Media Dordrecht 2014

Fig. 30.1 Location of the Göta River (*Swedish*: Göta älv) (© SGI, Lantmäteriet)

affected by landslides. The landslides are mostly small and shallow, but a number of large landslides have occurred during the past century, some of them with human casualties and extensive property damages (e.g. Alén et al. 2000; Hultén et al. 2007).

The valley has a long history of human activities resulting in residential areas, industries, shipping on the river and harbours. Roads and railroads of national importance follow the river banks, and environmentally hazardous activities and contaminated sites are not uncommon. The river is also the fresh water source for communities in the valley and the city of Göteborg. Due to the relatively high exploitation of the area, the consequences of a landslide in the valley can be severe.

Parts of the most populated areas in the valley have earlier been investigated for the risk of landslides. However, the predicted effects of the ongoing climate changes, such as increased precipitation and erosion and also longer dry periods, are expected to result in an increased frequency of landslides. In order to meet this challenge, the Swedish Government in 2008 commissioned the Swedish Geotechnical Institute to investigate and map landslide risks in the whole valley.

The mapping of the landslide risk (defined as probability × consequences), considered both present conditions and the estimated effects of a future climate change. The results showed that high level of risk is already prevailing in many areas in today's climate and that a climate change will increase the risk of landslides in the entire valley (SGI 2012).

A number of sub projects were initiated during the investigation process to study and improve different methods used in the stability risk analysis. To ensure a similar approach to be taken regarding quick clay in the various areas in the investigations along the river, common guidelines were developed for both mapping (Löfroth 2011) and handling (Åhnberg et al. 2011) of quick clay. The paper describes the methodology used in the investigations for mapping of quick clay, determining geotechnical properties and estimating possible landslide extent, in order to take this into account in the risk analysis.

30.2 The Geology of the Göta River Valley

A reason for the high frequency of landslides in the Göta River valley is its geological history. During the latest deglaciation of the area, which started around 14,500 years ago, the valley was part of an arctic ocean. Large amounts of soil particles were transported by the melt water from the ice and deposited, mainly as marine, glacial clays in the valley. The clay layers are underlaid by coarser soils, such as till and glacifluvial deposits, with a thickness mostly of 1–3 m, but deposits of more extensive thickness also cross the valley. The clay thickness increases towards the sea and reaches in Göteborg up to 100 m. A typical soil profile along the river is presented in Fig. 30.2. The isostatic uplift process has been extensive in the area, resulting in a marine limit located around 130 m above the present sea level in the inner part of the valley. The valley is surrounded by mountain outcrops which reach heights of 200–300 m above sea level. The layers of till and glacifluvial deposits are often exposed next to these outcrops. Runoff of rain water from the outcrops can thus easily infiltrate into these coarser deposits at high levels, providing favorable conditions for high artesian groundwater pressures in the valley.

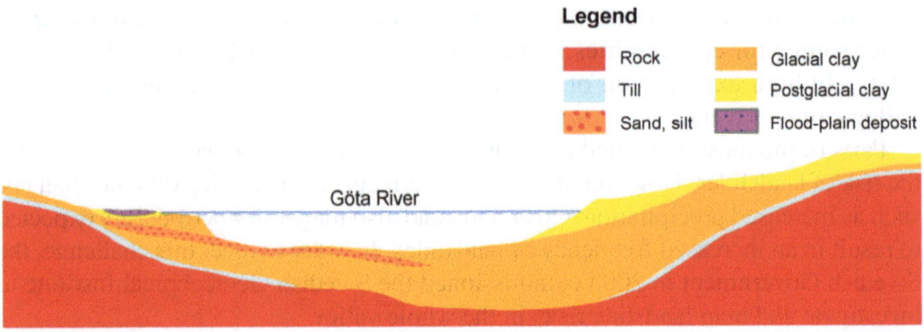

Fig. 30.2 A typical soil profile along the river (cross section) in the northern part of the valley (After SGI 2012)

30.3 Method of Quick Clay Mapping

Within the Göta River Commission, determinations of quick clay was made by taking undisturbed samples in the field and performing fall cone tests in the laboratory on the clay in its undisturbed and remoulded states. In addition, quick clay was estimated from CPTs and static pressure soundings by measurements of the total penetration force (Löfroth et al. 2011, 2013). While these methods primarily were used in the investigation for other purposes, as determination of undrained shear strength and depth to firm bottom layers respectively, they were also useful for mapping quick clay.

Estimation of quick clay from CPTs and static pressure soundings is based on measurements of the total penetration resistance of the soil, i.e. the measured thrust force on the rods (Möller and Bergdahl 1982). The measured penetration resistance together with the weight of the rods and, for the CPT, reduced with the tip resistance, corresponds to the total rod friction. This rod friction is compared to a rod friction of 1 kPa. Where the inclination of the curve for rod friction versus depth is less than the inclination of the curve for 1 kPa rod friction, the clay is assumed to be quick (Larsson and Lundström 2012; Rankka et al. 2004).

On the undisturbed samples, the sensitivity and the remoulded undrained shear strength were determined in the laboratory. For the clay to be classified as quick it should, according to Swedish praxis, have a sensitivity $S_t \geq 50$ and a remoulded undrained shear strength $\tau_R \leq 0.4$. If piston sampling, CPT and static pressure sounding were carried out at the same point, the laboratory determinations supersede the estimations from the CPT, which in turn supersede those from static pressure sounding.

Within the investigation an Excel spread sheet was developed for estimation of quick clay by measurement of penetration resistance in CPTs and static pressure soundings. In Fig. 30.3, striped markings in the diagram show the parts of the soil profile evaluated as quick clay in the spread sheet. An engineering judgement of the depth of quick clay has to be made based on this evaluation. From the laboratory determinations of sensitivity and quick clay the following cases are distinguished (see also Fig. 30.3);

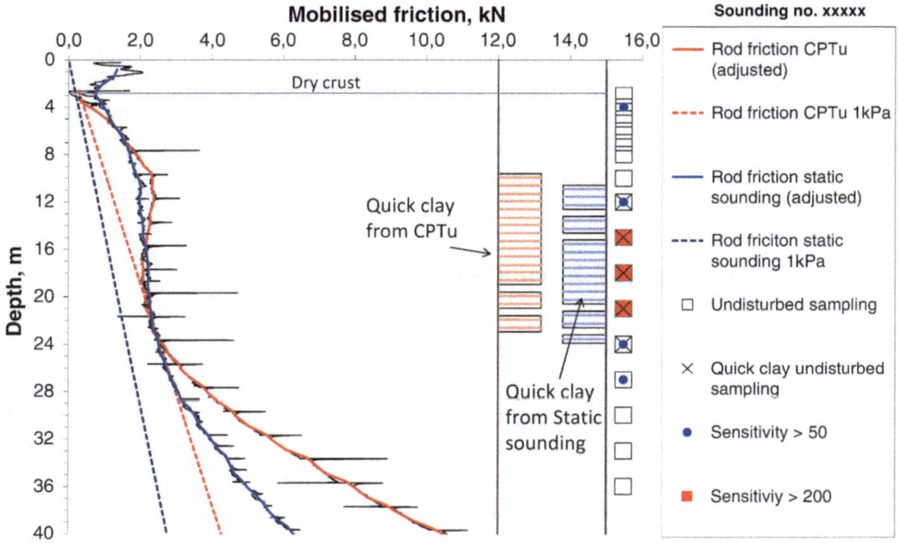

Fig. 30.3 Estimation of quick clay results from a CPTu and a static pressure sounding compared to laboratory results (After Löfroth 2011)

- sensitivity $S_t \geq 50$ and remoulded shear strength $\tau_R > 0.4$ kPa (highly sensitive clay, but not quick)
- sensitivity $S_t \geq 50$ and remoulded shear strength $\tau_R \leq 0.4$ kPa (quick clay)
- sensitivity $S_t \geq 200$ and remoulded shear strength $\tau_R \leq 0.4$ kPa (extremely quick clay)

In the diagram, peaks in the curves appear at regular intervals. These occur when the penetration is stopped to add new rods, due to reconsolidation of the clay. For the clay not to be wrongly classified as quick due to these peaks, the curves are smoothed by drawing mean curves where all values of half a standard deviation from the average are disregarded.

Based on the measured sensitivities in the laboratory, exported to ArcGIS, a GIS-based map showing the highest sensitivity in all sampling points was developed. For estimation of possible landslide extent in progressive landslides, a subdivision was made based on the sensitivity of the clays. On the GIS map, investigation points with sensitivities lower than 50 were marked with green, points with sensitivities between 50 and 99 were marked with yellow, points with sensitivities between 100 and 199 with orange and extremely quick clays with sensitivities above 200 were marked with red. From the interpreted plans, quick clay estimated from CPTs and static pressure soundings was marked manually on the same GIS map, cf. Fig. 30.6.

In addition, a separate study was carried out within the Göta River Commission with the purpose to investigate whether surface resistivity measurements and CPT-Rs (resistivity CPT) with measurement of the total rod friction, in combination with fall-cone tests and chemical analyses in the laboratory could give a more complete picture of the presence of quick clay (Löfroth et al. 2011; Dahlin et al. 2013).

Surface resistivity may be used as a screening tool in order to delimit leached areas, where further investigations are needed, from unleached areas which are not expected to contain quick clay. By combining surface resistivity with the CPT-R with measurement of the total rod friction, it is possible to acquire an indication of how much quick clay there is in the volume that from the surface resistivity measurements is estimated to be sufficiently leached to possibly form quick clay.

30.4 Sampling and Characterisation of Quick Clay

Determination of the geotechnical properties of quick clays were carried out as for other, less sensitive, clays with CPTs (including registration of the total penetration resistance, see above), vane tests, pore pressure measurements etc in the field. Sampling was made mainly with the 50 mm Swedish Standard piston sampler (StII) in accordance with recommendations for sampling in quick clay (SGF 2009), with a somewhat prolonged resting time compared to that normally used for ordinary clays. A special large-diameter tube sampler "block sampler" was constructed at SGI (Larsson et al. 2012; Löfroth 2012) as an alternative sampler that could be used in case the piston sampler would fail to bring up full samples in some quick clays or if the sample quality was deemed too low. However, the Standard piston sampler showed to be adequate for sampling throughout most of the sampling campaign along the Göta River, providing samples generally of very good quality. The large-diameter sampler was used successfully for research purposes at a number of locations in the valley (e.g. Åhnberg and Larsson 2012). In the laboratory the samples were investigated as soon as possible after sampling, since significant changes in properties may take place in quick clay with time after sampling (e.g. Åhnberg and Larsson 2012).

In estimates of the geotechnical properties for stability analyses, it was to be observed that the shear strength and the preconsolidation pressures in a chemically changed quick clay may be lower than empirically expected for normally sensitive clays (e.g. Bjerrum 1954; Torrance 1974; Larsson 2011).

30.5 Assessment of Landslide Extent

The presence of quick clay within a large number of areas along the Göta River entails a risk for local landslides to expand and affect large parts of the surrounding land. The possibility of initial landslides spreading backwards (uphill) or forward must be taken into account. However, calculations of the final land area involved in landslides are subject to large uncertainties. Development of advanced numerical methods for calculation of progressive failures have been of interest for many years and received increased attention in recent years (e.g. Nordal 2007; Locat et al. 2011; Bernander 2011; Gylland 2012). However, for the survey of landslide risk in the vast areas along the Göta River valley, these methods were deemed not yet adequately adapted for practical use. Instead a schematic method initially developed by

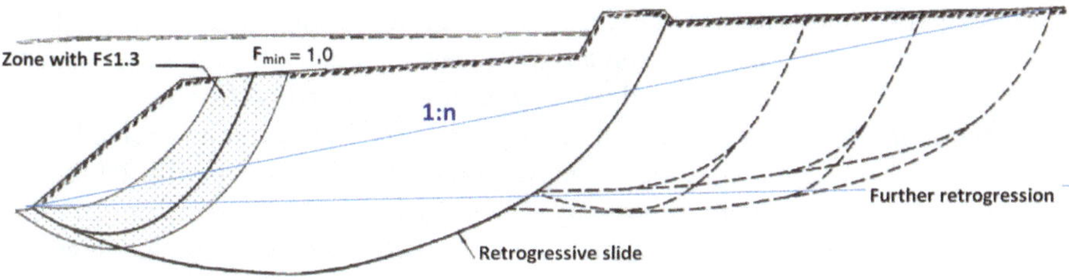

Fig. 30.4 Assessment of the probable extension of a landslide (After Larsson et al. 2008)

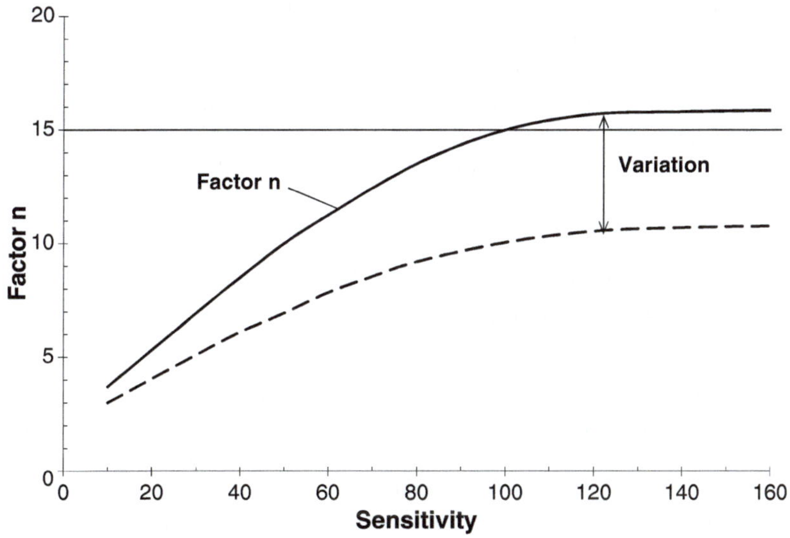

Fig. 30.5 Diagram for evaluation of the factor n by guidance of the sensitivity of the clay (After Åhnberg et al. 2011)

Bengtsson (2008) for assessment of potential retrogression of landslides along the Göta River was employed. This method is based on earlier experience of landslide extent coupled to the slope geometry and the sensitivity of the soils involved (Larsson et al. 2008). Further investigations of areas involved in various landslides in Sweden (Ahlbom et al. 2009) have provided additional support to the method. However, its applicability is geographically restricted, as indicated by results from similar landslide investigations in Norway (Thakur and Degago 2012).

Calculations of the initial stability of the slopes along the Göta River valley were performed in accordance with the guidelines of the Commission on slope stability (IVA 1995; IEG 2010). For potential initial slip surfaces at or below the river bank with a calculated factor of safety of 1.3 or less in undrained conditions, and 1.2 or less in combined conditions (drained and undrained), a further extension backwards was then calculated by a factor n times the height of the slope (Fig. 30.4). The factor n is a function of the sensitivity of the clay involved in the initial slide (Fig. 30.5). For clays with sensitivities higher than 100, n is assigned a value of 15. In cases of extreme quick clays with sensitivities of 200 and higher, an initial slide involving this type of clay is anticipated to spread over the entire area with extreme quick clay, which in some cases may reach all the way back to surrounding firm ground.

In the mapping of landslide risks along the Göta River, it was deemed more practical for some areas to adopt a stepwise increase in factor *n* with increasing sensitivity in the assessment of possibly affected areas.

The methods used refer to common static loads. Effects of cyclic loadings and enforced deformations possibly causing degradation of the quick clay were not part of the analyses. Estimates of such effects were assigned to be taken into consideration in connection to future construction works. Research regarding degradation of clay due to cyclic loadings and deformations was carried out partly within the Göta River Commission, highlighting the increased risks involved for e.g. highly sensitive clays (Åhnberg and Larsson 2012; Åhnberg et al. 2013).

Assessments of downhill progressive landslides were not a specific issue in the mapping of landslide risks along the Göta River where landslides due to natural causes, were assessed to start in close proximity to or below the river banks.

30.6 Management of the Quick Clay Areas in the Risk Assessments

The estimated stability with respect to an initial landslide, expressed as a specific probability class for the Göta River valley, was assumed to be valid also for areas potentially becoming involved in a secondary retrogressive landslide. The landslide probability was calculated based on estimates of the safety factor using conventional stability analysis, and of the variation in parameters used in the calculation of this safety factor, such as undrained shear strength, slope geometry and density of the soil. The probability of landslide was ranked from negligible to substantial, in five probability classes S1–S5 (e.g. Tremblay et al. 2013; SGI 2012). In areas with quick clay, the calculated probability class for an initial slip surface that affects an area with quick clay was extended backwards to a point reflecting an inclination of 1:10 (or 1:15) from the toe of the slope. This was to indicate that the area may be affected by retrogressive landslides as a result of local landslides close to the river.

The procedure used is presented in Fig. 30.6 where it can be seen that on the west side of the river geotechnical investigations and stability calculations have been performed in two sections, 37/250 and 37/850, marked with red lines. The interpreted sensitivity values are presented by dots in different colors and by stars showing probable quick clay according to CPTs and static pressure soundings. A line representing the inclination 1:10 is also shown. For section 37/250, the highest sensitivity close to the river is between 50 and 99 (yellow dots), and the probability class for the area has been moved to the point reflecting an inclination of about 1:10. For section 37/850, the highest sensitivity close to the river is below 50 (green dot). However, quick clay has been found in other boreholes more to the west with values above 200 (red dots) and also results from CPTs and static pressure soundings

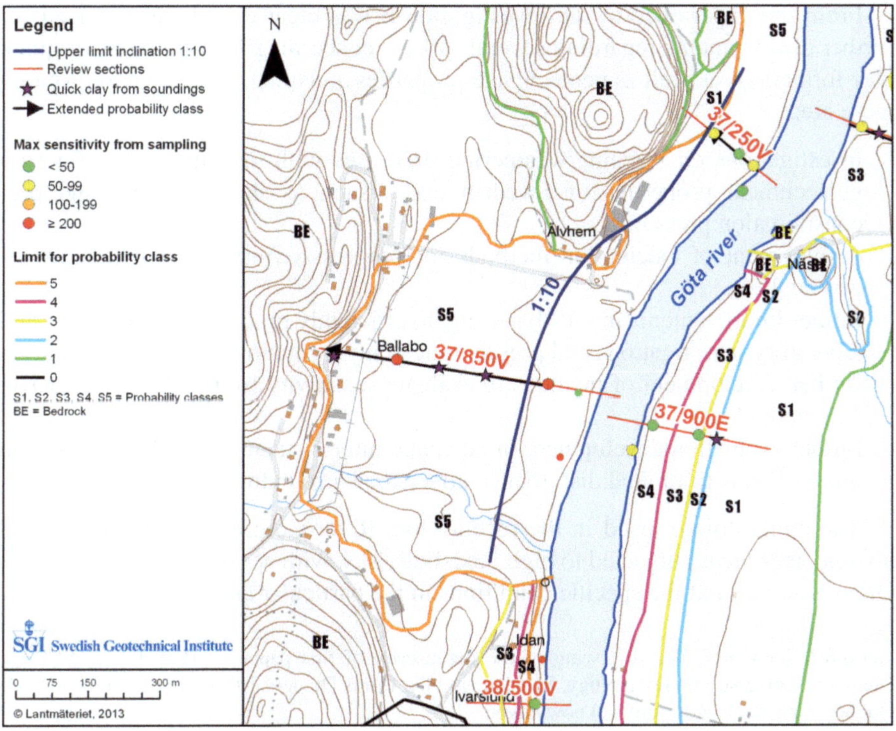

Fig. 30.6 Example of probability classes adjusted for presence of quick clay along the Göta River

indicate probable quick clay occurrence (stars). The probability for the first slip surface reaching the quick clay area has therefore been moved all the way back to firm ground. In this case the probability for slip surfaces close to the river was the same as for slip surfaces reaching quick clay areas.

The results from this procedure were summarized in a probability map that covers all the areas studied along the Göta River. Mapping of probability classes in this way in combination with mapping of consequence classes for the various areas was then used to establish landslide risk maps for the entire Göta River valley (Tremblay et al. 2013; SGI 2012).

30.7 Conclusions

In the mapping of landslide risks along the Göta River valley, the managing of areas containing quick clay was a central issue. The methodology used in the investigations was developed to provide a rational tool for the general evaluation of landslide risks in this particular area.

From the experience of the investigation it was clear that there is a need for further developments regarding several aspects concerning investigations of quick clay formations as well as geotechnical properties and quick clay behavior. Among these are:

- investigations on leaching/weathering processes with time and their effect on geotechnical properties of Swedish clays, such as shear strength and preconsolidation pressures.
- development of calculation methods for progressive/retrogressive failures for practical use.
- further improvements in GIS-handling, to efficiently process data from field and laboratory investigations and present quick clay formations on various plans.
- further development of methods to evaluate sensitivity from CPTs and other in situ methods.
- further evaluation/development of adequate improvement methods for slopes in quick clay, with limited disturbance effects on surrounding soil.

The methodology used in the recent Göta River Commission could be used also in other areas subjected to high landslide risks, with appropriate modifications taking into account the specific conditions in the particular area.

Acknowledgements The reviewing of the manuscript of this paper by Prof. Göran Sällfors, Chalmers University of Technology, Sweden, and Dr. Denis Demers, Ministère des Transports du Québec, Canada, is gratefully acknowledged.

References

Ahlbom E, Eriksson A, Storvall E, Strömqvist L (2009) Why does it slide? A digging investigation into landslides in clay. Candidate report 2009, Geology and Geotechnical Department, Chalmers University, Gothenburg (in Swedish)

Åhnberg H, Larsson R (2012) Strength degradation of clay due to cyclic loadings and enforced deformations. Report 75, Swedish Geotechnical Institute, SGI, Linköping. www.swedgeo.se

Åhnberg H, Bengtsson PE, Larsson R, Lundström K, Löfroth H, Tremblay M (2011) Management of quick clay formations in stability investigations for the Göta Älv valley. Guidelines. The Göta River Commision, Sub Report No. 32, Swedish Geotechnical Institute, SGI, Linköping. www.swedgeo.se (in Swedish)

Åhnberg H, Larsson R, Holmén M (2013) Degradation of clay due to cyclic loadings and deformations. In: Proceedings of the 18th international conference on soil mechanics and geotechnical engineering ICSMGE, Paris

Alén C, Bengtsson PE, Berggren B, Johansson L, Johansson A (2000) Landslide riskanalysis along the Göta River valley. Report 58, Swedish Geotechnical Institute, SGI, Linköping. www.swedgeo.se (in Swedish)

Bengtsson PE (2008) Proposed method to take into account secondary landslides. FB Engineering E45/Nordlänken, Bohus-Nödinge, Project Report, 080115 (in Swedish)

Bernander S (2011) Progressive landslides in long natural slopes. Formation, potential extension and configuration of finished slides in strain-softening soils. Theses, Department of Civil Mining and Environmental Engineering, Luleå University of Technology, Luleå

Bjerrum L (1954) Geotechnical properties of Norwegian marine clays. Geotechnique 4:49–69

Dahlin T, Löfroth H, Schälin D, Suer P (2013) Mapping of quick clay using geoelectrical imaging and CPTU-Resistivity. Geotechnical assessment and geoenvironmental engineering, special issues in near surface geophysics and J Environ Eng Geophys (in review)

Gylland AS (2012) Material and slope failure in sensitive clays. Doctoral thesis, Department of Civil and Transport Engineering, Norwegian University of Science and Technology, NTNU, Trondheim, 352pp

Hultén C, Andersson-Skold Y, Ottosson E, Edstam T, Johansson A (2007) Case studies of landslide risk due to climate change in Sweden. In: Proceedings of the international geotechnical conference on climate change and landslides, Ventnor, Isle of Wight, pp 149–157

IEG (2010) Classification of natural slopes and buildings and constructions slopes. Guidelines for application of IVA Commission on Slope Stability Reports 3:95 and 2:96, Implementation Commission for European Geotechnical standards (IEG) Report 4:2010 (in Swedish)

IVA (1995) Guidelines for slope stability investigations. Commission on Slope Stability, Report 3:95, Royal Swedish Academy of Engineering Sciences, Linköping (in Swedish)

Larsson R (2011) Effects of changes in pore water chemistry, particularly leaching of salts, on the properties of natural clays. A literature study, The Göta River Commission, Sub report No. 31, Swedish Geotechnical Institute, SGI, Linköping. www.swedgeo.se (in Swedish)

Larsson R, Lundström K (2012) On the use of penetration tests and geophysical methods for mapping of quick clay deposits. In: Proceedings of the 4th international conference on geotechnical and geophysical site characterization, vol 1, Porto de Galinhas, pp 743–748

Larsson R, Bengtsson PE, Edstam T (2008) Road construction with consideration to overall stability. Swedish Road Administration, Region West Dnr AL90 B 2007:27435, SGI final report 08-05-29, Linköping (in Swedish)

Larsson R, Åhnberg H, Löfroth H (2012) A new Swedish large-diameter sampler for soft and sensitive clays. In: Proceedings of the 4th international conference on geotechnical and geophysical site characterization (ISC'4), vol 1, Porto de Galinhas, Pernambuco, pp 737–742

Locat A, Leroueil S, Bernander S, Demers D, Jostad HP, Ouehb L (2011) Progressive failures in eastern Canadian and Scandinavian sensitive clays. Can Geotech J 48(11):1696–1712

Löfroth H (2011) Mapping of quick clay formations within the Göta River Commission – evaluation of proposed method and preliminary guidelines. The Göta River Commission, Sub report No. 29, Swedish Geotechnical Institute, SGI, Linköping. www.swedgeo.se (in Swedish)

Löfroth H (2012) Sampling in normal and high sensitive clay – a comparison of results from specimens taken with the SGI large-diameter sampler and the standard piston sampler St II. SGI Varia No. 637, Swedish Geotechnical Institute, Linköping. www.swedgeo.se

Löfroth H, Suer P, Dahlin T, Leroux V, Schälin D (2011) Quick clay mapping by resistivity – surface resistivity, CPTU-R and chemistry to complement other geotechnical sounding and sampling. The Göta River Commision, Sub report No. 30, Swedish Geotechnical Institute, SGI, Linköping. www.swedgeo.se

Löfroth H, Suer P, Schälin D, Dahlin T, Leroux V (2013) Mapping of quick clay using sounding methods and resistivity in the Göta River valley. In: Proceedings of the 4th international conference on geotechnical and geophysical site characterization (ISC'4), vol 1, Porto de Galinhas, Pernambuco, pp 1001–1008

Möller B, Bergdahl U (1982) Estimation of the sensitivity of soft clays from static and weight sounding tests. In: Proceedings of the European symposium on penetration testing, 2, ESOPT2, vol 1, Amsterdam, pp 291–295

Nordal S (2007) Material models in numerical simulations. Presentation at geotechnical R&D seminar, SGI, Linköping, 23–24 Oktober 2007 (in Norwegian)

Rankka K, Anderson-Sköld Y, Hultén C, Larsson R, Leroux V, Dahlin T (2004) Quick clay in Sweden. Report 65, Swedish Geotechnical Institute, SGI, Linköping. www.swedgeo.se

SGF (2009) Guidelines for sampling with standard piston sampler. Report 1:2009, Swedish Geotechnical Society, Linköping. www.sgf.net (in Swedish)

SGI (2012) The Göta River landslide risk investigation. Final report, part 2 – mapping, The Göta River Commission, Swedish Geotechnical Institute, SGI, Linköping. www.swedgeo.se

Thakur V, Degago SA (2012) Quickness of sensitive clays. Géotech Lett 2:87–95

Torrance JK (1974) A laboratory investigation of the effect of leaching on the compressibility and shear strength of Norwegian marine clays. Geotechnique 24(2):155–173

Tremblay M, Svahn V, Lundström K (2013) Landslide risk assessment in the Göta River valley: effect of climate changes. In: Proceedings of the 18th international conference on soil mechanics and geotechnical engineering, ICSMGE, Paris

Chapter 31
Safety Concepts for Slope Stability

Tim Länsivaara and T. Poutanen

Abstract The use of partial safety factors has, with the adaptation of the Eurocodes, become the dominant safety concept, also for slope stability in most European countries. The article discusses the partial safety factor approach based on considerations of uncertainties involved and consequences of failure. Partial safety factors are calculated based on reliability theory, following first the present ideas of the Eurocodes, and second an alternative approach aimed to improve the shortcomings in the code. In the alternative approach, all uncertainties are placed on the material safety factor, which are then calculated for different target reliability index values. Two simple examples are presented, where the differences of the two approaches are outlined.

Keywords Slope stability • Safety • Eurocode • Reliability based design

31.1 Introduction

Slope stability analysis has traditionally been performed using the total factor of safety approach. As it has been used for a long time, there exists a lot of experience based data about the implication of its values. For example, if the total factor of safety is below certain threshold in soft clay, this often implies that there is the potential for larger horizontal movements in the soil, as discussed e.g. by Leroueil et al. (1990). Although the required factor of safety might be case sensitive, it is still often selected rather subjectively based on previous experience or to follow codes and guidelines, while the uncertainties related to loads, material properties, calculation models and construction are usually not considered.

T. Länsivaara (✉) • T. Poutanen
Tampere University of Technology, Tampere, Finland
e-mail: tim.lansivaara@tut.fi

J.-S. L'Heureux et al. (eds.), *Landslides in Sensitive Clays: From Geosciences to Risk Management*, Advances in Natural and Technological Hazards Research 36,
DOI 10.1007/978-94-007-7079-9_31, © Springer Science+Business Media Dordrecht 2014

Reliability based design (RBD) offers a framework to account for all the uncertainties as well as consequences of failure. Although the method is advancing, a full RBD analysis is perhaps still not available for daily engineering purposes.

Many of the present design codes strive between these two options and try to maintain the experience based knowledge of the former and include a better evaluation of risk offered by the later. The partial factor of safety method can be considered as a first step towards RBD. In Europe, the Eurocodes have been adopted in most countries for both structural and geotechnical design. This article reviews Eurocodes way to apply partial safety factors to slope stability analysis, and suggests some improvements.

31.2 Slope Stability Analysis According to EN 1997

31.2.1 Application of Safety in the Eurocodes

The Eurocode consists of several different parts of which EN 1990 defines the principles and requirements for safety, serviceability and durability of structures for all materials. According to EN 1990 the partial safety factors should account for the possible unfavorable deviation of the property from its characteristic value and the uncertainties in the model used in the calculations. Three consequence classes are introduced to account for consequence of failure.

Eurocodes allows the determination of partial safety factors based on reliability theories. In that case, the target reliability index β is set to 3.8 for a 50 years reference period, corresponding to a probability of failure of approximately 1/15,000. A second approach is to calibrate the partial factors of safety to past experience. It is the understanding of the authors that the partial safety factors for material and resistance in geotechnical design is mainly based on calibration to older practice.

31.2.2 Slope Stability According to EN 1997

In EN 1997-1, three alternatives are presented on how safety can be factored in geotechnical design. These design approaches, named DA1, DA2 and DA3 result from the difficulty of finding a single way to combine factors between actions, ground properties and resistances. However, for slope stability most countries have chosen either DA3 or DA1. DA1 consists of checking two combinations, of which combination 2 is relevant for slope stability. As DA3 and DA1combination 2 are consistent (here after referred as DA3), there seems to be a consensus on how safety should be applied for stability problems.

According to DA3, the safety is placed on material properties (strength) and on actions. The recommended values for partial safety factors on soils resistance are

$\gamma^{\phi'} = \gamma_c' = 1.25$ for effective stress analysis and $\gamma_{cu} = 1.4$ for total stress analysis, where the subscripts ϕ', c' and c_u refer to effective friction angle, cohesion and undrained shear strength, and where γ stands for the partial safety factor. The largest permanent action on slope stability comes often from the soil weight itself. While it is quite difficult to divide soil weight into favorable and unfavorable actions, the permanent loads are left unfactored. Actions from variable loads, like e.g. traffic load, are on the other hand factored. Accordingly, the recommended value in EN-1997 for permanent actions is $\gamma_G = 1.0$ and for variable actions $\gamma_Q = 1.3$, where the subscripts G and Q refer to permanent and variable load.

31.3 Reliability Based Design Partial Safety Factors

31.3.1 Introduction

In the following, partial factors of safety will be calculated based on reliability theory. Firstly, the assumptions made regarding different uncertainties will be presented. Thereafter, the theoretical bases for reliability calculations are outlined. Two different calculation schemes are presented;

I. Approach I (EN 1997-1): the material partial safety factors are calculated by placing safety on both load and material factors.
II. Approach II (alternative approach): uncertainty related to loads and material properties are placed on the material factor.

In both of these calculations, a general uncertainty will also be included. The consequence of failure will be introduced into Approach II by setting different target reliability index values.

31.3.2 Assumptions

31.3.2.1 Material Properties

A vital part of reliability based design (RBD), including the partial factor of safety method, is that uncertainties are dealt in a proper way. There exits quite a lot of data about the variation of soil properties, especially on soil strength. The selection of appropriate values to use in RBD is however not obvious. It is not the intention of this article to review data on soil strength variation, but rather to outline some difficulties in addressing the issue.

Traditionally, codes tend to suggest clearly higher partial safety factors for undrained shear strength (s_u) than for effective strength parameters (ϕ', c'). It is quite correctly thought that there is generally a greater uncertainty in the determination of s_u from e.g. vane test than in the determination of the friction angle from a triaxial test. But there are many ways to determine undrained shear strength which are rather advanced. The reported coefficient of variation for s_u varies also highly from test to test, see e.g. Phoon and Kulhawy (1999b). It might be argued, that even though the triaxial test gives quite reliable values, the test is not run on a representative sample in the triaxial test, if one is dealing with coarse grained material. In addition if effective strength parameters are used for an undrained analysis of clays, the uncertainty is greater on the pore pressure than on the friction angle. So one might ask should the type of analysis, i.e. drained vs. undrained also be included in the material factor?

A thorough review of the sources of uncertainty can be found in Lacasse and Nadim (1996) and Phoon and Kulhawy (1999a). According to these sources, the uncertainty related to geotechnical properties can be divided into inherent (natural) variability and epistemic uncertainty. The inherent variability represents the natural variation of the property. The epistemic uncertainty is more complex. It includes statistical uncertainty, measurement error and transformation model uncertainty. It is often unclear how these different sources of uncertainty are dealt with. The coefficients of variation might be much higher than the inherent variability due to errors in processing the data. Following the ideas of Phoon and Kulhawy (1999a), a simple example is presented to demonstrate interpretation errors. Let us assume that for a normally consolidated clay layer there are three values for undrained shear strength, i.e. 10, 17.5 and 25 kPa. If the values are treated without further consideration about their origin, one would calculate that the mean as 17.5 kPa and the standard deviation 6.1 kPa, corresponding to a coefficient of variation (CoV) of 35 %. Let us then assume that the values correspond to the depths of 3, 8 and 13 m. Plotting the data versus depth reveals an obvious linear trend in data corresponding to $s_u = \alpha\sigma_v'$ with no fluctuation and that the inherent variability is zero. Phoon and Kulhawy (1999a) pointed out that some statistical analyses reported in the geotechnical literature probably ignore such trends. The resulting coefficient of variation would then be overestimated.

Although the information about the true variation of soil properties in continuously increasing, it is difficult, if not impossible, to fix a general coefficient of variation to a soil property that can be determined in various ways. Therefore the partial material factors in the present paper are calculated for three different CoV, namely 0.1, 0.2 and 0.3. This uncertainty could be included in codes by defining a minimum CoV-value the engineers should use, and adding that if there exist data from which the CoV can be determined with more reliability, the engineer should use the data.

In addition to CoV, one needs to choose the type of distribution for the material strength. Often there are not enough data to get a good statistical fit to a probability density function (PDF). Both normal and lognormal PDF are often considered as valid. For a normal PDF, the data population ranges practically between the mean $\pm$3 times the standard deviation. A CoV higher than 0.33 indicates thus that negative values are possible. Combining high values of CoV with a normal PDF for strength can lead to erroneous values of s_u in the probabilistic analyses. Lognormal PDFs are often used to characterize properties that have only positive values. While the authors prefer using lognormal PDFs for soil strength, it is sound to check the possible effect of each PDF assumption. The Approach I calculations are performed for both PDFs.

31.3.2.2 Loads

As mentioned earlier, the permanent load resulting mostly from soil weight is usually left unfactored. The uncertainty can, however, be accounted for by including it in the material factor. Herein a CoV set equal to 0.1 was used for the permanent load.

For the variable load, a CoV_Q of 0.4 might be applicable for structural design as the uncertainty in wind and snow loads is high. For geotechnical design however, the variability should be considered separately. In Approach I the CoV on the variable load was set to 0.4, but for Approach II the calculation was done with CoV-values of 0.4 and 0.25. A normal PDF was used for both the permanent and variable loads and the loads are combined dependently, see Poutanen (2011, 2013).

31.3.2.3 Additional Generic Random Variable

The partial factor of safety method implies indirectly that safety is placed where there is uncertainty. In practice this is not always the case. For example in Europe, the different safety factor approaches, i.e. design approach 1, 2 and 3, are mostly chosen based on previous tradition in each country, not based on the assessment of uncertainty. There is then a danger that all uncertainties are not considered. With a full RBD, it is, at least in theory, possible to directly account for all the uncertainties involved. For a partial safety factor approach, this is not the case. One should thus add a generic variable with an uncertainty to account for factors such as calculation model errors and poor construction.

31.3.2.4 Consequence of Failure

When evaluating a target probability of failure, one should consider not only the uncertainties involved, but also the consequences of a possible failure. It is not economical or logical to require the same safety level for slopes with only a pedestrian way a for a slope in a densely populated residential area. In the Eurocodes, this is addressed by three different consequence classes. A multiplication factor KFI is applied to unfavorable loads and its value depends on the consequence of failure. For slope stability problems, the effect of external actions on the stability varies from nil to rather substantial. It seems thus rather random to apply safety related to the consequence of failure. However, EN 1990 also stated that "Reliability differentiation may also be applied through the partial factors on resistance γ_M". In this study, the partial material factors will be calculated for different target reliability index (β) corresponding to the different reliability classes (RC) in the

Eurocode. The target β values were chosen according to EN 1990 as $\beta_{50}=4.3$ (for RC3), $\beta_{50}=3.8$ (for RC2) and $\beta_{50}=3.2$ (for RC1).

31.3.3 Theoretical Bases

The design point was set at unity and the target reliability in the reliability calculation chosen according to EN 1990. The permanent load distribution was assumed as normal, the coefficient of variation equal to 0.1. The cumulative distribution is $FG(x, \mu_G, \sigma_G)$ and the density distribution is $fG(x, \mu_G, \sigma_G)$. For the variable load, a normal distribution was also used. The distributions were $FQ(x, \mu_{QN}, \sigma_{QN})$, $fQ(x, \mu_{QN}, \sigma_{QN})$, and the 0.98 fractile was set at the design point according to the 1-year load. The n-year variable load distribution can be calculated according to:

$$FQ\left(x, \mu_Q, \sigma_Q\right)^n \tag{31.1}$$

The material property distribution was assumed as lognormal, the cumulative distribution was $FM(x, \mu_M, \sigma_M)$ and the density distribution $fM(x, \mu_M, \sigma_M)$. The characteristic value was a 5 % fractile value which was set at the design point. With the cumulative distribution of the load as $FL(x, \mu_L, \sigma_L)$, the density distribution of the material property $fM(x, \mu_M, \sigma_M)$, the load safety factor γ_L and the material safety factor γ_M, the failure probability P_f becomes:

$$1-\int_0^{\infty} \mathrm{FL}(x, \mu_L, \sigma_L) \cdot \mathrm{f\,M}(x, \mu_M \cdot \gamma_L \cdot \gamma_M, \sigma_M \cdot \gamma_L \cdot \gamma_M) dx = P_f \tag{31.2}$$

When the two loads $F_1(x, \mu_1, \sigma_1)$, $f_1(x, \mu_1, \sigma_1)$ and $F_2(x, \mu_2, \sigma_2)$, $f_2(x, \mu_2, \sigma_2)$ with the random variable $x_{1,i}$ and $x_{2,i}$ in fractile i are combined dependently in the proportion α and 1-α (where α is the proportion of the load 1 in the total load) to obtain random variable $x_{1,2,i}$ of the combination load in fractile i, the variable x_i can be calculated by adding up the partial random variables:

$$x_{1,2,i} = x_{1,i} \cdot \alpha + x_{2,i} \cdot (1-\alpha) \tag{31.3}$$

The distributions used for the actions and resistances are presented in Fig. 31.1.

31.3.4 Material Factors for Approach I (EN 1997-1)

The Approach I calculations resemble the partial safety factor approach DA3 in the Eurocodes. The loads were combined dependently and the partial safety factors for the loads are $\gamma_G=1$ and $\gamma_Q=1.3$. The load distribution for both the permanent and the variable loads was assumed as normal. The variable load was

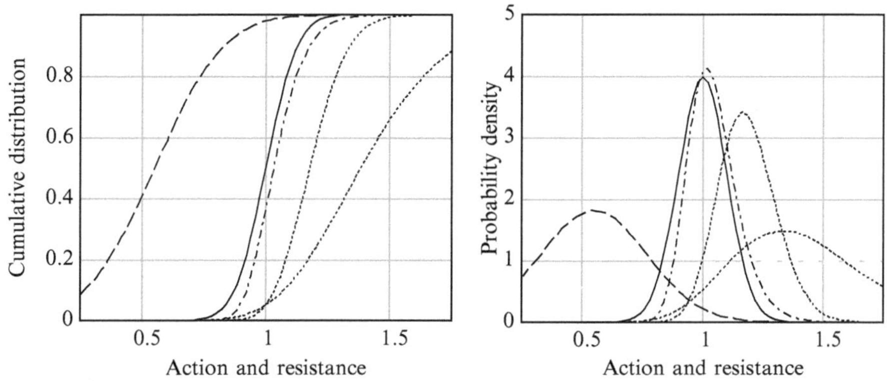

Fig. 31.1 Distributions set at the design point. *Solid line*: permanent load; *dashed line*: variable load 1-year location; *dash-dotted line*: variable load in 50-year location; *dotted lines*: material properties. The uncertainty distribution is equal to the permanent load distribution but it is located at the origin

selected as that with a 50-year return period, and the target reliability corresponds to RC2 of the Eurocodes, i.e. $\beta_{50}=3.8$.

The calculations were done for material CoVs of 0.1, 0.2 and 0.3. using both normal and lognormal PDFs. The generic random variable followed a normal distribution with zero mean and 0.1 CoV. The results of the calculations with Approach I are shown in Fig. 31.2 as a function of the load ratio (the proportion of the variable load to the total load, in %). The case with lognormal PDF for the material factor γ_M is on the left side, the normal PDF on the right side.

Figure 31.2 shows that with the given assumptions, the material factors vary for a given reliability index as a function of load proportion. If one considers a CoV of 0.1 for the effective strength parameters, ignores the generic random variable (as in the Eurocodes), one finds, that the recommended partial material factor $\gamma\phi$ in EN 1997-1 is 1.25 for a load proportion of 60 %. To achieve the same probability of failure for a case with no variable load, one would need a material factor of 1.4.

Figure 31.2 also shows that for a normal PDF for the material strength, the partial factors increase significantly with increasing CoV. For a CoV equal to 0.3, the calculation did not converge. This was expected, as a CoV of 0.3 indicates that the strength values can be close to zero.

31.3.5 Material Factors for Approach II, (Alternative)

In Approach II, all uncertainty is placed on the material partial safety factor, i.e. the partial safety factors for both permanent and variable load are set to 1.0. The material factors were calculated for three different reliability index values, corresponding to the three reliability classes in the Eurocodes. The calculations were done for CoVs on the variable load of 0.4 and 0.25. The results are presented in Fig. 31.3, again as

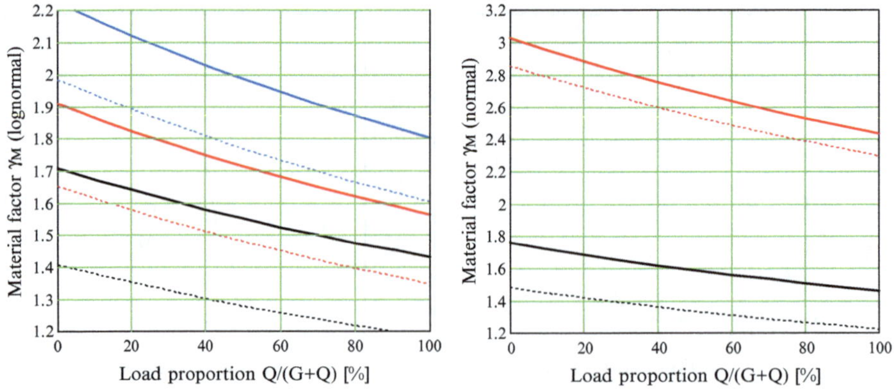

Fig. 31.2 Material factors according to approach I (EN 1997) to $\gamma_G = 1$, $\gamma_Q = 1.3$ and $\beta_{50} = 3.8$ as function of load ratio. *Black lines* denote $CoV_M = 0.1$; *red lines* $CoV_M = 0.2$ and *blue lines* $CoV_M = 0.3$. The *dotted lines* correspond to values without uncertainty, $CoV_U = 0$, and the *solid lines* values with uncertainty, $CoV_U = 0.1$. *Left side* figure uses lognormal distributions for material and *right side* uses normal distribution

a function of the load ratio. The case with CoV_Q of 0.4 is on the left side, the case with CoV_Q of 0.25 on the right side.

Comparing Figs. 31.2 and 31.3, the material safety factors for the same reliability index were equal when the load proportion was zero. For $CV_Q = 0.25$, the material factors were practically not affected by the load proportion. The same probability of failure would be reached with a constant material factor. In addition, the results were comparable, regardless if the safety was set on both the action and resistance or if set on the resistance only. For example, for a load proportion of 60 % and a CoV on material factor 0.1, Fig. 31.2 gives a γ_M of 1.53, resulting in a factor of [(0.4 + 0.6*1.3)*1.53] or 1.8, which is equal to the value of Fig. 31.3.

31.4 Examples of Application

31.4.1 Case 1, External Load with High Impact

Let us first study a case where the external load has a high impact on the safety. The geometry of the problem is given in Fig. 31.4. A 1-m thick embankment is laid upon a 1-m dry crust layer and a 10-m thick soft clay layer. The unit weight of the embankment material is 20 kN/m^3 and the characteristic friction angle is $\phi' = 38°$. The dry crust layer has a unit weight of 17 kN/m^3 and characteristic undrained shear strength of 30 kPa. The soft clay has unit weight of 16 kN/m^3 and a characteristic s_u of 10 kPa at the top of the layer increasing with 1.4 kPa/m with depth. A 5-m wide load of 40 kPa is placed two meters from the crest of the embankment. According to the recommendations in EN 1997-1, the design (d) values are: ϕ'_d of 32° for the

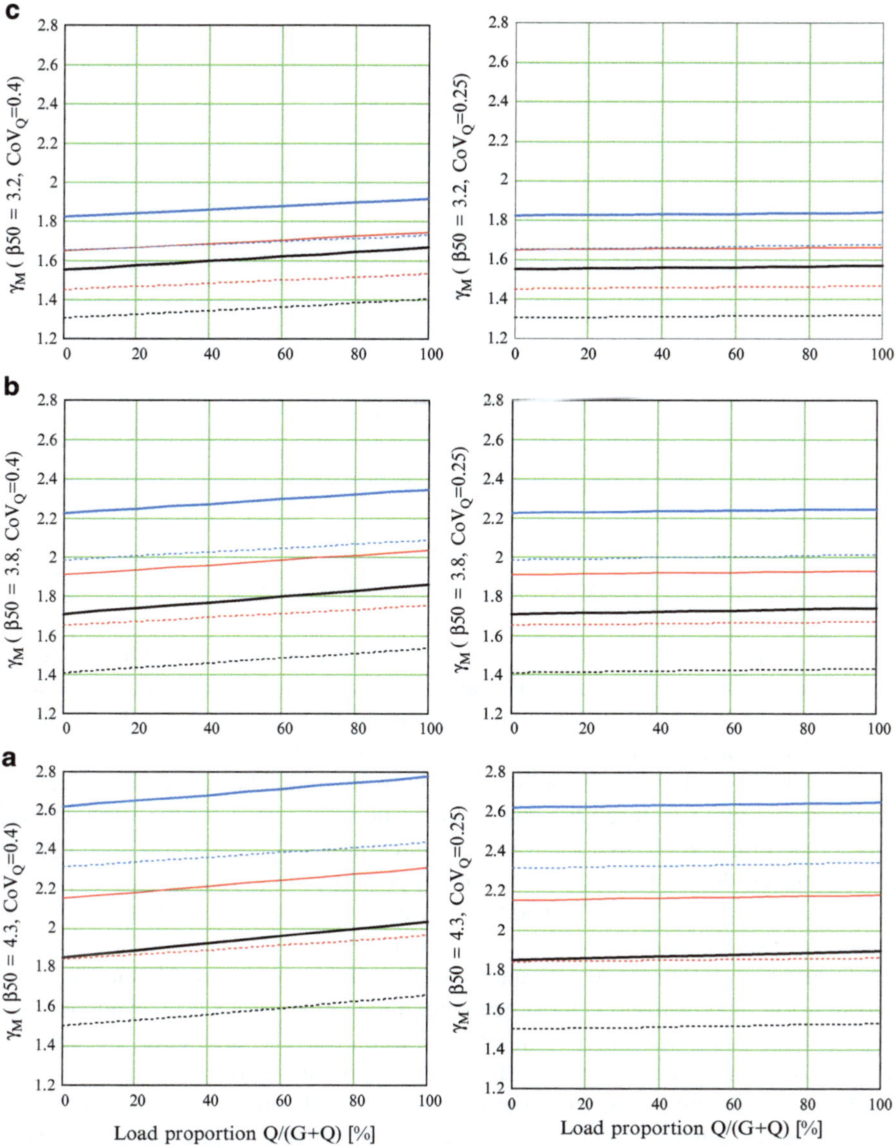

Fig. 31.3 Material factors corresponding to $\gamma_G=\gamma_Q=1.0$ and (**a**) $\beta_{50}=4.3$ (RC3), (**b**) $\beta_{50}=3.8$ (RC2) and (**c**) $\beta_{50}=3.2$ (RC1) as function of load ratio. *Black lines* denote $CoV_M=0.1$, *red lines* $CoV_M=0.2$ and *blue lines* $CoV_M=0.3$. The *dotted lines* correspond to values without uncertainty, $CoV_U=0$, and the *solid lines* values with uncertainty, $CoV_U=0.1$

embankment, s_{ud} of 21.4 kPa for the dry crust and $s_{ud}=7.1$ kPa + 1 kPa/m for the soft clay. For a permanent load G_d is 40 kPa while for a variable load Q_d is 52 kPa.

The total factor of safety calculated with the characteristic values was 1.46 for a circular failure surface applying the Bishop method. According to EN 1997-1 for the permanent load case, the resulting over dimensioning factor (ODF) is 1.04,

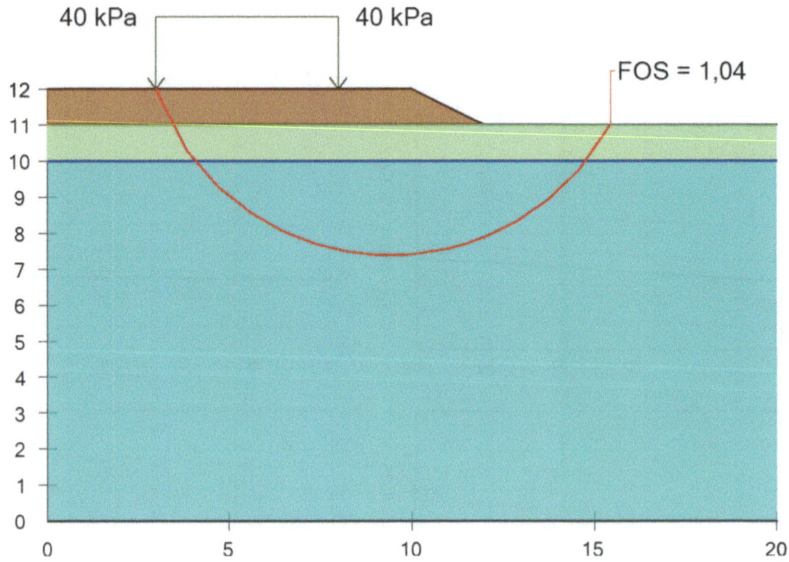

Fig. 31.4 Geometry for Case 1 and calculated ODF according to EN 1997-1

indicating that the situation is safe. For the variable load case, the ODF reduces to 0.88. For this case, to achieve an allowable design, the characteristic initial value of undrained shear strength of the soft clay should increase by about 30 %, up to 13 kPa. This corresponds to a total safety factor of 1.69.

Considering risk and probability of failure, it might be asked if higher soil strength should be required to achieve the same safety level for the variable load case? One might argue that there is more uncertainty in the variable load than in the permanent load. However, if the variable load is due to railway train load, the maximum allowable operation load is set by each track, so the characteristic load is rather a maximum load. It might then be more important to consider the possible consequences of failure. A permanent load might involve a residential building where lives could be lost. On the other hand, if the railway track is in an uninhabited area and the high load comes from a freight train, the risks are merely economical and the owner might consider allowing a higher probability of failure.

The alternative Approach II allows for such considerations. Assuming CoVs of 0.1 for the friction angle and 0.2 for the undrained shear strength, and $CoV_Q = 0.25$ for the variable load, Fig. 31.3 gives a partial factor γ^{ϕ} of 1.3 and γ_{su} of 1.45 for β_{50} of 3.2, and $\gamma^{\phi} = 1.41$ and $\gamma_{su} = 1.66$ for $\beta_{50} = 3.8$.

For a reliability index β_{50} of 3.2, the design values for approach II would be $\phi_d = 31°$ for the embankment, $s_{ud} = 20.6$ kPa for the dry crust and $s_{ud} = 6.9$ kPa + 0.96 kPa/m for the soft clay. The design load would be 40 kPa for both permanent and variable loads. These values give an ODF = 1.01 indicating that the situation is just acceptable. For β_{50} of 3.8, the design values would be $\phi_d = 29°$ for the embankment, $s_{ud} = 18$ kPa for the dry crust and $s_{ud} = 6.0$ kPa + 0.84 kPa/m for the soft clay and the design load would again be 40 kPa for both permanent and

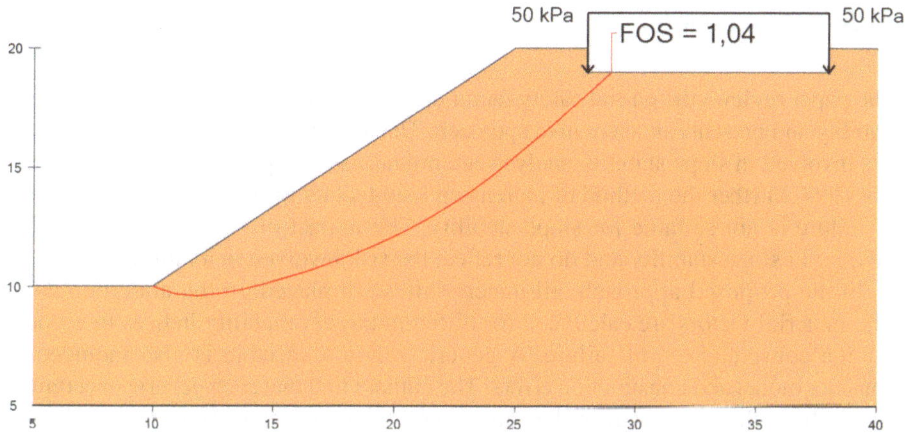

Fig. 31.5 Geometry for Case 2 and calculated ODF according to EN 1997-1

variable loads. These values result in an ODF of 0.88. To have an acceptable safety the initial undrained shear strength of the soft clay would need to increase by 26 % to 12.6 kPa. This corresponds to a total factor of safety requirement of F = 1.66.

31.4.2 Case 2, External Load with Minor Impact

Let us then study a case where the external loads have minor impact on the safety. The geometry of the problem is given in Fig. 31.5. The 10-m high slope is in a dry silty soil with a unit weight of 18 kN/m^3 and characteristic effective strength parameters of $\phi'_k = 26°$ and $c_k = 8$ kPa. At the top of the slope, there is a 10 m wide load corresponding to a characteristic stress of 50 kPa. According to the recommendations in EN 1997-1, the design values are $\phi'_d = 21.3°$ and $c_d = 6.4$ kPa, and $G_d = 50$ kPa if the load is permanent and $Q_d = 65$ kPa if the load is a variable load.

The total factor of safety calculated with the characteristic values was 1.29 for a non-circular failure surface applying the Morgenstern-Price method. In the case of a permanent load, the ODF according to EN 1997-1 was 1.04. In the case of a variable load of 65 kPa, the ODF was 1.03. In this case the load, even if rather high, does not influence significantly the safety factor. If one would then apply a consequence perspective based on loads as in the Eurocodes, it would not improve the reliability. For the alternative approach II, the consequence is on the other hand placed on the material safety factor. Then, for a CoV on the friction angle of 0.1 and a CoV on the variable load of 0.25, one obtains the partial material factors $\gamma^{\phi} = 1.3$ for $\beta_{50} = 3.2$, $\gamma^{\phi} = 1.41$ for $\beta_{50} = 3.8$ and $\gamma^{\phi} = 1.51$ for $\beta_{50} = 4.3$. As there is only one soil layer (only effective strength parameters are used), the material partial factors of safety are in practice also total safety factors.

31.5 Conclusions

The paper reviews the partial safety factor approach adopted in EN 1997-1 for slope stability and presents an alternative approach. The authors conclude that the uncertainties involved in slope stability analysis are not necessarily properly accounted for in EN 1997. Further the method of increasing load factors to account for consequence of failure is not suitable for slope stability. The impact of such factors are rather random in slope stability and do not reflect the risk involved in a failure.

In the proposed approach, all uncertainties are placed on the material factor. The material factors are calculated for different target reliability indices to account for the consequences of failure. A generic random variable is also included to consider model and construction errors. The authors find that such generic uncertainty should be included. However, the value used in the present analyses is probably on the high side.

The coefficient of variation for soil strength varies depending on whether undrained or effective strength parameters are used and which method is used for the determination of the parameters. Codes could account for this by setting a fixed minimum value cases where little or no data are available, but allowing the designer to use lower values if reliable site-specific data on the variability are available.

The authors suggest that, instead of using a normal probability density function for soil strength, a lognormal PDF should be used. Calculated material factors increased exponentially for high coefficient of variation when a normal distribution was used.

If probability theory is used for the determination of partial safety factors in codes, careful consideration is required in the evaluation of the coefficients of variation, PDfs and target reliability index values. For stability analysis, a coefficient of variation of 0.4 for the variable load might be too high and a lower value of 0.25 is suggested.

Acknowledgments The authors would like to thank Dr. Suzanne Lacasse for her valuable comments when reviewing this article.

References

European Committee for Standardization, CEN (2002) EN 1990 Eurocode: basis of structural design. European Committee for Standardization, Brussels

European Committee for Standardization, CEN (2004) EN 1997-1 Eurocode 7: geotechnical design. Part 1: general rules. European Committee for Standardization, Brussels

Lacasse S, Nadim F (1996) Uncertainties in characterizing soil properties. In: Theory to practice. Proceedings of uncertainty '96, Madison, Wisconsin 1996, Geotechnical special publication 58. American Society of Civil Engineers, New York, pp 49–75

Leroueil S, Magnan J-P, Tavenas F (1990) Embankments of soft clays. Ellis Horwood, Chichester

Phoon K-K, Kulhawy FH (1999a) Characterization of geotechnical variability. Can Geotech J 36:612–624

Phoon K-K, Kulhawy FH (1999b) Evaluation of geotechnical property variability. Can Geotech J 36:625–639

Poutanen T (2011) Calculation of partial safety factors. In: Faber MH, Köhler J, Nishijima K (eds) Applications of statistics and probability in civil engineering. Taylor & Francis Group, London

Poutanen T (2013) Load combination. In: IABSE workshop on safety, Helsinki, 14–15 February 2013

Author Index

A
Aabøe, R., 91–101
Aagaard, P., 63–73
Åhnberg, H., 383–392
Alpay, S., 253–263
Andersson-Sköld, Y., 51–61
Aunaas, K., 91–101, 343–352

B
Bastani, M., 159–175
Bazin, S., 193–203, 229–238
Bilodeau, C., 331–341
Brooks, G.R., 119–130

C
Crow, H.L., 253–263

D
Dahlin, T., 217–227
Dalsegg, E., 179–189
Degago, S.A., 91–101, 291–300
Demers, D., 1–9, 77–87, 133–143, 331–341
Dolva, B.K., 91–101, 343–352
Donohue, S., 159–175

E
Emdal, A., 63–73

F
Fabien-Ouellet, G., 241–251
Fornes, P., 305–315
Fortier, R., 241–251

G
Geertsema, M., 105–115
Giroux, B., 241–251
Gjelsvik, V., 145–156, 355–367
Gjengedal, I., 63–73
Gylland, A.S., 267–276

H
Hansen, L., 179–189, 369–380
Helle, T.E., 63–73, 193–203
Hinton, M., 253–263
Høgaas, F., 369–380
Høydal, Ø., 63–73
Hunter, J.A., 253–263

J
Jensen, O.A., 91–101, 343–352
Jostad, H.P., 279–288, 291–300, 305–315, 355–367

K
Kalscheuer, T., 159–175
Kalsnes, B., 355–367
Kornbrekke, H.A., 291–300

L
Lacasse, S., 355–367
Länsivaara, T., 317–327, 395–406
Lecomte, I., 159–175, 229–238
Lehtonen, V., 317–327
Leroueil, S., 1–9, 133–143, 279–288
L'Heureux, J.-S., 1–9, 91–101, 105–115, 145–156, 159–175, 229–238

J.-S. L'Heureux et al. (eds.), *Landslides in Sensitive Clays: From Geosciences to Risk Management*, Advances in Natural and Technological Hazards Research 36, DOI 10.1007/978-94-007-7079-9, © Springer Science+Business Media Dordrecht 2014

Locat, A., 1–9, 279–288
Locat, J., 1–9, 25–35, 133–143
Locat, P., 77–87, 133–143
Löfroth, H., 51–61, 383–392
Long, M., 159–175
Lundström, K., 383–392
Lyche, E., 91–101, 145–156, 343–352

M
Mansikkamäki, J., 317–327
Moholdt, R., 145–156

N
Nadim, F., 355–367
Nigussie, D., 91–101
Nyheim, T., 91–101, 343–352

O
O'Connor, P., 159–175
Olsen, L., 369–380
Oset, F., 91–101, 343–352

P
Persson, L., 159–175
Persson, M.A., 39–49
Pfaffhuber, A.A., 193–203
Potvin, J., 77–87, 331–341
Poutanen, T., 395–406
Pugin, A.J.M., 253–263
Pullan, S.E., 253–263

R
Rindstad, B.I., 369–380
Robitaille, D., 77–87
Robsrud, A., 91–101, 343–352
Rømoen, M., 159–175
Rønning, J.S., 179–189

S
Sandven, R., 205–215
Sæter, M.B., 91–101, 343–352
Sauvin, G., 159–175, 229–238
Schälin, D., 217–227
Solberg, I.-L., 159–175, 179–189, 205–215
St-Gelais, D., 25–35
Suer, P., 51–61
Sveian, H., 369–380

T
Thakur, V., 91–101, 291–300, 305–315, 343–352
Thibault, C., 331–341
Tornborg, J., 217–227
Torrance, J.K., 15–23

V
Vanneste, M., 229–238
Viklund, M., 91–101, 343–352

Subject Index

A
Abrasion marks, 374
Active failure, 281, 283, 288
Active strength, 282, 287, 288, 311
Activity, 3, 17–19, 28, 30, 31, 33, 34, 81, 107, 202, 206, 230, 242, 344, 358, 360
Activity coefficient, 30
Alaska, 1, 3, 33
Amelioration of the quick, 21, 23
Anisotropy, 197, 320, 321, 323, 327, 347, 348, 351
Apparent conductivity, 256
Apparent resistivity, 163, 165, 180, 209, 232
Ariake clay, 34
Artesian groundwater, 385
Assessment, 1, 2, 4, 7–9, 49, 92, 97, 120, 173, 184, 189, 195, 201, 213, 230, 237, 251, 263, 292, 294, 300, 334–337, 339, 345, 355–367, 369–380, 388–391, 399

B
Back-calculation, 306, 357, 365–366
Backscarp, 19–22, 110, 111, 185
Bedrock, 5, 26, 40–42, 44, 47, 48, 53, 56, 111, 114, 161, 163, 164, 166, 180, 184–189, 194–196, 198, 200–203, 211–214, 221, 222, 230, 231, 234, 236–238, 243–248, 250, 251, 257, 258, 263, 372–379
BERT, 196, 200, 201, 203
BIFURC, 285
BING, 140–143
Bingham, 140
Block sampler, 388
Block samples, 66, 69, 174, 271, 272, 295–298, 312, 313, 388
Borehole, 53, 54, 56, 61, 136, 137, 148–151, 153, 163, 167, 168, 193, 195–198, 200, 202, 203, 214, 230, 231, 237, 254–262, 349, 390
Borehole geophysical logging, 256
British Columbia, 3, 33, 105, 106, 110
Brittle materials, 1, 209, 306, 350, 365
Brittleness, 6, 272, 274–276, 286–288, 310, 312–314
Building Code of Canada, 242
Bulk density, 148, 169, 221, 254, 256, 257, 259–260

C
Calcium, 17, 52–61, 65, 68, 69, 72, 73
Carbonates, 18, 30, 48, 52, 54, 56, 67
Cementation, 7, 17–19
Cementing agents, 18
Cement piles, 64, 361
Chalk, 64, 361
Champlain Sea, 26, 28, 29, 35, 81, 105, 110, 121, 242, 253–263, 370
Chemical factors, 3, 16, 19, 23
Chemical weathering, 202
Chemistry, 15–23
Chloride, 67–71, 73
Chlorite, 18, 27, 29, 30, 32–34, 66, 197
Chronology, 129
Clay
  minerals, 3, 18, 19, 29, 31–35, 65, 66
  resistivity, 182, 194, 200–202, 226
Climate, 385
Climate change, 8, 385

J.-S. L'Heureux et al. (eds.), *Landslides in Sensitive Clays: From Geosciences to Risk Management*, Advances in Natural and Technological Hazards Research 36,
DOI 10.1007/978-94-007-7079-9, © Springer Science+Business Media Dordrecht 2014

CMP reflection processing, 246
Coastal bluffs, 142
Codes, 53, 242, 244, 256, 285, 395–398, 406
Coefficient of earth pressure at rest (Ko), 282
Cohesion and friction softening, 297–300
Cohesion softening, 293, 299
Conductivity, 27, 164, 175, 208, 254, 256–263, 286
Cone penetration test (CPT), 42, 149, 167, 173, 175, 181, 206, 208, 218–220, 227, 386, 387
Cone penetration test with resistivity module (CPT-R), 217–227, 388
Consequence, 2, 7, 8, 160, 185, 206, 306, 307, 338, 345–350, 352, 356, 358–360, 365, 366, 385, 391, 396, 397, 399–400, 404–406
Consolidation, 2, 5, 7, 16, 17, 20, 65, 153–155, 202, 247, 272, 273, 275, 296, 379
Constrained inversion, 194, 196, 200, 203, 223
Constraining resistivities, 200
Constraint map, 336
CPT. *See* Cone penetration test (CPT)
Cracks, 281, 338
Cyclic loadings, 390

D

Debris, 4, 7, 16, 20–23, 79, 82, 83, 85, 92, 94, 97, 101, 110–112, 121, 122, 124, 125, 128, 130, 134, 135, 138–143, 147, 148, 150, 280, 281, 284, 378
Debris flow, 21, 97, 140, 141, 143, 147
Decision making, 92, 334, 344, 360, 370
Deglaciation, 160, 221, 257, 372–374, 378, 379, 385
Diffusion, 22, 54–56, 60, 61, 65–68, 70–73, 221
  cells, 66–68, 72
  coefficient, 22, 55, 61, 70–73
Digital elevation models, 106, 370, 378
Direct simple shear tests (DSS), 284, 286–288, 306, 307, 309, 313
Dislocation, 6, 281–283, 288
Downhole seismic survey, 251
2D resistivity, 161, 179–189, 206–215, 232
  data, 181, 183–185, 187, 210, 212–214
  measurements, 179–189, 206–210, 214, 232
3D sensitivity, 44, 48, 49
DSS. *See* Direct simple shear tests (DSS)

E

Earthflows, 19, 20
Earthquakes, 3, 20, 120, 121, 128, 129, 242, 255, 258, 356
Eastern Ontario, 121, 253–263
Effective stress analysis, 318–321, 327, 345, 397
Effective stress stability analysis, 318
Electrical resistivity, 160, 162, 163, 165, 196, 218, 219, 249
Electrical resistivity tomography (ERT), 161–166, 168, 173–175, 189, 194–203, 217–227, 230–232, 234–238
Electrode arrays, 161, 180, 209
Electrode spacing, 165, 197, 198, 214, 222, 224, 232
Electromagnetic conductivity, 164, 175
Electromagnetic method, 165
Electromagnetic techniques, 164
Emergency preparedness, 360–361
Energy, 4, 18, 73, 92, 97–101, 107, 109–110, 134, 141, 185, 238, 247, 288, 306, 346, 348–349, 352, 356
Engineering properties, 160, 220, 295
Erosion, 6, 8, 79, 94, 96, 147, 160, 166, 187, 220, 238, 247, 251, 295, 317, 357, 367, 386, 396
ERT. *See* Electrical resistivity tomography (ERT)
Eurocode, 306, 346, 349–350, 352, 396, 399–401, 405

F

Factors of safety, 280, 318, 361, 396, 397, 405
Failure
  mechanism, 4, 6, 268, 270, 279–289, 293, 333, 345–348, 350
  probability, 400
  propagation, 282–284, 287, 288
  surface, 5, 79, 83, 85, 86, 113, 138, 139, 141, 148, 270, 273, 280–288, 306, 323, 326, 365, 403, 405
  zone, 21, 22, 92, 138–140, 150, 153–155
Field vane, 79, 109, 137, 138, 335, 365
Finite element, 6, 285, 306, 309, 313, 315, 318, 323, 324, 349, 352, 365
Fjord-marine deposits, 370–373, 376, 377
Fjord valleys, 370, 376
Flocculated structure, 17, 19
Flocculation, 3, 16–19

Flow(s), 3, 4, 19–21, 25, 26, 40, 41, 53, 54, 56, 82, 83, 85, 87, 92–97, 100, 106–108, 110, 114, 120, 134, 135, 138–141, 188, 218, 221, 230, 233, 245, 246, 284, 320, 340
  behavior, 92, 96, 97
  capacity, 4, 134
Flowslides, 4, 78–87, 141, 333
Friction softening, 293, 297–300

G
Geochemical data, 53, 68
Geological history, 3, 34, 385
Geophysical data, 5, 219, 229–238, 254
Geophysical inversion, 194, 203
Geophysical logs, 256, 257, 259
Geophysical measurements, 160, 175, 237, 238, 254
Geophysical methods, 5, 8, 160, 168, 174, 175, 195, 218, 219, 230, 238
Geophysical surveys, 160, 186
Geophysical techniques, 5, 160, 174
Geophysical testing, 170
Geotechnical/geophysical correlations, 259–262
Geotechnical properties, 5, 65, 81, 82, 87, 134, 136–138, 146, 168–175, 254, 355, 388, 392, 398
GIS, 370, 372, 387, 392
Glaciation, 3, 25, 26, 160, 221, 257, 268, 373, 374, 378, 379, 385
Glaciofluvial deposits, 161, 372, 374, 376
Glacioisostatic depression, 374
Goldthwait Sea, 26, 28, 29, 77, 81, 133, 137
Göta Älv river valley, 217–227
Göta River, 52–54, 56, 220, 221, 383–392
Gothenburg, 39–49, 53
GPR. *See* Ground penetrating radar (GPR)
Grabens, 6, 84, 85, 121, 141, 142, 280–284, 288
Ground improvement, 63–73
Ground penetrating radar (GPR), 230, 231, 233–238
Guidelines, 243, 306, 345–349, 352, 356–358, 365, 385, 389
Gulf of St. Lawrence, 26

H
Hardening soil model, 285, 286
Havre-Saint-Pierre, 134–135, 141–143
Havre-St-Pierre, 4, 133–143
Hazard
  assessment, 2, 4, 7–8, 92, 120, 195, 336, 377, 380
  classes, 357–358, 363
  mapping, 2, 4, 7, 345, 357, 378
  score, 357, 358
  zones, 7, 185–187, 194, 207, 356–357, 362
Herschel-Bulkley, 140
High-quality block samples, 295
High-salinity, 16
Horsts, 6, 84, 85, 138, 141, 142, 280–284, 288
Hvittingfoss, 170, 229–238

I
Illite, 18, 27, 29, 30, 32–34, 66, 197
Induced Polarisation (IP), 164, 226, 227
Induction logs, 256
Infrastructure for floods and slides (NIFS), 345, 351
Inorganic dispersants, 19, 202
Inter-particle contact, 18
Inversion model, 165, 166, 194
Ion(s), 3, 52–54, 57, 59, 65, 68, 73, 161, 184, 268
  composition, 65, 67
  concentration, 57, 60, 65, 209, 221
  exchange, 3, 51–61, 65, 68, 72, 73
Ionic composition, 196
Ionic content, 73
IP effects, 226, 227
Iron, 18, 19, 30, 52, 65, 67–69, 137

J
Japanese clay, 34
Joint acquisition, 241–251

K
Kaldvelladalen, 206–208, 210–214
Kinematic, 92, 107, 282

L
Labrador, 3, 25, 33
Landslide
  barrier, 185, 186
  extent, 385, 387–390
  hazard, 49, 92, 180, 206, 230, 254, 336, 338, 345, 348, 350, 357
  susceptibility, 369–380
Landstreamer, 168, 242, 255

Leached clay, 5, 47, 55, 161, 163, 167, 181, 183–189, 206, 209, 212–214, 221, 235
Leaching, 3, 16, 17, 19, 21, 22, 40, 42, 48, 49, 52, 53, 56, 59, 61, 107, 160, 161, 166, 170, 180, 184, 187–189, 194, 202, 214, 221, 230, 237, 238, 260, 268, 392
Light detection and ranging (LIDAR), 4, 79–81, 122, 335, 378, 379
Limestone, 26, 165, 166, 194, 198
Limit equilibrium, 6, 271, 280, 284, 315, 318, 322, 326, 349, 351, 367
Limit equilibrium method (LEM), 6, 280, 318–322, 326, 327, 349, 367
Limit state design, 345
Liquidity index, 4, 19, 28, 93–95, 97, 107, 109, 110, 114, 134, 137, 138, 143, 284, 295, 352
Liquid limit, 16–19, 22, 28, 30, 67, 68, 93, 110, 170, 195, 197, 202, 218, 221, 268
Local stability, 207, 345–347, 349, 351
Low activity, 3, 17–19, 31, 34, 107, 202
Lund-system, 161, 180, 209, 232

M

Magnesium, 52–61, 65, 68, 69, 72, 73
Magnetic susceptibility, 256
Management, 1–9, 19, 86, 331–341, 356, 383–392
Mapping, 2, 39, 160, 179–189, 193–203, 217–227, 331–341, 356, 370, 385
Mapping of quick clay, 160, 208, 215, 217–227, 385
Mapping program, 7, 335, 340, 370, 380
Marine clays, 17, 42, 66, 84, 160, 161, 164, 166–168, 170, 173, 174, 180, 181, 188, 194, 196, 198, 206–208, 213–215, 221, 227, 257, 268, 311, 348, 349, 356, 370, 376, 377, 380
Marine deposits, 80, 161, 179, 180, 194, 206, 221, 369–380
Marine limit (ML), 77, 78, 207, 332, 348, 370–377, 380, 385
Material coefficient, 306, 347, 350, 352, 362–365
Microfabric, 150
Microstructure, 16, 146, 149, 155
Mineral, 3, 17–19, 27, 29–34, 52–57, 61, 65, 72, 73, 182, 210
Mineralogical analyses, 26, 27
Mineralogical composition, 29–31, 33, 66, 379
Mineralogy, 2, 3, 17, 19, 26, 27, 31, 35, 65, 66, 73, 196, 197, 202
Mitigation measures, 1, 2, 4, 180
Mobility, 4, 7, 9, 23, 92, 106, 107, 110, 115, 134, 140–142
Modeling, 39–49, 92, 97, 140, 285, 286, 288, 318, 323, 326, 371, 377
Monitoring plan, 7, 338
Montmorillonite, 27, 29, 30, 32, 33
Morphology, 4, 48, 112, 134–139, 141, 142, 281
Multi-channel analysis of surface wave (MASW), 167, 168, 173, 175, 230, 242–248, 250, 251

N

NGI-ADPSoft, 309–310, 365
NIFS. *See* Infrastructure for floods and slides (NIFS)
Norway, 5, 47, 52, 65, 94, 107, 142, 146, 161, 179–189, 194, 206, 229–238, 262, 268, 295, 311, 343–352, 369–380, 389
Norwegian clays, 66, 71, 72, 97, 99, 100, 155, 174, 294, 297–300
Norwegian landslides, 92, 94–98, 100, 344, 345, 352
Norwegian National Railways Administration (NNRA), 344–348
Norwegian practice, 355–367
Norwegian Public Roads Administration (NPRA), 207, 344–347, 352
Norwegian standards, 208, 230, 345, 349, 371, 372
Norwegian Water Resources and Energy Directorate (NVE), 306, 346, 348–349, 356

O

OnsØy clay, 296–298
Ontario, 3, 25, 33, 77, 79, 105, 109, 121, 253–263
Organic compounds, 19
Ottawa, 22, 242, 254, 256–260
Ottawa Valley, 3, 119–130, 164, 258
Outaouais, 241–251
Oxides, 18, 19, 30, 32

P

Paleoearthquakes, 120, 121, 126–130, 255
Paleoseismic, 120, 121, 127–130

Partial safety factor, 8, 345–347, 349, 351, 362, 397–402, 406
Peak shear strength, 6, 272, 282, 284, 286, 287, 292, 293, 297, 306, 312, 314, 364, 365, 367
Percentual improvement, 345–347, 349–351
Physico-chemical properties, 27, 28, 30–32
Piezocone, 79, 83, 85, 148, 242, 248–250, 335
Piston sampler, 233, 388
Piston sampling, 149, 195, 386
Plastic limit, 18, 28, 67, 68, 70, 109
PLAXIS, 285, 286
PLAXIS 2D, 285, 365
Pore water chemistry, 2, 55, 56, 60, 65, 66, 73, 160, 182, 202, 210
Porewater conductivity, 256, 260–263
Porewater salinity, 22, 30, 31, 73, 194, 208
Porosity, 71, 72, 149, 151, 153, 155, 254, 256, 257, 259–260, 263, 273
Post-failure, 3, 4, 9, 15, 22, 91–101, 133–143, 148, 312, 352
Postglacial marine sediments, 26
Post-peak shear strength, 293, 297, 298
Post-peak states, 293, 297
Post-peak strength, 292
Potassium, 52, 55, 57–59, 70–73
Potassium chloride, 63–73
Potential energy, 44, 99–101, 107, 134, 141, 185
Precipitation, 52, 57, 385
Pre-failure, 3, 4, 78, 92, 112–114, 134, 148
Preventive measures, 337–339, 349, 363
Probability, 22, 40, 42, 43, 185, 237, 313, 315, 336, 337, 357, 358, 372, 385, 390, 391, 396, 398–402, 404, 406
  of failure, 315, 396, 399, 401, 402, 404
  of occurrence, 337
Progressive failure, 2, 6, 92, 266, 269–271, 276, 280, 282–288, 292, 299, 306, 310, 312, 348, 350, 365, 367, 388
Progressive slides, 345, 347, 351

Q
Quantitative mineralogy, 27
Quaternary geology, 231, 242, 369–380
Quaternary map, 370–372, 375–378
Québec, 2, 20, 25–35, 77–87, 105, 121, 133–143, 241–251, 280, 331–341
Québec clays, 33, 34
Quick clay
  development, 16, 17, 19, 41–42, 49, 52, 65, 187
  landslides, 15–23, 161, 185, 214, 230, 370, 372
  mapping, 39, 49, 179–189, 193–203, 229–238, 370, 372, 380
  susceptibility index (QCSI), 40, 43–46, 48, 49
  zones, 370, 372, 379
Quickness, 48, 94, 96, 352
Quickness tests, 96, 97

R
Radiocarbon ages, 124–126, 128–130
Rapidity, 109–111, 115
Rapidity number, 94, 96
Rayleigh waves, 242, 243, 246, 251
RCPT. *See* Resistivity cone penetrometer testing
RCPTU, 163–164, 182, 198, 230, 231, 234, 236
Reactivation, 3, 92, 134, 292
Reconsolidation, 146, 153–156, 387
Recovery measures, 340
Reflectivity, 237
Regional scale, 263, 371
Regression, 3, 42, 44, 100
Regulations, 2, 7–8, 336–338, 340, 345–352, 357
Reliability, 8, 49, 227, 396–402, 404–406
  index, 396, 397, 399, 401, 402, 404, 406
  theory(ies), 396–398
Reliability based design (RBD), 8, 396–402
Remolded shear strength, 40, 65, 67, 69, 93–95, 98, 99, 221, 306, 312
Remolding energy, 97–100
Remolding energy ratio, 99, 100
Remoulded shear strength, 2, 4, 7, 8, 79, 84, 86, 107, 109, 110, 114, 115, 142, 149, 161, 163, 170, 172, 173, 181, 183, 194, 196, 202, 209, 230, 256, 268, 284, 287, 288, 349, 352, 365, 387
Remoulding, 85, 109–110, 114, 115, 139, 141–143, 155, 206, 261, 283, 288, 352
Remoulding energy, 109–111, 288, 352
RES2DINV, 180, 196, 200, 209, 218, 232
Resistivity, 5, 39, 53, 160, 179–189, 193–203, 206, 218, 230, 249, 254, 387
  cone, 163–164
  mapping, 219
  measurements, 5, 163, 175, 179–189
  method, 206, 214, 215, 254
  moduli, 196
  profiles, 163, 164, 181, 184, 186, 188, 210–214, 221, 222

Resistivity, (*cont.*)
profiling, 39, 163, 186, 189, 206
values, 160, 161, 163, 168, 170, 181–182, 184, 186, 188, 189, 197, 198, 201, 202, 209, 210, 212, 213, 218, 223, 224, 234–236
Resistivity cone penetrometer testing (RCPT), 5, 163–166, 194–203, 223, 237
Resonance period, 248, 251
Response measures, 339–340
Response planning measures, 339
Retrogression, 4, 7, 9, 20, 40, 77, 78, 81, 82, 84, 86–87, 92–97, 100, 106–111, 113–115, 121, 134, 135, 138, 147, 184–186, 281, 283, 287, 288, 292, 312, 332, 333, 352, 389
Retrogressive, 4, 7, 19, 20, 77–87, 93, 94, 96, 114, 120, 121, 134, 269, 281, 333, 335, 337, 339, 340, 356, 365, 390, 392
Retrogressive landslides, 4, 20, 77–87, 96, 108, 121, 134, 333, 335, 339, 340, 390
Rheological properties, 97, 142, 143
Rheology, 94
Ridges, 135, 138, 139, 186, 281, 376
Risk analyses, 336–339, 385
Risk assessment, 7, 9, 334–337, 355–367, 390–391
Risk evaluation, 337
Risk identification, 334–336
Risk management, 1–9, 86, 331–341
Risk matrix, 355
Risk mitigation, 361–362
Risk score, 359, 360
Risk treatment, 337–340
Rissa, 1, 20, 64–69, 71–73, 93, 111, 112, 142, 185–187, 268, 294–296, 298, 344, 356
Rissa clay, 65, 294–296, 298
Rotary pressure sounding (RPS), 181, 183, 195, 198–201, 203
Rotational landslides, 20, 82, 83, 332, 333
Run-out, 4, 92–94, 97, 100, 106, 108–110, 115, 312, 349, 359, 366
Russia, 1, 3, 33

S

Safety, 8, 9, 40, 64, 195, 201, 231, 237, 280, 284, 306, 314, 318, 319, 335, 339, 344–349, 351, 357, 361, 362, 364, 365, 389, 390, 395–406
factors, 8, 9, 195, 201, 237, 284, 306, 345–347, 349, 351, 362, 390, 396–402, 404–406
measures, 64
regulations, 351
requirements, 346–348, 405
Sainte-Monique, 280, 283–288
Salinity, 3, 16, 17, 19, 22, 27, 28, 30, 31, 52, 65, 67, 73, 107, 137, 170, 175, 180, 194, 208, 215, 254, 256, 260–263
Salt, 17–19, 22, 52–55, 57, 64, 66–70, 72, 73, 107, 148, 149, 160, 161, 163, 169–171, 175, 181, 183, 194, 197, 198, 202, 208, 209, 213, 214, 221, 230, 238, 256, 258, 268, 295
content, 52, 53, 67, 68, 107, 148, 149, 160, 161, 163, 169–171, 175, 181, 183, 194, 197, 198, 202, 208, 209, 213, 221, 258, 295
well, 66, 68–70, 72
Sample disturbance, 92, 174, 210, 296–297
Scandinavia, 3, 4, 18, 19, 21, 33, 92, 105, 160, 181, 209, 268–269
Scanning electron microscopy, 146, 149, 273
Scar, 73, 78–80, 84–86, 122, 125, 127, 128, 130, 142, 335, 370–372, 378, 379
Seismic-CPT, 167, 230
Seismic data, 203, 232, 233, 254, 256, 257, 358
Seismic hazard, 129, 248
Seismic measurements, 232–233
Seismic methods, 160, 175, 242, 255
Seismic piezocone penetration test (SCPTu), 242, 243, 248–250
Seismic reflection, 112, 167–169, 232, 233, 235–238, 241–251, 254, 255, 263
Seismic reflection survey, 242–248, 251, 263
Seismic refraction, 164, 166–167, 230, 232, 236, 237, 247
Seismic refraction tomography (SRT), 230, 232, 234
Seismic sections, 247, 254, 257–259
Seismic tomography, 167, 234
Sensitivity, 2, 3, 5, 7, 8, 15–23, 27, 40, 42–49, 52, 54, 67, 79, 86, 93, 94, 98, 99, 106–109, 114, 115, 138, 142, 149, 161, 164, 170–173, 181, 183, 194, 196–198, 209, 213, 218, 221–224, 227, 230, 231, 238, 254, 256, 258, 260–263, 268, 286, 295, 306, 349, 358, 386, 387, 389, 390, 392
Sensitivity development, 16–19
Shale, 26, 32, 33, 166, 194, 198, 200
Shallow surface geophysical, 254
Shear bands, 6, 269–273, 275, 276, 286, 307–310, 312–314

Shear stress, 270, 271, 273, 274, 280, 282, 283, 285–288, 294–296, 306–308, 310, 319, 326, 346, 362
Shear wave, 168, 255
velocity, 167–169, 242, 246, 247, 254, 259–260
velocity profiles, 251
Shear zone, 145–156, 281–283, 286–288
Shelby tubes, 27
Short-term stability, 293, 345
Simple shear, 275, 276, 284, 309, 365, 367
Site investigations, 2, 5, 148, 195–196, 206–208, 218–222, 238, 344, 357
Slip surface, 4, 47, 93, 110, 144, 306, 321, 322, 326–327, 332, 346, 349, 358, 362–365, 389–391
Slope stability, 2, 6, 8, 64, 110, 145, 183, 185, 201, 214, 237, 268, 269, 273–275, 334, 365, 383–392, 395–406
Slurry, 66, 68–70
SmØrgrav, 165, 167, 194–198, 200
Sodium, 53–60, 65, 67–69, 72, 73
Softening modulus, 98, 312
Soil behaviour, 280, 285, 287, 351
Soil strength, 15, 155, 345, 397, 399, 404, 406
Specific surface area, 27, 30, 31, 33–35
Spreads, 4, 6, 20, 31, 61, 78, 79, 81–87, 96, 97, 107, 108, 113, 120, 134, 135, 138, 140–141, 143, 243, 251, 269, 279–289, 333, 339, 386, 388, 390
SRT. *See* Seismic refraction tomography (SRT)
Stability analysis, 6, 8, 280, 284, 294, 317–327, 351, 357, 360, 361, 388, 390, 395–397, 406
Stability investigations, 383–392
Stability number, 86, 94, 95, 100, 107, 109, 110, 115, 352
Stabilization works
Stepwise failure, 20
Strain localization, 6, 9, 269–272, 275, 276, 286, 288
Strain rate, 269, 274, 276, 295, 307, 310, 312
Strain softening, 6, 9, 98, 209, 268, 269, 272–275, 282, 288, 291–300, 305–315, 345, 351–352, 364, 365
Stratigraphic information, 168, 371, 378
Subaqueous, 112, 185, 370, 378, 380
Subaqueous failure, 185
Sulphide content, 221, 224–227
Surface seismic methods, 255
Surface wave, 167, 175, 195, 230, 241–251
Surface-wave analysis, 167, 175, 232, 244
Susceptibility, 7, 40, 256, 335–337, 370, 372, 375–380
assessment, 369–380
map, 335–337, 370, 372, 377, 378, 380
SV seismic reflection, 245–246
SV-wave reflections, 243, 246, 251
SV-waves, 242, 251
S-wave velocity, 233, 237, 250, 256–259, 263
Sweden, 1, 2, 7, 8, 20, 39–49, 53, 161, 163, 180, 182, 206, 215, 219, 221, 227, 281, 383, 389
Swedish Terrameter LS, 196
Swelling clays, 33, 34

T

Thermogravimetric method (TGV), 27
Tiller, 64–69, 71–73, 93, 268, 269, 272, 274, 298
Topographical aspects, 94, 99, 100
Topography, 46, 94, 105, 106, 111, 112, 115, 140, 141, 168, 187, 194, 199, 201, 203, 230, 238, 284, 349, 352, 355, 357, 358, 373, 374
Total stress analyses, 317–319, 326, 345, 346, 397
Triaxial, 6, 149, 153, 155, 208, 268–272, 274, 275, 292, 294, 295, 297, 299, 300, 307, 309, 310, 322, 327, 345, 364, 365, 367, 398
Triaxial tests, 153, 155, 208, 268, 269, 292, 294, 300, 345, 398

U

Uncertainty, 35, 129, 226, 318, 357, 364, 397–399, 401–404, 406
Undisturbed samples, 210, 263, 386
Undisturbed strength, 16–18, 22

V

Vålen, 194–201
Vane tests, 79, 220, 318, 335, 365, 388, 398
Velocity model, 166, 232, 233, 242, 246–249
Vermiculite, 27, 29, 30, 32, 33
Vertically polarized shear (SV) wave, 242
Vertical seismic profiling (VSP), 242, 250
Vestfossen slide, 365–366
Viscosity, 97, 110, 140, 272, 318

W

Weathered crust, 21, 30, 112
Weathered zone, 21
Weathering crust, 16, 19

X
X-ray diffraction,
27, 66, 196

Y
Yield strength, 97, 110
Yield stress, 140, 142, 143

MIX
Papier aus verantwortungsvollen Quellen
Paper from responsible sources
FSC® C105338

If you have any concerns about our products,
you can contact us on
ProductSafety@springernature.com

In case Publisher is established outside the EU,
the EU authorized representative is:
Springer Nature Customer Service Center GmbH
Europaplatz 3, 69115 Heidelberg, Germany

Printed by Libri Plureos GmbH
in Hamburg, Germany